Inorganic Chemistry: Concepts and Applications

Inorganic Chemistry: Concepts and Applications

Edited by **Warren Gibbs**

NY RESEARCH
P R E S S

New York

Published by NY Research Press,
23 West, 55th Street, Suite 816,
New York, NY 10019, USA
www.nyresearchpress.com

Inorganic Chemistry: Concepts and Applications
Edited by Warren Gibbs

© 2016 NY Research Press

International Standard Book Number: 978-1-63238-472-0 (Hardback)

Printed in the United States of America.

Contents

Preface

It is often said that books are a boon to mankind. They document every progress and pass on the knowledge from one generation to the other. They play a crucial role in our lives. Thus I was both excited and nervous while editing this book. I was pleased by the thought of being able to make a mark but I was also nervous to do it right because the future of students depends upon it. Hence, I took a few months to research further into the discipline, revise my knowledge and also explore some more aspects. Post this process, I began with the editing of this book.

Inorganic chemistry is the study of compounds that do not contain carbon-hydrogen bonds. These compounds can be categorized into acids, bases, salts and oxides. Their study incorporates examining their composition, analysis, reactions, structure and properties. There are many sub-fields of inorganic chemistry like descriptive inorganic chemistry which deals with classifying compounds based on their properties, theoretical inorganic chemistry which is the study of bonding simple and complex molecules, thermodynamics and inorganic chemistry which focuses on the energy released during a reaction and mechanistic inorganic chemistry which refers to the study of reaction pathways, etc. This book will discuss in detail the applications and concepts of this field. This book contains some path-breaking studies in the field of inorganic chemistry and unfolds the innovative aspects of this field. It includes the experiments performed across the globe. Therefore, it will serve as a valuable source of reference for students and researchers alike.

I thank my publisher with all my heart for considering me worthy of this unparalleled opportunity and for showing unwavering faith in my skills. I would also like to thank the editorial team who worked closely with me at every step and contributed immensely towards the successful completion of this book. Last but not the least, I wish to thank my friends and colleagues for their support.

Editor

Synthesis, characterization, spectroscopic and crystallographic investigation of metal complexes of N-benzyl-N-nitrosohydroxylamine

Olga Kovalchukova[1], Ali Sheikh Bostanabad[1], Vladimir Sergienko[2], Irina Polyakova[2], Igor Zyuzin[3], Svetlana Strashnova[1]

[1]Peoples' Friendship University of Russia, Moscow, Russia
[2]Institute of General and Inorganic Chemistry, Russian Academy of Sciences, Moscow, Russia
[3]Institute of Problems of Chemical Physics, Russian Academy of Sciences, Chernogolovka, Russia
Email: okovalchukova@mail.ru

ABSTRACT

New metal complexes of N-benzyl-N-nitrosohy-droxylamine (BNHA) are isolated and characterized by elemental analysis, IR and UV-VIS spectroscopy. Spectrophotometric titration of aqueous solutions of BNHA by salts of some transition metals allowed to calculate the composition and formation constants of the metal complexes. The crystal structure of Cu (BNHA)$_2$ is studied by X-ray diffraction. The Cu atom is coordinated by four O atoms of two bidentate ligands, which close 5-membered chelate rings. The N-O (1.306 Å - 1.320 Å) and N-N (1.274 Å and 1.275 Å) bond lengths indicate that π electrons are delocalized over the chelating groups. Complexes form stacks with intermolecular Cu...N contacts equal to 3.118 Å and 3.306 Å.

Keywords: N-Benzyl-N-Nitrosohydroxylamine; BNHA; Potassium Salt; Complex Compounds; IR; UV-Visible; Single Crystal X-Ray Diffraction; Formation Constants

1. INTRODUCTION

The feature of chelate-forming derivatives of N-nitrosohydroxylamines is the possibility to form stable complexes with the metallic ions of various natures (for example, Al and Pd).

The ammonium salt of N-nitroso-N-phenylhydroxylamine, is a well known (as cupferron) analytical reagent and it was very popular especially during the classical period of analytical chemistry [1-3]. This ligand is known to form complexes with many metals but only few of them have been structurally characterized [4-15].

Recently it was shown that many ortho-substituted N-nitroso-N-oxybenzenamines are good NO donors for both *in vitro* and *in vivo* assays [16]. As the major bio-

logical functions of NO include controlling blood pressure, smoothing muscle tone and platelet aggregation, and participating in neuronal synaptic transmission, Thus, the development of new substances which are potential NO donors and studies of their structure and properties are current challenges facing medicinal and synthetic chemists.

In addition, cupferron is a biologically active compound, known to display carcinogenic, genotoxic, mutagenic and DNA-damaging effects [17,18]. The knowledge of the chemistry of cupferron and its analogues containing N-nitrosohydroxylamine fragments could contribute to a better understanding of the interaction of nitrogen oxide with metal centers of biologically important species.

Here we report the isolation and characterization of some transition metal complexes with N-benzyl-N-nitrosohy-droxylamine (BNHA).

2. EXPERIMENTAL

2.1. Measurements

The electronic spectra in the 200 - 800 nm range were obtained on a Varian spectrophotometer "Cary 50 scan", using aqueous 10^{-4} M solutions. IR spectra were recorded in KBr on a Varian "Excalibur HE 3100" FT spectrophotometer. Single crystal was characterized by a Bruker APEX II diffractometer. All the chemicals are commercially available and they were used without further purification. All the solvents were dried using standard methods before use.

2.2. Synthesis of K(BNHA) and Transition Metal Complexes

K(BNHA) and Cu(BNHA)$_2$. The synthesis was performed according to the procedure described in [19]. A solution of benzylmagnesium chloride was prepared from magnesium (24.3 g, 1 mol) and benzyl chloride (126.7 g, 1 mol) in the dry Et_2O (800 mL). The NO gas was bubbled through the solution under vigorous stirring and cooling at such a rate that NO was almost entirely absorbed. The NO flow rate and the degree of its absorbance were controlled by the comparison of the rate of NO bubbling through the bubble vessel filled by NaOH at the outlet of the reactional vessel. The reaction mixture temperature was maintained in the range −20°C - −30°C. A pasty solid was separated and encumbered stirring. After the period of a rapid NO absorption (1 h), stirring was continued in an NO atmosphere for 0.5 h until the NO absorption was completed, with a gradual increase in temperature to 0°C. The reaction mixture was purged with Ar to avoid site reactions involving both NO and the Grignard reagent, cooled to −10°C, and then a mixture of MeOH (50 mL) and Et_2O (50 mL) was gradually added. The reaction mixture was poured into 700 mL of water, acidified with 5 M HCl, the ethereal layer was separated, and the aqueous layer was extracted with Et_2O (100 mL × 3). The combined extracts were washed with an excess of 10% KOH. The aqueous layer was neutralized with 5 M HCl until pH 4. The precipitate of N-benzyl-N-nitrosohydroxylamine was washed with water. The wet product was dissolved in the appropriate quantity of 10% KOH, so that the pH value reached 8 - 9. The mixture was concentrated in vacuo, the residue (108 g) was crystallized from EtOH (200 mL), and dried in air. Yield 79.1 g (39.7%), colorless plates, dehydration temperature 110°C - 115°C, m.p. 211°C - 214°C (dec.). Upon heating (1 h) in vacuo at 110°C - 145°C, a sample of the air-dried salt (10.5547 g) lost 0.4838 g, $i.e.$, 0.507 mol of H_2O per one mol of the hydrated salt was removed. Filter liquor was concentrated to 50 mL, and the additional air-dried product (14.2 g) was precipitated with Et_2O (250 mL). The overall yield of K(BNHA) was 93.3 g (46.8%). The treatment of the combined filtrate and wash liquors with $CuSO_4$ additionally gave 9.1 g (5%) of $Cu(BNHA)_2$ (m.p. 157°C - 158°C (dec.)). The overall yield of K(BNHA) and $Cu(BNHA)_2$ 51.8%.

2.3. Transition Metal Complexes of BNHA

The above mentioned complexes were prepared by the reaction of the corresponding metal chloride with K(BNHA). In a typical reaction, an aqueous solution containing 1 mmol of $MCl_2·xH_2O$ was slowly mixed with 2 mmol of the K(BNHA) under vigorous stirring. The obtained precipitates were collected and re-crystallized from ethanol. Yields 50% - 70%.

The formulae of the obtained complexes are derived from the elemental analysis and presented in the **Table 1**.

2.4. Electronic Spectroscopic Measurements

Spectrophotometric titration of solutions of K(BNHA) with the solutions of metal salts in water was carried out in neutral and alkaline media. Based on the results of titration, the saturation curves were drawn for several wavelengths, and the metal-to-ligand ratios were calculated. The formation constants of metal complexes were determined by the procedure described by Beck & Nadypal [20].

2.5. X-Ray Data Collection and Structure Determination

Light blue needles of $Cu(BNHA)_2$ suitable for X-ray data collection were obtained while re-crystallisation from ethanol. The diffraction data were collected on a Bruker APEX-II CCD diffractometer at room temperature and processed using the Bruker software [21]. The structure was solved by the direct method. The positions of the hydrogen atoms were calculated. The non-hydrogen atoms were refined in the anisotropic approximation, and the hydrogen atoms were refined within the riding model.

Table 1. Chemical analysis (calculated/observed) and empirical formulae of the isolated complexes.

No.	Structure	Formula	% M	% C	% H	% N	M
1	K(BNHA)	$C_7H_7KN_2O_2$		44.19/44.23	3.71/3.44	14.73/14.92	190.24
2	$Cu(BNHA)_2$[1]	$C_{14}H_{14}CuN_4O_4$	17.37/17.08	45.96/45.25	3.86/3.78	15.31/15.44	365.84
3	$Co(BNHA)_2·4H_2O$	$C_{14}H_{22}CoN_4O_8$	13.60/13.78	38.81/38.61	5.12/4.88	12.93/12.45	433.28
4	$Ni(BNHA)_2·3H_2O$	$C_{14}H_{20}N_4NiO_7$	14.14/14.45	40.52/41.08	4.86/4.39	13.50/13.37	415.03
5	$Pb(BNHA)_2$	$C_{14}H_{14}N_4O_4Pb$	40.67/40.33	33.00/33.19	2.77/2.57	11.00/10.98	504.49
6	$Sn(BNHA)_2·3H_2O$	$C_{14}H_{20}N_4O_7Sn$	24.99/25.12	35.40/35.59	4.24/3.90	11.29/11.61	475.03
7	$La(BNHA)_3·2H_2O$	$C_{21}H_{25}LaN_6O_8$	22.11/22.38	40.11/40.29	4.01/3.33	13.37/12.94	628.37
8	$Cr(BNHA)_3·2H_2O$	$C_{21}H_{25}CrN_6O_8$	9.60/9.41	46.58/46.08	4.65/4.21	15.52/15.87	541.46
9	$Zn(BNHA)Cl·4H_2O$[2]	$C_7H_{15}ClN_2O_6Zn$	20.18/20.55	25.95/25.36	4.67/4.82	8.65/8.24	324.04

[1]M.p. 157°C (dec.) [19]. [2]For Cl: 10.94/11.05.

Structure solution and refinement were performed using the SHELX97 program package [22]. Crystal data and details of data collection and structure refinement are given in **Table 2**. Selected bond lengths and angles are listed in **Table 3**.

3. RESULTS AND DISCUSSION

3.1. Electronic Spectra

The electronic absorption spectrum of K(BNHA) **(Figure 1)**

Table 2. Crystal data and details of data collection and structure refinement for Cu(BNHA)$_2$.

Empirical formula	C$_{14}$H$_{14}$CuN$_4$O$_4$
Formula weight	365.83
Temperature (K)	296 (2)
Radiation, λ (Å)	MoK_a, 0.71073
Crystal system, space group	Monoclinic, P2$_1$/n
a (Å)	11.521 (2)
b (Å)	4.6883 (8)
c (Å)	28.979 (5)
β (°)	100.486 (3)
Volume (Å^3)	1539.0 (5)
Z	4
D$_{calc}$ (Mg/m^3)	1.579
F(000)	748
Crystal size (mm)	0.64 * 0.03 * 0.02
Absorption coefficient (mm^{-1})	1.445
Theta range for data collection (deg)	2.54 – 28.46
Limiting indices	$-15 \leq h \leq 15, -6 \leq k \leq 6,$ $-37 \leq l \leq 38$
Reflections collected /unique [R$_{int}$]	15423/3892 [0.0461]
Completeness to θ = 28.46	99.9%
Absorption correction	Semi-empirical from equivalents
Max. and min. transmission	0.7457 and 0.6487
Refinement method	Full-matrix least-squares on F^2
Data/ restraints/parameters	3892/0/208
Goodness-of-fit on F^2	1.024
Final R indices [I > 2σ (I)]	R1 = 0.0432, wR2 = 0.0922
R indices (all data)	R1 = 0.0831, wR2 = 0.1067
Largest diff. peak and hole, e Å^{-3}	0.424 and –0.247

Table 3. Selected bond lengths d (Å) and angles ω (°) in Cu(BNHA)$_2$.

Bond, angle	d, ω	Bond, angle	d, ω
Cu(1)-O(1)	1.9138 (18)	Cu(1)-O(3)	1.896 (2)
Cu(1)-O(2)	1.921 (2)	Cu(1)-O(4)	1.905 (2)
Cu(1)-N(2)^a	3.306 (2)	Cu(1)-N(4)^b	3.118 (3)
O(1)-N(1)	1.319 (3)	O(3)-N(3)	1.312 (3)
O(2)-N(2)	1.306 (3)	O(4)-N(4)	1.320 (3)
N(1)-N(2)	1.275 (3)	N(3)-N(4)	1.274 (3)
N(1)-C(7)	1.469 (3)	N(3)-C(14)	1.476 (3)
O(1)Cu(1)O(2)	82.30 (8)	O(3)Cu(1)O(4)	82.59 (9)
N(2)^{a}Cu(1)N(4)^b	177.87 (6)		
N(1)O(1)Cu(1)	107.49 (15)	N(3)O(3)Cu(1)	107.83 (15)
N(2)O(2)Cu(1)	113.29 (16)	N(4)O(4)Cu(1)	112.72 (16)
O(1)N(1)N(2)	123.9 (2)	O(3)N(3)N(4)	123.5 (2)
N(1)N(2)O(2)	113.0 (2)	N(3)N(4)O(4)	112.6 (2)

^{a}x, y + 1, z; ^{b}x, y – 1, z.

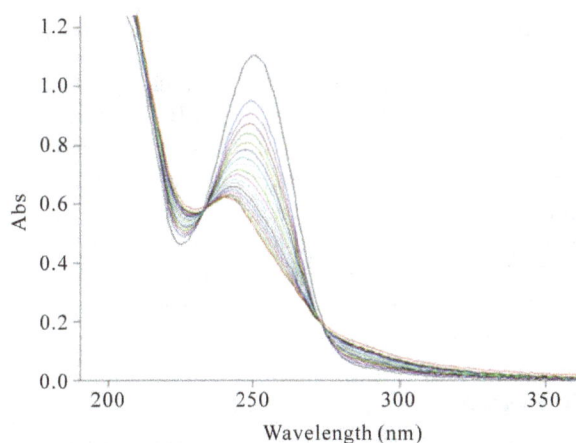

Figure 1. Spectrophotometric titration of an aqueous solution of K(BNHA) (10^{-4} M) by the solution of CuCl$_2$ (10^{-2} M).

is characterized by one wide absorption band in the range 230 nm - 260 nm with the λ_{max} 252 nm (log ε 4.06).

The addition of metal salts to the solutions of K(BNHA) provokes the hypsochromic shift and decrease in intensity of the ligand absorption band which indicates the change of the electronic structures of the organic anions due to their complexation and formation of coordinate bonds of a co-valent character. $\Delta\lambda_{max}$ 1 - 19 nm. The presence of isobestic points in the spectra indicates the reversible equilibria in the "ligand-complex" systems. The metal-to-ligand ratio and formation constants of the complexws in aqueous solutions are presented in the **Table 4**.

As the K-BNHA bonds may be considered to be of

Table 4. Composition (ML_n) and stability constants (β) of some metal complexes of BNHA in aqueous solutions and their absorbtion maxima (λ_{max}, nm).

Metal ion	n	β	λ_{max}, nm ($\log\varepsilon$)
Mn^{2+}	2	6.82×10^{11}	249 (3.87)
Co^{2+}	2	2.67×10^{10}	245 (3.84)
Ni^{2+}	2	7.51×10^{9}	245 (3.73)
Cu^{2+}	2	3.43×10^{8}	238 (3.80)
Zn^{2+}	2	2.3×10^{11}	242 (3.89)
Cd^{2+}	2	4.46×10^{11}	246 (3.96)
Pb^{2+}	2	3.62×10^{10}	248 (3.92)
Ba^{2+}	2	7.45×10^{10}	251 (3.93)
Al^{3+}	3	4.73×10^{14}	233 (3.73)
Cr^{3+}	3	3.22×10^{14}	243 (3.84)
La^{3+}	3	1.28×10^{17}	245 (3.94)
Sm^{3+}	3	1.03×10^{15}	243 (3.89)
Eu^{3+}	3	8.95×10^{15}	242 (3.92)
Er^{3+}	3	1.03×10^{15}	242 (3.92)

100% ionic character, the $\Delta\lambda_{max}$ value may indicate the increase in the degree of co-valency of M-BNHA interaction. Thus, from the data of the **Table 4** it is evident that among divalent ions of the metals of the 1st transition series, the degree of co-valency of metal-to-ligand interaction changes as following: $Mn^{2+} < Co^{2+} = Ni^{2+} < Cu^{2+} > Zn^{2+}$. The Al–BNHA interactions possess the most covalent character as well as Ba^{2+} forms the coordinate bonds of almost 100% of ionicity.

3.2. IR Spectra

The major IR absorption bands of K(BNHA) and its Cu(II) and Co(II) complexes are presented in the **Table 5**.

The assignments of the various absorption bands in K(BNHA) and the transitional metal complexes were performed according to N. Thakur *et al.* [23] for cupferron and metal cupferrates. It is shown that on complex formation NO stretching frequencies are lowered and the deformation frequencies increased while the C-N and N-N stretching frequencies are not very much altered, indicating that the complex formation takes place through the oxygen of the NO groups.

3.3. Crystal Structure of Cu(BNHA)₂

The copper atom has a nearly planar square coordination formed by oxygen atoms of two BNHA-ligands, which close 5-membered ONNOCu chelate rings (**Figure 2**).

The Cu(1) atom lies in the plane of the O(1)N(1)N(2)O(2) fragment (its deviation is 0.022 Å) and deviates from the O(3)N(3)N(4)O(4) plane by 0.194 Å. In both BGA-molecules, the planar $O_2N_2CH_2$ and $PhCH_2$ fragments lie in approximately perpendicular planes (the dihedral angles are 93.3° and 109.9°). The phenyl rings of the two NBHA-molecules are bent in opposite directions with respect to the coordination plane and form an angle of 139.4° with each other. The geometric parameters of the NBHA-ligands in Cu(BNHA)₂ are close to those in Zr(cp)2(BNHA) [24], which is the only compound of NBHA found in the Cambridge Structural Database (version 5.33). Bond lengths (**Table 3**) suggest that π electrons are delocalized over the ONNO chelating groups. The difference between the N-O bond lengths is insignificant, and the N-N bond lengths are close to the double-bond type [25]. This suggests an important contribution of the -O-N+(R)=N-O-canonical form. Cu(BNHA)₂ complexes are packed in stacks running along the b axis (**Figure 3**). In a stack, atoms N(2) and N(4) of neighboring complexes approach the Cu(1) atom by 3.306 (2) and 3.118 (3) Å, respectively, thus completing its coordination to an extremely elongated tetragonal bipyramid.

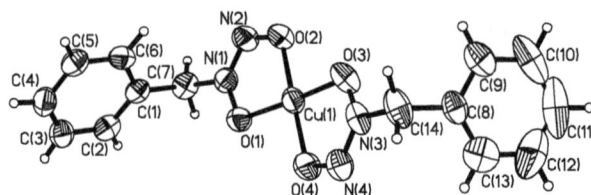

Figure 2. Molecular structure of Cu(BNHA)₂. The thermal ellipsoids are shown at the 50% probability level.

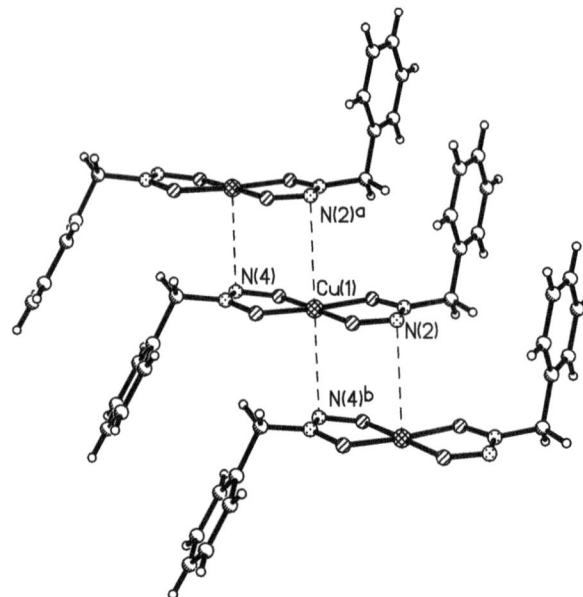

Figure 3. Mutual arrangement of neighboring Cu(BNHA)₂ complexes in a stack.

Table 5. Frequencies (cm^{-1}) and band assignments in K(BNHA) and some transition metal complexes.

K(BNHA)	Cu(BNHA)$_2$	Co(BNHA)$_2$·4H$_2$O	Assignment
3418	3417	3345	C-H stretching vibration
3150	3120	3243	N-H strtching vibration
1650	1633	1646	C=C ring stretching vibration
1434	1417	1425	C-N stretching vibration
1397	1344	1395	NN stretching vibration
1231	1214	1222	NO stretching vibration
1145	1149	1145	C-N ring in-plane vibration
961	940	929	Ring vibration
784	783	783	C-H ring out of plane vibration
705	717	707	C=C ring out of plane vibration

4. CONCLUSION

The N-benzyl-N-nitrosohydroxylamine (BNHA) ligand coordinates metal ions by two oxygen atoms. The bonding of the ligands to metal ions is confirmed by spectroscopic and single-crystal X-ray diffraction studies. The composition and the formation constants of the complexes in aqueous solutions were determined by UV spectroscopic titration.

6. ACKNOWLEDGEMENTS

This study was supported by the Russian Foundation for Basic Research (project 10-03-00003-a).

REFERENCES

[1]   Buscarron, F. and Canela, J. (1974) Analytical uses of some n-nitroso-n-alkyl (or -n-cyclo-alkyl)hydroxylamines: Part II. Solvent extraction of metal hexahydrocupferrates. *Analytica Chimica Acta*, **70**, 113-120. doi:10.1016/S0003-2670(01)82916-0

[2]   Buscarron, F. and Canela, J. (1974) Analytical uses of some N-nitroso-N-alkyl (or -N-cycloalkyl)-hydroxylamines: Part III. N-Nitroso-N-cycIooctylhydroxylamine, N-nitroso-N-cyclododecylhydroxyl-amine and N-nitroso-N-isopropyl-hydroxylamine. *Analytica Chimica Acta*, **71**, 468-471. doi:10.1016/S0003-2670(01)85455-6

[3]   Brown, J. (1917) The use of cupferron (phenylnitrosohydroxylamine ammonium) in: I. The qantitative separation of zirconium, titanium, iron, manganese and aluminum. II. The analysis of zircon and baddeleytte. *Journal of the American Chemical Society*, **39**, 2358-2366. doi:10.1021/ja02256a013

[4]   Párkányi, L., Kálmán, A., Deák, A., Venter, M. and Haiduc, I. (1999) A new inorganic (carbon-free) chelate ring: SnO$_2$N$_2$. Eight-coordinated tin (IV) in Sn(O$_2$N$_2$Ph)$_4$ and a self-assembled 20-membered macrocycle in [Me$_3$Sn(O$_2$N$_2$Ph)]$_4$. *Inorganic Chemistry Communications*, **2**, 265-268. doi:10.1016/S1387-7003(99)00062-3

[5]   Abraham, M.H., Bullock, J.I., Garland, J.H.N., Golder, A.J., Harden, G.J., Larkworthy, L.F., Povey, D.C., Riedl, M.J. and Smith, G.W. (1987) Copper and iron(III) complexes of N-nitroso-N-alkyl hydroxylamines, and the X-Ray crystal structures of bis(N-nitroso N-isopropylhydroxylaminato) copper(II) and tris(N-nitroso-N-isopropyl hydroxylaminato) iron(III). *Polyhedron*, **6**, 1375-1381. doi:10.1016/S0277-5387(00)80898-0

[6]   Basson, S.S., Ledipoldt, J.G., Roodt, A. and Venter, J.A. (1992) Structure of carbonyl(N-hydroxy-N-nitrosobenzenaminato-O,O')(4-methyl-2,6,7-trioxa-1-phosphabicyclo[2.2.2] octane)rhodium(I). *Acta Crystallographica Section C*, **48**, 171-173. doi:10.1107/S0108270191008223

[7]   Ahmed, M., Edwards, A.J., Jones, C.J., McCIeverty, J.A., Rothin, A.S. and Tate, J.P. (1988) Metal complexes of N-aryl-N-nitrosohydroxylamines: Cleavage of N-N bonds to give metal nitrosyl species and organonitrogen compounds, and the crystal structure of [RhCl$_2${ONN(C$_6$H$_4$Me-p)O}(H$_2$O)(PPh$_3$)]·0.5Me$_2$CO. *Journal of the Chemical Society, Dalton Transactions*, 257-263. doi:10.1039/dt9880000257

[8]   Najafi, E., Amini, M.M. and Ng, S.W. (2011) Bis (μ-N-nitroso-N-phenylhydroxylaminato)-κ^3O,O':O'; κ^3O':O,O'-bis[(Nnitroso-N-phenylhydroxylaminato-κ^2O,O')lead(II)]. *Acta Crystallographica Section E*, **67**, m377. doi:10.1107/S1600536811006775

[9]   Najafi, E., Amini, M.M. and Ng, S.W. (2011) Bis(N-nitroso-N-phenylhydroxylaminato-κ^2O,O')(1,10-phenanthroline-κ^2N,N')lead(II). *Acta Crystallographica Section E*, **67**, m378. doi:10.1107/S1600536811006787

[10] Tamaki, K. and Okabe, N. (1998) Diaquabis[N-(1-naphthyl)-N-nitrosohydroxylaminato-O,O']cobalt(II). *Acta Crystallographica Section C*, **54**, 195-197. doi:10.1107/S0108270197013620

[11] Okabe, N. and Tamaki, K. (1995) Dimethanolbis(N-nitroso-N-phenylhydroxylaminato-O,O')nickel(II). *Acta Crystallographica Section C*, **51**, 2004-2005. doi:10.1107/S0108270195005816

[12] Van der Helm, D., Merritt Jr., L.L., Degeilh, R. and MacGillavry, C.H. (1965) The crystal structure of iron cupferron Fe(O$_2$N$_2$C$_6$H$_5$)$_3$. *Acta Crystallographica*, **18**, 355-362. doi:10.1107/S0365110X65000816

[13] Bolboaca, M., Cîntă S., Venter, M., Deak, A., Haiduc, I., Cozar, O., Iliescu, T., Rösch, P. and Kiefer, W. (2000) Vi-

brational behavior of transition metal cupferronato complexes: Raman studies on cobalt(II) cupferronato derivatives. *Spectroscopy Letters*, **33**, 857-865. doi:10.1080/00387010009350162

[14] Pavel, I., Cîntă S., Venter, M., Deak, A., Haiduc, I., Rösch, P., Cozar, O. and Kiefer, W. (2000) Vibrational behavior of transition metal cupferronato complexes. Raman and SERS studies on nickel(II) cupferronato complexes. *Vibrational Spectroscopy*, **23**, 71-76. doi:10.1016/S0924-2031(99)00086-7

[15] Tamaki, K. and Okabe, N. (1996) Manganese cupferronate. *Acta Crystallographica Section C*, **52**, 1612-1614. doi:10.1107/S0108270196002661

[16] McGill, A.D., Zhang, W., Wittbrodt, J., Wang, J., Schlegel, H.B. and Wang, P.G. (2000) Para-substituted N-nitroso-N-oxybenzenamine ammonium salts: A new class of redox-sensitive nitric oxide releasing compounds. *Bioorganic & Medicinal Chemistry*, **8**, 405-412. doi:10.1016/S0968-0896(99)00300-4

[17] Shiino, M., Watanabe, Y. and Umezawa, K. (2001) Synthesis of N-substituted N-nitrosohydroxylamines as inhibitors of Mushroom Tyrosinase. *Bioorganic & Medicinal Chemistry*, **9**, 1233-1240. doi:10.1016/S0968-0896(01)00003-7

[18] Shiino, M., Watanabe, Y. and Umezawa, K. (2003) Synthesis and tyrosinase inhibitory activity of novel N-hydroxybenzyl-N-nitrosohydroxylamines. *Bioorganic Chemistry*, **31**, 129-135. doi:10.1016/S0045-2068(03)00026-9

[19] Zyuzin, I.N., Nechiporenko, G.N., Golovina, N.I., Trofimova R.F. and Loginova, M.V. (1997) Synthesis and structure of di(non-azoxy) formals and some related N-alkyl-N'-alkoxydiazene N-oxides. *Russian Chemical Bulletin*, **46**, 1421-1429. doi:10.1007/BF02505678

[20] Beck, M.T. and Nadypal, I. (1989) Chemistry of complex equilibria. Akad. Kiado, Budapest.

[21] APEX2 (Version 2008. 6-1), SAINT (V7.60A), SADABS (Version 2008/1), XPREP (Version 2008/2) (2005-2009) Bruker AXS Inc., Madison.

[22] Sheldrick G.M. (2008) A short history of *SHELX*. *Acta Crystallographica Section A*, **64**, 112-122. doi:10.1107/S0108767307043930

[23] Thakur, N.V., Kartha, V.B., Kaneka, C.R. and Marthe, V.R. (1972) Infrared spectra of cupferron and some rare earth cupferrates, *Journal of Inorganic and Nuclear Chemistry*, **34**, 2831-2836. doi:10.1016/0022-1902(72)80589-X

[24] Fochi, G., Floriani, C., Chiesi-Villa, A. and Guastini, C. (1986) Notes. Insertion of nitrogen oxide into a zirconium-carbon bond: Reaction of dialkylbis(cyclopentadienyl)zirconium(IV) complexes with nitrogen oxide. *Journal of the Chemical Society, Dalton Transactions*, 445-447. doi:10.1039/dt9860000445

[25] Allen, F.H., Kennard, O. and Taylor, R. (1983) Systematic analysis of structural data as a research technique in organic chemistry. *Accounts of Chemical Research*, **16**, 146-153. doi:10.1021/ar00089a001

2

Synthesis, characterisation and microbial studies of [bis(1,10-phenanthroline) (ethylenediamine) copper(II)] diperchlorate and its bromide analogue

Oluwayemi O. Esther Onawumi[*], Idowu O. Adeoye, Florence A. Oluwafunmilayo Adekunle

Department of Pure and Applied Chemistry, Ladoke Akintola University of Technology, Ogbomoso, Nigeria
Email: [*]estherdr@rocketmail.com

ABSTRACT

Two new tris-chelate-complexes have been synthesized and characterized with elemental and spectroscopic methods. IR and thermal studies correlate with the structures of the complex in the solid state. The structure of $[Cu(en)(phen)_2] \cdot 2ClO_4^-$ was determined with X-ray data using single crystal X-ray diffractometer while the molecular structure of $[Cu(en)(phen)_2] \cdot 2Br^- \cdot 2Phen \cdot 8H_2O$ was deduced from the used characterization methods. $[Cu(en)(phen)_2] \cdot ClO_4^-$ crystallizes as orthorhombic with space group Pbcn. Both complexes have distorted octahedral geometry. Microbial activities of these complexes against *Staphylococcus aureus, Escherichia coli, Pseudomonas aeruginosa, Streptoccocus pyogeneous, Candida albicans*, and *Aspergillus niger* were also reported.

Keywords: Tris-Chelate Complexes; X-Ray Data; Electronic Spectra; Microbial Studies

1. INTRODUCTION

Copper(II) ions and its metal complexes have continued to attract attention of coordination chemists due to its various interesting structural features, its usefulness as models of the active centers of various metallo-enzymes, magnetic, electronic, catalytic and biological properties [1-3].

In recent times, Copper(II) complexes from various ligand systems such as 1,10-phenanthroline and 1,2-diaminoethane have been synthesized and characterized with the aim of gaining a better understanding of the chemistry of copper and the various ligands in different environments.

We have previously described [4] the synthesis and spectral studied of tris chelate of mixed-ligand complexes of copper(II) ion of nitrate and that of chloride with 1,2-diaminoethane and 1,10-phenanthroline in which

the H-bonded network of lattice water molecules and the anions in both compounds are quite different from one another. The NO_3^-, H_2O, H-bonded network has a 1D chain structure while Cl^-, H_2O network has a 3D structure. The change in the structure is due to the change in the size of anions as Cl^- ions allow very close H-bonding because of its smaller size compare to NO_3^-. We are considering the effect of ClO_4^- ion upon these supramolecular structures. The activities of these two complexes on micro-organisms such as *Staphylococcus aureus*, *Escherichia coli*, *Pseudomonas aeruginosa*, *Streptoccocus pyogeneous*, *Candida albicans*, and *Aspergillus niger* being the micro-organisms that affect man and plants were also carried out.

In this paper, we therefore report the synthesis, characterization, structure, and microbial studies of $[Cu(en)(phen)_2](ClO_4)_2$-(1) and $[Cu(en)(phen)_2]2Br \cdot 2Phen \cdot 8H_2O$-(2).

2. EXPERIMENTAL

All the chemicals used, were analytical reagent from commercial sample. They were procured from Fluka and Merck. The reagents were used without further purification.

2.1. Synthesis of $[Cu(en)(phen)_2](ClO_4)_2$ (1)

$Cu(ClO_4)_2 \cdot 6H_2O$ (0.370 g, 1 mmol) and 1,2-diaminoethane (0.20 mL, 2.92 mmol) were dissolved in 25 mL water and heated to 80°C for 10 mins with stirring. To this, 5 mL ethanolic solution of 1,10-phenanthroline (0.198 g, 1.00 mmol) was added and stirred for another 5 mins and the solution was filtered. There was blue precipitate. The precipitate was recrystallized with methanol. Greenish blue block shaped X-ray type crystals were obtained after four days. Yield: 0.3576 g (52.35%). Anal. calcd. for $C_{26}H_{24}CuN_6Cl_2O_8$:

C 45.73, H, 3.54; N, 12.31, Found: C 45.72; H, 3.50; N, 12.27.

IR (KBr disk, cm^{-1}): 3393, 3312, 2953, 2853, 1622,

---
*Corresponding author.

1585, 1507, 1094, 843, 728, 422.

2.2. Synthesis of [Cu(en)(phen)$_2$]·2Br$^-$·2(phen)·8H$_2$O (2)

The procedure was similar to that of 1, except for the use of CuBr$_2$, (0.223 g, 1 mmole). Blue crystals in the form of hexagonal shaped were obtained from the filtrate after four days Similar crystals were obtained when the synthesis was carried out with the reagent in ratio 1:1:4. of Cu:en:phen in order to maximize the yield. Yield: 0.1857 g (64.68%). Anal. calcd. for C$_{50}$H$_{56}$Br$_2$CuN$_{10}$O$_8$: C, 52.29; H, 4.91; N, 12.19 Found: C, 52.19; H, 4.88; N, 12.14.

IR (KBr disk, cm^{-1}): 3403, 3226, 3093, 1623, 1585, 1506, 1420, 846, 727, 420.

2.3. Physical Measurements

IR spectra were recorded using KBr discs on a Perkin Elmer, GX-FTIR spectrometer in the range 4000 - 400 cm^{-1}. Elemental analysis for C, H and N was performed on a Perkin-Elmer, Series II, 2400. Electronic spectra measurements were obtained using a Shimadzu UV-3101 PC spectrophotometer. Thermogravimetric analysis was done using Mettler Toledo.

2.4. X-Ray Crystallography

X-ray data were collected on a Bruker SMART APEX CCD X-ray diffractometer, using graphite-monochromated Mo-Kα radiation (λ = 0.71073 Å) at 298 K. Data were reduced using SAINTPLUS [5] and a multi-scan absorption correction using SADABS [6] was performed. The structures were solved by direct method using SHELXS-97 and full-matrix least-square refinement was carried out using SHELXL-97. [7] All ring hydrogen atoms were assigned on the basis of geometrical considerations and were allowed to ride upon the respective carbon atoms. Drawings were made using Mercury [8].

2.5. Antimicrobial Screening

The antimicrobial screening was performed by paper diffusion method. The activities were expressed in terms of millimeter (mm) by diameter of zone of inhibition. Ampicilline, Chloramphenicol, Ciprofloxacin and Norfloxacin were used as standard drugs for bacteria. Greseofulvin and Nystatin were used for fungi.

2.5.1. Microorganisms Used

The following common standard strains were used for screening of antibacterial and antifungal activities. The strains, listed below, were procured from Institute of Microbial Technology, Chandigarh India. The bacteria used are 1) *Staphylococcus aureus* MTCC 96; 2) *Escherichia coli* MTCC 443; 3) *Pseudomonas aeruginosa* MTCC 1688; and 4) *Streptoccocus pyogeneous* MTCC 442, while fungi used are 5) Candida albicans MTCC 227; and 6) *Aspergillus niger* MTCC 282.

2.5.2. Culture Media

Nutrient agar medium was prepared by dissolving 7.0 g of agar powder in 250 mL of distilled water in a conical flask (for bacteria). Potato dextrose agar medium was prepared by dissolving 9.75 g of potato dextrose agar powder in 250 mL of distilled water in a conical flask (for fungi). The conical flask was capped with cotton wool and aluminium foil and then sterilized in an autoclave for 15 minutes at 121°C.

Using an inoculate wire loop, the micro organisms were collected from slant cultures of all the different micro-organisms and were inoculated separately into fresh sterile nutrient broths for bacteria and the resulting cultures were incubated at 37°C for 18 hrs and at room temperature to obtain new cultures of bacteria respectively.

2.5.3. Zone of Inhibition

The zone of inhibition of the complexes was carried out by dipping sterile paper disc into a range of metal complexes solutions using sterile forceps. These were then carefully and aseptically laid on the media flooded with cultures of test organisms. The concentrations used were 250, 100, 50, 25 and 5 μg/mL and finally, the Petri plates were incubated for 26 - 30 hrs, at 28°C ± 2°C. The zone of inhibition was measured in millimetre. The results are shown in **Table 3** for bacteria and **Table 4** for the fungi.

3. RESULTS AND DISCUSSION

3.1. Synthesis

The two complexes were prepared by a similar method of synthesis using the reagents in molar ratio of 1:3:1. of Cu^{2+}:en:phen. For the complex 2 it was resynthesized in ratio 1:1:4 to improve its yield. Greenish blue good crystals were isolated from the precipitate of 1 after re-crystallization from methanol while navy blue crystals can be isolated directly from the filtrate of 2 but its structure could not be refined. The complexes are non hygroscopic and thermally stable. They were obtained in fairly good yield and the C, H and N analysis calculated agrees with the experimental value for the two complexes. The anions of the complexes are different in that there are no free phenanthroline and lattice water in 1 as found in 2. This may be due to dissolution of the free phenanthroline and water during recrystallisation of precipitate of 1. The complexes are soluble in methanol, ethanol, DMF, DMSO, acetonitrile and water.

3.2. Infrared Spectra

The two complexes have almost similar IR absorption bands. Broad bands for ν (NH$_2$) stretching vibration in 1,2-diaminoethane appeared at 3393 cm^{-1}, 3312 cm^{-1} for complex 1. Both asymmetric and symmetric ν (C-H) bands in 1,2-diaminoethane could be seen at 2953 cm^{-1} and 2853 cm^{-1}. NH$_2$ deformation bands appeared at 1585 cm^{-1}, while aromatic ring stretching vibration for phenanthroline occurs at 1507 cm^{-1}. ClO$_4$ band is observed at 1094 cm^{-1}. Complex 2 also has similar bands as found in 1, ν (O-H), from uncoordinated water and ν (NH$_2$) stretching band from 1,2-diaminoethane could be seen at 3403 cm^{-1} and 3226 cm^{-1} respectively while ν (C-H) aromatic stretching vibrational band could be seen at 3093 cm^{-1}, aromatic ring stretching occurred at 1506 cm^{-1} for the coordinated and at 1623 cm^{-1} for the unco-ordinated phenanthroline rings. NH$_2$ deformation band is found at 1585 cm^{-1} [9]. The IR result shows that the two complexes coordinated via the N atoms from the 1,2-diaminoethane and from the N atoms of the aromatic rings. Molecules of water also present in complex 2 with free phenanthroline molecules. This is also confirmed from the C, H, N analysis. Cu-N band could not be observed for phenanthroline as its absorb at relatively low frequency [9]. The ν (Cu-N) from 1,2-diaminoethane for complexes 1 and 2 are 422 and 420 cm^{-1} respectively.

3.3. Electronic Spectra

The electronic absorption spectra of both complexes were measured in methanol solution. [Cu(en)(phen)$_2$] (ClO$_4$)$_2$ absorb at 15.3×10^3 cm^{-1} (652 nm), while [Cu(en)(phen)2]Br$_2$·2phen·8H$_2$O absorb at 15.3×10^3 cm^{-1} (646 nm). (**Figures 1(a)** and **(b)**). All the four d-d transitions of the rhombic distorted octahedral copper(II) chromophore may be placed within the broad envelop of the observed band. The energy band in this position for the two complexes is typically expected for a rhombic distorted octahedral configuration and may be assigned to $^2E_g \rightarrow {}^2T_{2g}$ transition [10].

3.4. Thermal Gravimetric Analysis

Thermal gravimetric studies and differential thermal analysis were carried out on the complexes in the temperature range of 50°C to 800°C at a heating rate of 10°C/min. The measurements were done primarily to determine the composition of materials and to predict their thermal stability and also to determine extent of weight loss or gain due to decomposition, oxidation or dehydration. The thermal analysis shows that there are two broad endothermic peaks in the differential thermal analysis curve of the bromide analogue which is typical of dehydration reaction [11], whereas the perchlorate analogue showed three sharp endothermic peaks which shows changes in the crystalline or fusion process in the complex. The two

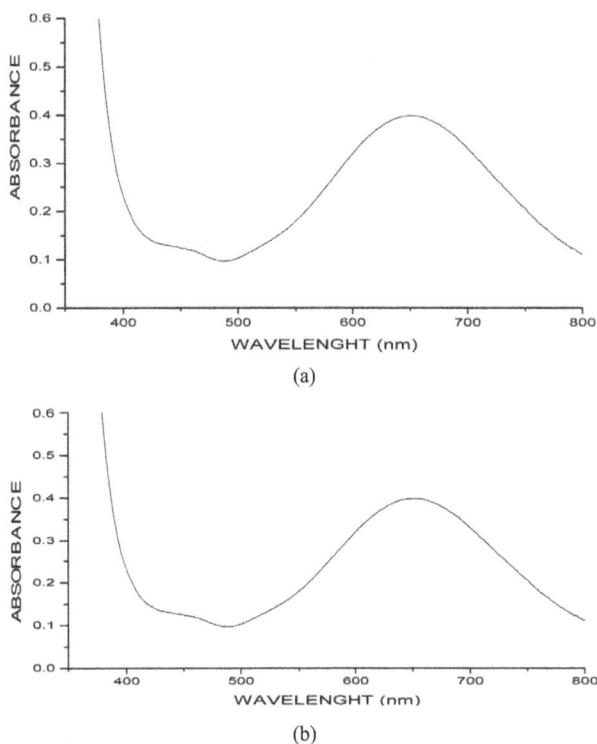

Figure 1. (a) UV-Visible spectra of [Cu(en)(phen)$_2$]$_2$phen·2Br·8H$_2$O; (b) UV-Visible spectra of [Cu(en)(phen)$_2$]·2ClO$_4$$^-$.

complexes showed no weight loss up to 170°C indicating the absence of uncoordinated water molecules in the complexes. The bromo complex decomposes in two endothermic peaks the first step occurs from 55°C - 300°C and this corresponds to the loss of the eight H$_2$O molecules and second from 300°C - 500°C represent the decomposition of the organic part. The perchlorate decomposes with two peaks the first from 150°C - 250°C then to 350°C. The remaining organic part decomposes with the peak occurring from 350°C - 650°C.

3.5. Structure of [Cu(en)(phen)$_2$](ClO$_4^-$)$_2$ (1)

The molecular structure of complex 1 was established by single crystal X-ray study. The crystal system in 1 is orthorhombic with space group Pbcn (**Table 1**). The coordination sphere of copper consists of one 1,2-diaminoethane and two phenanthroline molecules with distorted octahedral geometry. The equatorial plane of hexa coordinated copper consists of both nitrogen atoms of 1,2-diaminoethane molecule and nitrogen from the two phenanthroline molecules (**Figure 2(a)**). Two perchlorate anions also present in the lattice, there are no water molecules. The equatorial distances are Cu1-N1: 2.370 (2), Cu1-N2: 2.036 (2) Å and Cu1-N4: 2.367 (2) Å and Cu1-N5: 2.036 (2) Å, while the axial distances are Cu1-N3: 2.054 (2) Å and Cu1-N6: 2.042 (2) Å compared to the bond distances of its nitrate and chloride ions analogy

Table 1. Crystallographic data for [Cu(en)(phen)₂]·(ClO₄)₂ (1).

	1
Formula	$C_{26}H_{24}CuN_6Cl_2O_8$
Formula weight	682.95
Colour	Blue
Crystal system	Orthorhombic
a (Å)	18.4895 (12)
b (Å)	13.2926 (9)
c (Å)	22.9599 (15)
α (°)	90.00
β (°)	90.00
γ (°)	90.00
V (Å³)	5642.9 (6)
Z	8
Space group	Pbcn
T (°C)	110 (2)
λ (Å)	0.71073
D_{calcd} (g·cm⁻³)	1.608
μ (mm⁻¹)	1.025
F(000)	2792
Crystal size (mm)	0.45 × 0.35 × 0.25
θ range for data collection (°)	1.80 - 28.29
h/k/l	−23, 24/−17, 17/−30, 13
Reflections collected/unique [R_{int}]	32270/5695/0.0459
Completeness to $2\theta = 25.00$ (%)	99.7
Refinement method	Full-matrix least-squares on F^2
Data/restraints/parameters	5695/0/388
Goodness of fit on F^2	1.049
Final R indices [$I > 2\sigma (I)$]	
R1	0.0459
wR2	0.1200
R indices (all data)	
R1	0.0548
wR2	0.1269

[4], which have Cu-N₁: 2.390 (2), Cu-N₂ 2.0608 (17), Cu-N₃: 2.0210 (18) Å and 2.406 (3), 2.071 (3), 2.028 (3) Å respectively. Selected geometrical parameters are presented in **Table 2**. The packing diagram for 1 showing hydrogen bonding interactions as viewed along the b-axis is shown in **Figure 2(b)**. There is H-bonding interaction between the hydrogen of N5 from 1,2-diaminoethane and from O₂ and O₆ from perchlorate ions [N-H-O₂ = 2.936 Å, N-H-O₆: 3.064] Å.

3.6. Biological Activities

The antibacterial and antifungi activities of the com-

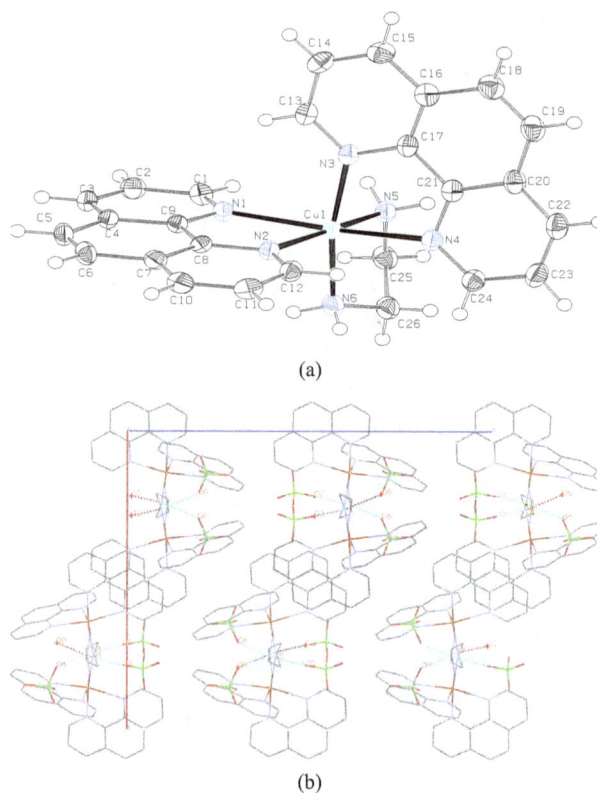

(a)

(b)

Figure 2. (a) ORTEP diagram depicting the cationic part of the copper complex with atom numbering scheme (50% probability factor for the thermal ellipsoids); (b) Packing structure of complex 1 showing the H-bonding between diaminoethane-H and O-from perchlorate as viewed along the b-axis.

Table 2. Bond lengths and bond angles for [Cu(en)(phen)₂] (ClO₄)₂.

Bond Lengths (Å)	Bond Angles (°)	
Cu1-N1 2.370 (2)	N2-Cu1-N5 173.02 (8)	C25 N5 Cu1 109.74 (16)
Cu1-N2 2.036 (2)	N2 Cu1 N6 92.92 (8)	C25 N5 H5A 109.7 (16)
Cu1-N3 2.054 (2)	N5 Cu1 N6 83.65 (9)	Cu1 N5 H5A 109.7 (16)
Cu1-N4 2.367 (2)	N2 Cu1 N3 92.33 (8)	C25 N5 H5B 109.7 (16)
Cu1-N5 2.036 (2)	N5 Cu1 N3 92.00 (8)	Cu1 N5 H5B 109.7 (16)
Cu1-N6 2.042 (2)	N6-Cu1-N3 169.91 (8)	H5A N5 H5B 108.2 (16)
O1-Cl1 1.428 (2)	N2 Cu1 N4 96.53 (8)	C26 N6 Cu1 107.89 (16)
O2-Cl1 1.4388 (19)	N5 Cu1 N4 89.82 (8)	C26 N6 H6A 110.1 (16)
O3-Cl1 1.434 (2)	N6-Cu1-N4 94.60 (8)	Cu1 N6 H6A 110.1 (16)
O4-Cl1 1.412 (2)	N3 Cu1 N4 76.23 (8)	C26 N6 H6B 110.1 (16)
O5-Cl2 1.429 (3)	N2 Cu1 N1 75.96 (7)	H6A N6 H6B 108.4 (16)
O6-Cl2 1.449 (3)	N5 Cu1 N1 98.08 (8)	N1 C1 C2 123.6 (2)
O7-Cl2 1.409 (3)	N6 Cu1 N1 92.84 (8)	N1 C1 H1 118.2 (2)
O8-Cl2 1.422 (2)	N3 Cu1 N1 96.79 (8)	C2 C1 H1 118.2 (2)
	N4 Cu1 N1 169.69 (7)	C3 C2 C1 118.6 (2)

plexes can easily be observed from **Tables 3** and **4** and from the bar chart as seen in **Figures 3-8**. The data were obtained for the bacteria and fungi which cause diseases in man and plants with that of the standard drugs normally used for these micro-organisms. From the results it can be observed that at higher concentrations the perchlorate analog have higher zone of inhibitions than their bromide counterpart when tested against the following bacteria: *E. coli*, *P. aeruginosa*, *S. aureus* and *P. pyogenes* compared to the zone of inhibitions observed with known antibiotics like Ampicillin and Chloralphenicol (**Figures 3-6**). The zones of inhibition are however lower when the values are compared to other antibiotics like Ciproflaxin and Norfloxacin. In general the results indicate that the complexes have potential and promising activity. Further investigations may however be required to illustrate the mechanisms by which the complexes exhibit its antibacterial effect [12].

Similarly, the perchlorate complex was more potent than the bromide complex as shown by the results of the zone of inhibition observed when tested with *A. niger* and *C. albicans* (**Figures 7** and **8**). Though the zones of inhibitions observed for standard antifungal drugs like Greseofulvin and Nystatin are higher than the ones observed for the complexes they are significant enough to indicate some antifungal properties inherent in the complexes.

Table 3. Zone of inhibition of complexes and the standard drugs against the bacteria (mm).

Bacteria species	*E. coli* MTCC 443					*P. aeruginosa* MTCC 1688					*S. aureus* MTCC 96					*S. pyogenes* MTCC 442				
Concn µg/ml	5	25	50	100	250	5	25	50	100	250	5	25	50	100	250	5	25	50	100	250
CCLOENP	-	15	17	19	24	-	14	15	18	21	-	11	14	18	21	-	13	15	17	18
CBRENP	-	17	18	19	21	10	13	15	18	20	-	12	15	18	19	-	12	14	16	18
Amplicilline	14	15	16	19	20	14	15	15	18	20	10	13	14	16	18	11	14	16	18	19
Chloramphenicol	14	17	23	23	23	14	17	18	19	21	12	14	19	20	21	10	13	19	20	20
Ciprofloxacin	20	23	28	28	28	20	23	24	26	27	17	19	21	22	22	16	19	21	21	22
Norfloxacin	22	25	26	27	29	18	19	21	23	23	19	22	25	26	28	18	19	20	21	21

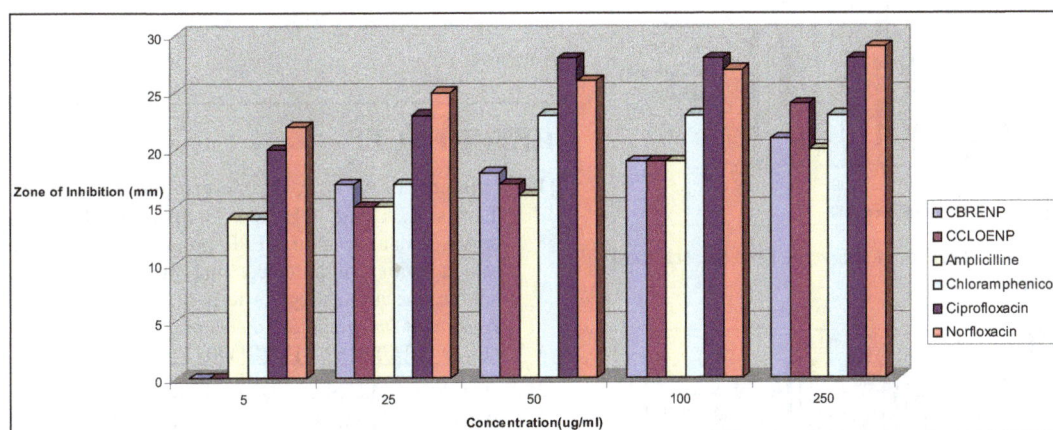

Figure 3. Zone of inhibition of *E. coli* against the concentration of the standard drugs and the complexes.

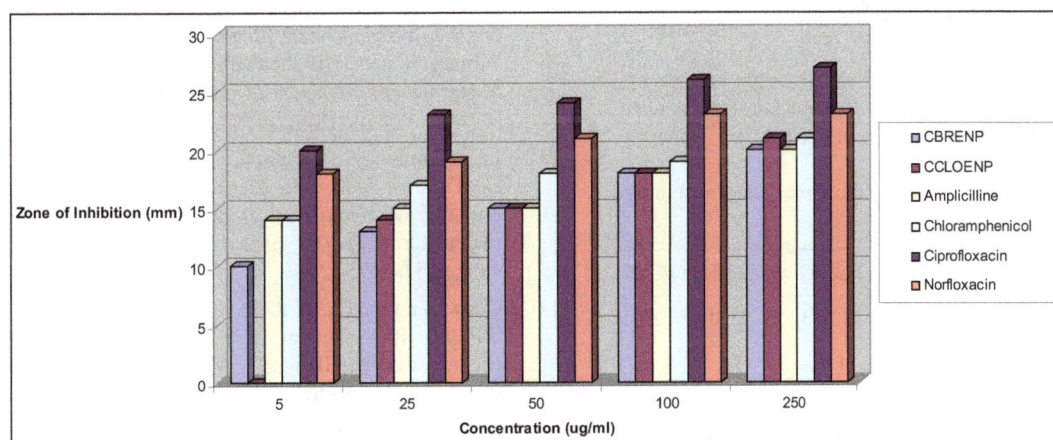

Figure 4. Zone of inhibition of *P. aeruginosa* against the concentration of the standard drugs and the complexes.

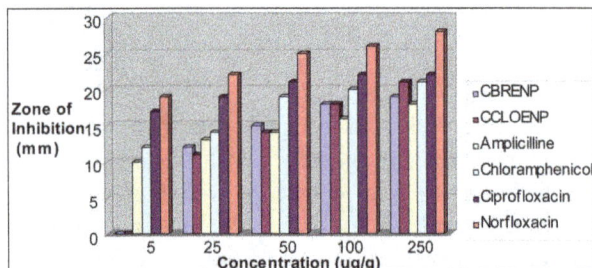

Figure 5. Zone of inhibition of *S. aureus* against the concentration of the standard drugs and the complexes.

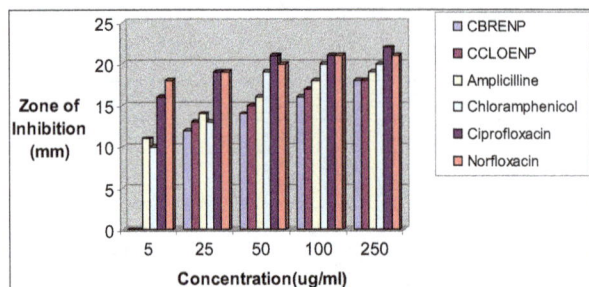

Figure 6. Zone of inhibition of *S. ptogenes* against the concentration of the standard drugs and the complexes.

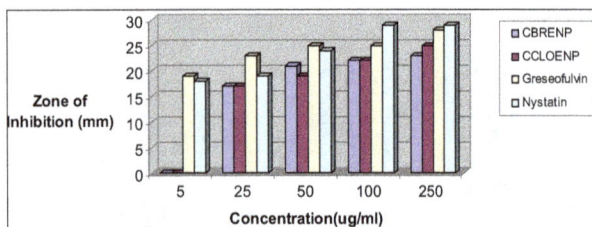

Figure 7. Zone of inhibition of *A. niger* against the standard drugs and the complexes.

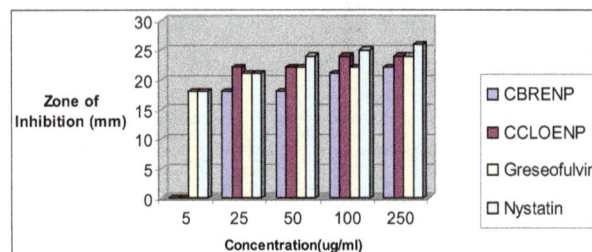

Figure 8. Zone of inhibition of *C. albican* against the concentration of the standard drugs and the complexes.

Table 4. Zone of inhibition of the complexes and the standard against the fungi (mm).

Fungi species	*A. niger* MTCC 282					*C. albicans* MTCC 227				
Concn µg/ml	5	25	50	100	250	5	25	50	100	250
CCLOENP	-	11	14	18	21	-	13	15	17	18
CBRENP	-	12	15	18	19	-	12	14	16	18
Greseofulvin	19	23	25	25	28	18	21	22	22	24
Nystatin	18	19	24	29	29	18	21	24	25	26

4. CONCLUSION

Two new mixed ligand complexes of copper(II) with two chelating ligands have been synthesized and characterized. Both complexes 1 and 2 have the same $[Cu(en)(phen)_2]^{2+}$ cation, but differ in anionic part, though complex 1 has no water molecule, hydrogen bonding exists between the H in 1,2-diaminoethane and O from chlorate ion which form network within the lattice. The two complexes exhibit antimicrobial activities which are comparable to the known standard drugs used for some of the micro-organisms considered.

5. ACKNOWLEDGEMENTS

Onawumi O. O. thanks TWAS (The Academy of Science For Developing World) and CSIR (Council of Scientific and Industrial Research), New Delhi for the award of postdoctoral fellowship. Ladoke Akintola University of Technology, Nigeria is gratefully acknowledged for study leave. Dr. Parimal Paul is highly acknowledged for being a good host at CSMCRI, Bharvnagar, Gujurat. India. Thanks to Vinod K. Agarwal for I.R analysis, E. Suresh for determined the crystal structure. A. S. Onawumi is highly appreciated for his support. Microcare Laboratory at Unapani Road, Surat, India is gratefully acknowledged for conducting microbial experiments. Thanks to Mr. H. Brahmbhatt for technical assistance.

REFERENCES

[1]   Lemoine, P., Viossat, B., Morgant, G., Greenaway, F.T., Tomas, A., Dung, N.H. and Sorenson, J.R.J. (2002) Synthesis, crystal structure, EPR properties, and anticonvulsant activities of binuclear and mononuclear 1,10-phenanthroline and salicylate ternary copper(II) complexes. *Journal of Inorganic Biochemistry*, **89**, 18-28. doi:10.1016/S0162-0134(01)00324-5

[2]   Devereux, M., Shea, D.O., Kellet, A., McCann, M., Walsh, M., Egan, D., Deggan, C., Kedziora, K., Rosair, G. and Muller-Bunz, H. (2007) Synthesis, X-ray crystal structures and biomimetic and anticancer activities of novel copper(II)benzoate complexes incorporating 2-(4'-thiazolyl)benzimidazole (thiabendazole), 2-(2-pyridyl)benzimidazole and 1,10-phenanthroline as chelating nitrogen donor ligands. *Journal of Inorganic Biochemistry*, **10**, 881-892. doi:10.1016/j.jinorgbio.2007.02.002

[3]   Lave-Cambot, A., Cantnuel, M., Leydet, Y., Jonususkas, G., Bassani, D.M. and McClenaghan, N.D. (2008) Improving the photophysical properties of copper(I) bis(phenanthroline) complexes. *Coordination Chemistry Reviews*, **252**, 2572-2584.

[4]   Onawumi, O.O.E., OFaboya, O.P., Odunola, O.A., Prasad, T.K. and Rajasekharan, M.V. (2011) Mixed ligand trischelate of copper(II) with *N*,*N*-donor ligands—Synthesis, structure and spectra of $[Cu(en)(phen)_2]X_2 \cdot 2phen \cdot 8H_2O$ (X=Cl$^-$, NO$_3^-$). *Polyhedron*, **30**, 725-729.

doi:10.1016/j.poly.2010.12.004

[5]   (2003) SAINTPLUS: Bruker AXS Inc., Madison.

[6]   Sheldrick, G.M. (1996) SADABS: Program for empirical absorption correction. Universityof Göttingen, Göttingen.

[7]   Sheldrick, G.M. (1997) SHELX-97: Programs for crystal structure analysis. University of Göttingen, Göttingen.

[8]   Macrae, C.F., Edgington, P.R., McCabe, P., Pidcock, E., Shields, G.P., Taylor, R., Towler, M. and van de Streek, J. (2006) Mercury: Visualization and analysis of crystal structures. *Journal of Applied Crystallography*, **39**, 453. doi:10.1107/S002188980600731X

[9]   Nakamoto K. (1963) Infrared and Raman spectra of inorganic and coordination compounds. John Willey and Sons, New York, 155.

[10]   Lever, A.B.P. (1984) Inorganic electronic spectroscopy. Elsevier, Amsterdam.

[11]   Kaur H. (2010) Instrumental methods of chemical analysis. 6th Edition, Pragati Prakashan Publishers, Meerut, 991.

[12]   Prakesh, A. and Singh, M.P.G.K. (2011) Synthesis spectroscopy and biological studies of nickel(II) complexes with tertradentate schiff bases having N_2O_2 donor group. *Journal of Developmental Biology and Tissue Engineering*, **3**, 13-19.

APPENDIX

CCDC 870713 contain the supplementary crystallo-graphic data for compound 1. These data can be obtained free of charge via

http://www.ccdc.cam.ac.uk/conts/retrieving.html, or from the Cambridge Crystallographic Data Centre, 12 Union Road, Cambridge CB2 1EZ, UK; fax: (+44) 1223-336-033; or e-mail: deposit@ccdc.cam.ac.uk.

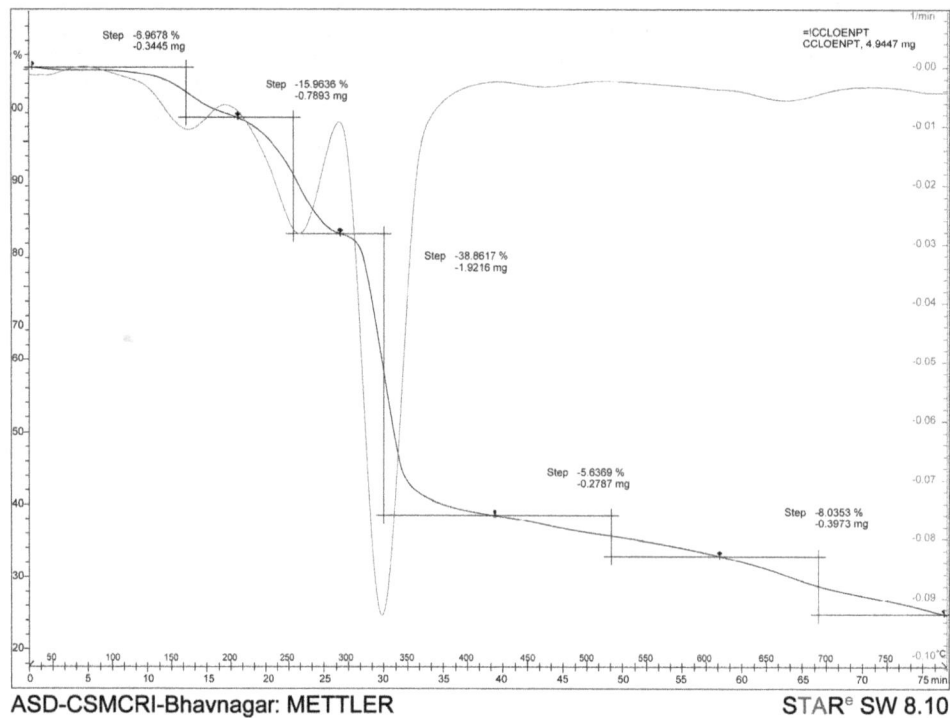

Thermogravimetric analysis of $[Cu(en)(phen)_2](ClO_4)_2$.

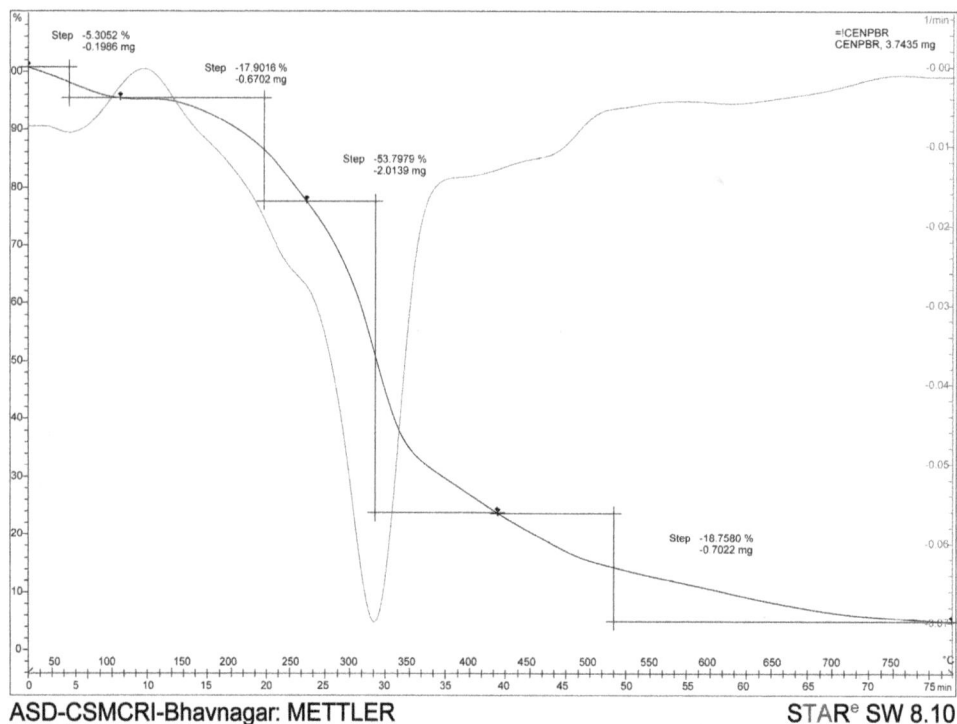

Thermogravimetric analysis of $[Cu(en)(phen)_2]Br_2 \cdot 2phen \cdot 8H_2O$.

3

Visible photoluminescence of polyoxoniobates in aqueous solution and their high electrocatalytic activities for water oxidation

Yichen Ye, Congcong Chen, Hui Feng*, **Jin Zhou, Juanjuan Ma, Jianrong Chen, Junhua Yuan, Lichun Kong, Zhaosheng Qian***

College of Chemistry and Life Science, Zhejiang Normal University, Jinhua, China
Email: *fenghui@zjnu.cn, *qianzhaosheng@zjnu.cn

ABSTRACT

The photoluminescence of four polyoxoniobates $[Nb_6O_{19}]^{8-}$, $[Nb_{10}O_{28}]^{6-}$, $[Ti_2Nb_8O_{28}]^{8-}$ and $[H_2Si_4Nb_{16}O_{56}]^{14-}$ was observed, and its origin was revealed in the view of molecular orbital by means of the computational method. The photoluminescence is originated from singlet transitions, and the calculated values agree well with the experimental data. The results indicate that the size of clusters and the foreigner atoms can affect the fluorescent properties of PONbs. The absorption and emission of these PONbs are originated molecular orbitals contributed mainly by μ_2-O and Nb atoms according to NBO analysis. These PONbs were also found as electrochemical catalysts with high performance for water oxidation, which can effectively split water into oxygen under basic condition with a high catalytic current, and pH values have remarkable influence on the electrocatalytic activities of these PONbs for water oxidation.

Keywords: PONbs; Photoluminescence; NBO Analysis; Electrocatalysis; Water Oxidation

1. INTRODUCTION

Polyoxoniobates (PONbs) have attracted significant interest for their fascinating properties and manifold potential applications in many fields such as photocatalytic water dissociation, medicine and radionuclide sequestration, as well as geochemistry [1-4]. Recent review has presented the progress of PONbs chemistry in the preparation and characterization of novel polyniobate compounds [5]. Furthermore, the PONbs family has also been successfully used as model complexes to explore

oxygen exchange dynamics for understanding aqueous reactions involving minerals and extended structures and dissolution of similar catalysts [6]. However, little attention has been paid on the optical and catalytic properties of the PONbs.

The hexaniobate $[Nb_6O_{19}]^{8-}$ Lindqvist ion is the first PONb, and has been used as an important precursor to prepare large isopolyoxoniobates such as $[Nb_{10}O_{28}]^{6-}$, $[Nb_{20}O_{54}]^{8-}$ and $[Nb_{24}O_{72}H_9]^{15-}$ [7-10]. Some other elements including Si, Ge, P and Ti were introduced to produce heteropolyoxoniobates in order to tailor the properties of niobate family [11-16]. The incorporation of transition metals into PONbs is able to control their solubility in alkaline solution. Several PONbs were selected to explore their oxygen exchange pathways and reaction activities since they can provide a molecular view of oxide dissolution and are accessible to both experimental approach and theoretical simulation. Oxygen-exchange rates were measured between sites in the Lindqvist ion $[Nb_6O_{19}]^{8-}$ and aqueous solution, and metastable polymorphs of the hexaniobate structure were proposed [17]. Reaction dynamics of the larger decaniobate ion $[Nb_{10}O_{28}]^{6-}$ revealed the possible mechanism for the base-promoted dissociation of the decaniobate ion, [18] and the comparison of two isostructrual polyoxoanions $[Nb_{10}O_{28}]^{6-}$ and $[Ti_2Nb_8O_{28}]^{8-}$ showed a profound difference in oxygen exchange rate [19]. Recently, a more complex ion $[H_2Si_4Nb_{16}O_{54}]^{14-}$ resembling colloidal oxides and minerals were explored by ^{17}O-NMR and computational approach [20]. In addition, $[Nb_{10}O_{28}]^{6-}$ as a model example of direct measurement of the formation of ion pairs in solution revealed remarkable similarity between solution and solid-state contact [21]. In spite of these progresses, these PONbs have not been found unique characteristics on redox chemistry, optical response or catalysis up to date.

---

*Corresponding authors.

Distinctly different reactivity of $[Nb_6O_{19}]^{8-}$ and $[Nb_{10}O_{28}]^{6-}$ with hydrogen peroxide was revealed, and the results showed the terminal niobium-bound oxygen atoms can be easily replaced by peroxy groups, implying the possible use of niobates as water-splitting catalysts [22]. Photoluminescence of amorphous niobium oxide was reported [23], and the unique visible luminescence of aluminum polyoxocations was found recently [24]. These results motivated us to explore the photoluminescence properties and electrocatalytic behaviour of decomposing water of these PONbs. In this work, we reported novel photoluminescence properties and high electrocatalytic activities for water oxidation of four PONbs including $[Nb_6O_{19}]^{8-}$, $[Nb_{10}O_{28}]^{6-}$, $[Ti_2Nb_8O_{28}]^{8-}$ and $[H_2Si_4Nb_{16}O_{56}]^{14-}$. These PONbs can exhibit relatively strong visible blue light, and the photoluminescence mechanism of these PONbs was revealed by means of theoretical approach. These PONbs were also found to efficiently split water to oxygen under basic condition, and there is no remarkable difference in electrocatatlytic activities between isoPONbs and heteroPONbs.

2. EXPERIMENTAL

2.1. Preparation of These PONbs

The synthesis of the four PONbs was according to the literature methods [7-16,20]. a) Preparation of $K_7HNb_6O_{19}$. 5 g KOH in a Ni crucible over a bunsen burner was melt and then 1.5 g Nb_2O_5 was very gently added in small portions. When all Nb_2O_5 has been added, the mixture was heated for 15 minutes before letting it cool. The solid melt was dissolved in 20 ml of water and then filtered to remove any solids. The crystals, $K_7HNb_6O_{19}$, grew when put in freezer overnight. b) Preparation of $K_6[Nb_{10}O_{28}]$. The $K_7HNb_6O_{19}$ crystals were dissolved in a minimum amount of water. Glacial acetic acid was gently added until the pH is below 6. The amorphous white precipitate on a cellulose filter was collected and washed repeatedly until the eluent is neutral. The solids on the frit were dried to remove as much trapped water as possible. The powder is hydrous niobium oxide, also known as niobic acid, which was the precursor for the synthesis of $K_6[Nb_{10}O_{28}]$ and $K_8[Ti_2Nb_8O_{28}]$.
$Nb_2O_5 \cdot nH_2O$ (1.00 g, 2.1 mmol; water 43% w/w) was mixed with a solution of KOH (3.3 mmol) in water (10 mL). The mixture was heated to 120°C in a Teflon-lined Parr vessel for 18 h, and then it is allowed to cool to room temperature. Supernatant by filtration becomes white and opaque on drying. c) Preparation of $K_8[Ti_2Nb_8O_{28}]$. Nb_2O_5 (0.35 g, 2.6 mmol Nb) and titanium isopropoxide (0.25 g, 0.88 mmol) were added in an 8mL KOH solution (0.34 M) in a 20 mL Teflon-lined Parr pressure vessel and stirred for approximately 20 min.

The closed vessel was placed in an oven at 200°C for 20 h. The products was the solution of $K_8[K_8Ti_2O_{28}]$. d) Preparation of $[K_{14}H_2Si_4Nb_{16}O_{56}]$. Nb_2O_5 (150 mg), tetraethyl orthosilicate (160 mg), and potassium hydroxide (110 mg) were mixed with 6 mL of water in a Teflon-lined Parr vessel. The vessel was then placed in a 200°C oven for two days.

2.2. Characterization of These PONbs

The prepared PONbs were firstly ^{17}O-enriched by dissolving in ^{17}O-enriched water (10% - 15%) and then standing night, and finally the ^{17}O-NMR spectroscopy was carried out on Bruker 400 MHz spectrometer to verify the structure of these PONbs. All the characteristic ^{17}O NMR signals were found for the four PONbs except their inert core oxygen atoms by oxygen-exchange with ^{17}O-enriched water, and corresponded to the data reported in the literature. The fluorescence spectra and lifetimes of these PONbs were preformed on a FLS 920 spectrophotometer, and their quantum yields were determined using quinine sulphate as the standard, whose quantum yield is 0.577. Determination of the quantum yields of these oxidized products was accomplished by comparison of the wavelength integrated intensity of these oxidized products to that of the standard quinine sulfate. The optical density is kept below 0.05 to avoid inner filter effects. The quantum yields of these oxidized products were calculated using

$$\Phi = \Phi_S \left[\left(I \cdot A_S \cdot n^2 \right) / \left(I_S \cdot A \cdot n_s^2 \right) \right]$$

The Φ is the quantum yield, I is the integrated intensity, A is the optical density and n is the refractive index of the solvent. The subscript S refers to the standard reference of known quantum yield. In this work, when determining PONbs, quinine sulfate was chosen as the standard, whose quantum yield is 0.577 for excitation wavelengths from 200 nm to 400 nm.

2.3. The Electrocatalytic Experiment of PONbs for Water Oxidation

During the electrocatalysis experiment, a 15 mL glass electrochemical cell containing a glassy Platinum disk working electrode (3 mm in diameter) embedded in Teflon, a platinum wire counter electrode (1 mm in diameter, 5 cm in length) and a saturated calomel electrode as the reference electrode was used. This three-electrode setup was configured with CHI 660C interfaced to a personal computer at room temperature for the electrochemical investigation of these PONbs for water oxidation. The electrolytic solution was bubbled with nitrogen for 15 minutes before the electrochemical experiment to remove oxygen. Homogeneity test of electrode surface from dilute $[Nb_6O_{19}]^{8-}$ solutions was conducted by firstly scanning the CVs for several cycles with Platinum disk electrode in the blank solution, and then getting the CVs in the $[Nb_6O_{19}]^{8-}$ solution, and finally repeating it in the blank solution after washing the Platinum disk electrode. The dependence of current on the concentration of $[Nb_6O_{19}]^{8-}$ was carried out using a series of $[Nb_6O_{19}]^{8-}$ solutions with different concentrations at the same electrochemical procedure. The influence of pH value on the electrochemical behavior was performed using $[Nb_6O_{19}]^{8-}$ solutions with different pH value.

2.4. Computational Details

The structures of these PONbs were firstly reproduced from their crystal structure and then optimized using DFT method with the B3LYP functional. The basis set LanL2DZ was performed for Nb and Ti, and 6-31+G (d,p) was used for the other elements [25,26]. The vibrational frequency was carried out to verify the stable structure. Solvent effects were evaluated by means of the integral equation formulation of the polarisable continuum model (IEFPCM) for all the calculations [27]. The electronic transition energies, oscillator strengths and excited-state compositions were computed by the time-dependent density functional theory method (TD-DFT) with B3LYP functional at the same basis sets [28,29]. The optimization of electronic transition structure and emission energies was also computed by TD-DFT method [30-32]. NBO analysis was also performed to obtain the molecular orbital [33]. All the calculations were carried out with the Gaussian 09 package [34]. All the calculations were performed for $[Nb_6O_{19}]^{8-}$, and the other calculations except the optimization of electronic transition structure and emission energies were performed for $[Nb_{10}O_{28}]^{6-}$ and $[Ti_2Nb_8O_{28}]^{8-}$. The ion $[H_2Si_4Nb_{16}O_{56}]^{14-}$ is so large that the present computational level can not deal with it at an acceptable accuracy.

3. RESULTS AND DISCUSSION

3.1. Photoluminescence of These PONbs

As shown in the **Figure 1**, the Lindqvist ion $[Nb_6O_{19}]^{8-}$ has a central μ_6-O site, 12 μ_2-O bridges and 6 terminal η-O sites. The decaniobate $[Nb_{10}O_{28}]^{6-}$ has 10 niobium atoms in one molecular, while $[Ti_2Nb_8O_{28}]^{8-}$ has eight niobium atoms and two Ti(IV) atoms replacing the two central Nb(V) atoms. The larger ion $[H_2Si_4Nb_{16}O_{56}]^{16-}$ has a lacunary, partial Keggin structure. The synthesis of $K_7[HNb_6O_{19}]$, $K_6[Nb_{10}O_{28}]$, $K_8[Ti_2Nb_8O_{28}]$ and $K_{14}H_2[Si_4Nb_{16}O_{56}]$ were according to the corresponding references, [7-11] and all the PONbs were characterized by solution-state ^{17}O-NMR technique. As shown in **Figure S1**, the $[Nb_6O_{19}]^{8-}$ shows the signals at 393 ppm and 623 ppm for μ_2-O and η-O respectively, but the signal for central oxygen atom is not detected since the central

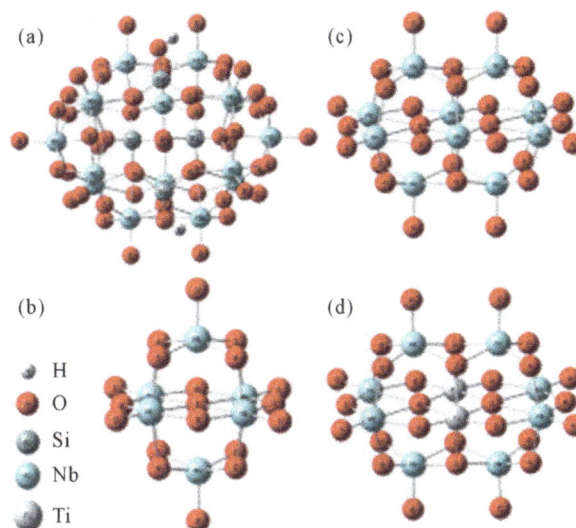

Figure 1. Structures of $[Nb_6O_{19}]^{8-}$ (b), $[Nb_{10}O_{28}]^{6-}$ (c), $[Ti_2Nb_8O_{28}]^{8-}$ (d) and $[H_2Si_4Nb_{16}O_{56}]^{14-}$ (a) from their crystal structures [7,8,14,15].

oxygen atom was not exchanged with ^{17}O water, which is consistent with the intertness of the central oxygen atom of $[Nb_{10}O_{28}]^{6-}$ found by Casey's group [17]. $[Nb_{10}O_{28}]^{6-}$ shows its characteristic chemical shifts from different sites (**Figure S2**), consistent with the result in the literature [18]. Major signals of ^{17}O-NMR for $[Ti_2Nb_8O_{28}]^{8-}$ was also detected (**Figure S3**), and the lack of the other signals may be induced by the low concentration and incomplete exchange with ^{17}O-enriched water [19]. The large cluster $[H_2Si_4Nb_{16}O_{56}]^{14-}$ was also confirmed by ^{17}O-NMR through comparing the chemical shifts at 50 ppm, 195 ppm, 394 ppm and 608 ppm (**Figure S4**) with the reported data [20].

It is the first time to observe the ultraviolet-visible luminescence of PONbs in aqueous solution at room temperature. **Figure 2** shows the excitation and emission spectra of the PONbs including $[Nb_6O_{19}]^{8-}$, $[Nb_{10}O_{28}]^{6-}$, $[Ti_2Nb_8O_{28}]^{8-}$ and $[H_2Si_4Nb_{16}O_{56}]^{14-}$. The isopolyoxobionates $[Nb_6O_{19}]^{8-}$ and $[Nb_{10}O_{28}]^{6-}$ have close central emission wavelengths at around 400 nm with a small difference of 10 nm under their excitation wavelength of 294 nm and 310 nm respectively. $[Ti_2Nb_8O_{28}]^{8-}$ exhibits the emission around 420 nm at the excitation wavelength of 306 nm, while the larger cluster $[H_2Si_4Nb_{16}O_{56}]^{14-}$ emits the light band at 390 nm, which suggests the role of introduced foreigner elements. The PONbs including $[Nb_6O_{19}]^{8-}$, $[Nb_{10}O_{28}]^{6-}$ and $[H_2Si_4Nb_{16}O_{56}]^{14-}$ exhibit a similar blue light under the UV lamp as shown in **Figure 2**, while $[Ti_2Nb_8O_{28}]^{8-}$ shows a green fluorescence. In order to explore the nature of the photoluminescence, we determined their lifetimes and quantum yields listed in **Table 1**. Their lifetimes are less than 10 ns, indicating that their photoluminescence is originated from fluorescence, similar with luminescence of aluminum polyoxo-

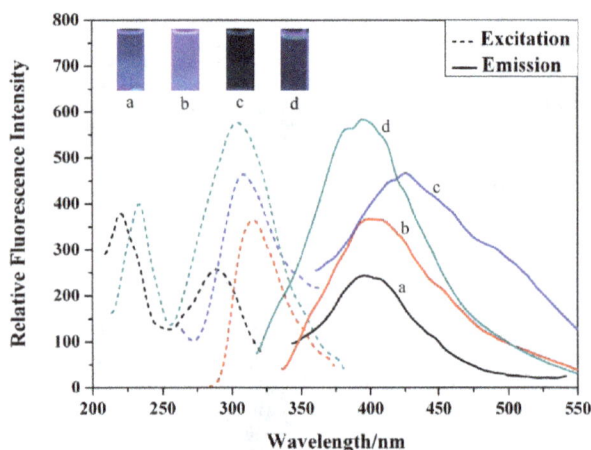

Figure 2. The photoluminescence spectra of $[Nb_6O_{19}]^{8-}$ (a), $[Nb_{10}O_{28}]^{6-}$ (b), $[Ti_2Nb_8O_{28}]^{8-}$ (c) and $[H_2Si_4Nb_{16}O_{56}]^{14-}$ (d) and their visible fluorescence photographs.

Table 1. The fluorescence quantum yields and lifetimes of $[Nb_6O_{19}]^{8-}$, $[Nb_{10}O_{28}]^{6-}$, $[Ti_2Nb_8O_{28}]^{8-}$ and $[H_2Si_4Nb_{16}O_{56}]^{14-}$.

Sample	$[Nb_6O_{19}]^{8-}$	$[Nb10O_{28}]^{6-}$	$[Ti_2Nb_8O_{28}]^{8-}$	$[H_2Si_4Nb_{16}O_{56}]^{14-}$
Φ^a (%)	0.067	0.51	0.13	1.1
τ^b (ns)	4.2	7.2	7.2	8.7

[a]The fluorescence quantum yields relative to quinine sulfate (0.577). [b]The average lifetimes.

cations [24]. The quantum yields are increased with their sizes except $[Ti_2Nb_8O_{28}]^{8-}$, and the incorporation of Ti into $[Nb_{10}O_{28}]^{6-}$ lowers the quantum yield.

In order to explore the nature of the fluorescence of these PONbs, we carried out DFT calculations on them. The geometries were optimized from their X-ray crystal structure with consideration of solvent effects. The electronic transition components were calculated at TD-DFT level of theory [30-32]. The calculations show that the computational absorption maximum wavelengths for $[Nb_6O_{19}]^{8-}$, $[Nb_{10}O_{28}]^{6-}$ and $[Ti_2Nb_8O_{28}]^{8-}$ are very close to their experimental data (Table S1). The largest contribution to actual transition for $[Nb_6O_{19}]^{8-}$ is the transition from HOMO−1 to LUMO+4, and five additional transitions also contribute to excitation wavelength 294 nm (**Figure S5**). The electronic transitions contributing largest to actual transitions for $[Nb_{10}O_{28}]^{6-}$ and $[Ti_2Nb_8O_{28}]^{8-}$ are from HUMO−3 to LOMO and from HOMO to LUMO+3 respectively (**Figures S6** and **S7**). All the transitions with largest contribution involve the electron transfer from μ_2-O to η-O, indicating their distinctive characteristics. We take $[Nb_6O_{19}]^{8-}$ as an example to demonstrate the whole absorption and emission process in the view of molecular orbital as shown in **Figure 3**. Six one-electron transitions including the transition from HOMO−1 to LUMO+3 with largest contribution result in the actual excitation (absorption) wavelength 294 nm.

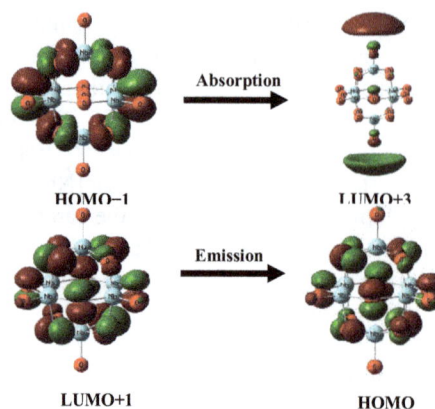

Figure 3. The electron-transitions with the largest contribution to actual absorption wavelength and emission wavelength of $[Nb_6O_{19}]^{8-}$ respectively.

The structure of the excited ion quickly changes after the transition occurs. Following internal conversion, the electron reaches the lowest energy vibrational state of S_1. The structure of $[Nb_6O_{19}]^{8-}$ at S_1 state was built with TD-optimization. The TD-calculations show the fluorescence emission is from the transition between LUMO+1 and HOMO. These results show the key role of μ_2-O and η-O atoms in the photoluminescence properties of these PONbs.

3.2. Electrocatalytic Activities of These PONbs for Water Oxidation

It was reported that manganese complexes, ruthenium complexes, cobalt complexes and iridium oxide colloid can be used as efficient catalysts for electrochemical water oxidation [35]. Except for cobalt complexes, it was found that cobalt polyoxometalates can also be used to splitting water, and the dominant water oxidation catalyst is actually heterogeneous cobalt oxides [36]. Ruthenium polyoxometalate cluster attached on carbon nanotubes was also shown as a very efficient and stable water oxidation catalyst.

These reports demonstrated that the efficient water oxidation catalysts are mainly composed of manganese, ruthenium, cobalt and iridium. However, herein we resented that the PONbs have high electrocatalytic activities for water oxidation under basic condition. **Figure 4** shows the electrocatalytic activities of these PONbs for water oxidation. An oxidation process is evident at 1.35 V and 1.48 V for $[Nb_{10}O_{28}]^{6-}$ and $[H_2Si_4Nb_{16}O_{56}]^{14-}$ respectively, and generates a significant rise in the current consistent with a catalytic process. The maximum peak current of $[Nb_{10}O_{28}]^{6-}$ is five times larger than that for the blank solution, while that for $[Si_4Nb_{16}O_{56}]^{16-}$ is nearly tenfold the blank solution. The rise in current around 1.4 V is attributed to oxidation of water catalyzed by

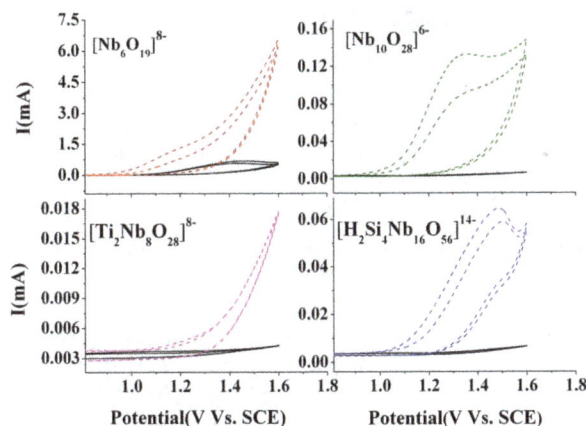

Figure 4. Cyclic voltammograms for $[Nb_6O_{19}]^{8-}$ at pH 13.7, $[Nb_{10}O_{28}]^{6-}$ at pH 9.1, $[Ti_2Nb_8O_{28}]^{8-}$ at pH 8.6, $[H_2Si_4Nb_{16}O_{56}]^{14-}$ at pH 10.3, and their corresponding blank solutions. Working electrode, platinum disk electrode; scan rate, 100 mV/s.

$[Nb_{10}O_{28}]^{6-}$ and $[H_2Si_4Nb_{16}O_{56}]^{14-}$, and the potentials at maximum current of water oxidation are close to those reported in the similar catalytic processes. The currents for $[Nb_6O_{19}]^{8-}$ and $[Ti_2Nb_8O_{28}]^{8-}$ solutions are alsoby several times relative to those for the blank although their CVs have no remarkable peak currents. This electrocatalytic behaviors of $[Nb_6O_{19}]^{8-}$ and $[Ti_2Nb_8O_{28}]^{8-}$ is similar to POM supported on multi-walled carbon nanotubes [37-39].

The possibility of a catalytically active niobium oxide film generated by the decomposition is excluded by homogeneity test of electrode surface from $[Nb_6O_{19}]^{8-}$ solution. As shown in **Figure 5**, the cyclic voltammetric response of working electrode in a blank electrolyte following two anodic cycles from 0 - 1.8 V vs SCE in 1 mM of $[Nb_6O_{19}]^{8-}$. The electrode was taken out of the solution, washed by distilled water and continued to the test. The current after removing $[Nb_6O_{19}]^{8-}$ is much lower than that for $[Nb_6O_{19}]^{8-}$ and close to that for the blank solution, which demonstrated that the true water oxidation catalyst is the ion $[Nb_6O_{19}]^{8-}$ in solution rather than the niobium oxide formed on the electrode surface. The dependence of the catalytic current on concentration of $[Nb_6O_{19}]^{8-}$ shown in **Figure 6** also supported the key role of these PONbs in the catalysis. From the figure one can see clearly the current of $[Nb_6O_{19}]^{8-}$ was increased as the concentration rises, demonstrating the dependence of the catalytic current on concentration of $[Nb_6O_{19}]^{8-}$ in aqueous solution. We also tested the influence of pH values on the electrocatalytic behavior of $[Nb_6O_{19}]^{8-}$ for water oxidation. As shown in **Figure 7**, the catalytic current is increased as pH values, but electrocatalytic behaviors of these PONbs for water oxidation were not remarkably affected by the variety of pH values, which indicated that the number of protons and charges of the PONbs would affect their electrocatalytic activities.

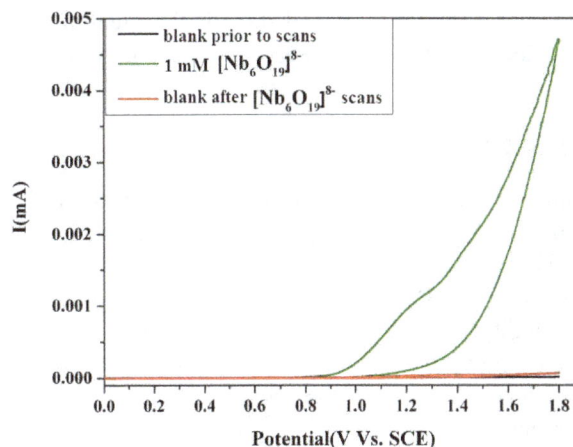

Figure 5. Homogeneity test of electrode surface from dilute $[Nb_6O_{19}]^{8-}$ solutions. The cyclic voltammetric response of glassy carbon working electrode in distilled water (black) and $[Nb_6O_{19}]^{8-}$ (green) at the same pH of 13 were detected (scan rate = 10 mV/s), and after gently washing the electrode with distilled H_2O then scanning in blank electrolyte (red).

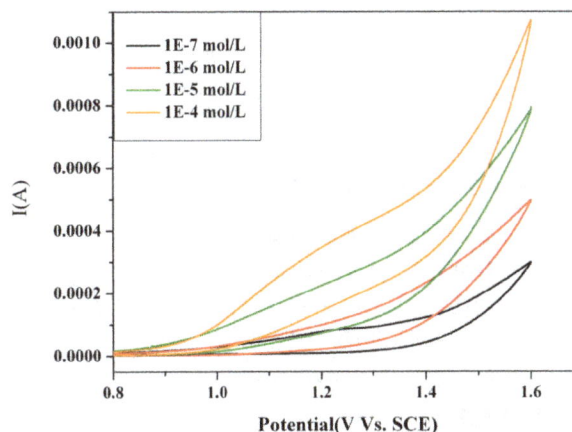

Figure 6. Cyclic voltammograms of $[Nb_6O_{19}]^{8-}$ with different concentrations recorded in aqueous solution (pH = 12, scan rate = 10 mV/s).

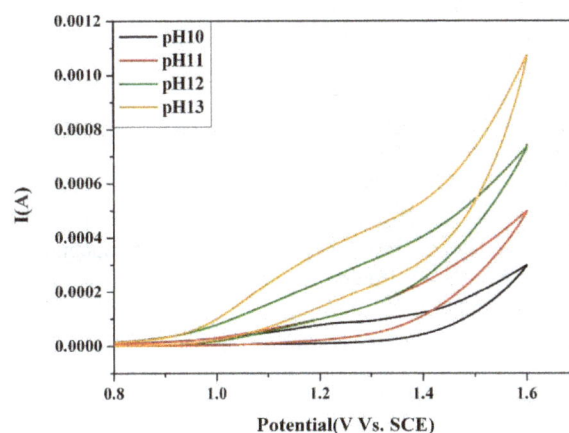

Figure 7. Cyclic voltammograms of $[Nb_6O_{19}]^{8-}$ under different pH conditions recorded in aqueous solution (scan rate = 10 mV/s).

However, the redox chemistry of these PONbs underlying the electrocatalytic processes is needed to further explore.

4. CONCLUSION

In summary, we provided the examples of PONbs for their novel photoluminescence and potential redox chemistry. The photoluminescence of $[Nb_6O_{19}]^{8-}$, $[Nb_{10}O_{28}]^{6-}$, $[Ti_2Nb_8O_{28}]^{8-}$ and $[H_2Si_4Nb_{16}O_{56}]^{14-}$ was observed, the tunable ability of their fluorescence by foreign atoms was demonstrated, and the origin of the fluorescence was revealed in the view of molecular orbital theory. The luminescence of these PONbs gives a good example for optical response of the other polyoxometalates. The high performance of these PONbs on water oxidation was also demonstrated, which implies their huge potential application for splitting water, and opens the door to redox chemistry of PONbs.

6. ACKNOWLEDGEMENTS

We are thankful for the supports by Chinese National Natural Science Foundation (No. 21005073) and Foundation of Zhejiang Province (No. LY13B050001, LQ13B050002 and Y201225601) and Xin Miao Talents Planning of Zhejiang Province (2012R404038).

REFERENCES

[1] Rhule, J.T., Hill, C.L. and Judd, D.A. (1998) Polyoxometalates in medicine. *Chemial Review*, **98**, 327-357. doi:10.1021/cr960396q

[2] Nyman, M., Bonhomme, F., Alam, T.M., Rodriguez, M. A., Cherry, B.R., Krumhansl, J.L., Nenoff, T.M. and Sattler, A.M. (2002) A general synthetic procedure for heteropolyniobates. *Science*, **297**, 996-998.

[3] Kudo, A. and Miseki, Y. (2009) Heterogeneous photocatalyst materials for water splitting. *Chemical Society Review*, **38**, 253-278. doi:10.1039/b800489g

[4] Rustad, J.R. and Casey, W.H. (2012) Destruction of the Kondo effect in the cubic heavy-fermion compound $Ce_3Pd_{20}Si_6$. *Nature Material*, **11**, 189-194. doi:10.1038/nmat3214

[5] Nyman, M. (2011) Polyoxoniobate chemistry in the 21st century. *Dalton Transactions*, **40**, 8049-8058. doi:10.1039/c1dt10435g

[6] Casey, W.H., Rustad, J.R. and Spiccia, L. (2009) Minerals as molecules-use of aqueous oxide and hydroxide clusters to understand geochemical reactions. *Chemical Europe Journal*, **15**, 4496-4515.

doi:10.1002/chem.200802636

[7] Lindqvist, I. (1952) The structure of the hexaniobate ion in $7Na_2O \cdot 6Nb_2O_5 \cdot 32H_2O$. *Ark Kemi*, **5**, 247-250.

[8] Graeber, E.J. and Morosin, B. (1977) The molecular configuration of the decaniobate ion ($Nb_{10}O_{28}^{6-}$). *Acta Crystallographica, Section B*, **33**, 2137-2143. doi:10.1107/S0567740877007900

[9] Maekawa, M., Ozawa, Y. and Yagasaki, A. (2005) Icosaniobate: A new member of the isoniobate family *Inorganic Chemistry*, **45**, 9608-9609.

[10] Ranko, P. and Nyman, M. (2006) Evolution of polyoxoniobate cluster anions. *Angewandte Chemie International Edition*, **45**, 6670-6672. doi:10.1002/anie.200602200

[11] Nyman, M., Bonhomme, F., Alam, T.M., Parise, J.B. and Vaughan, G.M.B. (2004) $[SiNb_{12}O_{40}]^{16-}$ and $[GeNb_{12}O_{40}]^{16-}$: Highly charged keggin ions with sticky surfaces. *Angewandte Chemie International Edition*, **43**, 2787-2792. doi:10.1002/anie.200353410

[12] Nyman, M., Celestian, A.J., Parise, J.B., Holland, G.P. and Alam, T.M. (2006) Solid-state structural characterization of a rigid framework of lacunary heteropolyniobates. *Inorganic Chemistry*, **45**, 1043-1052. doi:10.1021/ic051155g

[13] Ohlin, C.A., Villa, E.M., Fettinger, J.C. and Casey, W.H. (2008) The $[Ti_{12}Nb_6O_{44}]^{10-}$ ion—A new type of polyoxometalate structure. *Angewandte Chemie International Edition*, **47**, 5634-5636. doi:10.1002/anie.200801883

[14] Ohlin, C.A., Villa, E.M., Fettinger, J.C. and Casey, W.H. (2009) A new titanoniobate ion-completing the series $[Nb_{10}O_{28}]^{6-}$, $[TiNb_9O_{28}]^{7-}$ and $[Ti_2Nb_8O_{28}]^{8-}$. *Dalton Transactions*, **15**, 2677-2678. doi:10.1039/b900465c

[15] Johnson, R.L., Villa, E.M., Ohlin, C.A., Rustad, J.R. and Casey, W.H. (2011) ^{17}O NMR and computational study of a tetrasiliconiobate ion, $[H^{2+x}Si_4Nb_{16}O_{56}]^{(14-x)-}$. *Chemical Europe Journal*, **17**, 9359-9367. doi:10.1002/chem.201100004

[16] Huang, P., Qin, C., Wang, X.L., Sun, C.Y., Xing, Y., Wang, H.N., Shao, K.Z. and Su, Z.M. (2012) A new organic-inorganic hybrid based on the crescent-shaped polyoxoanion $[H_6SiNb_{18}O_{54}]^{8-}$ and copper-organic cations. *Dalton Transactions*, **41**, 6075-6077. doi:10.1039/c2dt30265a

[17] Black, J.R., Nyman, M. and Casey, W.H. (2006) Rates of oxygen exchange between the $[H_xNb_6O_{19}]^{8-x}$(aq) lindqvist ion and aqueous solutions. *Journal American Chemical Society*, **128**, 14712-14720.

[18] Villa, E.M., Ohlin, C.A., Balogh, E., Anderson, T.M., Nyman, M.D. and Casey, W.H. (2008) Reaction dynamics of the decaniobate ion $[H_xNb_{10}O_{28}]^{(6-x)-}$ in Water. *Angewandte Chemie International Edition*, **47**, 4844-4846. doi:10.1002/anie.200801125

[19] Villa, E.M., Ohlin, A., Rustad, J.R. and Casey, W.H. (2009) Isotope-exchange dynamics in isostructural decametalates with profound differences in reactivity. *Journal of American Chemical Society*, **131**, 16488-16492. doi:10.1021/ja905166c

[20] Johnson, R.L., Villa, E.M., Ohlin, C.A., Rustad, J.R. and Casey, W.H. (2011) ^{17}O NMR and computational study

of a tetrasiliconiobate ion, $[H^{2+x}Si^4Nb^{16}O^{56}]^{(14-x)-}$. *Chemical Europe Journal*, **17**, 9359-9367. doi:10.1002/chem.201100004

[21] Antonio, M.R., Nyman, M. and Anderson, T.M. (2009) Direct observation of contact ion-pair formation in aqueous solution, *Angewandte Chemie International Edition*, **48**, 6136-6140. doi:10.1002/anie.200805323

[22] Ohlin, C.A., Villa, E.M., Fettinger, J.C. and Casey, W.H. (2008) Distinctly different reactivities of two similar polyoxoniobates with hydrogen peroxide. *Angewandte Chemie International Edition*, **47**, 8251-8254. doi:10.1002/anie.200803688

[23] Zhou, X.F., Li, Z.C., Wang, Y.Q., Sheng, X. and Zhang, Z.J. (2008) Photoluminescence of amorphous niobium oxide films synthesized by solid-state reaction. *Thin Solid Films*, **516**, 4213-4216. doi:10.1016/j.tsf.2007.12.112

[24] Qian, Z.S., Chen, C.C., Chen, J.R., Kong, L.C., Wang, C., Zhou, J. and Feng, H. (2011) Unusual visible luminescence of aluminium polyoxocations in aqueous Solution. *Chemical Communication*, **47**, 12652-12654. doi:10.1039/c1cc15823f

[25] Musa, K.A.K. and Eriksson, L.A. (2009) Photodegradation mechanism of nonsteroidal anti-inflammatory drugs containing thiophene moieties: Suprofen and tiaprofenic acid. *Journal of Physical Chemistry B*, **113**, 11306-11313. doi:10.1021/jp904171p

[26] Cardenas-Jiron, G.I., Barboza, C.A., Lopez, R. and Menendez, M.I. (2011) Theoretical study on the electronic excitations of a porphyrin-polypyridyl ruthenium (II) photosensitizer. *Journal of Physical Chemistry A*, **115**, 11988-11997. doi:10.1021/jp202377d

[27] Tomasi, J., Mennucci, B., Cammi, R. (2005) Quantum mechanical continuum solvation models. *Chemical Review*, **105**, 2999-3094. doi:10.1021/cr9904009

[28] Runge, E. and Gross, E.K.U. (1984) Density-functional theory for time-dependent systems. *Physical Review Letters*, **52**, 997-1000. doi:10.1103/PhysRevLett.52.997

[29] Scalmani, G., Frisch, M.J., Mennucci, B., Tomasi, J., Cammi, R. and Barone, V. (2006) Geometries and properties of excited states in the gas phase and in solution: Theory and application of a time-dependent density functional theory polarizable continuum model. *Journal of Chemical Physics*, **124**, 94-107.

[30] Dreuw, A., Plotner, J., Lorenz, L., Wachtveitl, J., Djanhan, J.E., Bruning, J., Metz, T., Bolte, M. and Schmidt, M.U. (2005) Molecular mechanism of the solid-state fluorescence behavior of the organic pigment yellow 101 and its derivatives. *Angewandte Chemie International Edition*, **44**, 7787-7786. doi:10.1002/anie.200501781

[31] Fortage, J., Peltier, C., Perruchot, C., Takemoto, Y., Teki, Y., Bedioui, F., Marvaud, V., Dupeyre, G., Pospisil, L., Adamo, C., Hromadova, M., Ciofini, I. and Laine, P.P. (2012) Single-step versus stepwise two-electron reduction of polyarylpyridiniums: Insights from the steric switching of redox potential compression. *Journal of Chemical Physics*, **134**, 2691-2705.

[32] Bereau, V., Duhayon, C., Sournia-Saquet, A. and Sutter, J.-P. (2012) Tuning of the emission efficiency and HOMO-LUMO band gap for ester-functionalized {Al(salophen)(H$_2$O)$_2$}$^+$ blue luminophors. *Inorganic Chemistry*, **51**, 1309-1318. doi:10.1021/ic201208c

[33] Reed, A.E., Curtiss, L.A. and Weinhold, F. (1988) Intermolecular interactions from a natural bond orbital, donor-acceptor viewpoint. *Chemical Review*, **88**, 899-926. doi:10.1021/cr00088a005

[34] Frisch, M.J., Trucks, G.W., Schlegel, H.B., Scuseria, G.E., Robb, M.A., Cheeseman, J.R., Scalmani, G., Barone, V., Mennucci, B., Petersson, G.A., Nakatsuji, H., Caricato, M., Li, X., Hratchian, H.P., Izmaylov, A.F., Bloino, J., Zheng, G., Sonnenberg, J.L., Hada, M., Ehara, M., Toyota, K., Fukuda, R., Hasegawa, J., Ishida, M., Nakajima, T., Honda, Y., Kitao, O., Nakai, H., Vreven, T., Montgomery, J.A., Peralta, J.E., Ogliaro, F., Bearpark, M., Heyd, J.J., Brothers, E., Kudin, K.N., Staroverov, V. N., Kobayashi, R., Normand, J., Raghavachari, K., Rendell, A., Burant, J.C., Iyengar, S.S., Tomasi, J., Cossi, M., Rega, N., Millam, J.M., Klene, M., Knox, J.E., Cross, J.B., Bakken, V., Adamo, C., Jaramillo, J., Gomperts, R., Stratmann,, Yazyev, O., Austin, A.J., Cammi, R., Pomelli, C., Ochterski, J.W., Martin, R.L., Morokuma, K., Zakrzewski, V.G., Voth, G. A., Salvador, P., Dannenberg, J.J., Dapprich, S., Daniels, A.D., Farkas, O., Foresman, J. B., Ortiz, J.V., Cioslowski, J. and Fox, D.J. (2009) Gaussian 09, Revision A.02, Gaussian, Inc., Wallingford CT.

[35] Stracke, J.J. and Finke, R.G. (2011) Electrocatalytic Water Oxidation Beginning with the Cobalt Polyoxometalate $[Co_4(H_2O)_2(PW_9O_{34})_2]^{10-}$: Identification of heterogeneous CoO$_x$ as the dominant catalyst. *Journal of the American Chemical Society*, **133**, 14872-14875. doi:10.1021/ja205569j

[36] Toma, F.M., Sartorel, A., Lurlo, M., Carraro, M., Parisse, P., Maccato, C., Rapino, S., Gonzalez, B.R., Amenitsch, H., DaRos, G.T., Casalis, L., Goldoni, A., Marcaccio, M., Scorrano, G., Scoles, F., Paolucci, M. and Prato, B.M. (2010) Efficient water oxidation at carbon nanotubepolyoxometalate electrocatalytic interfaces. *Nature Chemistry*, **2**, 826-831.

[37] Concepcion, J.J., Jurss, J.W., Hoertz, P.G. and Meyer, T.J. (2009) Catalytic and surface-electrocatalytic water oxidation by redox mediator-catalyst assemblies. *Angewandte Chemie International Edition*, **48**, 9473-9476. doi:10.1002/anie.200901279

[38] Stracke, J.J. and Finke, R.G., (2011) Electrocatalytic Water Oxidation Beginning with the Cobalt Polyoxometalate $[Co_4(H_2O)_2(PW_9O_{34})_2]^{10-}$: Identification of heterogeneous CoO$_x$ as the dominant catalyst. *Journal of the American Chemical Society*, **133**, 14872-14875. doi:10.1021/ja205569j

[39] Wasylenko, D.J., Palmer, R.D., Schott, E. and Berlinguette, C.P. (2012) Interrogation of electrocatalytic water oxidation mediated by a cobalt complex. *Chemical Communications*, **48**, 2107-2109. doi:10.1039/c2cc16674g

SUPPORTING INFORMATION

1. The ^{17}O NMR spectrum of $[Nb_6O_{19}]^{8-}$ in aqueous solution.

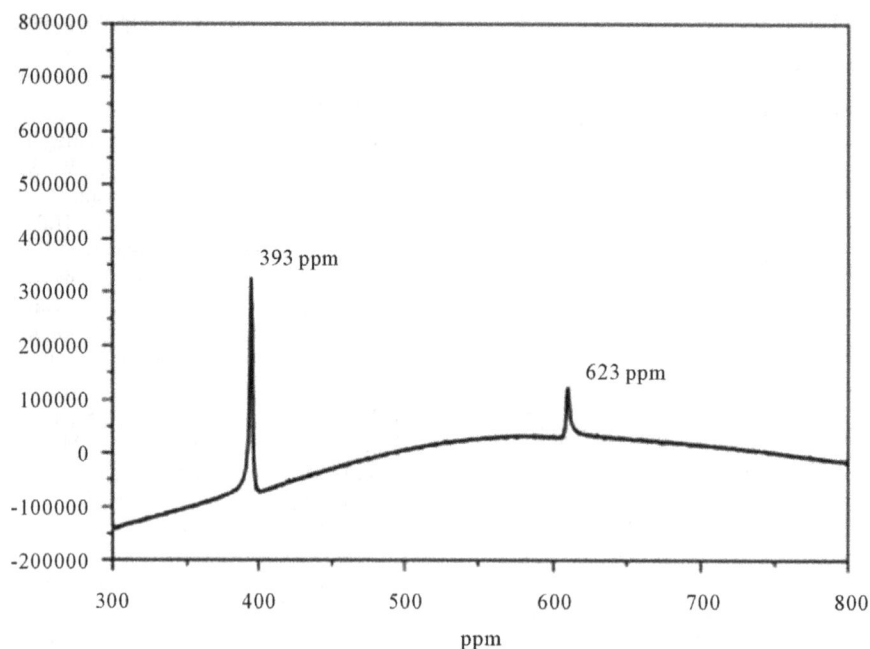

Figure S1. The ^{17}O NMR spectrum of $[Nb_6O_{19}]^{8-}$ in aqueous solution (pH = 13, T = 306 K). The signal at 393 ppm corresponds to μ_2-O, and the signal at 623 ppm corresponds to η-O. The signal from central oxygen atom is too weak to detect because its inert hinder the exchange with ^{17}O-enriched water.

2. The ^{17}O NMR spectrum of $[Nb_{10}O_{28}]^{6-}$ in aqueous solution.

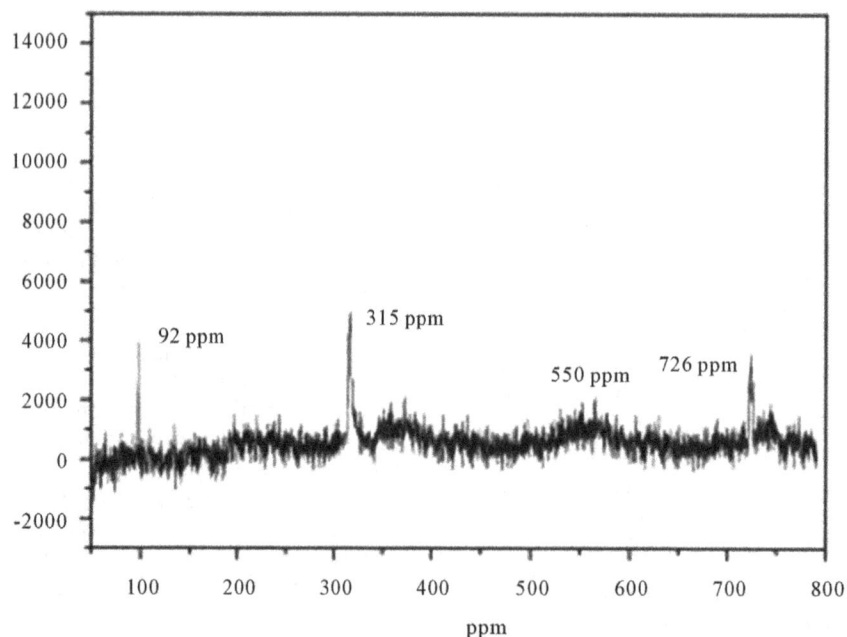

Figure S2. The ^{17}O NMR spectra of $[Nb_{10}O_{28}]^{6-}$ in aqueous solution (pH = 10, T = 306 K). The characteristic signals at 92 ppm, 315 ppm, 550 ppm and 726 pmm are in accordance with the literature.

3. The ^{17}O NMR spectrum of $[Ti_2Nb_8O_{28}]^{8-}$ in aqueous solution.

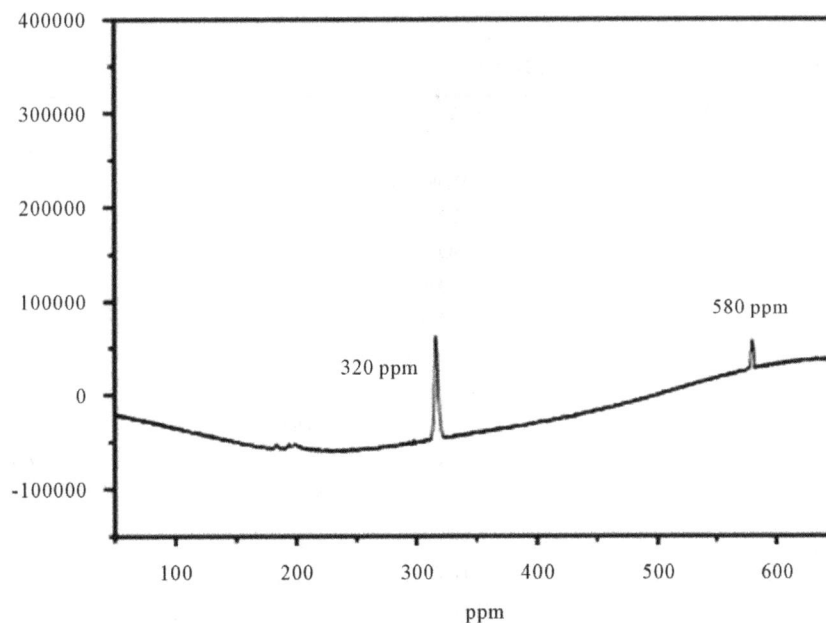

Figure S3. The ^{17}O NMR spectra of $[Ti_2Nb_8O_{28}]^{8-}$ in aqueous solution (pH = 8.3, T = 306 K). The characteristic signals at 320 ppm and 580 ppm are shown and consistent with the literature. The lack of the other signals may be induced by the low concentration and incomplete exchange with ^{17}O-enriched water.

4. The ^{17}O NMR spectrum of $[H_2Si_4Nb_{16}O_{56}]^{14-}$ in aqueous solution.

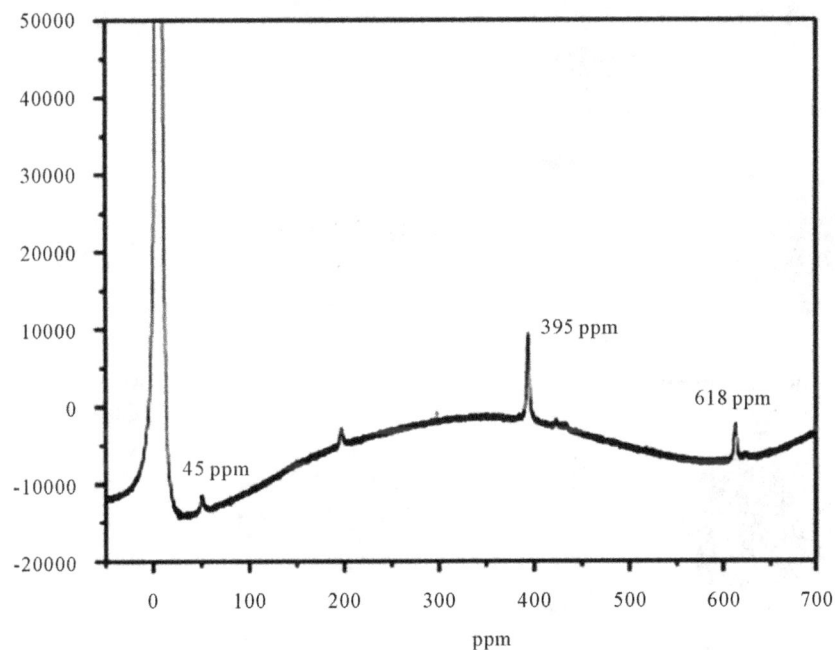

Figure S4. The ^{17}O NMR spectrum of $[H_2Si_4Nb_{16}O_{56}]^{14-}$ in aqueous solution (pH = 10.2, T = 306 K). The characteristic signals at 45 ppm, 440 ppm and 618 ppm are shown and consistent with the literature. The lack of the other signals may be induced by the low concentration and incomplete exchange with ^{17}O-enriched water.

5. The experimental and calculated absorption maximum wavelengths, oscillator strength (f_{calcd}), transitions and CI expansion coefficients (contributions) of $[Nb_6O_{19}]^{8-}$, $[Nb_{10}O_{28}]^{6-}$ and $[Ti_2Nb_8O_{28}]^{8-}$.

Table S1. The experimental and calculated absorption maximum wavelengths, oscillator strength (f_{calcd}), transitions and CI expansion coefficients (contributions) of $[Nb_6O_{19}]^{8-}$, $[Nb_{10}O_{28}]^{6-}$ and $[Ti_2Nb_8O_{28}]^{8-}$.

	$\lambda_{ex\text{-}expl}$ (nm)	$\lambda_{ex\text{-}cal}$ (nm)	f_{calcd}	Transitions	CI expansion coefficients
$[Nb_6O_{19}]^{8-}$	294	282	0.2334	HOMO − 1 -> LUMO + 4	0.52512
				HOMO -> LUMO + 5	0.24608
				HOMO -> LUMO + 4	0.23348
				HOMO − 1 -> LUMO + 5	−0.20839
				HOMO -> LUMO + 3	−0.11372
				HOMO − 1 -> LUMO + 1	0.10778
$[Nb_{10}O_{28}]^{6-}$	310	283	0.0011	HOMO − 3 -> LUMO	0.61008
				HOMO − 4 -> LUMO + 3	−0.25025
				HOMO − 6 -> LUMO + 1	0.15390
				HOMO − 7 -> LUMO + 2	−0.11470
$[Ti_2Nb_8O_{28}]^{8-}$	306	287	0.0259	HOMO -> LUMO + 3	0.64577
				HOMO − 6 -> LUMO + 1	0.14930
				HOMO − 3 -> LUMO	−0.12456
				HOMO -> LUMO + 9	−0.10122

6. The molecular orbitals of one-electron transitions contributing to actual transition for $[Nb_6O_{19}]^{8-}$.

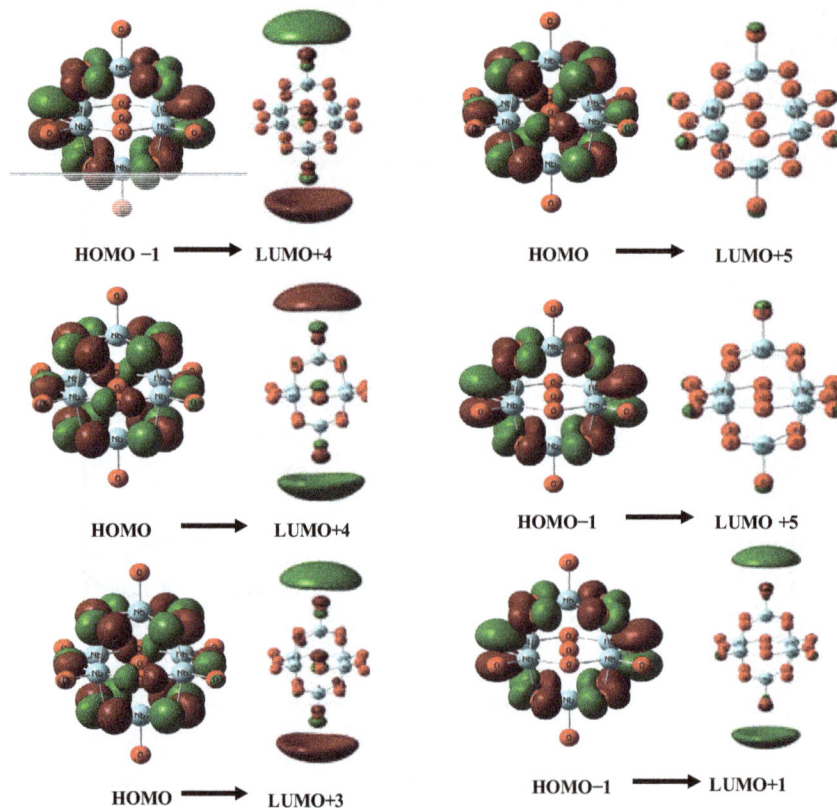

Figure S5. The molecular orbitals of one-electron transitions contributing to actual transition for $[Nb_6O_{19}]^{8-}$. These one-electron transitions contributes to the actual transition at aborption wavelength of 294 nm. The transition from HOMO − 1 to LUMO + 4 contributes the largest component, in which the electron was transferred to η-O from μ_2-O.

7. The molecular orbitals of one-electron transitions contributing to actual transition for $[Nb_{10}O_{28}]^{6-}$.

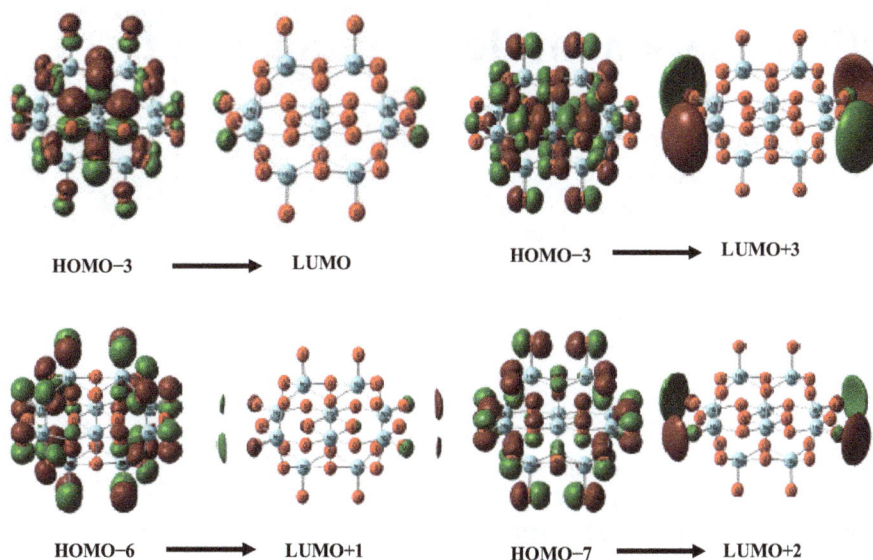

Figure S6. The molecular orbitals of one-electron transitions contributing to actual transition for $[Nb_{10}O_{28}]^{6-}$. The one-electron transition from HOMO − 3 to LUMO contributes the largest component to actual transition at absorption wavelength of 310 nm.

8. The molecular orbitals of one-electron transitions contributing to actual transition for $[Ti_2Nb_8O_{28}]^{8-}$.

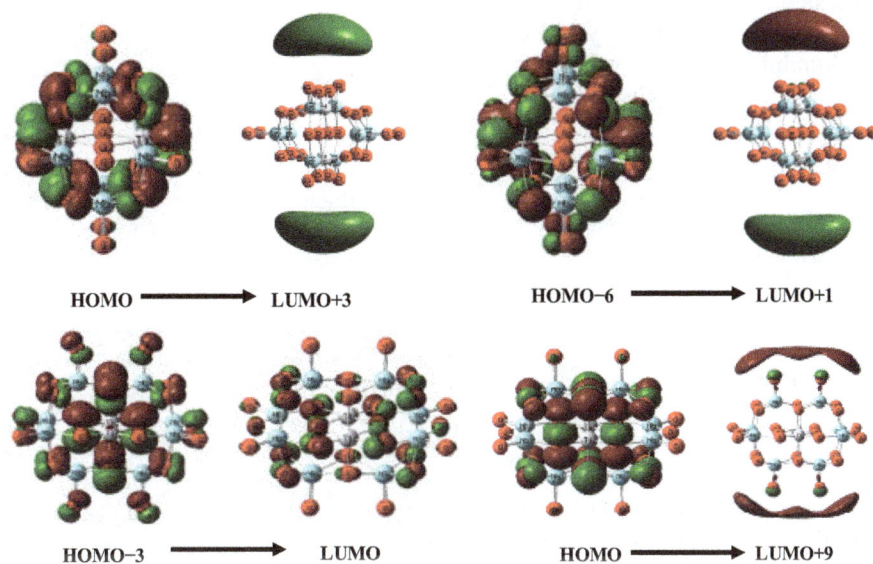

Figure S7. The molecular orbitals of one-electron transitions contributing to actual transition for $[Ti_2Nb_8O_{28}]^{8-}$. The one-electron transition from HOMO to LUMO + 3 contributes the largest component to actual transition at absorption wavelength of 306 nm.

4

Photocatalytic oxidation for degradation of VOCs

Lin Lin[1], Yuchao Chai[2], Bin Zhao[1], Wei Wei[3], Dannong He[1,2*], Belin He[4], Qunwei Tang[4*]

[1]National Engineering Research Center for Nanotechnology, Shanghai, China
[2]School of Material Science and Engineering, Shanghai Jiao Tong University, Shanghai, China
[3]State Key Laboratory of Bioreactor Engineering, New World Institute of Biotechnology, East China University of Science and Technology, Shanghai, China
[4]Institute of Materials Science and Engineering, Ocean University of China, Qingdao, China
Email: *hdnbill@sh163.net, *tangqunwei@ouc.edu.cn

ABSTRACT

Volatile organic compounds (VOCs) are the major group of indoor air pollutants, which significantly impact indoor air quality (IAQ) and influence human health. Photocatalytic oxidation (PCO) is a cost-effective technology for VOCs removal, compared with adsorption, biofiltration, or thermal catalysis method. Development of active photocatalyst systems is crucial for the PCO reaction. In this paper, the catalyst systems for photocatalysis under UV and visible light were discussed and the kinetics of photocatalytic oxidation was presented in order that some key influencing factors (relative huminity, light intensity, initial contaminant concentration and mass of catalyst) had also been studied. In addition, the future research directions were also presented in this paper.

Keywords: Indoor VOCs; Photocatalytic Oxidation; Synergistic Photocatalysts; Titanium Dioxide

1. INTRODUCTION

In today's life, indoor air pollutions have drawn a global attention regarding the imporvment of indoor air quality (IAQ). Indoor air pollutants mainly include nitrogen oxides (NO_x), carbon oxides (CO and CO_2), volatile organic compounds (VOCs) and particulates. Among the pollutions, VOCs are prominent representative indoor pollutants which commonly include trichloroethylene (C_2HCl_3) [1], acetone (C_3H_6O), 1-butanol ($C_4H_{10}O$), butyraldehyde (C_4H_8O) [2], m-xylene (C_8H_{10}) [2], 1,3-Butadiene (C_4H_6) [3], toluene ($C_6H_5CH_3$) [3], and formaldehyde (CH_2O) [2,3]. VOCs are known to have long-term effects on humans and considered to be carcinogenic, mutagenic, or teratogenic [4-6]. Moreover, some VOCs emissions can contribute to the formation of urban smog and ozone, stratospheric ozone depletion and the greenhouse effect [7].

With the increasing concerns on the indoor air quality, the degradation of VOCs has easily become the main point of research efforts in today's scientific world. Three methods are commonly used to control the emission of VOCs, such as adsorption by activated carbon [8] which merely transfers pollutants from gaseous phase to solid phase instead of destroying them, biofiltration which is slow and no obvious effect, and thermal oxidation destruction [9-11] which requires high temperatures of 200°C - 1200°C for efficient operation and expensive. Moreover, all these techniques have inherent limitations, and none of them is decisively cost-effective for gas stream with low to moderate concerntion and large numbers of compounds, as the recovery and reuse of the compounds is not economically feasible. Therefore, there is a great demand for a more cost-effective, more efficient and environmentally benign technology [9]. Photocatalytic oxidation (PCO) does not have the above mentioned problems and are cost-effective for treating low concerntion pollutions. PCO has received considerable attention regarding the removal of air pollutants during the last years [12,13]. It has been demonstrated that organics can be oxidized to harmless carbon dioxide, and water which makes PCO especially attractive for treating air indoor pollutants. Furthermore, PCO requires a low temperature and pressure, employs inexpensive semiconducting catalysts, and is suitable for the oxidation of a wide range of pollutants, both of organic and inorganic nature. Researchers have already focused on this promising technique and many beneficial advancements have been made in the field of VOCs [14-27].

Removal of VOCs from indoor air can be achieved using photocatalytic oxidation (PCO), an efficient, promising and cost-effective approach. This behavior is due to the employment of highly efficient photocatalysts. Development of active catalyst systems is crucial for this technology. TiO_2 is the most popular photocatalyst currently employed due to the hydrophilic properties of TiO_2 and its ability to degrade a wide range of inorganic and organic compounds under irradiation of UV or near UV-light [28-32].

In this work, the recent progress from various photo-

---

*Corresponding authors.

catalyst systems for on PCO reaction under UV and visible light is summarized. The key factors influencing the catalytic behaviros and the kinetics will be discussed. The future research directions and a larger scale application of the technology will also be presented.

2. MECHANISM OF PCO REACTION

Figure 1 presents the schematic of photocatalytic oxidation using TiO_2 as the photocatalyst. During the photocatalytic oxidation, the most important step of photoreaction is the formation of hole-electron pairs which need energy to overcome the band gap between the valence band (VB) and conduction band (CB) [34]. When the photon energy is equal to or exceeded the band gap energy Eg, the electron-hole pairs are created in the semiconductor, dissociating into free photoelectrons in the conduction band and photoholes in the valence band, respectively.

Simultaneously, the photo-oxidation and reduction reactions occur in the presence of air, oxygen and pollutant molecules. During the reactions, the hydroxyl radical (OH), coming from the oxidation of adsorbed water or OH^-, is highly reactive. In addition, the reducing power of the electrons can induce the reduction of molecular oxygen (O_2) to superoxide O_2^-. The highly reactive species OH and O_2^- show strong ability to degrade micro-organisms [35,36] as well as organic [36,37] and inorganic pollutants [37,38].

TiO_2 photocatalysis is capable of destructing many organic contaminants completely, the activation can be written as follows [39]:

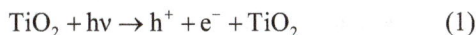

$$TiO_2 + h\nu \rightarrow h^+ + e^- + TiO_2 \qquad (1)$$

In this reaction, the photonic excitation of the catalyst appears as the initial step of the whole catalytic system, the produced h^+ and e^- are powerful oxidizing and reducing agents, respectively. The oxidation reaction:

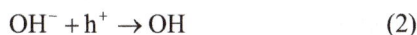

$$OH^- + h^+ \rightarrow OH \qquad (2)$$

The reduction reaction:

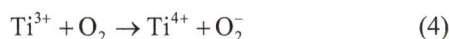

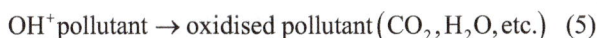

$$Ti^{4+} + e^- \rightarrow Ti^{3+} \qquad (3)$$

$$Ti^{3+} + O_2 \rightarrow Ti^{4+} + O_2^- \qquad (4)$$

$$OH^+ pollutant \rightarrow oxidised\ pollutant\left(CO_2, H_2O, etc.\right) \qquad (5)$$

For a complete PCO reaction, the final products of reactions are harmless CO_2 and H_2O, as well as few amounts of sulfates and chlorides. However, intermediates have been detected in some PCO processes. For example, Benoit-Marquie *et al.* [40] found that both butanal and 1-butanoic acid were the intermediates of 1-butanol. Nimolos *et al.* [41] used ethanol as the reactant and found the following intermediates: acetaldehyde, acetic acid, formaldehyde and formic acid.

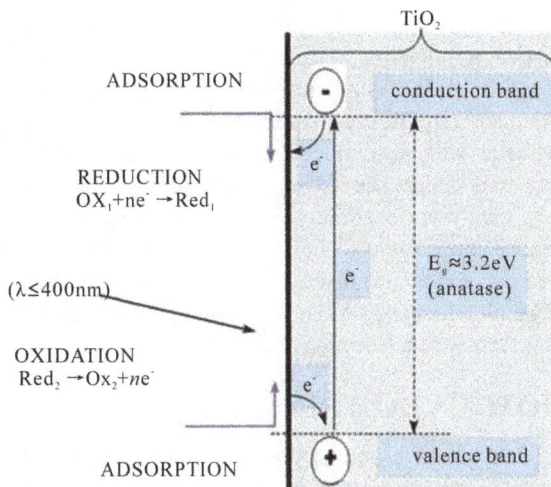

$$h^+ + e^- \rightarrow thermal\ heat\ or\ luminescence \qquad (6)$$

Figure 1. The schematic of TiO_2 UV photo-excitation process (R = reduction; O = oxidation) [33].

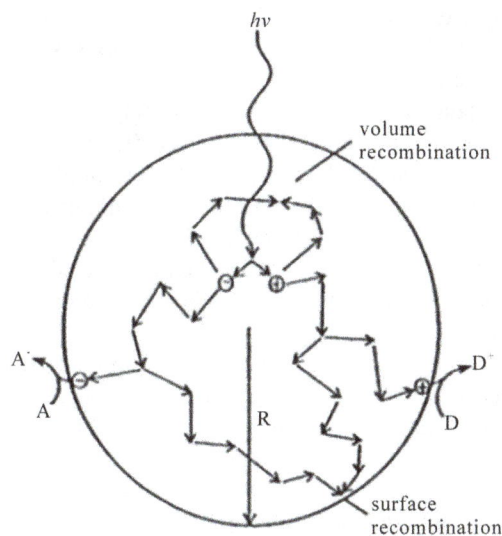

Figure 2. Fate of electrons and holes within a spherical particle of titania in the presence of acceptors (A) and (D) molecules [42].

Except the above reactions, there is also the electron-hole recombination (**Figure 2**). The PCO photoefficiency can be reduced by the electron-hole recombination, which corresponds to decreased electron and hole density as well as separation and the presence of oxygen can prevent the recombination of hole-electron pairs.

In short, the overall photocatalytic reactions can be decomposed into several steps, such as mass transfer of reactants from a fluidum (gas or liquid) to the photocatalyst surface, production of electron-hole, separation of the photogenerated electrons and holes, redox reactions between the trapped electrons and holes and adsorbed reactants, as well as desorption products and reconstruction of the surface.

Reaction mechanism of PCO reaction is a key issue for practical application. It is important to reveal the reactions of the PCO. However, few investigations have been conducted in this area. During the reactions, various intermediates will be produced and some of these intermediates will poison the active sites resulting in deactivation of catalysts. In addition, the produced intermediates can be more toxic to human health and should be removed or further oxidised to CO_2. Future research should be also focused on the detection of intermediates and their further re-adsorption and oxidation.

3. PHOTOCATALYSTS

The solids that can promote reactions in the presence of light but are not consumed in the overall reaction are referred as photocatalysts [28]. A photocatalyst for the reaction process needs to meet the following criteria: 1) photo-active; 2) utilizing near-UV light; 3) biologically and chemically inert; 4) photo-stable; 5) inexpensive and 6) non-toxic. Development of active photocatalyst systems is crucial for the PCO reaction.

For a semiconductor to be photochemically active as a sensitizer for the aforementioned reaction, the redox potential of the photogenerated valence band hole should be sufficiently positive to generate OH radicals that can subsequently oxidize the organic pollutants. The redox potential of the photogenerated conductance band electron must be sufficiently negative to be able to reduce the adsorbed O_2 to a superoxide O_2^-.

To date, semiconductors such as TiO_2, ZnO, ZnS, CdS, Fe_2O_3, SnO_2 have been found to have sufficient band-gap energies for promoting photocatalytic activities. Among plentiful of semiconductors, TiO_2 nanostructures are regarded as the most promising photocatalysts for the degradation of pollutants in water and air. The band-gap of several semiconductors is presented in **Figure 3**.

Of all the various semiconductors photocatalysts tested, TiO_2 is the most suitable semiconductors for

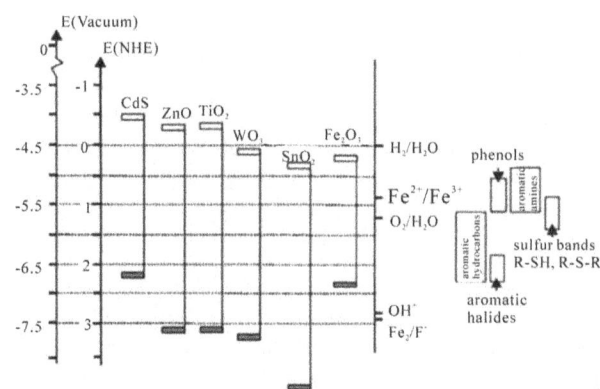

Figure 3. Band positions (top of valence band and bottom of conduction band) of several semiconductors together with some selected redox potentials [13].

photocatalytic reactions due to its superior characteristics [43]. Moreover, TiO_2 promotes ambient temperature oxidation of most indoor air pollutants and does not need any chemical additives.

In the following, UV-light irradiated catalysts and visible light irratiated catalysts are introduced respectively.

3.1. UV-Light Irradiated Catalysts

At the early stage of photocatalyst development, a commercially available Degussa P25 TiO_2 has been used as a standard photocatalyst for oxidation of gaseous or aqueous pollutants under UV or near-UV light due to its high photoactivity, excellent stability and low cost. Most of the investigations for indoor air photocatalytic oxidation used this catalyst.

Obee and Brown [44] pioneered reported an investigation of photocatalytic oxidation of VOCs using TiO_2 for indoor air. In the investigation, an important finding was that competitive adsorption between water and trace contaminants of VOCs has a significant effect on the oxidation rate. Subsequently, the photooxidation of toluene and formaldehyde titania on glass-plate reactor and compared the results of two reactor designs were reported by the same group [45]. A design model was also developed the proposed two reactors.

The particle size affects the catalyst activity. Cao *et al.* [46] developed nanoscaled TiO_2 photocatalysts. The photocatalyst nanostructures, composed mainly of mesopores with pore size in the range of 35 - 44 nm showed higher photoactivity than P25 TiO_2. Then they used the nanoscale TiO_2 catalysts for the photocatalytic oxidation of toluene [47]. The nanoscale TiO_2 samples were found to be highly active for the photodegradation of toluene. However, performing the reaction at room temperature resulted in rapid deactivation of TiO_2 catalysts due to the chemisorption of intermediates, such as benzaldehyde and benzoic acid.

A TiO_2-coated fibre glass mesh composed of anatase TiO_2 and SiO_2 was first employed for photocatalytic oxidation of benzene, toluene and xylenes in indoor air [48]. The average concentrations of benzene, toluene and xylenes were indeed reduced by a factor of 2 - 3 in an ordinary nonairtight room (**Figure 4**). O_3 addition in O_2 markedly increases the mineralisation percentage of n-octane, under otherwise identical conditions, in the laboratory photoreactor without photoexcitation of O_3.

Jo *et al.* [6] evaluated the application of P25 TiO_2 photocatalyst for the removal of five target VOCs, benzene, ethyl benzene, and o-, m-, p-xylenes in-vehicle air. Under the experimental conditions, the PCO destruction efficiencies were close to 100%. They further extended the investigation to chlorinated VOCs, trichloroethylene (TCE) and perchloroethylene (PCE) under non-occupa-

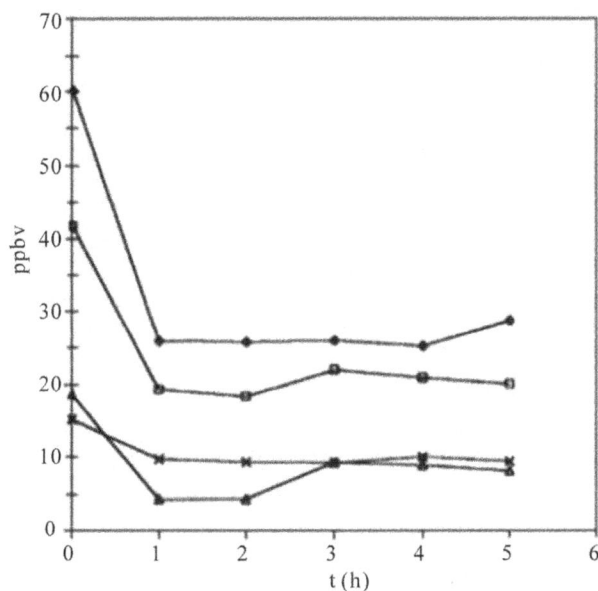

Figure 4. Decreases in the concentrations of benzene (crosses), toluene (diamonds), o-xylene (triangles) and both m- and p-xylene (squares) in a non-air tight roomcaused by the proto-type air purifier; t = 0 refers to the concentrations before the purifier UV lamps were switched on [48].

tional indoor air quality. The PCO destruction of VOCs was up to nearly 100%, and the CO generated during PCO was a negligible addition to indoor CO levels [49].

Shiraishi et al. [50] developed a high-performance photocatalytic reactor with a parallel array of nine light sources for photocatalytic decomposition of gaseous formaldehyde at a very low concentration using amorphous TiO_2. The experimental result indicateed that this photocatalytic reactor could rapidly decompose formaldehyde completely.

The application of employing artificial UV-light has the potential of being made more economical if the artificial UV is replaced by solar UV.

3.2. Visible-Light Irradiated Catalysts

The commonly used TiO_2 is an effective catalyst in photocatalytic oxidation. However, this type of catalysts only exhibit high catalytic activity in UV-light span, the usage indoors or inside vehicles is still a challenge. Few TiO_2 catalysts exhibit high activity for VOCs degradation under visible-light irradiation which hinder their practical application. In order to extend the applicable wavelength range and widen the practical applications, recent efforts focus on the doping of TiO_2 by metal ions like iron and tungsten and non-metal species like carbon, nitrogen and sulphur in order to reduce the band gap energy [51].

Anpo and Takeuchi [52] reported that the metal ion implanted catalysts were able to absorb the light in wavelengh of 400 - 600 nm and thus worked well in the

visible-light region. And the order of effectiveness in the red shift was found to be V > Cr > Mn > Fe > Ni. It was further observed that both the depth the metal ions inserted and the number of metal ions implanted influenced the photocatalytic activity. The results showed that there were optimal conditions in the depth and the number of metal ions implanted to achieve high photocatalytic activity under visible-light irradiation.

Yamashita et al. [53] have reported on the usage of metal ions implanted TiO_2 catalysts (V^+, Mn^+ and Fe^+) for the decomposition of propanol in the visible-light medium. The implanted metal ions were located at the lattice positions of Ti^{4+} in TiO_2 after the calcination, which was improtant to modify TiO_2 to be able to adsorb visible-light and operate as robust visible-light photocatalysts.

Wu and Chen [54] have also reported that vanadium doping provided a promising strategy to improve the photocatalytic activity of TiO_2 under visible light. The increase in vanadium doping increased the absorption of the light towards the visible region because vanadium was present in the V^{4+} state either by substituting Ti^{4+} in its site or embedded in the vacancy of the TiO_2 structure.

Fuerte et al. [55] reported the preparation and application of titanium-tungsten mixed oxides in the photocatalytic degradation of toluene using sunlight-type excitation. The photocatalytic activity of Ti-W mixed oxides increased with W content and is much better than TiO_2 itself and TiO_2 P25.

However, Chapuis et al. [56] found that doping of TiO_2 by metal oxides such as Ag_2O, CuO, NiO, CoO, and Cr_2O_3 gave much lower catalytic conversion. And nanosized doping TiO_2 photocatalysts were more effective in the degradation of toluene than TiO_2 itself and P25 TiO_2

Another new approach of synthesizing visible-light activated TiO_2 photocatalysts is by doping with anions, such as N^{3-}, C^{4-}, S^{4-} or halides (F^-, Cl^-, Br^-, I^-) [57]. It was suggested that these species substitute the oxygen lattice on TiO_2 and lead to a band gap narrowing, resulting in high visible absorption.

Doping with non-metallic species also causes the photosensitization of TiO_2 semiconductor in the visible light like C, N and S.

For carbon modified TiO_2 photocatalysts, Lettmann et al. [58] reported the synthesis of photocatalysts based on TiO_2 by modified sol-gel process using various alkoxide precursors. It was seen that the employment of alcohols pyrolysis in the sol-gel process leaded to carbonaceous species embedded within the TiO_2 matrix which were responsible for the observed visible light sensitization of TiO_2 semiconductor.

Many investigations have described the enhanced photocatalytic properties of TiO_2 by N doping. Asahi et *al.* first reported the visible-light photocatalysis in N-doped TiO_2 [59]. They tested various TiO_2-xNx catalysts for VOCs decomposition. These catalysts were found to photodegrade gaseous acetaldehyde [59], acetone [60] and 2-propanol [61] upon irradiation of visible-light. Recently, Irokawa *et al.* [62] have reported the photo-oxidation of aromatics over N-doped TiO_2 (TiO_2-xNx) under visible-light irradiation. They used toluene as test compound. Results indicated that toluene, weakly adsorbed on the catalyst surface, was initially photooxidised to benzaldehyde which firmly adsorbed onto the TiO_2-xNx surface, leading to the formation of ring-opening products such as carboxylic acids and aldehydes. Major intermediates adsorbed at the catalyst surface were gradually photodegraded to CO_2 and H_2O under visible-light irradiation.

Mesoporous nanocrystalline TiO_2-xNx and TiO_2-xNx/ZrO_2 visible-light photocatalysts have been prepared by a sol-gel method. The photocatalytic activity of the samples was evaluated by the decomposition of ethylene in air under visible-light (lambda N450 nm) illumination. The introduction of ZrO_2 into TiO_2-xNx considerably inhibited the undesirable crystal growth during the calcinations and improved the efficiency of degradation of ethylene [63].

A series of nanosized N-containing TiO_2-based materials were also tested in the photocatalytic degradation of methylcyclohexene, a representative example of volatile organic compounds (VOCs) present in urban atmospheres. It was found that the samples contained substitutional and interstitial N-containing impurities and a significant number of oxygen vacancies. Photocatalytic activity was correlated with an optimum of oxygen vacancies, above and below which a decrease of the steady state reaction rate was observed. The physico-chemical bases of this behaviour were discussed on the light of the above-mentioned experimental results [64].

Reports on band gap narrowing of TiO_2 by sulfur doping are also available in the literature [65]. Degradation of methylene blue using S-doped TiO_2 photocatalyst under visible light irradiation has been reported [66,67]; However, the degradation of catalytic activity, coming from catalytic poison caused by oxidation of SO_2 to SO_4^{2-}, has also been reportedly noticed in the decomposition of 2-proponal [68].

Recently, a novel nanocomposites of Ag-AgBr-TiO_2 were prepared by a deposition-precipitation method [69]. The results showed that the nanocomposites exhibited much higher photocatalytic activity and stability under both UV light and visible-light irradiation as compared with that of P25 TiO_2 toward aromatic benzene and non-aromatic acetone due to the synergetic effects between Ag-AgBr and TiO_2.

Many efforts have been made on the application of PCO degradation VOCs under visible light, the results would be of great significance in developing visible-light or sunlight responsive photocatalytic materials.

4. KINETIC PARAMETERS

The reaction rate is an important parameter to evaluate the effciency of PCO. Lots of experimental results showed that the overall photocatalytic degradation rate was governed by physicochemical as well as produced related parameters, such as the effects of humidity, pollutant concentration, temperature, the UV-light intensity, the weight of catalysts, etc [18,20,26,70].

In the following, a short and descriptive explanation of some key influencing parameters is given.

4.1. Process Related Parameters

4.1.1. Adsorption Rate

Adsorption of pollutants on the titania surface is one of the most important parameter affecting the catalyst performance. The adsorption trends of these pollutants may determine their conversion rate [71]. As shown in **Figure 5**, ethylene undergoes a lower adsorption phenomenon compared to the other hydrocarbons tested, whereas propylene has slightly higher adsorption than ethylene on the optimized TiO_2 nano particles surface. Compared to both ethylene and propylene, toluene has very high adsorption which attributes a multilayer adsorption. The adsorption of pollutants is proportional to the enrichment of pollutants on the surface of catalyst and the faster the degradation rate.

4.1.2. Light Spectrum

As to the PCO reaction, the photonic excitation of the catalyst appears as the initial step of the activation of the whole catalytic system. When photons of a certain wavelength hit its surface, electrons are promoted from the valence band and transferred to the conductance band. This leaves positive holes in the valence band, which then reacts with the hydroxylated surface to produce OH radicals, the true oxidizing agents. Thus the spectrum of the light source has the dramatic effect on the photo-degradation of the reactants.

The variation of the reaction rate as a function of the wavelength follows the absorption spectrum of the catalyst (**Figure 6(b)**), with a threshold corresponding to its band gap energy.

4.1.3. Initial Concentration of VOCs

Another important factor that should be discussed is the concentration of the VOCs. Generally, the influence of the pollutant concentration plays an important role during the PCO reaction. Different contaminant concentrations lead to different reaction rates. The influence of the pollutant concentration in the lower range results in a

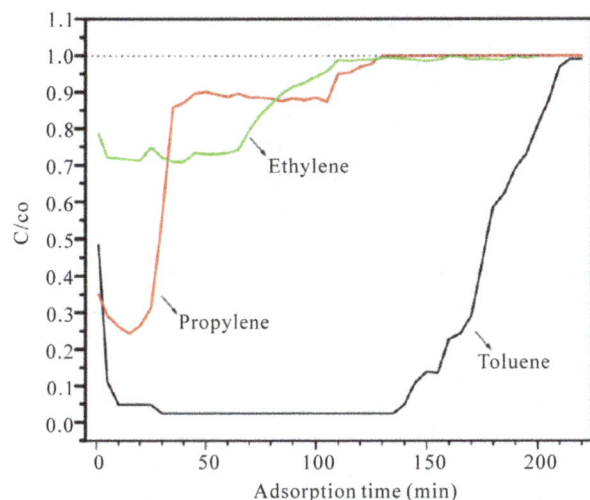

Figure 5. Adsorption of ethylene, propylene and toluene at optimized TNP photocatalyst.

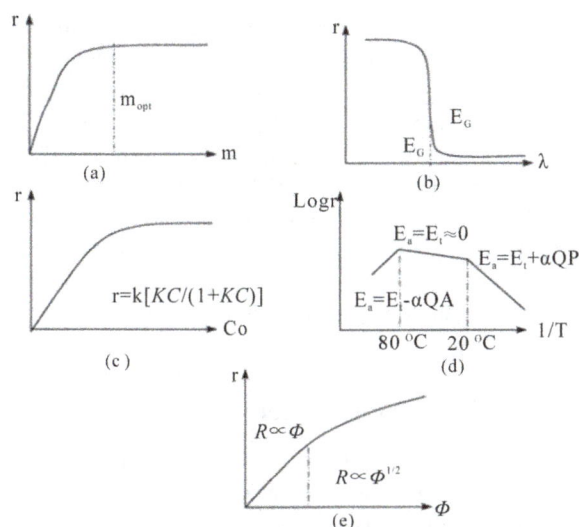

Figure. 6 Influence of different physical parameters governing the reaction rate [72]. (a) Mass of catalysts; (b) Wavelength; (c) Initial concentration of reactants; (d) Temperature; (e) Radiant flux.

remarkable higher changes of the degradation rate than that of in the range of higher concentrations.

Obee and Hay [73] found that the reaction rate of ethylene was enhanced by increasing concentrations of pollutants. Noguchi *et al.* [74] and Cao [75] employed formaldehyde and 1-butene as target reactants and found that the reaction rate was enhanced by increasing the initial contaminant concentration. In Cao's research, the reaction rate did not change largely when concentration of 1-butene was higher than 7 ppmv. In Noguchi *et al.*'s research, the increasing rate of the reaction rate slowed down when the initial concentration of formaldehyde was higher than 600 ppmv. They further gave the following results with regard to formaldehyde and acetal-

dehyde: in the low contaminant concentration region (less than 1200 ppmv), the degradation rate of formaldehyde was higher than that of acetaldehyde because the adsorption strength of formaldehyde on the catalyst surface was higher.

The rate equations can be used to explain the related results. The rate equations follow a Langmuir-Hinshelwood mechanism confirming the catalytic characteristic of the system with the rate r varying proportionally with the coverage θ as below:

$$r = k\theta = k\left(KC/(1+KC)\right) \qquad (7)$$

For diluted solutions (C < 10^{-3} M), KC becomes $\ll 1$ and the reaction is of the apparent first order, whereas for concentrations > 5×10^{-3} M, $(KC \ll 1)$, the reaction rate is maximum and of the apparent order (**Figure 6(c)**). In the gas phase, similar Langmuir-Hinshelwood expressions have been found including partial pressures P instead of C.

4.1.4. Reaction Temperature

Because of the photonic activation, the photocatalytic systems do not require heating and usually operate at room temperature. The true activation energy Et is nil, whereas the apparent activation energy Ea is often very small (a few kJ/mol) in the medium temperature range (20°C $\leq \theta \leq$ 80°C). However, at very low temperatures (−40°C $\leq \theta \leq$ 0°C), the activity decreases and the activation energy Ea becomes positive (**Figure 6(d)**).

By contrast, for various types of photocatalytic reactions at "high" temperatures ($\theta \geq$ 70°C - 80°C), the activity decreases and the apparent activation energy becomes negative (**Figure 6(c)**). This behavior can be easily explained within the frame of the Langmuir-Hinshelwood mechanism described above. The decrease in temperature favors adsorption (which is a spontaneous exothermic phenomenon) with in particular that of the final product, which becomes the inhibitor of the reaction. By contrast, when θ increases above 80°C and tends to the boiling point of water, the exothermic adsorption of reactant A becomes disfavored and tends to become the rate limiting step of the whole reaction.

As a consequence, the optimum temperature is generally between 20°C and 80°C. This explains why solar devices which use light concentrators instead of light collectors require coolers [76].

Hussain *et al.* [71] also investigated the affects of the operating temperature. The results showed that a small increase in temperature to 35°C, suggesting an increase in the ethylene conversion. Conversely, when a higher temperature of 80°C was reached, a decrease was observed. A similar trend was observed for propylene and toluene, and the highest conversion was shown at 40°C. This was due to the best compromise between the increase in reaction kinetics and the decrease in adsorption entailed by the temperature rise.

4.1.5. Radiant Flux

Not only the wavelength is one of the influencing factors of the system's efficiency but also the intensity of the radiation or so-called irradiance E has an effect on the degradation rate [77].

The activity of the photocatalyst depends strongly on the light-irradiation E or the photon flux on the surface of the catalyst. It has been shown that for all types of photocatalytic reactions, the reaction rate r is proportional to the radiant flux Φ (**Figure 3(e)**). This confirms the photo-induced nature of the activation of the catalytic process, with the participation of photo-induced electron-hole pairs to the reaction mechanism. However, the linear behavior for values could not be verified under a low irradiance [51], during which, the reaction rate r becomes proportional to $\Phi 1/2$, indicating strong electron-hole recombination. These two regimes have been independently demonstrated by Egerton et al. [78] and the author [79,80].

4.1.6. Relative Humility

The influence of the relative humidity depends to a large extent on the characteristics of contaminants [81]. It has been observed that the surface OH groups and/or physisorbed H_2O, as well as the anion radicals, can play a significant role as key active species in the PCO reaction of various VOCs [71]. In the absence of water vapor, the photocatalytic degradation of some chemical compounds (e.g., toluene, formaldehyde) is seriously retarded and the total mineralization to CO_2 does not occur. However, excessive water vapor on the catalyst surface will lead to the decrease of reaction rate because water molecules can occupy the active sites of the reactants on the surface and the hydrophilic effect at the surface prevails over the oxidizing effect [12]. According to Beeldens [82], the hydrophilic effect at the surface prevails over the oxidizing effect when high values of relative humidity are applied.

TiO_2 surface carries weakly or strongly bound molecular water, as well as hydroxyl groups created by the dissociative chemisorption of water.

Obee and Brown [3] investigated the effect of humility on photooxidation rate of formaldehyde, toluene, and 1, 3-butadiene on titania. The data showed that competitive adsorption between water and trace contaminants had a significant effect on the oxidation rate.

However, Jo et al. [6,49] found that the humidity had little effect on photocatalytic oxidation of benzene, ethyl benzene, and o-, m-, p-xylenes, trichloroethylene, and perchloroethylene. The PCO destruction efficiencies were close to 100% for four different RH ranges from 18

to 78%. Ao and Lee [83] compared the effect of humidity on photocatalytic oxidation of BTEX on TiO_2 and TiO_2/ AC systems and the results indicated that TiO_2/AC exhibited higher resistance to humidity effect. This negative effect of water vapour on TOL removal was in agreement with results obtained by other authors investigating photocatalytic TOL degradation by TiO_2/UV [47].

The practical evaluation by installation of the TiO_2/AC filter in an air cleaner was also conducted. The use of TiO_2/AC not only increased the target pollutant removal efficiency but also reduced the amount of intermediate exiting the system. The study showed that the enhancement effect of the TiO_2/AC shown in the laboratory scale using the photoreactor was also achieved by installing it into an air cleaner available in the commercial market [84].

Except the factors, the volatile organic compound structure is also an important feature. O'Malley and Hodnett [85] found that a primary factor in determining the reactivity of volatile organic compounds in oxidation reaction was the strength of the weakest C-H bond in the structure, which implicted that destruction of volatile organic compounds over oxidation catalysts proceeded by initial rupture of the weakest C-H bond in the compound, followed by further steps that may involve free radical chemistry. The research showed that the reactivity of VOCs with different functional groups for total oxidation varied as alcohols > aromatics > ketones > carboxylic acids > alkanes.

4.2. Product Related Parameters

Besides varying experimental parameters, the influence of product related parameters like crystalline structure of TiO_2, the mass of the TiO_2 and surface of the catalyst is also investigated [86,87].

4.2.1. Mass of Catalysts

Herrmann [77] investigated the relationship between the initial rates of photocatalytic reaction and the mass of catalysts. The results indicated that the initial rates of reactions were found to be directly proportional to the mass m of catalysts (**Figure 6(a)**). This indicated a true catalytic regime. Due to the increased weight, photocatalyst has higher surface area and more porosity, which offers larger specific surface areas and caused the higher adsorption and consequently the higher conversion. However, above a certain value of m, the reaction rate levels off and becomes independent of m. This limit depends on the mass transfer and light penetration limitations, which entailed that the increase in the available catalyst mass did not actually lead to an increase in terms of increased pollutants conversion.

For higher amounts of catalysts, a screening effect of

multilaters occurs, which masks part of the photosensitive surface and were not accessible to the reaction. For applications, this optimum mass of catalysts has to be chosen in order 1) to avoid an unuseful excess of catalysts and 2) to ensure total absorption of efficient photons.

4.2.2. Crystalline Forms of TiO_2

In solid state, TiO_2 can appear in three different crystalline modifications namely: rutile (tetragonal), anatase (tetragonal) and the seldom brookite (orthorhombic).

Among these crystalline forms, rutile is thermodynamically the most stable, whereas anatase and brookite are metastable and transform to rutile on heating. In spite of this similarity between anatase and brookite, the last one occurs rarely compared to the anatase form of TiO_2 and exhibits no significant photocatalytic activity under daylight irradiation. Rutile and anatase have more industrial applications.

Most of the studies have shown that the photocatalytic activity of titanium dioxide is influenced to a great extent by the crystalline form, although controversial results have also been reported in the literature. Some authors have stated that anatase works better than rutile [24], others have found the best photocatalytic activity for rutile [88], and some others have detected synergistic effects in the photocatalytic activity for anatase-rutile mixed phases [89]. It has recently been demonstrated that photo-activity towards organic degradation depends on the phase composition and on the oxidizing agent; for example, when the performance of different crystalline forms was compared, it was discovered that rutile shows the highest photocatalytic activity with H_2O_2, whereas anatase shows the highest with O_2 [90].

It is generally accepted that anatase demonstrates a higher activity than rutile, for most photocatalytic reaction systems, and this enhancement in photoactivity has been ascribed to the fact that the Fermi level of anatase is higher than that of rutile [91]. The precursors and the preparation method both affect the physicochemical properties of the specimen. In recent years, Degussa P25 TiO_2 has set the standard for photoreactivity in environmental VOC applications. Degussa P25 is a non-porous 70%:30% (anatase to rutile) material. Despite the presence of the rutile phase, this material has proved to be even more reactive than pure anatase [26]. Therefore, a mixed anatase-rutile phase seems to be preferable to enable some synergistic effects for photocatalytic reactions since the conduction band electron of the anatase partly jumps to the less positive rutile part, thus reducing the recombination rate of the electrons and positive holes in the anatase part.

4.2.3. Effects Surface Areas

The surface area is another important factor that affect

VOCs degradation in catalytic oxidation reactions [28]. The high surface area results in more adsorption sites for pollutants to be oxidized.

Zou et al. [92] reported that the catalyst with higher surface area (421.1 m^2/g) achieved 100% conversion efficiency for a substantial period of about 4 h, whereas the catalyst with a low surface area only had a very brief period of 100% conversion efficiency. This can be explained by the higher surface area of the porous catalyst material, increasing the adsorption capacity. These experimental results demonstrated that it was possible to achieve desirable degradation efficiency using the innovative TiO_2-SiO_2 photocatalyst. Within the regeneration limits, the adsorption enhanced the photocatalysis for better overall degradation. The surface structure was, in this respect, found to have a major influence. Surfaces with high roughness are providing more active surface area than smooth surfaces.

Moreover, the ability of TiO_2 particles to degrade organic compounds also depended on the size of the particles, since the smaller the particle size, the larger specific surface areas and the better catalyst activity.

It has also been found that photoformed OH species, as well as O_2^- and O_3^- anion radicals, play a significant role as a key active species in the complete photocatalytic oxidation of ethylene with oxygen into carbon dioxide and water [10].

Attempts should be made to obtain catalyse with a high surface area and a mixed crystalline phase with more anatase and small amounts of rutile. Instead, in the present work, an attempt has been made to optimize catalyst by optimizing the synthesis operating conditions.

5. SUMMARY AND PERSPECTIVES

Indoor air quality is currently a public intensive concern. VOCs are important indoor air pollutions. The most promising approach of improving the indoor quality is the PCO. Great efforts have been made on the PCO. However, the technique is still in the initial stage for VOCs removal from indoor air.

TiO_2 is widely used as photocatalyst. However, it merely exhibits high catalytic activity under UV-light, how to make TiO_2 photosensitized in the visible by doping or other methods or to develop novel photocatalyst, different from TiO_2 and directly active in the visible is highly demanded. More research should be directed in the development of effective visible-light or solar light irradiated photocatalysts and also in testing the photodegradation of indoor pollutions. Furthermore, nanoscaled photocatalysts exhibited higher activity on the PCO, great efforts should be made to fabricate nano photocatalysts at a large scale by adjusting conditions and the synthesis conditions. The development of nano technologies make the PCO promising for future research and further applications.

For a complete PCO reaction, the final products of reactions are harmless CO_2 and H_2O. However, intermediates have been detected in some PCO processes. Ring the reaction, various intermediates will be produced and some of these intermediates will poison the active sites resulting in deactivation of catalysts. In addition, the produced intermediates can be more toxic to human health and should be removed or further oxidised to CO_2. Thus, how to avoid the intermediates is also a hard task. Future research should be focused on the detection of intermediates and their further readsorption and oxidation. Modelling of catalytic reactor performance should also be developed for a pilot scale practice.

A variety of products containing TiO_2 are already available on the European market and their working mechanism under laboratory conditions is proven. However, there is still a lack of transforming the experimental results obtained under laboratory conditions to practical applications considering real world conditions. The potential applications strongly depend on the future development of photocatalytic engineering.

Novel TiO_2 nanopaticles showed a superior catalytic performance of VOCs than that of P25 TiO_2. Thus, the development of nano-scale TiO_2 photocatalysts with anatase-rutile mixed phase, high specific surface area, good porosity is promising for future research and further applications.

A variety of novel photocatalysts, such as Fe_2O_3, TiO_2, Cu_2O, WO_3, ZnS and CdS can be prepared with the synergetic photocatalytic effect between Ag-AgBr and semiconductors. This would advance the application of nanocomposites as photocatalyst for the degradation of harmful VOCs and widely optimize the photocatalytic performance.

Although development and application of PCO still need improvement, the potential for a wider use of this technology is huge and promising.

6. ACKNOWLEDGEMENTS

This work was supported by the National Nature Science Foundations of China (21071098), the China International Science and technology Cooperation Project (2011DFA50530), Shandong Provincial Natural Science Foundation (ZR2011BQ017), China, Seed Fund from Ocean University of China, and Fundamental Research Funds for the Central Universities (201313001).

REFERENCES

[1]   Jacoby, W.A., Blake, D.M., Noble, R.D., *et al.* (1995) Kinetics of the oxidation of trichloroethylene in air via heterogeneous photocatalysis. *Journal of Catalysis*, **157**, 87-96. doi:10.1006/jcat.1995.1270

[2]   Peral, J. and Ollis, D.F. (1991) Heterogeneous photocatalytic oxidation of gas-phase organics for air purification: Acetone, 1-butanol, butyraldehyde, formaldehyde, and mxylene oxidation. *Journal of Catalysis*, **136**, 554-565. doi:10.1016/0021-9517(92)90085-V

[3]   Obee, T.N. and Brown, R.T. (1995) TiO_2 photocatalysis for indoor air applications: Effects of humidity and trace contaminant levels on the oxidation rates of formaldehyde, toluene, and 1,3-butadiene. *Environmental Science and Technology*, **29**, 1223-1231. doi:10.1021/es00005a013

[4]   Alberici, R.M. and Jardim W.F. (1997) Photocatalytic destruction of VOCs in the gas-phase using titanium dioxide. *Applied Catalysis B: Environmental*, **14**, 55-68.

[5]   Tsoukleris, D.S., Maggos, T., Vassilakos, C., *et al.* (2007) Photocatalytic degradation of volatile organics on TiO_2 embedded glass spherules. *Catalyst Today*, **129**, 96-101. doi:10.1016/j.cattod.2007.06.047

[6]   Jo, W.K., Park, J.H. and Chun H.D. (2002) Photocatalytic destruction of VOCs for in-vehicle air cleaning. *Journal of Photochemistry and Photobiology A: Chemistry*, **148**, 109-119. doi:10.1016/S1010-6030(02)00080-1

[7]   Derwent, R.G., Jenkin, M.E., Saunders, S.M., *et al.* (2003) Photochemical ozone formation in north west Europe and its control. *Atmospheric Environment*, **37**, 1983-1991. doi:10.1016/S1352-2310(03)00031-1

[8]   Yu, H., Zhang, K. and Rossi C. (2007) Theoretical study on photocatalytic oxidation of VOCs using nano-TiO_2 photocatalyst. *Journal of Photochemistry and Photobiology A: Chemistry*, **188**, 65-73. doi:10.1016/j.jphotochem.2006.11.021

[9]   Ray, M.B. (2000) Photodegradation of the volatile organic compounds in the gas phase: A review. *Development and Chemistry Engineering Mineral Process*, **8**, 405-439. doi:10.1002/apj.5500080502

[10] Kumar, S., Fedorov, A.G. and Gole, J.L. (2005) Photodegradation of ethylene using visible light responsive surfaces prepared from titania nanoparticle slurries. *Applied Catalysis B: Environmental*, **57**, 93-107.

[11] Everaert, K. and Baeyens, J. (2004) Catalytic combustion of volatile organic compounds. *Journal of Hazardous Materials*, **109**, 113-139. doi:10.1016/j.jhazmat.2004.03.019

[12] Zhao, J. and Yang, X.D. (2003) Photocatalytic oxidation for indoor air purification: A literature review. *Building and Environment*, **38**, 645-654. doi:10.1016/S0360-1323(02)00212-3

[13] Carp, O., Huisman, C.L. and Reller, A. (2004) Photoinduced reactivity of titanium dioxide. *Progress in Solid State Chemistry*, **32**, 33-177. doi:10.1016/j.progsolidstchem.2004.08.001

[14] Benoit-Marquié, F., Wilkenhoner, U., Simon, V., *et al.* (2000) VOC photodegradation at the gas-solid interface of a TiO_2 photocatalyst. Part I. 1-butanol and 1-butylamine. *Journal of Photochemistry and Photobiology A: Chemistry*, **132**, 225-232. doi:10.1016/S1010-6030(00)00196-9

[15] Biomorgi, J., Oliveros, E., Coppel, Y., *et al.* (2010) Effect of V-UV-radiation on VOCs-saturated zeolites. *Journal of Photochemistry and Photobiology A: Chemistry*, **214**,

194-202. doi:10.1016/j.jphotochem.2010.06.026

[16] Martra, G., Coluccia, S., Marchese, L., et al. (1999) The role of H$_2$O in the photocatalytic oxidation of toluene in vapour phase on anatase TiO$_2$ catalyst: A FTIR study. Catalyst Today, 53, 695-702. doi:10.1016/S0920-5861(99)00156-X

[17] Muggli, D.S. and Ding, L. (2001) Photocatalytic performance of sulfated TiO$_2$ and Degussa P-25 TiO$_2$ during oxidation of organics. Applied Catalysis B: Environmental, 32, 181-194.

[18] Zhang, P.Y., Liang, F.Y., Yu, G., et al. (2003) A comparative study on decomposition of gaseous toluene by O$_3$/UV, TiO$_2$/UV and O$_3$/TiO$_2$/UV. Journal of Photochemistry and Photobiology A: Chemistry, 156, 189-194. doi:10.1016/S1010-6030(02)00432-X

[19] Ao, C.H. and Lee, S.C. (2004) Combination effect of activated carbon with TiO$_2$ for the photodegradation of binary pollutants at typical indoor air level. Journal of Photochemistry and Photobiology A: Chemistry, 161, 131-140. doi:10.1016/S1010-6030(03)00276-4

[20] Sleiman, M., Conchon, P., Ferronato, C., et al. (2009) Photocatalytic oxidation of toluene at indoor air levels (ppbv): Towards a better assessment of conversion, reaction intermediates and mineralization. Applied Catalysis B: Environmental, 86, 159-165.

[21] Blount, M.C. and Falconer, J.L. (2002) Steady-state surface species during toluene photocatalysis. Applied Catalysis B: Environmental, 39, 39-50.

[22] Fresno, F., Hernandez-Alonso, M.D., Tudela, D., et al. (2008) Photocatalytic degradation of toluene over doped and coupled (Ti,M)O$_2$ (M= Sn or Zr) nanocrystalline oxides: Influence of the heteroatom distribution on deactivation. Applied Catalysis B: Environmental, 84, 598-606.

[23] Zou, L., Luo, Y., Hooper, M., et al. (2006) Removal of VOCs by photocatalysis process using adsorption enhanced TiO$_2$-SiO$_2$ catalyst. Chemical Engineering and Processing, 45, 959-964. doi:10.1016/j.cep.2006.01.014

[24] Zuo, G.M., Cheng, Z.X., Chen, H., et al. (2006) Study on photocatalytic degradation of several volatile organic compounds. Journal of Hazardous Materials, 128, 158-163. doi:10.1016/j.jhazmat.2005.07.056

[25] Deveau, P.A., Arsac, F., Thivel, P.X., et al. (2007) Different methods in TiO$_2$ photodegradation mechanism studies: Gaseous and TiO$_2$-adsorbed phases. Journal of Hazardous Materials, 144, 692-697. doi:10.1016/j.jhazmat.2007.01.097

[26] Augugliaro, V., Coluccia, S., Loddo, V., et al. (1999) Photocatalytic oxidation of gaseous toluene on anatase TiO$_2$ catalyst: Mechanistic aspects and FT-IR investigation. Applied Catalysis B: Environmental, 20, 15-27.

[27] Hussain, M., Ceccarelli, R., Marchisio, et al. (2010) Synthesis, characterization, and photocatalytic application of novel TiO$_2$ nanoparticles. Chemical Engineering Journal, 157, 45-51. doi:10.1016/j.cej.2009.10.043

[28] Bhatkhande, B.S., Pangarkar, V.G. and Beenackers, A.A.C.M. (2001) Photocatalytic degradation for environmental applications—A review. Journal of Chemical Technology and Biotechnology, 77, 102-116.

doi:10.1002/jctb.532

[29] Fu, X., Zeltner, W.A. and Anderson, M.A. (1996) Applications in photocatalytic purification of air. Science, 445-461.

[30] Peral, J., Domenech, S. and Ollis, D.F. (1997) Heterogeneous photocatalysis for purification, decontamination and deodorization of air. Journal of Chemical Technology and Biotechnology, 70, 117-140. doi:10.1002/(SICI)1097-4660(199710)70:2<117::AID-JCTB746>3.0.CO;2-F

[31] Mills, A. and Le Hunte, S. (1997) An overview of semiconductor photocatalysis. Journal of Photochemistry and Photobiology A: Chemistry, 108, 1-35. doi:10.1016/S1010-6030(97)00118-4

[32] Demeestere, K., Dewulf, J. and Van Langenhove, H. (2007) Heterogeneous photocatalysis as an advanced oxidation process for the abatement of chlorinated, monocyclic aromatic and sulfurous volatile organic compounds in air: State-of-the-art. Critical Reviews in Environmental Science and Technology, 37, 489-548. doi:10.1080/10643380600966467

[33] Herrmann, J.M. (2010) Environmental photocatalysis: Perspectives for China. Science China Chemistry, 53, 1831-1843. doi:10.1007/s11426-010-4076-y

[34] Hufschmidt, D., Liu, L., Selzer, V., et al. (2004) Photocatalytic water treatment: Fundamental knowledge for its practical application. Water Science and Technology, 49, 135-140.

[35] Kikuchi, Y., Sunada, K., Iyoda, T., et al. (1997) Photocatalytic bactericidal effect of TiO$_2$ thin films: Dynamic view of the active oxygen species responsible for the effect. Photochemistry and Photobiology A: Chemistry, 106, 51-56.

[36] Agrios, A.G. and Pichat, P. (2005) State of the art and perspectives on materials and applications of photocatalysis over TiO$_2$. Reviews in Applied Electrochemistry, 58, 655-663. doi:10.1007/s10800-005-1627-6

[37] Demeestere, K., Dewulf, J. and Van Langenhove, H. (2007) Heterogeneous photocatalysis as an adavanced oxidation process for the abatement of chlorinated, monocyclic aromatic and sulfurous volatile organic compounds in air: State of the art. Critical reviews in Environmental Science and Technology, 37, 489-538. doi:10.1080/10643380600966467

[38] Toma, F.L., Bertrand, G., Klein, D., et al. (2004) Photocatalytic removal of nitrogen oxides via titanium dioxide. Environmental Chemistry Letters, 2, 117-121. doi:10.1007/s10311-004-0087-2

[39] Cao, L.X., Spiess, F.J., Huang, A.M., et al. (1999b) Heterogeneous photocatalytic oxidation of 1-butene on SnO$_2$ and TiO$_2$ films. Journal of Physical Chemistry B, 103, 2912-2917. doi:10.1021/jp983860z

[40] Benoit-Marquie, F., Wilkenhoner, U., Simon, V., et al. (2000) VOC photodegradation at the gas-solid interface of a TiO$_2$ photocatalyst part I: 1-butanol and 1-butylamine. Journal of Photochemistry and Photobiology A: Chemistry, 132, 225-232. doi:10.1016/S1010-6030(00)00196-9

[41] Nimlos, M.R., Wolfrum, E.J., Brewer, M.L., *et al.* (1996) Gas-phase heterogeneous photocatalytic oxidation of ethanol: Pathways and kinetic modeling. *Environmental Science and Technology*, **30**, 3102-3110. doi:10.1021/es9602298

[42] Ollis, D.F. and Al-Ekabi, H. (1993) Photocatalytic purification and treatment of water and air. *Science*, 511-532.

[43] Hager, S., Bauer, R. and Kudielka, G. (2000) Photocatalytic oxidation of gaseous chlorinated organics over titanium oxide. *Chemosphere*, **41**, 1219-1225. doi:10.1016/S0045-6535(99)00558-5

[44] Obee, T.N. and Brown, R.T. (1995) TiO_2 photocatalysis for indoor air applications- effects of humidity and trace contaminant levels on the oxidation rates of formaldehyde, toluene, and 1,3-butadiene. *Environmental Science and Technology*, **29**, 1223-1231. doi:10.1021/es00005a013

[45] Obee, T.N. (1996) Photooxidation of sub-parts-per-million toluene and formaldehyde levels an titania using a glass-plate reactor. *Environmental Science and Technology*, **30**, 3578-3584. doi:10.1021/es9602713

[46] Cao, L.X., Huang, A.M., Spiess, F.J., *et al.* (1999a) Gas-phase oxidation of 1-butene using nanoscale TiO_2 photocatalysts. *Journal of Catalysis*, **188**, 48-57. doi:10.1006/jcat.1999.2596

[47] Cao, L.X., Gao, Z., Suib, S.L., *et al.* (2000) Photocatalytic oxidation of toluene on nanoscale TiO_2 catalysts: Studies of deactivation and regeneration. *Journal of Catalysis*, **196**, 253-261. doi:10.1006/jcat.2000.3050

[48] Pichat, P., Disdier, J., Hoang-Van, C., *et al.* (2000) Purification/deodorization of indoor air and gaseous effluents by TiO_2 photocatalysis. *Catalyst Today*, **63**, 363-369. doi:10.1016/S0920-5861(00)00480-6

[49] Jo, W.K. and Park, K.H. (2004) Heterogeneous photocatalysis of aromatic and chlorinated volatile organic compounds (VOCs) for non-occupational indoor air application. *Chemosphere*, **57**, 555-565. doi:10.1016/j.chemosphere.2004.08.018

[50] Shiraishi, F., Toyoda, K. and Miyakawa, H. (2005) Decomposition of gaseous formaldehyde in a photocatalytic reactor with a parallel array of light sources-Reactor performance. *Chemical Engineering Journal*, **114**, 145-151. doi:10.1016/j.cej.2005.09.008

[51] Bloβ, S.P. and Elfenthal, L. (2007) Doped titanium dioxide as a photocatalyst forUVand visible light. *Proceedings International RILEM Symposium on Photocatalysis, Environment and Construction Materials*, Florence, 8-9 October 2007, 31-38.

[52] Anpo, M. and Takeuchi, M. (2003) The design and development of highly reactive titanium oxide photocatalysts operating under visible light irradiation. *Journal of Catalysis*, **216**, 505-516. doi:10.1016/S0021-9517(02)00104-5

[53] Yamashita, H., Harada, M., Misaka, J., *et al.* (2002) Degradation of propanol diluted in water under visible light irradiation using metal ion-implanted titanium dioxide photocatalysts. *Journal of Photochemistry and Photobiology A*: *Chemistry*, **148**, 257-261.

[54] Wu, J.C.S. and Chen, C.H. (2004) A visible-light response vanadium-doped titania nanocatalyst by sol-gel method. *Journal of Photochemistry and Photobiology A*: *Chemistry*, **163**, 509-515. doi:10.1016/j.jphotochem.2004.02.007

[55] Fuerte, A., Hernandez-Alonso, M.D., Maira, A.J., *et al.* (2002) Nanosize Ti-W mixed oxides: Effect of doping level in the photocatalytic degradation of toluene using sunlight-type excitation. *Journal of Catalysis*, **212**, 1-9. doi:10.1006/jcat.2002.3760

[56] Chapuis, Y., Kivana, D., Guy, C., *et al.* (2002) Photocatalytic oxidation of volatile organic compounds using fluorescent visible light. *Journal of the Air & Waste Management Association*, **52**, 845-854. doi:10.1080/10473289.2002.10470816

[57] Belver, C., Bellod, R., Fuerte, A., *et al.* (2006) Nitrogen-containing TiO_2 photocatalysts: Part 1. Synthesis and solid characterization. *Applied Catalysis B*: *Environmental*, **65**, 301-308.

[58] Lettmann, C., Hildenbrand, K., Kisch, H., *et al.* (2001) Visible light photodegradation of 4-chlorophenol with a coke containing titanium dioxide photocatalyst. *Applied Catalysis B*: *Environmental*, **32**, 215-227.

[59] Asahi, R., Morikawa, T., Ohwaki, T., *et al.* (2001) Visible-light photocatalysis in nitrogen-doped titanium oxides. *Science*, **293**, 269-271. doi:10.1126/science.1061051

[60] Ihara, T., Miyoshi, M., Iriyama, Y., *et al.* (2003) Visible-light-active titanium oxide photocatalyst realized by an oxygen-deficient structure and by nitrogen doping. *Applied Catalysis B*: *Environmental*, **42**, 403-409.

[61] Miyauchi, M., Ikezawa, A., Tobimatsu, H., *et al.* (2004) Zeta potential and photocatalytic activity of nitrogen doped TiO_2 thin films. *Physical Chemistry Chemical Physics*, **6**, 865-870. doi:10.1039/b314692h

[62] Irokawa, Y., Morikawa, T., Aoki, K., *et al.* (2006) Photodegradation of toluene over TiO_2-xNx under visible light irradiation. *Physical Chemistry Chemical Physics*, **8**, 1116-1121. doi:10.1039/b517653k

[63] Wang, X.C., Yu, J.C., Chen, Y.L., *et al.* (2006) ZrO_2-modified mesoporous manocrystalline TiO_2-xNx as efficient visible light photocatalysts. *Environmental Science and Technology*, **40**, 2369-2374. doi:10.1021/es052000a

[64] Belver, C., Bellod, R., Stewart, S.J., *et al.* (2006b) Nitrogencontaining TiO_2 photocatalysts: Part 2. Photocatalytic behavior under sunlight excitation. *Applied Catalysis B*: *Environmental*, **65**, 309-314.

[65] Umebayashi, T., Yamaki, T., Itoh, H., *et al.* (2002) Band gap narrowing of titanium dioxide by sulfur doping. *Applied Physics Letters*, **81**, 454-456. doi:10.1063/1.1493647

[66] Umebayashi, T., Yamaki, T., Itoh, H., *et al.* (2003) Visible Light-Induced Degradation of Methylene Blue on S-doped TiO_2. *Chemistry Letters*, **32**, 330-331. doi:10.1246/cl.2003.330

[67] Ohno, T., Mitsui, T. and Matsumara, M. (2003) Photocatalytic activity of S-doped TiO_2 photocatalyst under visible light. *Chemistry Letters*, **32**, 364-365.

doi:10.1246/cl.2003.364

[68] Irie, H., Watanabe, Y. and Hashimoto, K. (2003) Carbon-doped anatase TiO_2 powders as a visible-light sensitive photocatalyst. *Chemistry Letters*, **32**, 772-773. doi:10.1246/cl.2003.772

[69] Zhang, Y.H., Tang, Z.R., Fu, X.Z., *et al.* (2011) Nanocomposites of Ag-AgBr-TiO_2 as a photoactive and durable catalyst for degradation of volatile organic compounds in the gas phase. *Applied Catalysis B: Environmental*, **106**, 445-452. doi:10.1016/j.apcatb.2011.06.002

[70] Yu, H., Zhang, K. and Rossi, C. (2007) Theoretical study on photocatalytic oxidation of VOCs using nano-TiO_2 photocatalyst. *Photochemistry and Photobiology A: Chemistry*, **188**, 65-73.

[71] Hussain, M., Russo, N. and Saracco, G. (2011) Photocatalytic abatement of VOCs by novel optimized TiO_2 nanoparticles. *Chemical Engineering Journal*, **166**, 138-149. doi:10.1016/j.cej.2010.10.040

[72] Herrmann, J.M. (1999) Heterogeneous photocatalysis: Fundamentals and applications to the removal of various types of aqueous pollutants. *Catalyst Today*, **53**, 115-129. doi:10.1016/S0920-5861(99)00107-8

[73] Obee, T.N. and Hay, S.O. (1997) Effects of moisture and temperature on the photooxidation of ethylene on titania. *Environmental Science and Technology*, **31**, 2034-2038. doi:10.1021/es960827m

[74] Noguchi, T., Fujishima, A., Sawunyama, P., *et al.* (1998) Photocatalytic degradation of gaseous formaldehyde using TiO_2 film. *Environmental Science and Technology*, **32**, 3831-3833. doi:10.1021/es980299+

[75] Cao, L. (1999) Gas-phase oxidation of 1-butene using nanoscale TiO_2 photocatalysts. *Journal of Catalysis*, **188**, 48-57. doi:10.1006/jcat.1999.2596

[76] Mehos, M.S. and Turchi, C.S. (1993) Field testing solar photocatalytic detoxification on TCE-contaminated groundwater. *Environmental Progress*, **12**, 194-199. doi:10.1002/ep.670120308

[77] Herrmann, J.M., Peruchon, L., Puzenat, E., *et al.* (2007) Photocatalysis: From fundamentals to self-cleaning glass application. *Proceedings International RILEM Symposium on Photocatalysis, Environment and Construction Materials*, Florence, 8-9 October 2007, 41-48.

[78] Egerton, T.A. and King, C.J. (1979) The influence of light intensity on photoactivity in titanium dioxide pigmented systems. *Journal of the Oil and Colour Chemists Association*, **62**, 386-391.

[79] Herrmann, J.M. (2006) Water Treatment by Heterogeneous Photocatalysis. *Kirk-Othmer Encyclopedia of Chemical Technology*, **19**, 73-106.

[80] Herrmann, J.M. (2006) From catalysis by metals to bifunctional photo catalysis. *Topics in Catalysis*, **33**, 421-431.

[81] Luo, Y. and Ollis, D.F. (1996) Heterogeneous photocatalytic oxidation of trichloroethylene and toluene mixtures in air: Kinetic promotion and inhibition, time-dependent catalyst activity. *Journal of Catalysis*, **163**, 1-11. doi:10.1006/jcat.1996.0299

[82] Beeldens, A. (2007) Air purification by road materials: Results of the test project in Antwerp. *Proceedings International RILEM Symposium on Photocatalysis, Environment and Construction Materials*, Florence, 8-9 October 2007, 187-194.

[83] Ao, C.H., Lee, S.C., Mak, C.L., *et al.* (2003) Photodegradation of volatile organic compounds (VOCs) and NO for indoor air purification using TiO_2: Promotion versus inhibition effect of NO. *Applied Catalysis B: Environmental*, **42**, 119-129.

[84] Ao, C.H. and Lee, S.C. (2005) Indoor air purification by photocatalyst TiO_2 immobilized on an activated carbon filter installed in an air cleaner. *Chemical Engineering Science*, **60**, 103-109. doi:10.1016/j.ces.2004.01.073

[85] O'Malley, A. and Hodnett, B.K. (1999) The influence of volatile organic compound structure on conditions required for total oxidation. *Catalysis Today*, **54**, 31-38. doi:10.1016/S0920-5861(99)00166-2

[86] Lin, H., Huang, C.P, Li, W., *et al.* (2006) Size dependency of nanocrystalline TiO_2 on its optical property and photocatalytic reactivity exemplified by 2-chlorophenol. *Applied Catalysis B: Environmental*, **68**, 1-11. doi:10.1016/j.apcatb.2006.07.018

[87] Bakardjieva, S., Stengl, V., Szatmary, L., *et al.* (2006) Transformation of brookite-type TiO_2 nanocrystals to rutile: Correlation between microstructure and photoactivity. *Materials Chemistry*, **16**, 1709-1716. doi:10.1039/b514632a

[88] Watson, S.S., Beydoun, D., Scott, J.A., *et al.* (2003) The effect of preparation method on the photoactivity of crystalline titanium dioxide particles. *Chemical Engineering Journal*, **95**, 213-220. doi:10.1016/S1385-8947(03)00107-4

[89] Bacsa, R.R. and Kiwi, J. (1998) Effect of rutile phase on the photocatalytic properties of nanocrystalline titania during the degradation of p-coumaric acid. *Applied Catalysis B: Environmental*, **16**, 19-29. doi:10.1016/S0926-3373(97)00058-1

[90] Testino, A., Bellobono, I.R., Buscaglia, V., *et al.* (2007) Optimizing the photocatalytic properties of hydrothermal TiO_2 by the control of phase composition and particle morphology. *Journal of the American Chemical Society*, **129**, 3564-3575. doi:10.1021/ja067050+

[91] Porkodi, K. and Arokiamary, S.D. (2007) Synthesis and spectroscopic characterization of nanostructured anatase titania: A photocatalyst. *Materials Characterization*, **58**, 495-503. doi:10.1016/j.matchar.2006.04.019

[92] Zoua, L., Luo, Y.G., Hooper M., *et al.* (2006) Removal of VOCs by photocatalysis process using adsorption enhanced TiO_2-SiO_2 catalyst. *Chemical Engineering and Processing*, **45**, 959-964. doi:10.1016/j.cep.2006.01.014

5

Synthesis of NaA zeolite using PTMAOH (phenyltrimethylammoniumhydroxide): Hydrothermal and microwave heating methods and comparison of their XRD patterns

S. N. Azizi[*], A. R. Samadi-Maybodi, M. Yarmohammadi

Analytical Division, Faculty of Chemistry, University of Mazandaran, Babolsar, Iran
Email: [*]azizi@umz.ac.ir

ABSTRACT

Ion exchanging is one of the characteristics of the zeolites. Zeolites have octahedral and tetrahedral holes to trap ions and molecules. They also can exchange many ions in solution because of the size of the attendant ions. As a matter of fact, the property of the ion-exchanged zeolites depends on the ligands involved in the ion exchanging solutions. Ion exchanged zeolites are used as catalyst for studying the anodic oxidation of methanol in an acidic medium to investigate their suitability for use in direct methanol fuel cells (DMFCs). Some of the zeolites that have exchanged ions are shown to have redox and catalytic properties [1-3]. As an example A. Itadani *et al.*, have reported the preparation of copper ion exchanged ZSM-5 for calorimetric study of N_2 adsorption on Cu-ZSM-5 zeolite [4]. In another study, A. Ribera *et al.* have reported the characterization of redox properties and application of Fe-ZSM-5 catalysts [5]. In this research we prepared silicate solutions by dissolving silica in sodium hydroxide. Aluminosilicate solutions with different Al/Si ratios were prepared by mixing appropriate quantities of sodium silicate solutions with freshly prepared sodium aluminate solutions and the NaA zeolites were made by hydrothermal method. Then, their XRD patterns and IR spectra were also considered. Obviously, those zeolites which have Al-OH and Si-OH groups can lose their protons in basic solutions. In this way zeolites can adsorb many ions with positive charge. We investigated ion exchange property of Fe^{3+}, Cu^{2+}, Ni^{2+} and Hg^{2+} in systems with pH equals to 2, 4, 6 and 8. We found that the aluminosilicate with Si/Al = 1, has greatest exchange capacity for all of the ions studied in this work.

Keywords: Zeolite NaA; PTMAOH; Microwave Heating; Hydrothermal

1. INTRODUCTION

Wastewater contaminated with heavy metals is produced in many industrial activities, such as tanneries, metal-plating facilities, and mining operations [6]. Hg^{2+}, Cu^{2+}, Fe^{3+}, and Ni^{2+} are among the most common metals found in many industrial wastewater. These heavy metals which are not biodegradable and tend to accumulate in organisms cause numerous diseases and disorders.

The understanding of the nature of micro porous solid acid catalysts, which have many important applications in commercial processes, is of great interest. Two of the most commonly studied types of materials in this category are zeolites (aluminosilicates) and silicoaluminophosphates (SAPOs). Such negatively charged defect sites can be counterbalanced by the presence of a proton leading to the formation of Bronsted-acid sites which are believed to be the active sites for catalysis [7].

Zeolites are micro porous crystalline aluminosilicates with frameworks made of SiO_4 and AlO_4 tetrahedral (these tetrahedral atoms are often referred to as T atoms). Insertion of trivalent Al^{3+} in place of tetrahedral coordinated Si^{4+} creates negative charge on the lattice, which is compensated by extra framework cations. If the charge compensating cation is H^+ a bridged hydroxyl group (Si-O(H)-Al) is formed which functions as a strong Bronsted-acid site. Due to these acid sites, zeolites are solid acids and are used as catalysts. The catalytic activity of zeolites is often related to strength of the acid sites, which depends on chemical composition and topology of zeolite frameworks [8]. The formation of many different zeolites is shown in **Figure 1**.

The crystallization process usually proceeds via the solution phase. Therefore, the composition and the structure of the silicate and silicoaluminate species in the so-

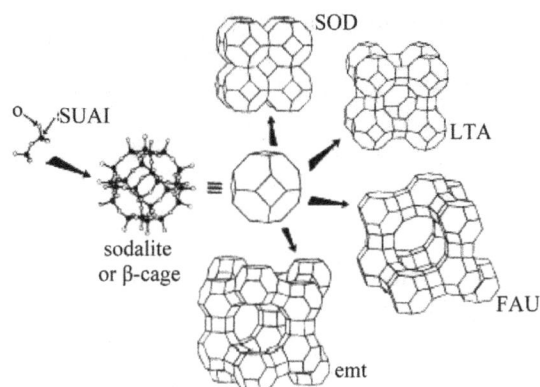

Figure 1. Building up zeolite structures.

lution phase are of importance in the crystallization process. In many experiments, the Si/Al ratio and the rate constant of the growth were studied based on the composition of the solution phases. The Si/Al ratio as an important parameter in application of zeolites is responsible for their thermal and hydrothermal stability as well as acidity. For the planning of a synthesis procedure for a special zeolite, exact information on the possibilities of adjustment of the Si/Al ratio using the experimental conditions is, therefore, of vital importance. Among the Various available treatment processes, ion exchange is considered to be more cost effective if low cost ion exchangers such as zeolites are used [6].

The selectivity of the exchanger for the higher valence ions increases with increasing dilution. The SiO_4 and AlO_4 are shown in **Figure 2**.

The present study is an attempt to investigate the selectivity series of Pb^{2+}, Fe^{3+}, Cu^{2+} and Ni^{2+} using single solutions, for three sample of zeolites with different Si/Al ratios.

2. EXPERIMENTAL

2.1. Materials

Sodium hydroxide; Aluminum foil and $NiCl_2 \cdot 6H_2O$ were purchased from Merck; SiO_2 was synthesized in our laboratory; silicon tetrachloride (99.8%) was purchased from Janssen chemical Co; ferric chlorid, $CuSO_4 \cdot 5H_2O$ and $Hg(OAc)_2$ were obtained from Fluka; and water was purified using a homemade instrument.

2.2. Preparation of Pure SiO_2

Pure silica was produced by hydrolysis of silicon tetrachloride using doubly distilled water. The precipitate was filtered and washed many times with doubly distilled water to remove the whole acid. It was then dried at $105^{\circ}C$ for 48 h. The chemical reaction is quite simple:

$$SiCl_4 + 2H_2O \rightarrow SiO_2 + 4HCl$$

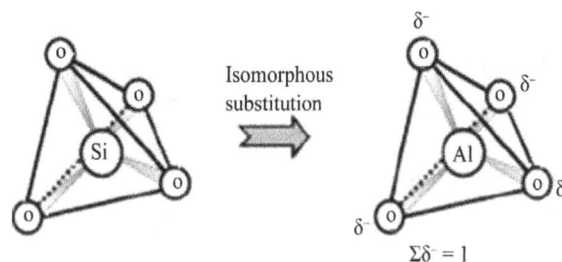

Figure 2. Figure of silicate and aluminate groups.

2.3. Making Silicate Solution

Aqueous silicate solution was prepared in plastic bottle by dissolving 1.5 gr. SiO_2 in sodium hydroxide solution 2 M. The dissolution of the silica was very slow at room temperature, therefore the sample was heated in an oven at $70^{\circ}C$ to help the dissolution process.

2.4. Making Aluminosilicate Solutions

Aqueous sodium aluminate solution was prepared by dissolving of 2 gr. aluminum foil in NaOH solution (5 M). Aluminosilicate solutions were then obtained by mixing the freshly prepared sodium aluminate and sodium silicate solution with the following initial ratios:

Sample No. 1, $SiO_2:Al_2O_3:4Na_2O:2PTMAOH:100H_2O$
Sample No. 2, $2SiO_2:Al_2O_3:4Na_2O:2PTMAOH:100H_2O$
Sample No. 3, $4SiO_2:Al_2O_3:4Na_2O:2PTMAOH:100H_2O$

2.5. Synthesis of Zeolite

The above mixtures were placed in hydrothermal vessel and then placed in an oven at $90^{\circ}C$ for 42 hours. The precipitate was filtered and washed with doubly distilled water and then dried at $105^{\circ}C$ for 48 h.

3. RESULTS AND DISCUSSION

3.1. XRD Pattern

In zeolite systems, perturbations in the framework structure, crystal morphology, extra framework material, phase purity, crystallite size, and the setting and occupation of cation sites can produce differences in the X-ray diffraction patterns. The first requirement is having a good and clean crystalline material that yields very high-resolution patterns. This results in a considerable improvement of the X-ray diffraction pattern, intensity and resolution. It also increases the ability to characterize the zeolite.

The characteristic of XRD patterns for the prepared zeolites are presented in **Figure 3** for three samples. The X-ray diffraction analysis indicated that all of the zeolites have crystalline structure.

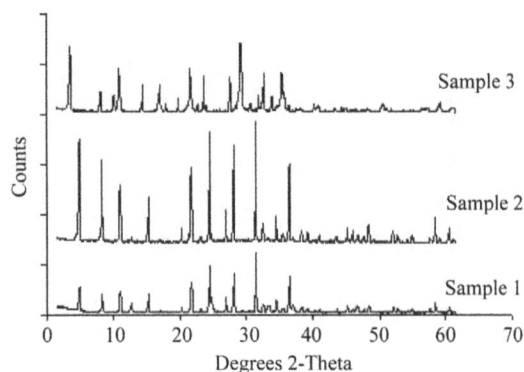

Figure 3. The XRD patterns of the obtained samples.

3.2. IR Spectra

Complementary information concerning the structure of the prepared zeolites was obtained through further research using IR spectroscopy. The bands occurring in the spectra of zeolites can be divided into two groups: bands connected with the internal vibrations of either SiO_4 and AlO_4 tetrahedra or Si-O-Si and Si-O-Al bridges, depending on the model used. Bands arising from the external vibrations of tetrahedra form the SBU and fragments of aluminosilicate framework. Each of the zeolites exhibits a typical IR pattern, which can be observed in **Figure 4**.

The results obtained from IR spectroscopy which present the characteristic spectral frequencies of zeolites are in agreement with other data reported in the literature [9].

The band at ca. 635 cm^{-1} observed in the spectra of the minerals of the D6R group (double six-membered rings) can be due to the presence of six-membered rings in the

three-dimensional zeolites structure (**Figure 2**).

According to the literature, the increase of the number of ring members leads to the shift of the characteristic ring band to the higher wave numbers [10]. Thus, one can expect the presence of the 4-membered ring band in the region of 700 - 740 cm^{-1}. According to the previous paper [11], the two bands related to the Si-O-Si and Al-O-Si bridges can be expected in the region of 690 - 800 cm^{-1}. At the same time, the position of the most intensive band in the region of 950 - 1200 cm^{-1} arising from the Si(Al)-O stretching vibrations depends on the Al:Si ratio in zeolites. It shifts to the lower wave numbers with the increase of Al contents [12]. The comparison of the intensity of pseudo lattice bands in the region of 550 - 650 cm^{-1} with the intensity of internal bending bands in the region of 410 - 470 cm^{-1} can be used to determine the degree of amorphization in zeolites [13]. The bands occurring in the first region are connected to the inter tetrahedral bonds typical for the ordered crystal structure, while the intensity of the band at about 410 - 470 cm^{-1} does not depend on the degree of crystallization.

3.3. Selectivity Determination

Uptake of metals (q) is calculated with the equation:

$$q(t) = (C_0 - C_t)\frac{V}{m}$$

where C_0 and C_t are the initial and final metal concentration in solution (in $mol\cdot l^{-1}$) after time t, V is solution volume (in l), and m is zeolite mass (in g). **Figures 5(a)-(d)** show metal uptake for three samples of synthesized zeolites.

The above charts show the decrease of the concentra-

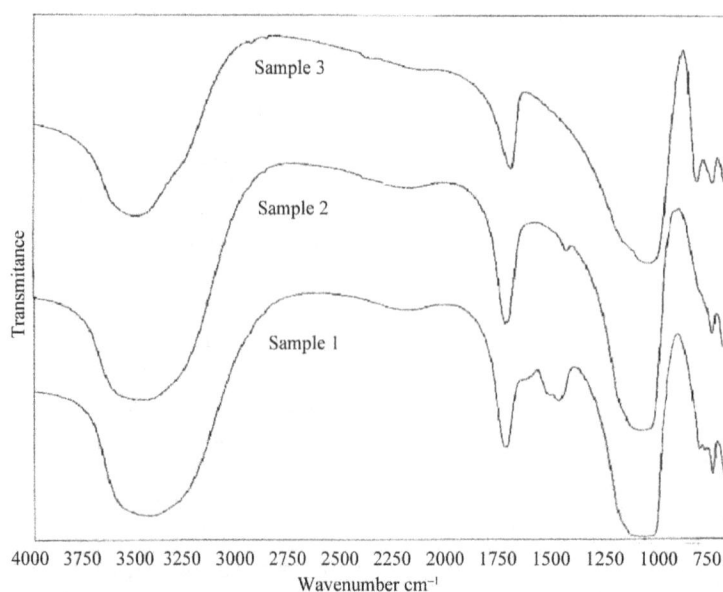

Figure 4. The IR spectra of obtained zeolites.

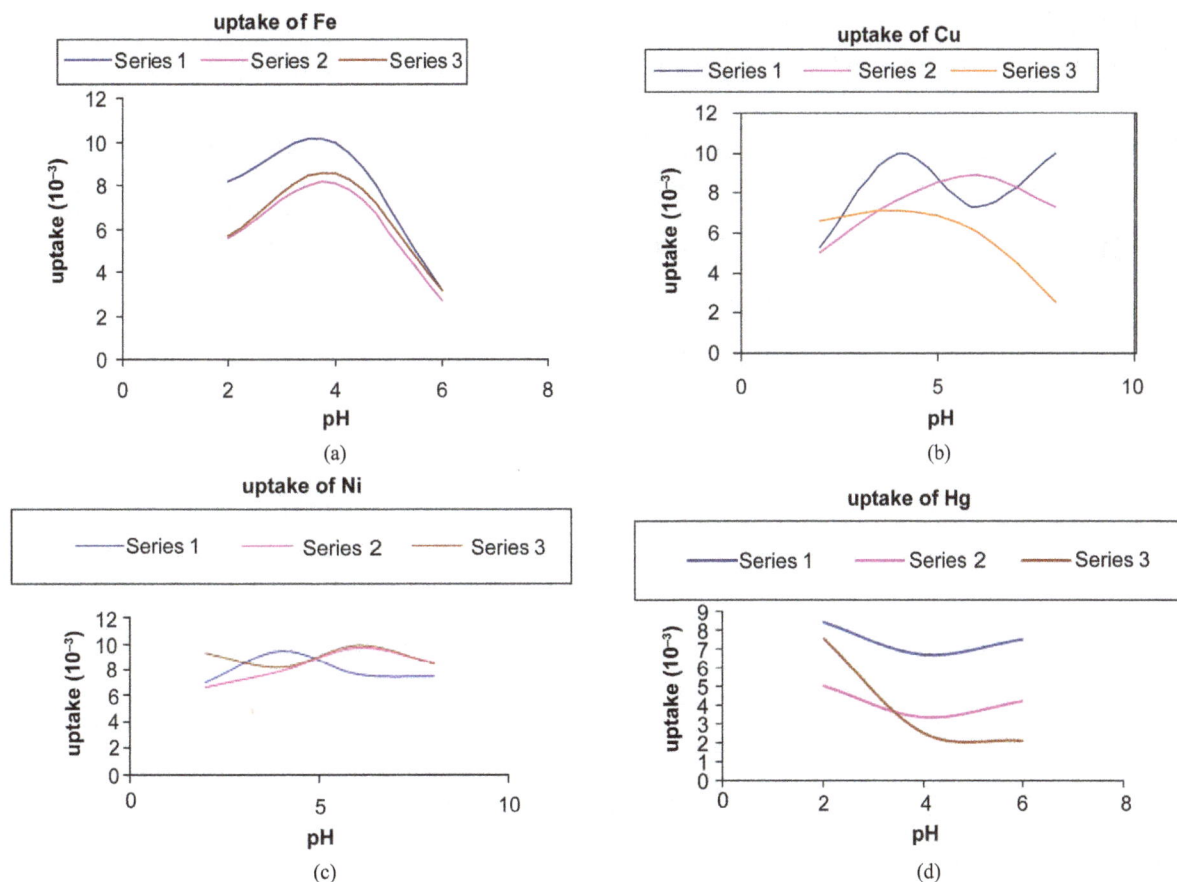

Figure 5. The uptake of metal ions in the different pH = 2 - 8 for three synthesized zeolite. In these charts series 1, 2 and 3 refer to samples 3, 2 and 1 respectively.

tion of metal ions.

4. CONCLUSIONS

The results indicate that sample 2 has higher exchange capacity than the other ones. It is mainly due to its higher Al substitution of Si that provides a negative framework favorable to superior ion exchange capability. It Also indicates that exchange capacity of samples 1 and 2 are approximately similar.

5. ACKNOWLEDGEMENTS

This project was carried out at the analytical laboratory in University of Mazandaran. Authors acknowledge Dr. Haghighi and Mr. Feyzi for their cooperation in preparing IR spectra.

REFERENCES

[1] Samant, P.V. And Fernandes, J.B. (2004) Enhanced activity of Pt(HY) and Pt-Ru(HY) zeolite catalysts for electrooxidation of methanol in fuel cells. *Power Source*, **125**, 172-177. doi:10.1016/j.jpowsour.2003.07.013

[2] Hadjiivanov, K., Ivanova, E. and Dimitrov, L. (2003) FTIR spectroscopic study of CO adsorption on Rh-ZSM-5: Detection of Rh^+-CO species. *Molecular Structure*, **459**, 661-662.

[3] Yamashita, H. and Takada S. (2003) Experimental study and *ab initio* molecular orbital calculation on the photolysis of *n*-butyrophenone included within the alkali metal cation-exchanged ZSM-5 zeolite. *Journal of Photochemistry and Photobiology A*: *Chemistry*, **160**, 37-42. doi:10.1016/S1010-6030(03)00218-1

[4] Itadani, A., Kumashiro, R., Kuroda, Y. and Nagao M. (2004) Calorimetric study of N_2 adsorption on copper-ion-exchanged ZSM-5 zeolite. *Thermochimica Acta*, **416**, 99-104. doi:10.1016/j.tca.2002.12.001

[5] Ribera A., Arends, I.W.C.E., de Vries, S., Pérez-Ramirez, J. and Sheldon, R.A. (2000) Preparation, characterization, and performance of FeZSM-5 for the selective oxidation of benzene to phenol with N_2O. *Journal of Catalysis*, **195**, 287-297. doi:10.1006/jcat.2000.2994

[6] Inglezakis, V.J., Loizidou M.D. and Grigoropoulou, H.P. (2003) Ion exchange of Pb^{2+}, Cu^{2+}, Fe^{3+} and Cr^{3+} on natural clinoptilolite: Selectivity determination and influence of acidity on metal uptake. *Journal of Colloid and Interface Science*, **261**, 49-54. doi:10.1016/S0021-9797(02)00244-8

[7] Shah, R., Gale, J.D. and Paynea, M.C. (1997) Comparing the acidities of zeolites and SAPOs from first principles. *Chemical Communications*, **1**, 131-132.

doi:10.1039/a605200b

[8]   Bekkum H., Flanigen, E.M. and Jansen, J.C. (1991) Introduction to zeolite science and practice. Elsevier, Amsterdam.

[9]   Azizi, S.N. and Yousefpour, M. (2010) Synthesis of zeolites NaA and analcime using rice husk ash as silica source without using organic template. *Journal of Materials Science*, **45**, 5692-5697. doi:10.1007/s10853-010-4637-7

[10] Sitarz, M., Mozgawa, W. and Handke, M. (1997) Synthesis of zeolites NaA and analcime using rice husk ash as silica source without using organic template. *Journal of*

Molecular Structure, **404**, 193-197. doi:10.1016/S0022-2860(96)09381-7

[11] Handke, M. and Mozgawa, W. (1993) Vibrational spectroscopy of the amorphous silicates. *Vibrational Spectroscopy*, **5**, 75-84. doi:10.1016/0924-2031(93)87057-Z

[12] Mozgawa, W., Sitarz, M. and Rokita, M. (1999) Spectroscopic studies of different aluminosilicate structures. *Journal of Molecular Structure*, **251**, 511-512.

[13] Ciciszwili, G.W., Andronikaszwili, T.G., Kirov, G.N. and Filizowa, L.D. (1990) Zeolity Naturalne. WNT, Warszawa.

6

Crystallization of amorphous boron by the calorimetric method

Tengiz Machaladze[1], Madona Samkharadze[2], Nino Kakhidze[2], Maia Makhviladze[2]

[1]Rafiel Agladze Institute of Inorganic Chemistry and Electrochemistry, I. Javakhishvili Tbilisi State University, Tbilisi, Georgia
[2]Akaki Tsereteli State University, Kutaisi, Georgia
Email: nino.kakhidze@gmail.com

ABSTRACT

The kinetics of formation of crystal boron was studied by X-ray structure and differential-thermal analyses. The enthalpy of transition of amorphous boron to crystal boron with formation of an intermediate non-equilibrium phase was determined by the calorimetric method. Studies were carried out by using the differential scanning calorimeter SETARAM. Amorphous boron received by diboron cracking was used as the starting material. The test sample is characterized by stability even at a high temperature. When heated, amorphous boron first transforms to crystal boron of α-modification, and during a further heating, there occur several phase transitions, which means the transition of α-rhombohedric crystal boron to the β' and β'' metastable states. Studies of various modifications were carried out by radiographic and electro-optic methods. The high sensitiveness of the calorimeter made it possible to control temperature in the oven, make records and detect even the smallest thermal effects.

KEYWORDS

Amorphous boron; α,β-Rombohedric Crystalline Phase; Calorimeter; Thermogram

1. INTRODUCTION

Elementary boron exists in the form of various crystalline α,β-rhombohedric and tetragonal modifications as well as in amorphous state. The heating of boron in vacuum or in an inert gas medium brings about its crystallization. Boron undergoes some transformations until the stable crystalline β-rhombohedric phase is formed. The transformation process depends on various factors: sample heating conditions, content of impurities and the presence of defects in the structure of the initial amorphous boron [1].

The kinetics of crystalline boron formation was studied by X-ray diffraction and differential-thermal analyses in [2-4], but thermal effect values were not determined. The $D_\alpha \rightarrow D_\beta$ transformation was studied by the electro-optical method in [5].

2. EXPERIMENTAL

The enthalpy of transformation of amorphous boron to the crystalline state with formation of an intermediate non-equilibrium state was measured by the calorimetric method. Experimental studies were carried out using the differential electronic scanning microcalorimeter HT-1500. This high-quality calorimetric apparatus made by SETARAM company is able to control the conditions in the kiln, to record signals automatically and to measure even the minutest thermal effects about 20 micro J (0.0005 cal).

The calorimeter was calibrated against the known metal melting heat values. Calorimetric and thermal signals were processed through the amplifier which delivered signals simultaneously to the recorder and to the numerical printing integrator which, upon the completion of transformation, gave out the corresponding numerical values. The integrator provides the high accuracy of integration since it directly processes the signal and not the signal copy, registered at another device.

The standard compounds were weighed to an accuracy of 0.0005 g. After putting the melting pot with tested samples into the calorimeter, they were subjected to degassing. Experiments were run in the argon steam medium at the pressure $P = 1.0 \times 10^{-5}$ Pa.

The calorimeter constant K for the melting point of the

standard compound was determined by the formula:

$$K = \frac{\Delta H}{A} \times \frac{G}{M},$$

where ΔH is the melting heat of the standard compound, cal/mole; A is the number of pulses produced by the generator of the amplifier for the electromotive force of the differential thermal battery which corresponds to the transformation area, in mV; G is the weight of the standard compound; M is the molecular mass of the standard compound.

The standard compounds used for the calorimeter calibration were tin, lead, zinc, aluminum, potassium chloride, silver, gadolinium, silicon and iron of "c·p" qualification.

The calorimeter constant K was calculated for each standard compound by averaging the values of 8 - 10 measurements. The standard error was determined by the formula:

$$\Delta S = \pm 2 \sqrt{\frac{\sum (\Delta i)^2}{n(n-1)}}$$

where n is the number of experiments; Δi is a deviation from the average value.

3. RESULT AND DISCUSSIONS

The dependence of the calorimeter constant on temperature is shown in **Figure 1**.

Experiments with the investigated compounds were run similarly to experiments with the standard compounds. The phase transition heat was calculated by the formula

$$\Delta H = K \times A \frac{M}{G},$$

where K is the calorimeter constant shown by the calibration curve, A is the number of pulses produced by the generator of the amplifier for the electromotive force of the differential thermal battery, corresponds to the transformation area, in mV; M is the molecular mass of the investigated compound, G is the investigated compound weight.

To determine the experiment accuracy, we studied the transformation $F_{e_\beta} \to F_{e_\gamma}$. The difference between our measurements and those given in [6] is approximately 1%. Amorphous boron received by diboron cracking was used as the starting material. The test sample is characterized by stability even at a high temperature. When heated, amorphous boron first transforms to crystal boron of α-modification, and during a further heating there occur several phase transitions, which means the transition of α-rhombohedric crystal boron to the β' and β'' metastable states.

Studies of various modifications were carried out by

radiographic and electro-optic methods. The high sensitiveness of the calorimeter made it possible to control temperature in the oven, make records and detect even the smallest thermal effects.

Amorphous boron selected as the initial material was prepared by cracking diboron. Its purity was 99.5%.

No thermal effect was observed when heating the boron sample up to 1300°C at the rate of 200, 400, 600°C per hour. However, with a further increase of the heating temperature by 200° per hour there occur two different exothermal peaks: the first peak is observed in the range 1360-1400°C (**Figure 2**) and the second occurs immediately upon the completion of the first peak and attains its maximum at 1450°C.

The X-ray and electro-optical analysis confirmed the transformation of amorphous boron to crystalline boron of α-modification. The enthalpy change during the transformation process was $\Delta H = -4.45$ kcal/mole. The standard error was equal to $\Delta S = \pm 1.5\%$. A further heating of the sample revealed several phases of transformations with a maximum 1550°C, 1662°C and 1712°C (**Figure 3**). The X-ray study showed that as a result of these transformations α-rhombohedral crystalline boron transforms to the metastable phase β' and β''. Transformation at 1712°C is associated with the final formation of the β-rombohedric phase. The thermal effect, correspond-

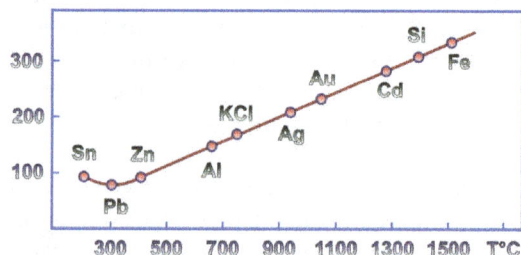

Figure 1. Calibration curve for calorimeter HT-1500.

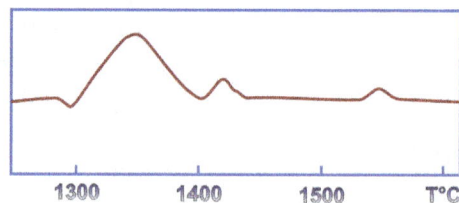

Figure 2. Thermogram of phase transformation of amorphous boron to α-boron.

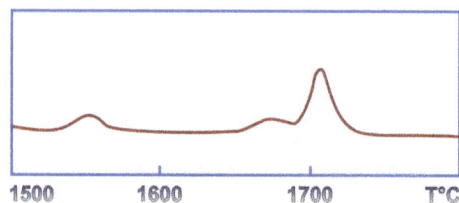

Figure 3. Thermogram of phase transformation of α-boron to β-boron.

ing to the variation of system enthalpy, comprises $\Delta H = -1.05$ kcal/mole.

4. CONCLUSIONS

Amorphous boron received by diboron cracking was used as the starting material. The test sample is characterized by stability even at a high temperature. When heated, amorphous boron first transforms to crystal boron of α-modification, and during a further heating, there occur several phase transitions, which means the transition of α-rhombohedric crystal boron to the β' and β'' metastable states.

Studies of various modifications were carried out by radiographic and electro-optic methods. The high sensitiveness of the calorimeter made it possible to control temperature in the oven, make records and detect even the smallest thermal effects.

REFERENCES

[1] Lavrinenko, V.A., Gogotsi, Yu.G. and Fratzevich, I.N. (1984) High-temperature oxidation of hot pressed boron. *Transactions of Academy of Sciences of USSR*, **275**, 114-117. (in Russian)

[2] Bagramashvili, I.A. and Pirtskhalaishvili, R.M. (1974) Study of the transfer of amorphous boron to crystalline state. In: Tavadze, F.N., Ed., *Boron Preparation, Structure, Properties*, Nauka, Moscow, 23.

[3] Ugai, L.A. and Soloviev, N.F. (1977) Methods of preparations of α-rhombohedral boron. In: Matkovich, V.I., Ed., *Boron and Refractory Borides*, Springer Verlag, Berlin, Heidelberg, 227-240. http://dx.doi.org/10.1007/978-3-642-66620-9_14

[4] Krutski, Ju.L., Galevski, G.V. and Kornilov, A.A. (1983) Oxidation of the ultra-disperse powders of carbide, boron, vanadium and chromium. *Powder Metallurgy*, **2**, 47-50. (in Russian)

[5] Kervalishvili, P.D., Kutelia, E.R., Dzigrashvili, T.A., Dekanosidze, R.., and Petrov, V.I. (1985) Electron-microscopic research of the structure of the amorphous boron. *Solid State Physics*, **27**, 1414-1418. (in Russian)

[6] Kelly, K.K. (1960) High-temperature heat-contents, heat capacity and entropy data for element and inorganic compounds. *Bureau of Mines Bulletin*, **584**, 232 p.

7

Reaction *in situ* found in the synthesis of a series of lanthanide sulfate complexes and investigation on their structure, spectra and catalytic activity

Zhaoyan Deng[1], Fengying Bai[2], Yongheng Xing[1*], Na Xing[1], Liting Xu[1]

[1]College of Chemistry and Chemical Engineering, Liaoning Normal University, Dalian, China
[2]College of Life Science, Liaoning Normal University, Dalian, China
Email: *yhxing2000@yahoo.com

ABSTRACT

A series of lanthanide sulfates coordination complexes, $Ln_2(SO_4)_3(H_2O)_8$ (Ln = Pr (1), Nd (2), Tb (3), Sm (4), Dy (5), Gd (7), Ho (8)), and $EuK(SO_4)_2$ (6), were constructed by the reaction in situ of lanthanide ions (Ln^{3+}) with flexible dodecanedioic acid and rigid aromatic 5-sulfosalicylic acid under hydrothermal conditions. All of them were characterized by elemental analysis, IR spectroscopy, and single-crystal X-ray diffraction. The crystal structures and coordination modes of metal centers and sulfate ions, as well as the novel reaction mechanism and different conditions of lanthanide ions and 5-sulfosalicylic acid to form the series of lanthanide sulfate complexes, were discussed in detail. Solid-state properties for these crystalline materials, such as thermal stability and powder X-ray diffraction have been investigated. Additionally, the photoluminescent characterizations of the complexes 3, 4, 5 and 6, and the catalytic properties of all the complexes about cyclohexane being oxidized into cyclohexanone/cyclohexanol were investigated and compared.

Keywords: Lanthanide Sulfates; Hydrothermal Synthesis; Crystal Structure; Photoluminescent Properties; Catalytic Reaction

1. INTRODUCTION

Metal-organic frameworks (MOFs) [1], as a relatively new class of crystalline coordination polymers, have in the past decades become one of the fastest growing fields in chemistry, which is due to the significance both in academia and industry not only for their structural varie-

ties but also for their fascinating potential applications as functional crystalline materials, including gas storage, carbon dioxide capture and renewable catalysts [2-10]. Compared with transition metal, lanthanide-organic frameworks (LMOFs) possess more unique advantages because of the better optics and magnetism characters from the lanthanide themselves and more coordination numbers which is due to the larger radii of the lanthanide atoms [11-16]. In addition, the selection and accommodation of ligands play a key role for the construction of LMOFs, and therefore, mixed organic ligands, particularly in rigid-flexible ligands, have been proven to be effective and useful to enrich the varieties of lanthanide-organic frameworks because of their specific features: 1) Rigid ligands play important roles in constructing a stable framework and enhance the fluorescence emissions of complexes; 2) The rotation of the flexible ligands increases the variety of configurations of the coordination polymers [17-33].

As we know, a great number of examples about lanthanide-organic framework with rigid aromatic multicarboxylate have been reported, such as benzoic acid [34,35], 1,2-benzene dicarboxylic acid (o-H_2BDC) [36], 1,3-benzene dicarboxylic acid (m-H_2BDC) [37,38], 1,4-benzene dicarboxylic acid (p-H_2BDC) [39,40], benzene-1,3,5-tricarboxylic acid (H_3BTC) [41], and 1,2,4,5-benzenetetracarboxylic acid (H_4BTEC) [42,43]. However, ligands with the interesting functional sulfonic group to construct lanthanide complexes have been rarely explored and only the complexes [Eu(p-Tos) $(H_2O)_7][p$-Tos$]_2$ $(H_2O)_2$ (p-Tos = Toluene-4-sulfonate) [44]; $[Ln_2(ad)_{2.5}(BSA)(H_2O)_2]_n$ (Ln = Sm, Nd; H_2ad = adipic acid; BAS = benzenesulfonate) [45]; [Eu(BSA)(glu)($H_2O)_2$]·H_2O (Ln = Eu, Sm, Ce, Pr, Nd; H_2glu = glutaric acid, HBSA = benzene sulfonic acid); Ln(SSA)($H_2O)_2$ (SSA = 5-sulfosalicylic acid, Ln = Ce, Pr, Nd and Dy) [46]; and

---

*Corresponding author.

$[Ln(SSA)(H_2O)_2]_n \cdot nH_2O$ (SSA = 5-sulfosalicylic acid, Ln = Gd, Sm, Nd, Tb, Eu, Yb and Dy) [47] appear in the present literatures. 5-sulfosalicylic acid, possessing three potential coordinating groups, -COOH, -SO$_3$H and -OH, can act as the preferential ligand of the lanthanide complexes [48-51]. In addition, although a number of the lanthanide complexes with flexible linkers have been reported, complexes with aliphatic diacid with more than ten carbon atoms have never been reported [52].

Therefore, in our previous work, we selected 5-sulfosalicylic acid and at the same time chose dodecanedioic acid in synthesizing the rigid-flexible lanthanide coordination polymers. Nevertheless, it is not predictable at all to receive the established products, but surprising that there exist two kinds of 2-D or 3-D lanthanide sulfates after the hydrothermal reaction when these two different acids and lanthanide chloride were dissolved in water/ethanol system, and the resultant reaction has never before been reported. Although there are a great number of reports on organic amine template complexes of the lanthanide sulfates, they generally used lanthanide sulfates as the starting material and employed a special organic amine as the structure-directing agent (SDA) or the pH-adjusting agent [53-56]. Hence, one recognizes that organic amine was directly inserted into the structure of the product, and inevitably, as a result, it is hard to produce some single sulfato lanthanide complexes. Moreover, reports on a series of complexes of lanthanide sulfates are very few. Recently, we used a reaction in situ to synthesize a series of the sulfato lanthanide complexes, $Ln_2(SO_4)_3(H_2O)_8$ (Ln=Pr(**1**), Nd(**2**), Tb(**3**), Sm(**4**), Dy(**5**), Gd(**7**), Ho(**8**)), EuK(SO$_4$)$_2$ (**6**) (shown in **Scheme 1**).

2. EXPERIMENTAL SECTION

2.1. Materials and Methods

All chemicals purchased were of reagent grade or better and were used without further purification. Lanthanide chloride salts were prepared via dissolving 10 g praseodymium oxides with 100 ml 12 M HCl and then evaporating at 100°C until the crystal film formed. The infrared spectra were recorded on a JASCO FT/IR-480 PLUS Fourier Transform spectrometer with pressed KBr pellets in the range 400 - 4000 cm^{-1}. The luminescence spectra were reported on F-7000 FL Spectrophotometer (200 - 800 nm). The elemental analyses were carried out on a Perkin Elmer 240C automatic analyzer. Lanthanide con-

tents were analyzed on a Plasma-Spec(I)-AES model ICP spectrometer. X-ray powder diffraction (XRD) data were collected on a Bruker Advance-D8 with Cu-Ka radiation, in the range $5° < 2\theta < 60°$, with a step size of 0.02° (2θ) and an acquisition time of 2 s per step.

2.2. Synthesis of the Complexes

$[Pr_2(SO_4)_3(H_2O)_8]$ (**1**). Complexes **1** were prepared by hydrothermal reaction. In a typical synthesis, solution I was prepared by dissolving PrCl$_3$·6H$_2$O (0.107 g, 0.3 mmol) and 5-sulfosalicylic acid (0.076 g, 0.30 mmol) into 5.0 ml ethanol under stirring for 1 - 2 h, dodecanedioic acid (0.069 g, 0.3 mmol) was added to 5.0 ml ethanol to make solution II. 5.0 ml deionized water was added after solution I was mixed with solution II under stirring for a minimum of 1-2 h. Then, one drop of saturated KOH (aq) was added into the mixture solution under stirring. The final mixture was transferred to a 25 ml Teflon-lined stainless steel vessel under autogenous pressure and heated at 160°C for 3 days. Colorless single crystals of **1** for X-ray diffraction analysis were obtained in ca. 74% yield based on Pr(III) after two weeks. Elemental analysis for H$_{16}$O$_{20}$S$_3$Pr$_2$ (M_r = 714.16), calcd: Pr, 39.46; H, 2.26%. Found: Pr, 39.41; H, 2.50%. IR data (KBr pellet, v[cm^{-1}]): 3450.51 (s), 2923.64 (w), 2855.35 (w), 1637.26 (s), 1123.90 (s), 980.17 (w), 651.43 (m), 597.03 (m).

$[Nd_2(SO_4)_3(H_2O)_8]$ (**2**). This complex was synthesized by a procedure similar to that used for **1** but changing the PrCl$_3$·6H$_2$O to NdCl$_3$·6H$_2$O (0.108 g, 0.30 mmol), purple crystals of **2** were obtained in ca. 69% yield based on Nd (III). Elemental analysis for H$_{16}$O$_{20}$S$_3$Nd$_2$ (M_r = 720.82): Nd: 40.02; H, 2.24%. Found: Nd: 40.00; H, 2.36%. IR data (KBr pellet, v [cm^{-1}]): 3423.01(s), 1642.38 (s), 1130.83 (s), 998.28 (w), 658.03 (m), 599.73 (m).

$[Tb_2(SO_4)_3(H_2O)_8]$ (**3**). This complex was synthesized by a procedure similar to that used for **1** but changing the PrCl$_3$·6H$_2$O to TbCl$_3$·6H$_2$O (0.112 g, 0.30 mmol), colorless crystals of **3** were obtained in ca. 78% yield based on Tb(III). Elemental Anal. Calc. for H$_{16}$O$_{20}$S$_3$Tb$_2$ (Mr = 750.20): Tb, 42.37; H, 2.15%. Found: Tb, 42.31; H, 2.42%. IR data (KBr pellet, m[cm^{-1}]): 3487.70 (s), 3371.48 (s), 1642.38 (s), 1143.62 (s), 1092.09 (s), 1001.07 (m), 806.98 (w), 606.12 (s), 489.52 (s), 424.82 (w).

$[Sm_2(SO_4)_3(H_2O)_8]$ (**4**). This complex was synthesized by a procedure similar to that used for **1** but changing the PrCl$_3$·6H$_2$O to SmCl$_3$·6H$_2$O (0.109 g, 0.30 mmol), primrose yellow crystals of **4** were obtained in ca. 76% yield based on Sm (III). Elemental analysis for H$_{16}$O$_{20}$S$_3$Sm$_2$ (M_r = 733.06): Sm, 41.02; H, 2.20%. Found: Sm, 39.96; H, 2.51%. IR data (KBr pellet, v [cm^{-1}]): 3448.96 (s), 1642.38 (s), 1124.44 (s), 1007.84 (w), 651.64 (w), 599.73 (s), 489.52 (w).

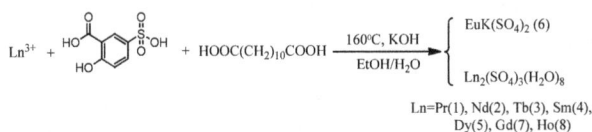

Scheme 1. Synthesis method of complexes 1 - 8.

$[Dy_2(SO_4)_3(H_2O)_8]$ (**5**). This complex was synthesized by a procedure similar to that used for **1** but changing the $PrCl_3 \cdot 6H_2O$ to $DyCl_3 \cdot 6H_2O$ (0.113 g, 0.30 mmol), colorless crystals of **5** were obtained in ca. 71% yield based on Dy (III). Elemental analysis for $H_{16}O_{20}S_3Dy_2$ (M_r = 757.34): Dy, 42.91; H, 2.13%. Found: Dy, 42.83; H, 2.16%. IR data (KBr pellet, ν [cm^{-1}]): 3481.31 (s), 3371.48 (s), 3235.32 (m), 1642.38 (s), 1150.02 (w), 1198.49 (s) , 1001.07 (s), 813.37 (m), 748.68 (m), 690.38 (w), 651.64 (w), 606.12 (s), 489.52 (m), 431.22 (m).

$EuK(SO_4)_2$ (**6**). This complex was synthesized by a procedure similar to that used for **1** but changing the $PrCl_3 \cdot 6H_2O$ to $EuCl_3 \cdot 6H_2O$ (0.110 g, 0.30 mmol), primrose yellow crystals of **6** were obtained in ca. 67% yield based on Eu (III). Elemental analysis for $EuK(SO_4)_2$ (M_r = 383.21): Eu, 39.66; K, 10.20%. Found: Eu, 39.61; K, 10.12%. IR data (KBr pellet, ν [cm^{-1}]): 3432.24 (w), 2926.42 (w), 2857.74 (w), 1743.66 (w), 1265.24 (w), 1121.91 (s) , 636.34 (s), 602.19 (s), 451.72 (w).

$[Gd_2(SO_4)_3(H_2O)_8]$ (**7**). This complex was synthesized by a procedure similar to that used for **1** but changing the $PrCl_3 \cdot 6H_2O$ to $GdCl_3 \cdot 6H_2O$ (0.111g, 0.30 mmol), colorless crystals of **7** were obtained in ca. 66% yield based on Gd (III). Elemental analysis for $H_{16}O_{20}S_3Gd_2$ (M_r = 746.84): Gd, 42.11%; H, 2.16%. Found: Gd, 42.06; H, 2.25%. IR data (KBr pellet, ν [cm^{-1}]): 3481.31 (s), 3371.48 (s), 3235.32 (m), 1648.77 (s), 1143.62 (s), 1092.09 (s), 1001.07 (s), 806.98 (w), 742.28 (w), 651.64 (w), 599.73 (s), 489.52 (m), 431.22 (m).

$[Ho_2(SO_4)_3(H_2O)_8]$ (**8**). This complex was synthesized by a procedure similar to that used for **1** but changing the $PrCl_3 \cdot 6H_2O$ to $HoCl_3 \cdot 6H_2O$ (0.114 g, 0.30 mmol), pink crystals of **8** were obtained in ca. 61% yield based on Ho (III). Elemental analysis for $H_{16}O_{20}S_3Ho_2$ (M_r = 762.20): Ho, 43.28%; H, 2.12%. Found: Ho, 43.15%; H, 2.32%. IR data (KBr pellet, ν [cm^{-1}]): 3486.64 (s), 3384.20 (s), 3240.87 (w), 2331.67 (w), 1641.23 (s), 1395.07 (w), 1148.91 (s), 1101.26 (s), 1005.58 (s), 608.94 (m), 479.11 (w), 438.22 (w).

2.3. Initial Characterization

Initial characterizations were carried out by elemental analysis, PXRD and IR studies.

PXRD patterns were recorded in the 2θ range 5-50°using Cu-Kα radiation (Bruker Advance-D8), with a step size of 0.02°(2θ) and a count time of 2s per step. As shown in **Figures S1-S8**, all the peaks presented in the measured patterns closely match the simulated patterns generated from single crystal diffraction data, which confirm the phase purity of the bulk samples.

The IR spectra for the complexes were recorded as KBr pellets (Supporting Information, **Figure S9** and **S10**). The IR spectra of complexes **1**, **2**, **3**, **4**, **5**, **7** and **8** exhibit strong and broad absorption bands in the range of 3009 –3679 cm^{-1} and 1640 cm^{-1}, indicating the presence of

coordination water moleculars. In **1**, **2**, **3**, **4**, **5**, **7** and **8**, the SO_4^{2-} ions adopt μ_3 or μ_2 coordination modes and lead to low site symmetry C_{3v} or C_{2v}. The bands show medium strong intensity at 994 and 607 cm^{-1}. These may be attributable to the symmetric S-O stretching mode (ν_1) and the symmetric SO_4^{2-} bending mode(ν_2). The strong band around 1117 cm^{-1} splitting into two bands, 1142 and 1093 cm^{-1}, may be assigned to the ν_3 mode because of the coordination of the free sulfate group to the metals. Compared with that of the above complexes, the IR spectra of complex **6** shows some difference in that there were only two main characteristic bands of 1139 cm^{-1} and 601 cm^{-1} ascribed to the vibration of the sulfate group [57-59].

2.4. Single Crystal Structural Determinations

A suitable single crystal of each compound was carefully selected under a cubic microscope and glued to a thin glass fiber for X-ray measurement. Of these reflection data of the complexes, **1-5**, **7** and **8** were collected on a Bruker AXS SMART APEX II CCD diffractometer with graphite monochromatized Mo Kα radiation (λ = 0.71073 Å), and that of **6** was collected on a Xcalibur, Atlas, Gemini ultra CCD diffractometer. The data were reduced using SAINTPLUS, and an empirical absorption correction was applied using the SADABS program. The structure was solved and refined using SHELXL-97. All the hydrogen positions were initially located in the different Fourier maps. Final refinement included atomic positions for all the atoms, anisotropic thermal parameters for all the non-hydrogen atoms, and isotropic thermal parameters for all the hydrogen atoms. Details of the structure solution and final refinements for the compounds are given in **Tables 1** and **2**. The selected bond lengths and bond angles of complexes **3** are listed in **Table 3**, that of **6** are represented in **Table 4**, and that of **1**, **2**, **4**, **5**, **7** and **8** are shown in the supplement materials (**Table S1-S6**).

2.5. Experiment Set up for Catalytic Oxidation

The oxidation reactions were carried out under air condition (atmospheric pressure) in Schlenk tubes. In a typical experiment, 0.0004 g of the catalysts (complex **1-8**) was dissolved in 3.00 ml of desired solvent. Then the required amounts of H_2O_2 (30% H_2O_2 solution) and HNO_3 were added according to this order. Finally, 0.68 g of cyclohexane was added into the solution to make the cyclohexane/catalyst molar ratio equal to 15,000. The reaction solution was stirred for some time at the given temperature.

For the product analysis, 0.03 g of methylbenzene (internal standard) and 1.5 ml of diethyl ether (to extract the substrate and the organic products from the reaction mixture) were added. The obtained mixture was stirred

Table 1. Crystallographic data for complexes **1-4**[*].

Complexes	1	2	3	4
Empirical Formula	$H_{16}O_{20}Pr_2S_3$	$H_{16}O_{20}Nd_2S_3$	$H_{16}O_{20}Tb_2S_3$	$H_{16}O_{20}Sm_2S_3$
M (g·mol^{-1})	714.13	720.79	750.15	733.01
Temperature(K)	273(2)	296(2)	296(2)	296(2)
Crystal system	Monoclinic	Monoclinic	Monoclinic	Monoclinic
Space group	C2/c	C2/c	C2/c	C2/c
a (Å)	13.7004(13)	13.6617(8)	13.5006(17)	13.552(5)
b (Å)	6.8613(7)	6.8366(4)	6.7165(8)	6.757(3)
c (Å)	18.4632(18)	18.4353(12)	18.247(2)	18.272(7)
α (deg)	90	90	90	90
β (deg)	102.7970(10)	102.6390(10)	102.100(2)	102.320(6)
γ (deg)	90	90	90	90
V (Å^3)	1692.5(3)	1680.13(18)	1617.8(3)	1634.8(11)
Z	4	4	4	4
D_{calc} (g·cm^{-3})	2.803	2.850	3.080	2.978
Crystal size (mm)	0.17 × 0.14 × 0.09	0.18 × 0.14 × 0.09	0.18 × 0.14 × 0.09	0.18 × 0.14 × 0.09
F(000)	1368	1376	1416	1392
μ(Mo-Kα)/mm^{-1}	6.158	6.584	9.162	7.599
θ (deg)	2.26 - 24.97	2.26 - 24.99	2.28 - 28.43	2.28 - 25.00
Reflections collected	4108	4034	4886	3914
Independent reflections	1479	1473	1976	1423
R int	0.0231	0.0249	0.0244	0.0231
Parameters	122	123	130	131
$\Delta(\rho)$ (e Å^{-3})	0.633, −0.847	0.889, −0.781	0.776, −1.297	1.012, −0.956
Goodness of fit	1.090	1.125	1.073	1.121
R^a	0.0207 (0.0228)[b]	0.0209 (0.0219)[b]	0.0253 (0.0311)[b]	0.0180 (0.0186)[b]
wR$_2$^{a}	0.0499 (0.0509)[b]	0.0550 (0.0556)[b]	0.0565 (0.0590)[b]	0.0448 (0.0451)[b]

[*][a]$R = \Sigma||Fo|-|Fc||/\Sigma|Fo|$, $wR_2 = [\Sigma(w(Fo^2-Fc^2)^2/[\Sigma(w(Fo^2)^2)^{1/2}; [Fo > 4\sigma(Fo)]$. [b]Based on all data.

for 10 min and then a sample (0.8 μL) was taken from the organic phase and analyzed by a GC equipped with a capillary column and a flame ionization detector by the internal standard method. Blank experiments confirmed that no cyclohexanol or cyclohexanone were formed in the absence of the metal catalyst under the same conditions.

3. RESULTS AND DISCUSSION

3.1. Synthesis

As shown in **Table 5**, in the investigation we tried four methods in order to let the lanthanide ions coordinate with 5-sulfosalicylic acid and a flexible ligand, or to coordinate with 5-sulfosalicylic acid and another rigid ligand. The details are as follows. At first, we designed the experimental method (II), for the purpose of the coordination of lanthanide ions and 5-sulfosalicylic acid and chain-like aliphatic dicarboxylic acid to achieve novel complexes with 5-sulfosalicylic acid-dicarboxylic acid as rigid-flexible ligands. But unfortunately, every experimental result demonstrated that lanthanide ions only coordinated with aliphatic dicarboxylic acid. In this case, in order to explore the experimental conditions of the coordination of 5-sulfosalicylic acid and lanthanide

Table 2. Crystallographic data for complexes **5-8**[*].

Complexes	5	6	7	8
Empirical Formula	$H_{16}O_{20}Dy_2S_3$	O_8KEuS	$H_{16}O_{20}Gd_2S_3$	$H_{16}O_{20}Ho_2S_3$
M (g·mol^{-1})	757.34	383.20	746.81	762.20
Temperature(K)	296(2)	293(2)	296(2)	296(2)
Crystal system	Monoclinic	Triclinic	Monoclinic	Monoclinic
Space group	C2/c	P1	C2/c	C2/c
a (Å)	13.5034(15)	5.3589(3)	13.574(2)	13.4516(17)
b (Å)	6.7192(7)	6.8831(6)	6.7622(12)	6.6885(8)
c (Å)	18.253(2)	8.9525(11)	18.337(3)	18.171(2)
α (deg)	90	97.419(9)	90	90
β (deg)	102.049(2)	92.338(7)	102.173(2)	102.000(2)
γ (deg)	90	91.025(6)	90	90
V (Å^3)	1619.6(3)	327.09(5)	1645.3(5)	1599.1(3)
Z	4	2	4	4
D_{calc} (g·cm^{-3})	3.106	3.891	3.015	3.166
Crystal size (mm)	0.18 × 0.14 × 0.09	0.13 × 0.10 × 0.09	0.18 × 0.14 × 0.09	0.18 × 0.14 × 0.09
F(000)	1424	356	1408	1432
μ(Mo-Kα)/mm^{-1}	9.646	10.868	8.474	10.320
θ (deg)	3.09 - 24.99	2.99 - 24.99	3.38 - 25.00	2.29 - 28.47
Reflections collected	3792	2042	3814	4543
Independent reflections	1408	1158	1430	1910
R int	0.0392	0.0382	0.0324	0.0458
Parameters	115	109	123	115
$\Delta(\rho)$ (e Å^{-3})	5.885, −6.613	6.535, −1.757	2.758, −2.049	7.564, −6.235
Goodness of fit	1.258	1.052	1.150	1.180
R[a]	0.0477 (0.0478)[b]	0.0511 (0.0535)[b]	0.0384 (0.0397)[b]	0.0584 (0.0589)[b]
wR$_2$[a]	0.1077 (0.1078)[b]	0.1351 (0.1379)[b]	0.1101 (0.1113)[b]	0.1413 (0.1420)[b]

[*][a]$R = \Sigma||Fo|-|Fc||/\Sigma|Fo|$, $wR_2 = [\Sigma(w(Fo^2 - Fc^2)^2/[\Sigma(w(Fo^2)^2)]^{1/2}$; [Fo > 4$\sigma$(Fo)]. [b]Based on all data.

ions, the experiment of (I) was devised. That is, there were only the lanthanide chloride and 5-sulfasalicylic acid in the reaction systems. However, after the normal hydrothermal reaction and cooling, there was still only the clear solution and no desired crystal of the complexes, which lead us to reach a conclusion that 5-sulfosalicylic acid and lanthanide chloride could not directly react with each other in this condition. Then, we designed the experiments of (III) and (IV), that is, the weak acid of dodecanedioic acid or the weak base of 2,4,6-tis (3,5-dimethyl-1H-pyrazol-1-yl)-1,3,5-triazine were added in addition to the lanthanide ions and 5-sulfosalicylic acid because neither dodecanedioic acid nor 2,4,6-tis(3,5-dimethyl-1H-pyrazol-1-yl)-1,3,5-triazine can easily coordinate with lanthanide ions in this reaction condition. Although lanthanide-3pz-CQ has been reported in the past literature, a lanthanide-dodecanedioic acid complex has never been reported. For 3pz-CQ compound, it is possible that lanthanide ions tend to coordinate with oxygen atoms, and coordinate with nitrogen only in an appropriately polar non-aqueous solvent such as ethanol. For dodecanedioic acid, since it possesses a too long

Table 3. Selected bond distances (Å) and angles (deg) of complex 3.

Bond distances					
Tb-O(5)	2.253(8)	Tb-O(2)	2.316(8)	Tb-O(6)	2.317(9)
Tb-O(4)	2.326(9)	Tb-O(1)	2.366(9)	Tb-O(8)	2.429(8)
Tb-O(3)	2.438(10)	Tb-O(7)	2.487(9)		
Bond angles					
O(5)-Tb-O(2)	143.0(3)	O(5)-Tb-O(6)	80.4(3)	O(2)-Tb-O(6)	126.1(3)
O(5)-Tb-O(4)	88.3(3)	O(2)-Tb-O(4)	79.2(3)	O(6)-Tb-O(4)	70.8(3)
O(5)-Tb-O(1)	147.2(3)	O(2)-Tb-O(1)	69.2(3)	O(6)-Tb-O(1)	79.7(3)
O(4)-Tb-O(1)	109.1(4)	O(5)-Tb-O(8)	99.5(3)	O(2)-Tb-O(8)	75.8(3)
O(6)-Tb-O(8)	141.3(3)	O(4)-Tb-O(8)	147.7(3)	O(1)-Tb-O(8)	80.6(3)
O(5)-Tb-O(3)	69.9(3)	O(2)-Tb-O(3)	73.2(3)	O(6)-Tb-O(3)	134.4(3)
O(4)-Tb-O(3)	74.4(3)	O(1)-Tb-O(3)	140.6(3)	O(8)-Tb-O(3)	78.9(3)
O(5)-Tb-O(7)	73.4(3)	O(2)-Tb-O(7)	133.4(3)	O(6)-Tb-O(7)	74.8(3)
O(4)-Tb-O(7)	143.2(3)	O(1)-Tb-O(7)	76.4(3)	O(8)-Tb-O(7)	68.4(3)
O(3)-Tb-O(7)	125.0(3)				

Table 4. Selected bond distances (Å) and angles (deg) of complex 6[*].

Bond distances					
Eu-O(4)	2.325(8)	Eu-O(7)#1	2.330(8)	Eu-O(8)#2	2.377(8)
Eu-O(3)	2.438(7)	Eu-O(5)	2.443(9)	Eu-O(1)	2.445(8)
Eu-O(6)#3	2.469(7)	Eu-O(2)	2.566(8)	O(6)-Eu#6	2.469(7)
O(7)-Eu#1	2.330(8)	O(8)-Eu#2	2.377(8)		
Bond angles					
O(4)-Eu-O(7)#1	83.7(3)	O(4)-Eu-O(8)#2	79.3(3)	O(7)#1-Eu-O(8)#2	82.9(3)
O(4)-Eu-O(3)	121.0(3)	O(7)#1-Eu-O(3)	78.6(3)	O(8)#2-Eu-O(3)	150.3(3)
O(4)-Eu-O(5)	70.5(3)	O(7)#1-Eu-O(5)	121.9(3)	O(8)#2-Eu-O(5)	137.1(3)
O(3)-Eu-O(5)	72.7(3)	O(4)-Eu-O(1)	136.4(3)	O(7)#1-Eu-O(1)	139.5(3)
O(8)#2-Eu-O(1)	106.5(3)	O(3)-Eu-O(1)	74.8(3)	O(5)-Eu-O(1)	78.0(3)
O(4)-Eu-O(6)#3	149.2(3)	O(7)#1-Eu-O(6)#3	74.0(3)	O(8)#2-Eu-O(6)#3	77.1(3)
O(3)-Eu-O(6)#3	75.6(2)	O(5)-Eu-O(6)#3	139.7(3)	O(1)-Eu-O(6)#3	70.2(3)
O(4)-Eu-O(2)	85.3(3)	O(7)#1-Eu-O(2)	158.8(3)	O(8)#2-Eu-O(2)	77.4(3)
O(3)-Eu-O(2)	122.6(3)	O(5)-Eu-O(2)	70.6(3)	O(1)-Eu-O(2)	55.6(2)
O(6)#3-Eu-O(2)	108.3(3)				

[*]Symmetry transformations used to generate equivalent atoms: #1 −x + 2, −y, −z + 2 #2 −x + 1, −y, −z + 2 #3 x, y + 1, z #6 x, y−1, z.

carbon chain and large steric hindrance, it maybe not coordinate with lanthanide ions to form stable skeleton of complexes. According to the experimental case above, we can use their insolubility and inertia of coordination with lanthanide ions to promote the coordination of lanthanide ions and 5-sulfosalicylic acid. Based on these points, we chose the mixed solvents of water-ethanol. On one hand, it makes the inert ligand dissolve well in the

Table 5. Experimental strategies demonstrating the reaction mechanism.

Methods	Materials			Conditions	Products
(I)	LnCl$_3$·6H$_2$O	H$_3$SSA		Hydrothermal (160°C, H$_2$O/EtOH)	Solution (No crystal)
(II)	LnCl$_3$·6H$_2$O	H$_3$SSA	HOOC(CH$_2$)$_n$COOH (0 ≤ n ≥ 3)	Hydrothermal (160°C, H$_2$O/EtOH)	Ln-dicarboxylate complexes
(III)	LnCl$_3$·6H$_2$O	H$_3$SSA	HOOC(CH$_2$)$_n$COOH (10 ≤ n ≥ 12)	Hydrothermal (160°C, H$_2$O/EtOH)	Ln-SO$_4$ complexes
(IV)	LnCl$_3$·6H$_2$O	H$_3$SSA	3Pz-CQ	Hydrothermal (160°C, H$_2$O/EtOH)	Ln(SSA)$_3$(H$_2$O)$_{1.5}$ complex

H$_3$SSA:5-sulfosalicylic acid 3Pz-CQ:2,4,6-tis(3,5-dimethyl-1H-pyrazol-1-yl)-1,3,5-triazine

*H3SSA: 5-sulfosalicylic acid. 3Pz-CQ: 2,4,6-tis(3,5-dimethyl-1H-pyrazol-1-yl)-1,3,5-triazine.

ethanol and strengthen the possibility of coordination with lanthanide ions. On the other hand, because of the addition of ethanol, it increases the mole concentration of 5-sulfosalicylic acid which easily dissolves in water, and as a result, enhances the molecular collision possibilities and coordination opportunities of 5-sulfosalicylic acid and lanthanide ions. All these promote the lanthanide ions to coordinate with the inert ligand as well as 5-sulfosalicylic acid. But in fact, the results of the experiment of (III) indicated that when a weak acidic ligand was added in the reaction system a decomposition reaction of 5-sulfosalicylic acid occurred, and at the same time, it gave rise to a new reaction in situ, namely, the decomposed product (SO_4^{2-}) coordinated to lanthanide directly. This shows that in the reaction above both dodecanedioic acid and 2,4,6-tis(3,5-dimethyl-1H-pyrazol-1-yl)-1,3,5-triazine played a certain role as a pH-adjusting agent and a structure-directing agent for the coordination of 5-sulfosalicylic acid and lanthanide ions. Compared to 2,4,6-tis(3,5-dimethyl-1H-pyrazol-1-yl)-1,3,5-triazine, it seems that dodecanedioic acid played a more important role for coordination. It might offer some protons for the system to be acidic. Then these protons attached at the juncture of the sulfo-group and phenyl and made the unstable 5-sulfosalicylic to decompose and form the rather stable salicylic acid, and the sulfo-group, which has fallen off from the phenyls, then coordinated with lanthanide ions. The detailed reaction mechanism was speculated in **Scheme 2**.

3.2. Structural Description of Complexes

X-ray diffraction determination results reveal that coordination polymers 1 - 5, 7, 8 are isomorphous, therefore complexes 3 and 6 are taken as the examples to present and describe the structures in detail.

Structure of Ln$_2$(SO$_4$)$_3$(H$_2$O)$_8$, (Ln = Pr(**1**), Nd(**2**), Tb(**3**), Sm(**4**), Dy(**5**), Gd(**7**), Ho(**8**))

The structure of **3** reveals that it is a 2-D framework complex. The asymmetric unit of **3** contains one Tb^{3+} atom, one and a half SO$_4^{2-}$ anions and four coordination water molecules. The Tb atom is eight-coordinated to

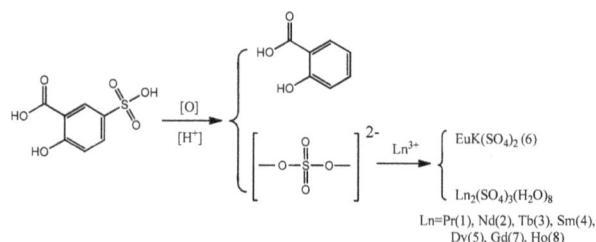

Scheme 2. Reaction mechanism of complexes **1-8**.

four sulfate ions in the monodentate fashion and four water molecules (**Figure 1(a)**, **Figures S11(a)-S16(a)**). Of the two framework sulfate ions, one binds to two Tb atoms in a monodentate fashion leaving two terminal S-O bonds, whereas another binds to three metal atoms in a monodentate manner leaving one terminal S–O bond (**Scheme 3(d) and (c)**). The linkage of the TbO$_8$ polyhedra and the SO$_4$ tetrahedra by sharing vertices gives rise to neutral inorganic layers parallel to the bc-plane, containing four-membered and eight-membered rings (**Figure 1(b)**), (**Figures S11(b)-S16(b)**).

SO_4^{2-} can be regarded as 2-connected and 3-connected linkers, and Tb^{3+} is surrounded by four SO_4^{2-} ligands and can be regarded as a 4-connected node. Topology analysis by the TOPOS 4.0 software package suggests that complex **3** possesses a 2D (2, 3, 4)-connected 3-nodal network with a Schläfli symbol of $(4^2.6.8^2.10)_2$ $(4^2.6)_2(8)$ (**Figure 1(c)**, **Figures S11(c)-S16(c)**).

In the complex Tb$_2$(SO$_4$)$_3$(H$_2$O)$_8$, there exists seven kinds of intermolecular hydrogen bonding interactions of O1-H1A··O9, O1-H1B··O10, O3-H3A··O7, O3-H3B··O8, O4-H4A··O9, O4-H4B··O10, O7-H7B··O8, in which they are respectively formed by water molecular and SO_4^{2-} ligands, except O3-H3A··O7 which is formed by water molecules. In addition, O1-H1A··O9, O4-H4A··O9, O4-H4B··O10 and O7-H7B··O8 linked the adjacent bc planes to generate a 3D framework. At the same time, O1-H1B··O10, O3-H3A··O7, O3-H3B··O8, as the intermolecular hydrogen bondings, further strengthen the whole structure of complexes (**Figure 1(d)**, **Figures S11(d)-S16(d)**). As shown in **Table 6**, the hydrogen

(a)

(b)

(c)

(d)

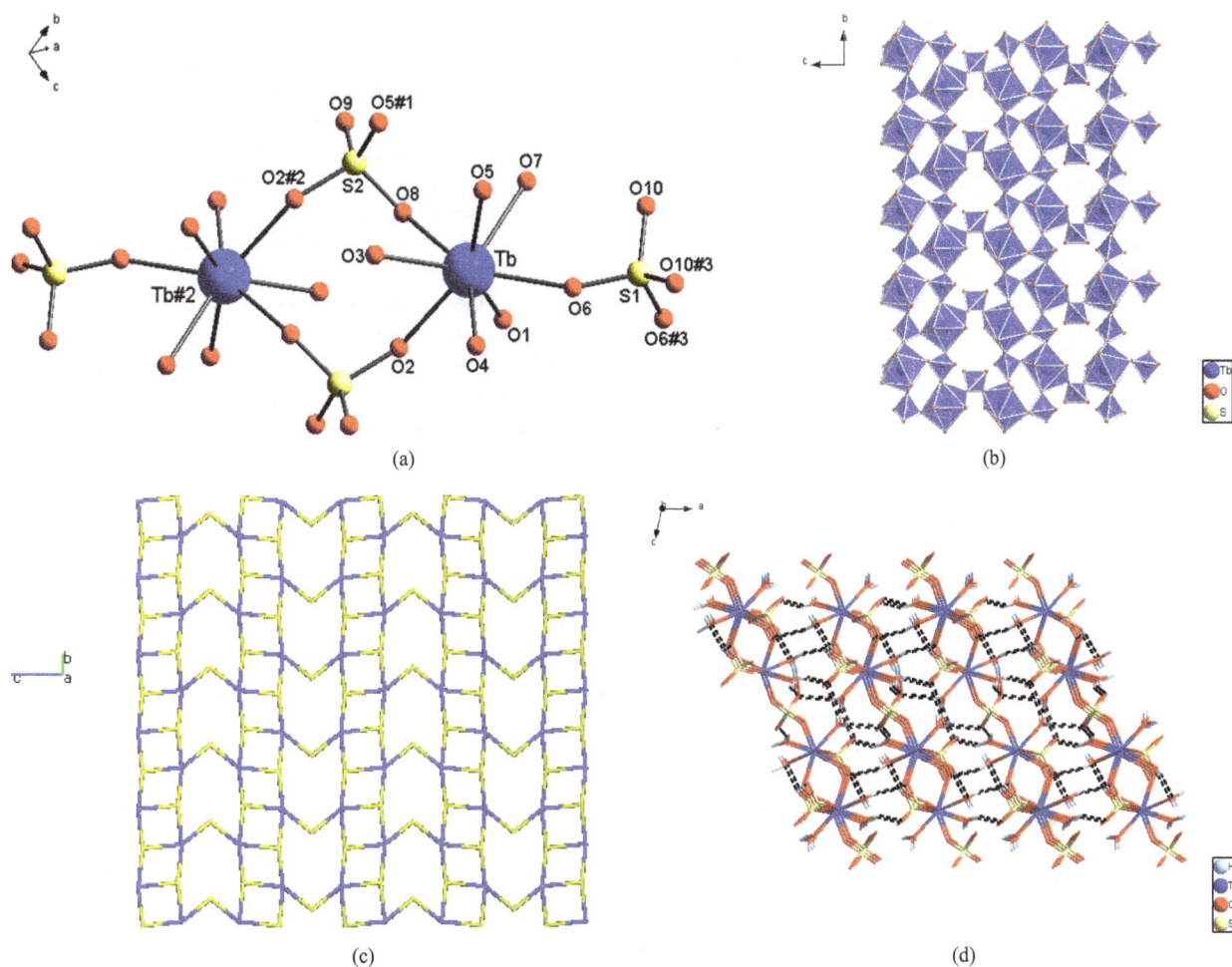

Figure 1. (a) The coordination environment of Tb in complex 7; (Symmetry codes: #1: −x+3/2, −y−1/2, −z + 1; #2: −x + 3/2, −y + 1/2, −z + 1; #3: −x + 2, y, −z + 3/2); (b) The 2D layer in complex 3, viewed along the a-axis (c) Schematic representation of the 2D $(4^2.6.8^2.10)_2(4^2.6)_2(8)$ topology network in complex 3. Blue, Tb; Yellow, SO_4^{2-}. (d) A view of the 3D hydrogen bonding network structure of complex 3 (black dotted lines representing hydrogen bonding).

Table 6. Hydrogen bonds distances (Å) and angles (°) of complex **3**[*].

D-H···A	D-H	H···A	D···A	<D-H···A
O1—H1A···O9[g]	0.9000	1.8777	2.7151	153.93
O1—H1B···O10[b]	0.9000	1.9754	2.8213	155.95
O3—H3A···O7[e]	0.7911	2.1652	2.9413	166.93
O1—H3B···O8[d]	0.7423	2.1431	2.7885	145.81
O4—H4A···O9[h]	0.8999	2.1300	2.7584	126.16
O4—H4B···O10[i]	0.9000	2.2402	2.8019	120.09
O7—H7B···O8[k]	0.9044	2.1543	3.0366	164.89

[*]Symmetry codes: g: 1/2 + x, −1/2 + y, z; b: x, 1 + y, z; e: 1/2−x, 1/2−y, −z; d: 1/2−x, 3/2−y, −z; h: 1/2−x, −1/2 + y, 1/2−z ; i: −1/2 + x, 1/2 + y,z; k:1−x, 1 + y, 1/2−z.

bonding distances of D···A vary from 2.7151 to 2.9413 Å, and the angles of D-H···A are in the rage of 120.09° - 166.93°. The detail hydrogen bondings of the complexes are listed in **Tables S7-S12**. All these are comparable with those found in other reported lanthanide sulfate complexes [55,60-62].

Structure of EuK(SO$_4$)$_2$ (**6**)

The structure of **6** reveals that it is a 3-D framework complex. The asymmetric unit of **6** contains one Eu^{3+} cation, one K$^+$ cation and two SO$_4^{2-}$ anions. The metal atom is eight-coordinated by the oxygen atoms from seven sulfate ions as can be seen in **Figure 2(a)**. Thus, two of the sulfur atoms in the asymmetric unit S1 forms four S-O-Ln bonds to four crystallographically distinct Ln atoms, thereby sharing corners with four metal-oxygen polyhedra. The S2 forms four S-O-Ln bonds to three unique metal atoms, sharing the corners with two metal-oxygen polyhedra and the edge with another polyhedron (**Scheme 3(a)** and **(b)**). The Eu-O bond distances are in the range of 2.325(5) - 2.566(8) Å (av.

2.416(1) Å). The O-Eu-O bond angles are in the range of 55.6(2) - 158.8(3)°. The selected bond distances and angles are given in **Table 4**. LnO$_8$ polyhedra are linked by the S(1)O$_4$ tetrahedra in two-dimensions to form four-membered rings, the rings connected by S(2)O$_4$ to form layers parallel to the bc-plane of the unit cell. Such a connectivity between the four-membered rings results in the formation of six four-membered rings around each eight-membered ring (**Figure 2(b)**). The layers are stacked over one another in AAA... fashion, with two adjacent layers separated by a unit cell length along the a-axis of the unit cell, thus forming four and eight-membered channels. The S(1)O$_4$ tetrahedra share corners and the S(2)O$_4$ tetrahedra share corners and edges with

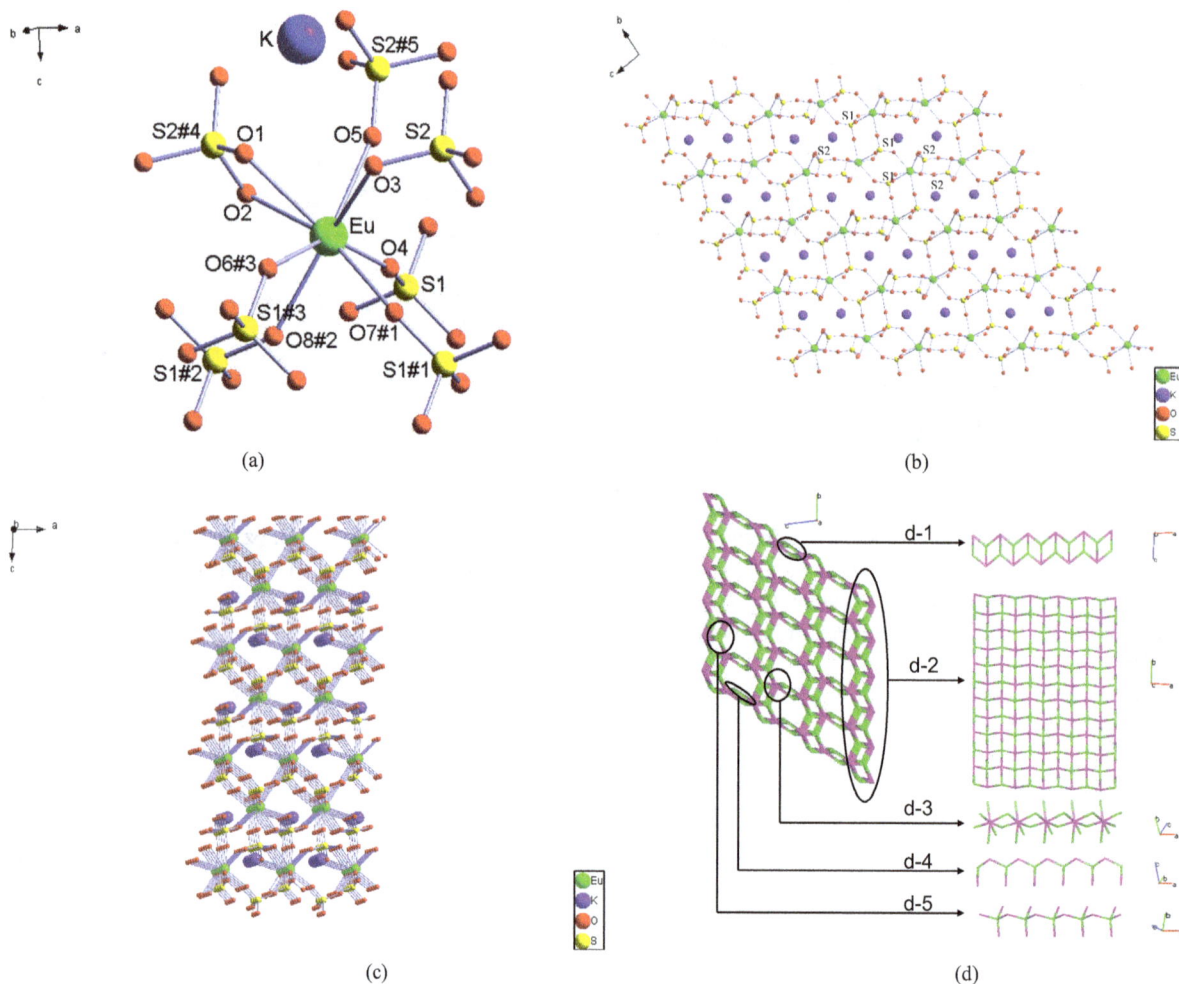

(a)

(b)

(c)

(d)

Figure 2. (a) The ORTEP view of the coordination environment of Eu in complex 6. (Symmetry codes: #1: −x+2, −y, −z + 2; #2: −x + 1, −y, −z + 2; #3: x, y + 1, z; #4:x−1, y, z; #5:−x + 2,−y, −z + 1); (b) The layer, parallel to the bc-plane of the unit cell in I, formed by connecting EuO$_8$ polyhedra with SO$_4$ tetrahedra, by sharing edges and vertices, and thereby forming four-membered and eight-membered rings. Note the arrangement of the four-membered rings around the eight-membered rings. The kalium cations are shown in one eight-membered rings. (c) The three-dimensional framework formed by the linking of the layers, stacked one over another, along the a-axis of the unit cell by the sulfate groups. (d) Schematic representation of the 3D $(4^3)(4^5.6)(4^8.6^9.8^4)$ topology network in complex 6. Fuchia, Eu; Lime, SO$_4^{2-}$. (d-1) The 1D zig-zig chains in the ac plane. (d-2) The 2D wavelike layer in the ab plane. (d-3) A 7-connected metal mode containing one europium center with seven SO$_4^{2-}$. (d-4) A 3-connected ligand node containing one SO$_4^{2-}$ with three samarium center. (d-5) A 4-connected ligand node containing one SO$_4^{2-}$ with four europium center.

the metal-oxygen polyhedra respectively from adjacent layers, thereby connecting the layers and forming a three-dimensional framework (**Figure 2(c)**). This connectivity gives rise to a square grid of intersecting four membered channels running perpendicular to one an other, and intersecting the channels along the a-axis of the unit cell. The kalium cations are located in the cages formed by the eight-membered channels.

Each $S(1)O_4$ links four adjacent Eu atoms to afford an infinite wavelike 2D layer in an ab plane (**Figure 2(d)-2**). Then $S(2)O_4$ between the adjacent layers links three Eu atoms in zig-zig 1D chains in ac planes to construct a 3D framework (**Figure 2(d)-1**). In this case, $S(1)O_4$ ligands can be regarded as 4-connected nodes (**Figure 2(d)-5**), $S(2)O_4$ ligands can be regarded as 3-connected nodes (**Figure 2(d)-4**), and Eu atoms surrounded by seven SO_4^{2-} ligands can be considered as 7-connected nodes (**Figure 2(d)-3**). Topology analysis by the TOPOS4.0 software package suggests that $EuK(SO_4)_2$ possesses a 3D (3, 4, 7)-connected 3-nodel network with a $(4^3) (4^5.6) (4^8.6^9.8^4)$ Schläfli symbol (**Figure 2(d)**).

3.3. Photoluminescent Properties

Lanthanide coordination polymers often exhibit intense luminescence and thereby are particularly interesting for luminescent materials. Owing to the excellent luminescent properties of Tb (III), Sm (III), Dy (III) and Eu (III) ions, the photoluminescent behaviors of the coordination polymers **3, 4, 5, 6** were investigated in the form of a solid state at room temperature.

The luminescent spectrum of complex **3** was investigated under excitation of 368 nm with a slit width (2.5:2.5). In **Figure 3** it can be seen that emission peaks at 491 nm, 545 nm, 587 nm and 622 nm are attributed to the characteristic emissions of Tb emissive state $^5D_4$ to the ground state ^7F_J (J = 6→3), respectively. The spectra are dominated by the $^5D_4 \rightarrow {}^7F_5$ transitions at 545 nm which create the intense green luminescence output for the solid sample. In addition, there exists a weak luminescence emission at 417 nm assigned to intraligand $\pi \rightarrow \pi^*$ transition, implying that the energy is not fully transferred from the sulfate ion to the Tb ion.

The emission spectrum involving complex **4** was determined on excitation at 367 nm with a slit width (2.5:2.5), which is depicted in **Figure 4**. As expected, the

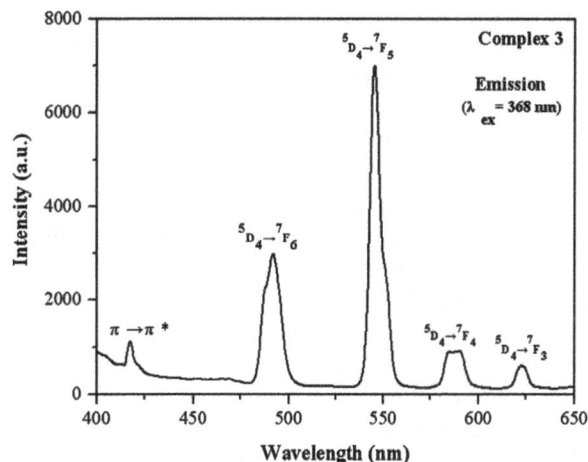

Figure 3. Photoluminescence emission spectra for complex **3** in the solid state at room temperature.

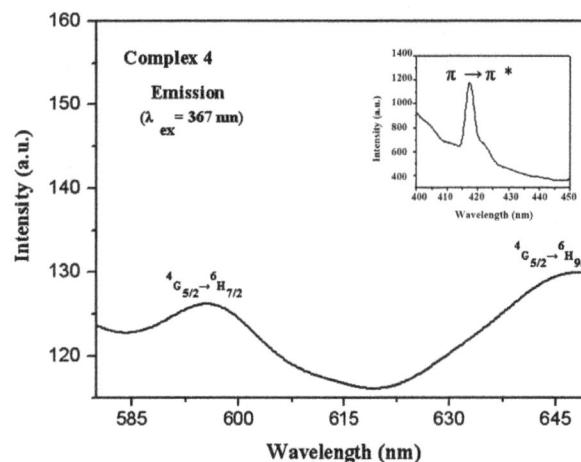

Figure 4. Photoluminescence emission spectra for the complex **4** in the solid state at room temperature.

two luminescence emission peaks at 595 and 647 nm correspond to the characteristic emissions of Sm emissive state $^4G_{5/2}$ to the $^4H_{7/2}$ and $^4H_{9/2}$ levels, respectively. Similar to complex **3**, there also exists an intraligand $\pi \rightarrow \pi^*$ transition at 417 nm, implying that the energy is not fully transferred from the sulfate ion to the Sm ion. Compared to those of the other three complexes, the emission spectrum of complex **3** showed very weak and broad.

With regard to the luminescent characteristic of the dysprosium complex, the luminescent spectra of complex **5** was determined under excitation of 240 nm with slit width (2.5:2.5). As shown in **Figure 5**, it presents good luminescent properties with a narrow, sharp and strong emission peak at 482 nm assigned to the blue insensitive transitions $^4F_{9/2} \rightarrow {}^6H_{15/2}$. Additionally, it should be mentioned that no further emission peaks were observed, indicating that the energy is fully transferred from the sulfate ion to the Dy ion. This is different from the other three complexes in which the characteristic emission

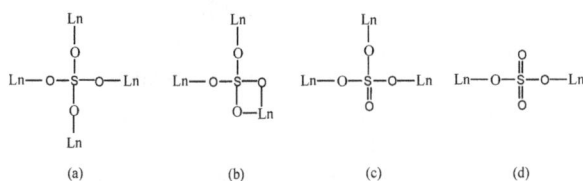

Scheme 3. The four kinds of coordination modes of sulfate ions coordinated to lanthanide ions.

bands of the sulfates ions are found, suggesting that the more efficient ligand-to-metal energy transfers occur in dysprosium coordination polymer. Furthermore, the one strong emission band suggests that this homochromy effect possesses good optical application, which is rare in other dysprosium coordination polymers.

The luminescent properties regarding complex **6** were studied at the excitation wavelength of 370 nm with a slit width (2.5:2.5). **Figure 6** gives the emission spectra of this complex. The characteristic $^5D_0 \rightarrow ^7F_J$ ($J = 1 \rightarrow 4$) transitions of the Eu(III) ions at 595, 619, 653, and 689 nm show well an efficient ligand-to-Eu energy transfer. The quite weak emission peak $^5D_0 \rightarrow ^7F_0$ at 580 nm are attributed to the symmetry-forbidden emission of the Eu(III) ions in these coordination polymers. The $^5D_0 \rightarrow ^7F_1$ emission bands pertain to the prominent magnetic dipole transitions, which are almost unin- fluenced by the coordination environment. On the other hand, the outstanding $^5D_0 \rightarrow ^7F_2$ emission bands, possessing a strong electric dipole character, are hypersensitive to

the coordination environment. Therefore, we can take advantage of the relative intensity disparity of these transitions to probe the nature of the linker environment and herein Eu luminescence can act as a sensitive probe of the lanthanide coordination environment [63,64]. In particular, the ratio of the intensities of the ($^5D_0 \rightarrow ^7F_2$): ($^5D_0 \rightarrow ^7F_1$) transition is very sensitive to the symmetry of the Eu(III) center. In the spectra, it can be obviously seen that the intensity of the electric dipole transitions $^5D_0 \rightarrow ^7F_2$ are much stronger than that of the magnetic dipole transition $^5D_0 \rightarrow ^7F_1$, which implies that the Eu(III) ions in these complexes are located in lower symmetric co-ordination environments. This is in agreement with the result of the single X-ray analysis. Additionally, the strongest emission peaks in the $^5D_0 \rightarrow ^7F_1$ and $^5D_0 \rightarrow ^7F_2$ transition region generate splitting of energy, which can be also ascribed to the Eu(III) centers in asymmetric sites in these complexes. Among these emission lines, $^5D_0 \rightarrow ^7F_2$ transitions are most striking, indicating intense red luminescence of complex **6**. Additionally, there also exits a $\pi \rightarrow \pi^*$ transition, which is similar to that of the complex **3** and **4**.

3.4. TG Properties

Thermal analysis shows that the complexes are consid-erably stable, especially complex **6**. As shown in **Figure 7(a)**, the weight loss of 20.01% in the range 91°C - 294°C can be attributed to the loss of the coordinated water molecules (cal. 21.07%). Then the complexes stayed at a very stable stage until 800°C. The collapse of the fragment in the second stage may be due to the loss of SO_3, and the final residual is Ln_2O_3. In comparison, the thermal stability of complex **6** is better, and it didn't decompose until 700°C because of the absence of water in the structure (**Figure 7(b)**).

3.5. Catalysis Study

In the primary stage of our work, the oxidation reaction of cyclohexane to produce cyclohexanone and cyclo-hexanol was employed as a model reaction and complex **1** was examined as a catalyst. The reaction was carried out in a solvent of CH_3CN at 40 centigrade. It showed small TON value, indicating that the capability of the catalysts for the reaction of cyclohexane conversion is weak. A blank experiment conducted in the absence of the catalyst, under the above reaction conditions, found that no products were detected. For the sake of compare-son, complexes **2-8** were also used as catalysts. As shown in **Table 7**, similarly, there is very low catalytic activity with TON values in the range of 0.68 - 4.83, although cyclohexane can reach high conversion after the reaction for some time. It is found that the order of the catalytic activity is **8 > 2 > 5 > 3 > 7 > 4 > 6 > 1** when

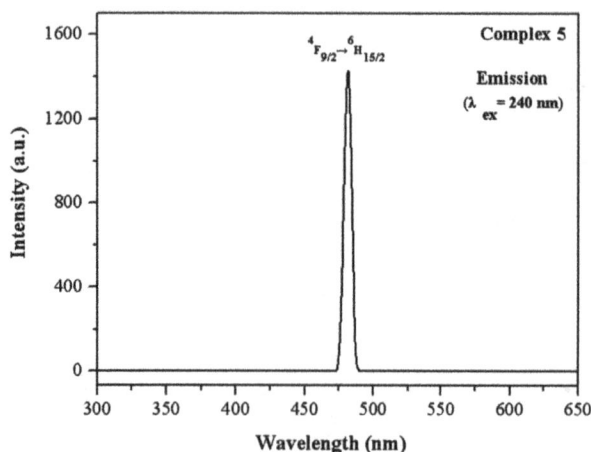

Figure 5. Photoluminescence emission spectrum for complex **5** in the solid state at room temperature.

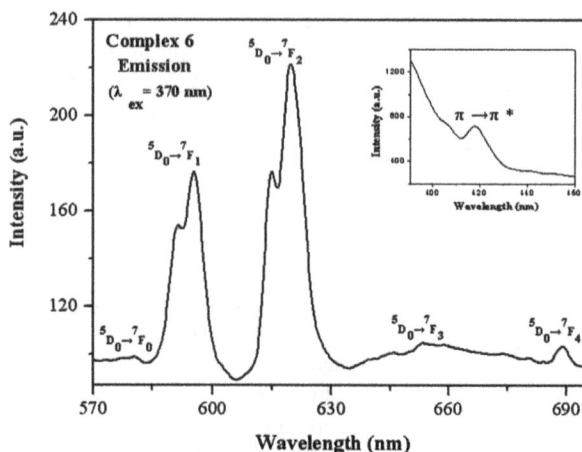

Figure 6. Photoluminescence emission spectra for **complex 6** in the solid state at room temperature.

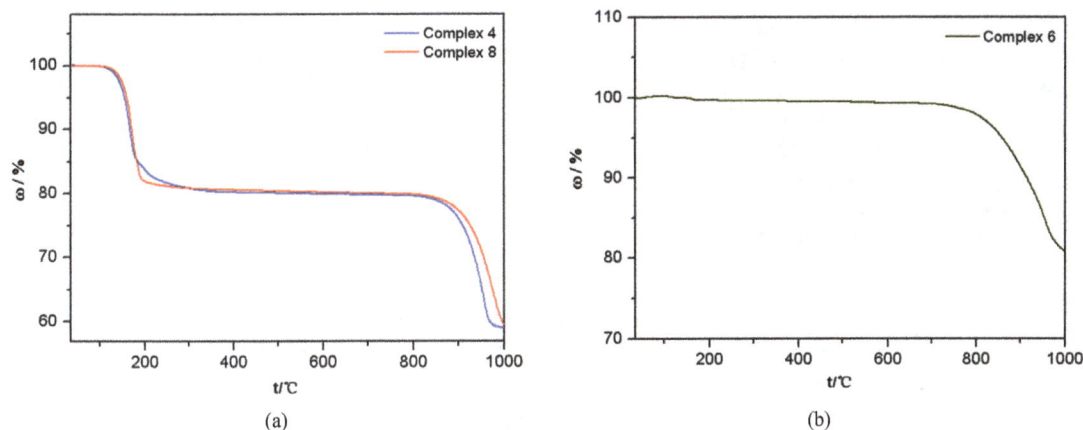

(a)

(b)

Figure 7. (a) TG curve of $Ln_2(SO_4)_3(H_2O)_8$ (Ln = Sm(4), Ho(8)); (b) TG curve of $EuK(SO_4)_2(6)$.

Table 7. Data of oxidation for cyclohexane with complexes **1-8** as catalysts in the system of $/H_2O_2/HNO_3/CH_3CN$ at 40°C*.

Catalysts	Mole ratio			Time (h)	Reaction medium	Cyclohexane Conversion (%)	TON (Cyclohexanone)	TON (Cyclohexanol)	TON (Cyclohexanol and Cyclohexanone)
	Catalyst/ Cyclohexane (10^{-5})	Catalyst/H_2O_2 (10^{-6})	Catalyst /HNO_3 (10^{-4})						
1	6.52	1	2.5	16	CH_3CN	99.88	0	0.68	0.68
2	6.52	1	2.5	16	CH_3CN	99.95	0.27	4.56	4.83
3	6.52	1	2.5	16	CH_3CN	99.92	0.06	3.43	3.49
4	6.52	1	2.5	16	CH_3CN	99.94	0	2.09	2.09
5	6.52	1	2.5	16	CH_3CN	99.95	0	4.19	4.19
6	6.52	1	2.5	16	CH_3CN	99.95	0	1.67	1.67
7	6.52	1	2.5	16	CH_3CN	99.94	0.13	2.51	2.64
8	6.52	1	2.5	16	CH_3CN	99.94	0.18	4.65	4.83

*1: $Pr_2(SO_4)_3(H_2O)_8$; 2: $Nd_2(SO_4)_3(H_2O)_8$; 3: $Tb_2(SO_4)_3(H_2O)_8$; 4: $Sm_2(SO_4)_3(H_2O)_8$; 5: $Dy_2(SO_4)_3(H_2O)_8$; 6: $Eu(SO_4)_2K$; 7: $Gd_2(SO_4)_3(H_2O)_8$; 8: $Ho_2(SO_4)_3(H_2O)_8$. TON: moles of products (cyclohexanol and cyclohexanone)/moles of catalysts.

comparing the catalytic activity of the complexes **1-8**. In particular, the catalytic efficiency of the complexes **2** and **8** is found to be much better than other complexes. The results also showed that in the cases of **1, 4, 5** and **6**, the final products were only cyclohexanol, which indicated that the selectivity of those complexes is very high, and therefore, complexes of **1, 4, 5** and **6** may possess a value for potential application.

4. CONCLUSION

In summary, we have reported two kinds of simple, high-yield and pure phase synthesis of 2-D and 3-D framework lanthanide sulphate coordination complexes, $Ln_2(SO_4)_3(H_2O)_8$ (Ln = Pr (**1**), Nd (**2**), Tb (**3**), Sm (**4**), Dy (**5**), Gd (**7**), Ho (**8**)) and $EuK(SO_4)_2$ (**6**). The synthesis of all complexes have demonstrated that the lanthanide sulfate complexes were designed and prepared by using appropriate solvent and SDAs. On the other hand, all

these complexes represent a certain activity in catalysis. These Tb, Sm, Dy and Eu complexes exhibit different luminescence characterizations of Tb^{3+}, Sm^{3+}, Dy^{3+} and Eu^{3+}, respectively. The luminescence in the solid state indicates the complex **5** is an excellent candidate for pure fluorescent materials.

5. ACKNOWLEDGEMENTS

This work was supported by grants from the National Natural Science Foundation of China (Grant No. 21071071) and the State Key Laboratory of Inorganic Synthesis and Preparative Chemistry, College of Chemistry, Jilin University, Changchun 130012, P. R. China (Grant No. 2013-05).

REFERENCES

[1]   Zhou, H.C., Long, J.R. and Yaghi, O.M. (2012) Introduction to metal-organic frameworks. *Chemical Reviews*,

112, 673-674. http://dx.doi.org/10.1021/cr300014x

[2] Rosi, N.L., Ecket, J., Eddaoudi, M., Vodak, D.T., Kim, J., O'Keeffe, M. and Yaghi, O.M. (2003) Microporous metal-organic frameworks. *Science*, **300**, 1127-1129. http://dx.doi.org/10.1126/science.1083440

[3] Dincă, M., Dailly, A., Liu, Y., Brown, C.M., Neumann, D.A.and Long, J.R. (2006) Hydrogen storage in a microporous metal-organic framework with exposed Mn^{2+} coordination sites. *Journal of the American Chemical Society*, **128**, 16876-16883. http://dx.doi.org/10.1021/ja0656853

[4] Collins, D.J. and Zhou, H.C. (2007) Hydrogen storage in metal-organic frameworks. *Journal of Materials Chemistry*, **17**, 3154-3160. http://dx.doi.org/10.1039/b702858j

[5] Lin, J.D., Cheng, J.W. and Du, S.W. (2008) Five d^{10} 3D metal-organic frameworks constructed from aromatic polycarboxylate acids and flexible imidazole-based ligands. *Crystal Growth and Design*, **8**, 3345-3353. http://dx.doi.org/10.1021/cg8002614

[6] Sun, Y.Q., Zhang, J. and Yang, G.Y. (2006) A series of luminescent lanthanide-cadmium-organic frameworks with helical channels and tubes. *Chemical Communications*, 4700-4702. http://dx.doi.org/10.1039/b610892j

[7] Kuppler, R.J., Timmons, D.J., Fang, Q.R., Li, J.R., Makal, T.A., Young, M.D., Yuan, D.Q., Zhao, D., Zhuang, W.J. and Zhou, H.C. (2009) Potential applications of metal-organic frameworks. *Coordination Chemistry Reviews*, **253**, 3042-3066. http://dx.doi.org/10.1016/j.ccr.2009.05.019

[8] Li, J.R., Kuppler, R.J. and Zhou, H.C. (2009) Selective gas adsorption and separation in metal-organic frameworks. *Chemical Society Reviews*, **38**, 1477-1504. http://dx.doi.org/10.1039/b802426j

[9] Férey, G. (2008) Hybrid porous solids: Past, present, future. *Chemical Society Reviews*, **37**, 191-214. http://dx.doi.org/10.1039/b618320b

[10] Furukawa, H., Ko, N., Go, Y.B., Aratani, N., Choi, S.B., Choi, E., Yazaydin, A.O., Snurr, R.Q., O'Keeffe, M., Kim, J. and Yaghi, O.M. (2010) Ultra-high porosity in metal-organic frameworks. *Science*, **239**, 424-428. http://dx.doi.org/10.1126/science.1192160

[11] Kuriki, K., Koike, Y. and Okamoto, Y. (2002) Plastic optical fiber lasers and amplifiers containing lanthanide complexes. *Chemical Reviews*, **102**, 2347-2356. http://dx.doi.org/10.1021/cr010309g

[12] Bünzli, J.C.G. (2006) Benefiting from the unique properties of lanthanide ions. *Accounts of Chemical Research*, **39**, 53-61. http://dx.doi.org/10.1021/ar0400894

[13] Plečnik, C.E., Liu, S.M. and Shore, S.G. (2003) Lanthanide-transition-metal complexes: From ion pairs to extended arrays. *Accounts of Chemical Research*, **36**, 499-508. http://dx.doi.org/10.1021/ar010050o

[14] Winpenny, R.E.P. (1998) The structures and magnetic properties of complexes containing 3d- and 4f-metals. *Chemical Society Reviews*, **27**, 447-452. http://dx.doi.org/10.1039/a827447z

[15] Li, G.M., Akitsu, T., Sato, O. and Einaga, Y. (2003) Pho-

toinduced Magnetization for Cyano-bridged 3d-4f Hetero-bimetallic Assembly Nd(DMF)$_4$(m-CN)$_5$Fe(CN)$_5$·H$_2$O (DMF=N,N-dimethylformamide). *Journal of the American Chemical Society*, **125**, 12396-12397. http://dx.doi.org/10.1021/ja037183k

[16] Zhao, B., Chen, X.Y., Cheng, P., Liao, D.Z., Yan, S.P. and Jiang, Z.H. (2004) Coordination Polymers Containing 1D Channels as Selective Luminescent Probes. *Journal of the American Chemical Society*, **126**, 15394-15395. http://dx.doi.org/10.1021/ja047141b

[17] Qi, Y., Che, Y.X. and Zheng, J.M. (2008) A zinc(II) coordination polymer constructed from mixed-ligand 1,2-bis(2-(1H-imidazol-1-yl)ethoxy)ethane and 1,4-benzenedicarboxylic acid. *CrystEngComm*, **10**, 1137-1139. http://dx.doi.org/10.1039/b801387j

[18] Henninger, S.K., Habib, H.A. and Janiak, C. (2009) MOFs as adsorbents for low temperature heating and cooling applications. *Journal of the American Chemical Society*, **131**, 2776-2777. http://dx.doi.org/10.1021/ja808444z

[19] Ma, L.F., Wang, L.Y., Wang, Y.Y., Du, M. and Wang, J.G. (2009) Synthesis, structures and properties of Mn(II) coordination frameworks based on R-isophthalate (R = –CH$_3$ or –C(CH$_3$)$_3$) and various dipyridyl-type co-ligands. *CrystEngComm*, **11**, 109-117. http://dx.doi.org/10.1021/ja808444z

[20] Habib, H.A., Sanchiz, J. and Janiak, C. (2009) Magnetic and luminescence Properties of Cu(II), Cu(II)$_4$O$_4$ core and Cd(II) mixed-ligand metal-organic frameworks constructed from 1,2-bis(1,2,4-triazol-4-yl)ethane and benzene-1,3,5-tricarboxylate. *Inorganica Chimica Acta*, **362**, 2452-2460. http://dx.doi.org/10.1016/j.ica.2008.11.003

[21] Habib, H.A., Hoffmann, A., Hoppe, H.A. and Janiak, C. (2009) Crystal structures and solid-state CPMAS ^{13}C NMR correlations in luminescent zinc(II) and cadmium(II) mixed-ligand coordination polymers constructed from 1,2-bis(1,2,4-triazol-4-yl)ethane and benzenedicarboxylate. *Dalton Transactions*, 1742-1751. http://dx.doi.org/10.1039/b812670d

[22] Habib, H.A., Sanchiz, J. and Janiak, C. (2008) Mixed-ligand coordination polymers from 1,2-bis(1,2,4-triazol-4-yl)ethane and benzene-1,3,5-tricarboxylate: Trinuclear nickel or zinc secondary building units for three-dimensional networks with crystal-to-crystal transformation upon dehydration. *Dalton Transactions*, 1734-1744. http://dx.doi.org/10.1039/b715812b

[23] Zhang, L.P., Yang, J., Ma, J.F., Jia, Z.F., Xie, Y.P. and Wei, G.H. (2008) A series of 2D and 3D metal-organic frameworks based on different polycarboxylate anions and a flexible 2,2'-bis(1H-imidazolyl)ether ligand. *CrystEngComm*, **10**, 1410-1420. http://dx.doi.org/10.1039/b804578j

[24] Wisser, B., Lu, Y. and Janiak, C. (2007). Chiral Coordination Polymers with Amino Acids: $^2_\infty$[Cu$_2$(μ-L-tryptophanato)$_2$(μ-4,4'-bipyridine)(H$_2$O)$_2$](NO$_3$)$_2$. *Zeitschrift für Anorganishe und Allgemeine Chemie*, **633**, 1189-1192.

[25] Manna, S.C., Okamoto, K.I., Zangrando, E. and Chaudhuri, N.R. (2007) Conformational morphosis in azocalix [4]arenes. *CrystEngComm*, **9**, 199-122.

[26] Chen, Z.F., Zhang, S.F., Luo, H.S., Abrahams, B.F. and Liang, H. (2007) Ni₂(R*COO)₄(H₂O)(4,4'-bipy)₂—A robust homochiral quartz-like network with large chiral channels. *CrystEngComm*, **9**, 27-29. http://dx.doi.org/10.1039/b613047j

[27] Pichon, A., Fierro, C.M., Nieuwenhuyzen, M. and James, S. (2007) A pillared-grid MOF with large pores based on the Cu₂(O₂CR)₄ paddle-wheel. *CrystEngComm*, **9**, 449-451. http://dx.doi.org/10.1039/b702176c

[28] Pasán, J., Sanchiz, J., Lloret, F., Julve, M. and Ruiz-Pérez, C. (2007) Crystal engineering of 3-D coordination polymers by pillaring ferromagnetic copper(II)-methylmalonate layers. *CrystEngComm*, **9**, 478-487. http://dx.doi.org/10.1039/b701788j

[29] Stephenson, M.D. and Hardie, M.J. (2006) Coordination and hydrogen bonded network structures of Cu(II) with mixed ligands: A hybrid hydrogen bonded material, an infinite sandwich arrangement, and a 3-D net. *Dalton Transactions*, 3407-3417. http://dx.doi.org/10.1039/b600357e

[30] Carballo, R, Covelo, B, Vázquez-López, E.M., García-Martínez, E., Castiñeiras, A., Janiak, C. (2005) *Zeitschrift für Anorganishe und Allgemeine Chemie*, **631**, 1929-1931.

[31] Yao, J., Lu, Z.D., Li, Y.Z., Lin, J.G., Duan, X.Y., Gao, S., Meng, Q.J. and Lu, C.S. (2008) Three-dimensional metal-organic frameworks constructed from bix and 1,2,4-benzenetricarboxylate. *CrystEngComm*, **10**, 1379-1383. http://dx.doi.org/10.1039/b805263h

[32] Wei, G.H., Yang, J., Ma, J.F., Liu, Y.Y., Li, S.L. and Zhang, L.P. (2008) Syntheses, structures and luminescent properties of zinc(II) and cadmium(II) coordination complexes based on bis(imidazole) and different carboxylate ligands. *Dalton Transactions*, 3080-3092. http://dx.doi.org/10.1039/b716657e

[33] Zhang, J.Y., Ma, Y., Cheng, A.L., Yue, Q., Sun, Q. and Gao, E.Q. (2008) A manganese(II) coordination polymer with mixed pyrimidine-2-carboxylate and oxalate bridges: synthesis, structure, and magnetism. *Dalton Transactions*, 2061-2066. http://dx.doi.org/10.1039/b717837a

[34] Roh, S.G., Nah, M.K., Oh, J.B., Baek, N.S., Park, K.M. and Kim, H.K. (2005) Synthesis, crystal structure and luminescence properties of a saturated dimeric Er(III)-chelated complex based on benzoate and bipyridine ligands. *Polyhedron*, **24**, 137-142. http://dx.doi.org/10.1016/j.poly.2004.10.014

[35] Qin, C., Wang, X.L., Wang, E.B. and Su, Z.M. (2005) A series of three-dimensional lanthanide coordination polymers with rutile and unprecedented rutile-related topologies. *Inorganic Chemistry*, **44**, 7122-7129. http://dx.doi.org/10.1021/ic050906b

[36] Wan, Y.H., Jin, L.P. and Wang, K.Z. (2003) Hydrothermal synthesis and structural characterization of two novel lanthanide supramolecular coordination polymers with nano-chains. *Journal of Molecular Structure*, **649**, 85-93. http://dx.doi.org/10.1016/S0022-2860(03)00021-8

[37] Wan, Y., Zhang, L., Jin, L., Gao, S. and Lu, S. (2003) High-dimensional architectures from the self-assembly of lathanide ions with benzenedicarboxylates and 1,10-

phenanthroline. *Inorganic Chemistry*, **42**, 4985-4994. http://dx.doi.org/10.1021/ic034258c

[38] Eddaoudi, M., Kim, J., Wachter, J.B., Chae, H.K., O'Keeffe, M. and Yaghi, O.M. (2001) Porous Metal-organic polyhedra: 25 Å cuboctahedron constructed from 12 Cu₂(CO₂)₄ paddle-wheel building blocks. *Journal of the American Chemical Society*, **123**, 4368-4369. http://dx.doi.org/10.1021/ja0104352

[39] Guo, X.D., Zhu, G.S., Sun, F.X., Li, Z.Y., Zhao, X.J., Li, X.T., Wang, H.C. and Qiu, S.L. (2006) Synthesis, structure, and luminescent properties of microporous lanthanide metal-organic frameworks with inorganic rod-shaped building units. *Inorganic Chemistry*, **45**, 2581-2587. http://dx.doi.org/10.1021/ic0518881

[40] Thirumurugan, A. and Natarajan, S. (2004) Terephthalate bridge frameworks of Nd and Sm phthalates. *Inorganic Chemistry Communications*, **7**, 395-399. http://dx.doi.org/10.1016/j.inoche.2003.12.023

[41] Zhang, Z.H., Okamura, T.A., Hasegawa, Y., Kawaguchi, H., Kong, L.Y., Sun, Y.W. and Ueyama, N. (2005) Syntheses, structures, and luminescent and magnetic properties of novel three-dimensional lanthanide complexes with 1,3,5-benzenetriacetate. *Inorganic Chemistry*, **44**, 6219-6227. http://dx.doi.org/10.1021/ic050453a

[42] Cao, R., Sun, D., Liang, Y., Hong, M., Tatsumi, K. and Shi, Q. (2002) Syntheses and characterizations of three-dimensional channel-like polymeric lanthanide complexes constructed by 1,2,4,5-benzenetetracarboxylic acid. *Inorganic Chemistry*, **41**, 2087-2094. http://dx.doi.org/10.1021/ic0110124

[43] Cañadillas-Delgado, L., Fabelo, Ó., Ruíz-Pérez, C., Delgado, F.S., Julve, M., Hernndez-Molina, M., Laz, M.M. and Lorenzo-Luis, P. (2006) Zeolite-like nanoporous gadolinium complexes incorporating alkaline cations. *Crystal Growth and Design*, **6**, 87-93. http://dx.doi.org/10.1021/cg050170t

[44] Tang S.F. and Mudring A.V. (2009) The missing link crystallized from the ionic liquid 1-ethyl-3-methylamonium tosylate: Bis-aqua-(p-toluenesulfonato-o)-europium(III)-bis-toluenesulfonate dehydrate. *Crystal Growth and Design*, **9**, 2549-2557.

[45] Wang, Z., Bai, F.Y., Xing, Y.H., Xie, Y., Ge, M.F. and Niu, S.Y. (2010) Two New 3D Lanthanide Coordination Polymers with Benzenesulfonic and Adipic Acids: Synthesis, Structure and Luminescent Properties. *Zeitschrift für Anorganishe und Allgemeine Chemie*, **636**, 1570-1575.

[46] Zhou, R.S., Ye, L., Ding, H., Song, J.F., Xu, X.Y. and Xu, J.Q. (2008) Sytheses, structures, luminescence, and magnetism of four 3D lanthanide 5-sulfosalicylates, *Journal of Solid State Chemistry*, **181**, 567-575. http://dx.doi.org/10.1016/j.jssc.2007.12.027

[47] Lu, Z., Wen, L., Yao, J., Zhu, H. and Meng, Q. (2006) Two types of novel layer framework structures assembled from 5-sulfosalicylic acid and lanthanide ions. *CrystEngComm*, **8**, 847-853. http://dx.doi.org/10.1039/b612147k

[48] Ma, J.F., Li, J.Y., Zheng G.L. and Liu J.F. (2003) The first ladder structure containing three different squares: The structure of barium 3-carboxy-4-hydroxybenzene-

sulfonate. *Inorganic Chemistry Communications*, **6**, 581-583. http://dx.doi.org/10.1016/S1387-7003(03)00044-3

[49] Ma, J.F., Yang, J., Li, S.L., Song, S.Y., Zhang, H.J., Wan, H.S. and Yang, K.Y. (2005) Two Coordination polymers of Ag(I) with 5-sulfosalicylic acid. *Crystal Growth and Design*, **5**, 807-812. http://dx.doi.org/10.1021/cg049723a

[50] Gao, S., Zhu, Z.-B, Huo, L.-H. and Ng, S.W. (2005) Poly[[triaquabis(μ_4-3-carboxy-4-hydroxybenzenesulfonato)disilver(I)] monohydrate]. *Acta Crystallographica*, **61**, m279-m281. http://dx.doi.org/10.1107/S1600536805000504

[51] Gao, S., Zhu, Z.-B., Huo, L.-H. and Ng, S.W. (2005) Poly[di-μ-aqua-bis(μ_8-3-carboxylato-4-hydroxybenzenesulfonato)tetrasilver(I)]. *Acta Crystallographica*, **61**, m282-m284. http://dx.doi.org/10.1107/S1600536805000516

[52] Borkowski, L.A. and Cahill, C.L. (2004) Poly [[aquaneodymium(III)]-μ_3-decane-1,10-dicarboxylato-μ_3-9-carboxynonanecarboxylato]. *Acta Crystallographica C*, **60**, m159-m161. http://dx.doi.org/10.1107/S0108270104003427

[53] Cheetham, A.K., Férey, G. and Loiseau, T. (1999) Open-Framework Inorganic Materials. *Angewandte Chemie International Edition*, **38**, 3268-3292. http://dx.doi.org/10.1002/(SICI)1521-3773(19991115)38:22<3268::AID-ANIE3268>3.0.CO;2-U

[54] Chesnut, D.J., Hagrman, D., Zapf, P.J., Hammond, R.P., LaDuca, R. Jr., Haushalter, R.C., Zubieta, J. (1999) Organic/inorganic composite materials: The roles of organoamine ligands in the design of inorganic solids. *Coordination Chemistry Reviews*, **190-192**, 737-769. http://dx.doi.org/10.1016/S0010-8545(99)00119-8

[55] Bataille, T. and Louër, D. (2002) Two new diamine templated lanthanum sulfates, $La_2(H_2O)_2(C_4H_{12}N_2)(SO_4)_4$ and $La_2(H_2O)_2(C_2H_{10}N_2)_3(SO_4)_6 \cdot 4H_2O$, with 3D and 2D crystal structures. *Journal of Materials Chemistry*, **12**, 3487-3493. http://dx.doi.org/10.1039/b207212m

[56] Rao, C.N.R., Behera, J.N. and Dan, M. (2006) Organically-templated metal sulfates, selenites and selenates. *Chemical Society Reviews*, **35**, 375-387. http://dx.doi.org/10.1039/b510396g

[57] Nakamoto, K., Fujita, J., Tanaka, S. and Kobayashi, M. (1957) Infrared spectra of metallic complexes. IV. Comparison of the infrared spectra of unidentate and bidentate metallic complexes. *Journal of the American Chemical Society*, **79**, 4904-4908. http://dx.doi.org/10.1021/ja01575a020

[58] Anbalagan, G., Mukundakumari, S., Murugesan, K.S. and Gunasekaran, S. (2007) Preparation, crystal structure and infrared spectroscopy of the new compound rubidium beryllium sulfate dihydrate, $Rb_2Be(SO_4)_2 \cdot 2H_2O$. *Vibrational Spectroscopy*, **44**, 226-272.

[59] Wang, T.W., Liu, D.S., Huang, C.C., Sui, Y., Huang, X.H., Chen, J.Z. and You, X.Z. (2010) Syntheses, crystal structures, and magnetic properties of two Mn(II) coordination polymers based on the 5-aminotetrazole ligand: Effect of sources of ligand on construction of topological networks. *Crystal Growth and Design*, **10**, 3429-3435. http://dx.doi.org/10.1021/cg100127w

[60] Bataille, T. and Louër, D. (2004) New linear and layered amine-templated lanthanum sufates. *Journal of Solid State Chemistry*, **177**, 1235-1243. http://dx.doi.org/10.1016/j.jssc.2003.10.031

[61] Zhu, Y., Sun, X., Zhu, D. and Xu, Y. (2009) Solvothermal synthesis, crystal structure and luminescence of the first organic amine templated europium sulfate. *Inorganica Chimica Acta*, **362**, 2565-2568. http://dx.doi.org/10.1016/j.ica.2008.11.015

[62] Xing, Y., Shi, Z., Li, G. and Pang, W. (2003) Hydrothermal synthesis and structure of $[C_2N_2H_{10}][La_2(H_2O)_4(SO_4)_4]$ $2H_2O$, a new organically templated rare earth sulfate with a layer structure. *Dalton Transactions*, 940-943. http://dx.doi.org/10.1039/b211076h

[63] Richardson, F.S. (1982) Terbium(III) and europium(III) ions as luminescent probes and stains for biomolecular systems. *Chemical Reviews*, **82**, 541-552. http://dx.doi.org/10.1021/cr00051a004

[64] Allendorf, M.D., Bauer, C.A., Bhakta, R.K. and Houk, R.J.T. (2009) Luminescent metal-organic frameworks. *Chemical Society Reviews*, **38**, 1330-1352. http://dx.doi.org/10.1039/b802352m

Supporting Information Available

Tables S1-S6 show the selected bond lengths and bond angles of the complexes **1, 2, 4, 5, 7, 8**; **Tables S7-S12** list the hydrogen bonds of the complexes **1, 2, 4, 5, 7, 8**; **Figures S1-S8** present the PXRD patterns for the simulations based on the X-ray single crystal diffraction and the experimental samples of complexes **1-8**; **Figures S9** and **S10** show the infrared spectra of complexes **1-8**, and **Figures S11-S16** unfurl the coordination environments of lanthanide atoms, the 2D polyhedron lays, the 2D topology network and the 3D hydrogen bonding network structures in complexes **1, 2, 4, 5, 7** and **8**. Tables of atomic coordinates, isotropic thermal parameters, complete bond distances and angles are deposited in the Cambridge Crystallographic Data Center. Copies of those tables can be obtained free of charge by contacting the Director of the CCDC, 12 Union Road, Cambridge, CB2 1EZ, UK (fax + 44-1223-336033; e-mail deposit@ccdc.cam.ac.uk; http://www.ccdc.cam.ac.uk), and quoting the publication citations and deposition numbers CCDC 904695 (**1**), 904696 (**2**), 904697 (**3**), 904698 (**4**), 904699 (**5**), 904700 (**6**), 904701 (**7**) and 904702 (**8**).

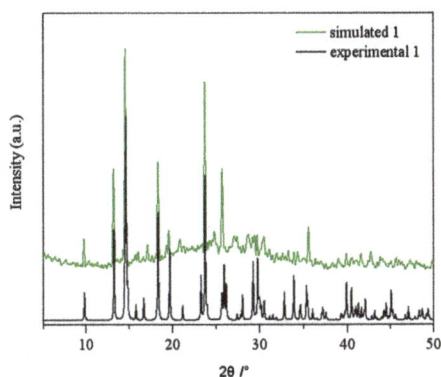

Figure S1. PXRD patterns for the simulated based on the x-ray single crystal diffraction and the experimental samples of complexes **1**.

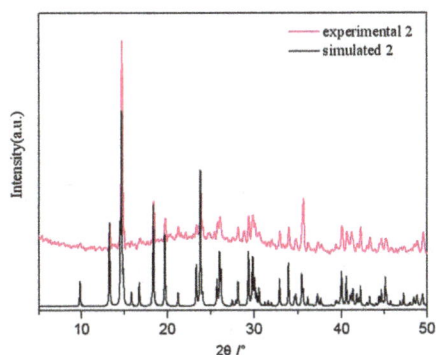

Figure S2. PXRD patterns for the simulated based on the x-ray single crystal diffraction and the experimental samples of complexes **2**.

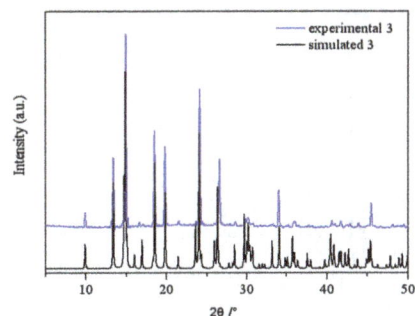

Figure S3. PXRD patterns for the simulated based on the x-ray single crystal diffraction and the experimental samples of complexes **3**.

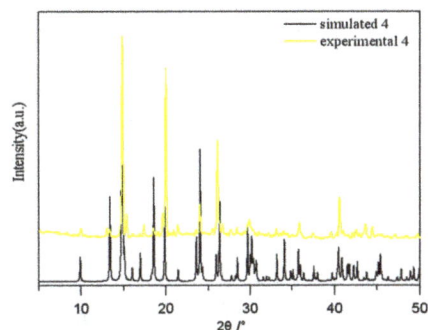

Figure S4. PXRD patterns for the simulated based on the x-ray single crystal diffraction and the experimental samples of complexes **4**.

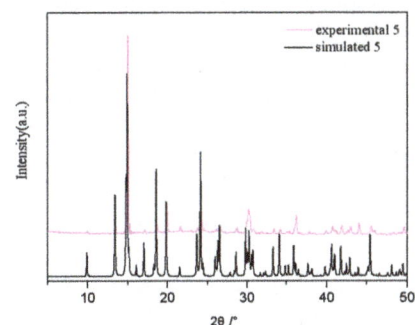

Figure S5. PXRD patterns for the simulated based on the x-ray single crystal diffraction and the experimental samples of complexes **5**.

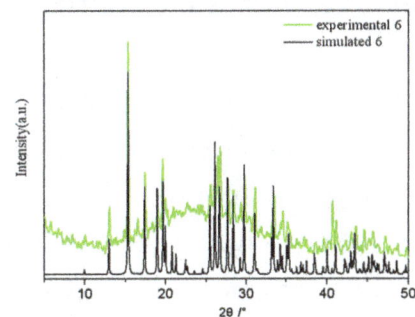

Figure S6. PXRD patterns for the simulated based on the x-ray single crystal diffraction and the experimental samples of complexes **6**.

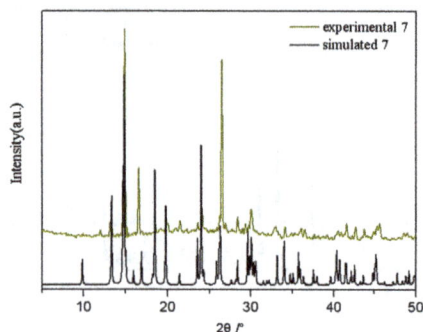

Figure S7. PXRD patterns for the simulated based on the x-ray single crystal diffraction and the experimental samples of complexes **7**.

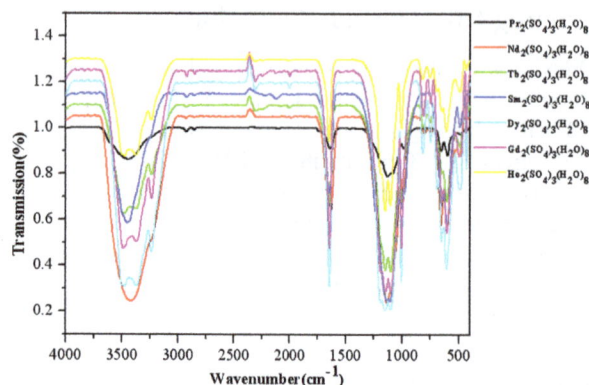

Figure S9. The IR spectra of complexes **1 - 5, 7 - 8**.

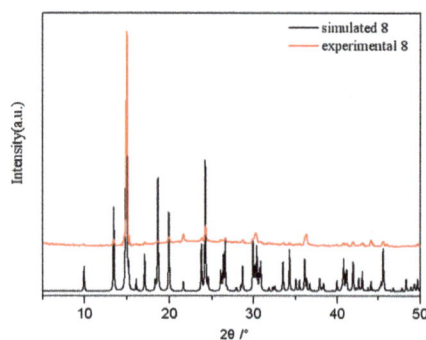

Figure S8. PXRD patterns for the simulated based on the x-ray single crystal diffraction and the experimental samples of complexes **8**.

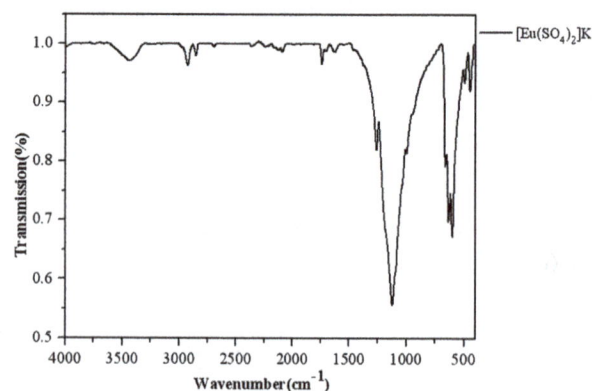

Figure S10. The IR spectra of complex **6**.

Table S1. Selected bond distances (Å) and angles (deg) of complex 1.

Bond distances					
Pr–O(8)	2.389(3)	Pr–O(3)	2.400(3)	Pr–O(1)	2.422(3)
Pr–O(7)	2.447(3)	Pr–O(6)	2.459(3)	Pr–O(4)	2.518(3)
Pr–O(5)	2.528(3)	Pr–O(2)	2.568(3)		
Bond angles					
O(8)–Pr–O(3)	80.18(11)	O(8)–Pr–O(1)	70.75(10)	O(3)–Pr–O(1)	88.19(11)
O(8)–Pr–O(7)	126.00(10)	O(3)–Pr–O(7)	143.31(10)	O(1)–Pr–O(7)	79.26(10)
O(8)–Pr–O(6)	79.70(11)	O(3)–Pr–O(6)	147.44(10)	O(1)–Pr–O(6)	108.79(11)
O(7)–Pr–O(6)	68.77(10)	O(8)–Pr–O(4)	134.27(11)	O(3)–Pr–O(4)	70.17(11)
O(1)–Pr–O(4)	74.21(11)	O(7)–Pr–O(4)	73.26(10)	O(6)–Pr–O(4)	140.32(11)
O(8)–Pr–O(5)	139.61(9)	O(3)–Pr–O(5)	101.70(10)	O(1)–Pr–O(5)	148.96(10)
O(7)–Pr–O(5)	75.33(9)	O(6)–Pr–O(5)	78.20(10)	O(4)–Pr–O(5)	81.56(10)
O(8)–Pr–O(2)	73.28(9)	O(3)–Pr–O(2)	73.71(9)	O(1)–Pr–O(2)	141.93(10)
O(7)–Pr–O(2)	133.85(9)	O(6)–Pr–O(2)	76.12(10)	O(4)–Pr–O(2)	126.66(10)
O(5)–Pr–O(2)	68.83(9)				

(a)

(b)

(c)

(d)

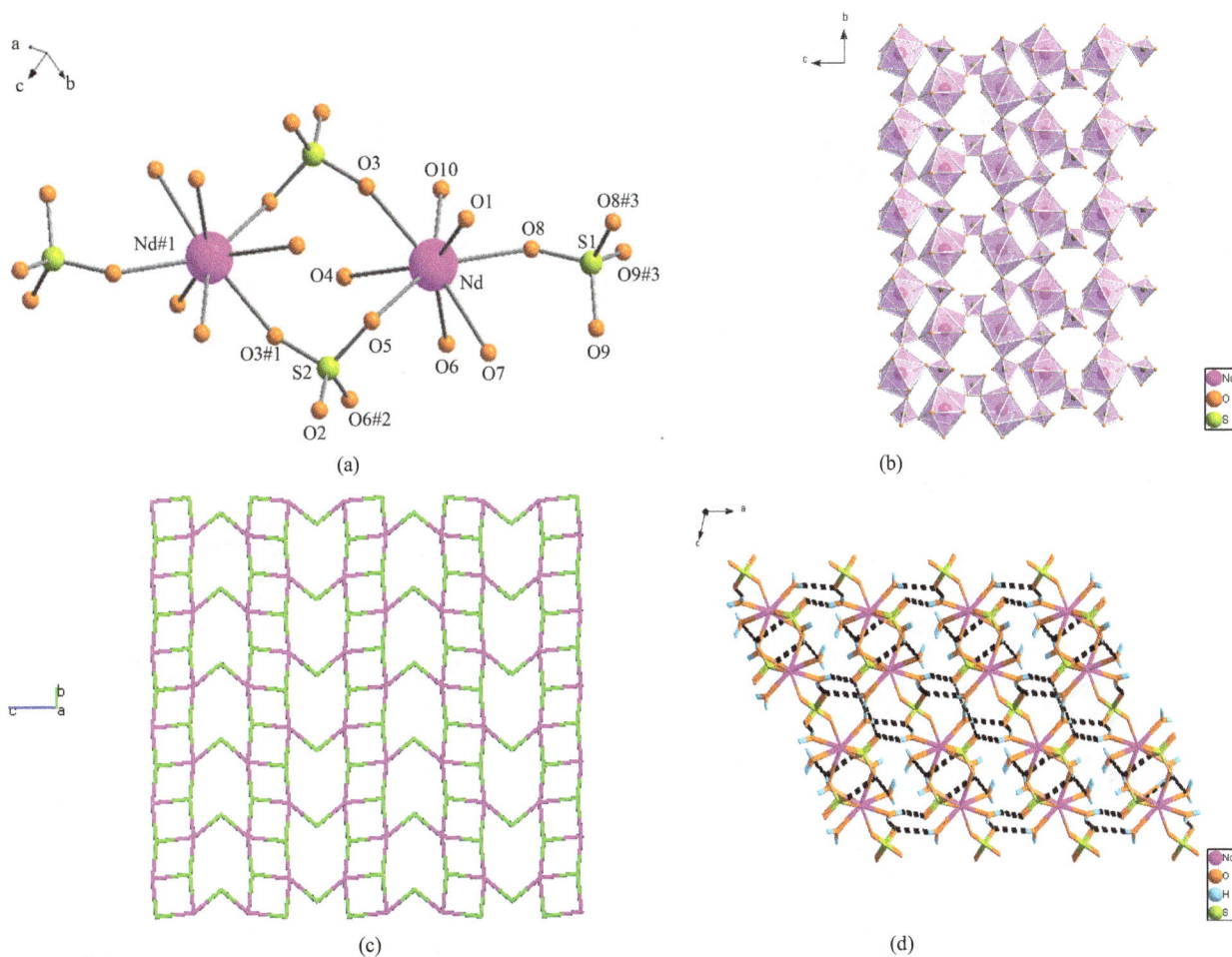

Figure S12. (a) The coordination environment of Nd in complex 2; (Symmetry codes: #1 −x + 1/2,−y−1/2,−z + 1 #2 −x + 1/2, −y + 1/2,−z + 1 #3 −x, y, −z + 1/2) (b) The 2D polyhedron layer in complex 2, viewed along the a-axis; (c) Schematic representation of the 2D $(4^2.6.8^2.10)_2(4^2.6)_2(8)$ topology network in complex 1. Fuchsia, Nd; Lime, SO_4^{2-} ; (d) A view of the 3D hydrogen bonding network structure of complex 2 (black dotted lines representing hydrogen bonding).

Table S2. Selected bond distances (Å) and angles (deg) of complex **2**.

Bond distances					
Nd-O(8)	2.373(3)	Nd-O(6)	2.382(3)	Nd-O(10)	2.398(3)
Nd-O(3)	2.434(3)	Nd-O(1)	2.447(3)	Nd-O(4)	2.502(3)
Nd-O(5)	2.510(3)	Nd-O(7)	2.555(3)		
Bond angles					
O(8)-Nd-O(6)	80.18(10)	O(8)-Nd-O(10)	70.76(10)	O(6)-Nd-O(10)	88.24(11)
O(8)-Nd-O(3)	126.04(10)	O(8)-Nd-O(3)	126.04(10)	O(6)-Nd-O(3)	143.34(8)
O(10)-Nd-O(3)	79.30(10)	O(8)-Nd-O(1)	79.80(10)	O(6)-Nd-O(1)	147.43(9)
O(10)-Nd-O(1)	108.87(11)	O(3)-Nd-O(1)	68.73(9)	O(8)-Nd-O(4)	134.31(10)
O(6)-Nd-O(4)	69.93(9)	O(10)-Nd-O(4)	74.48(11)	O(3)-Nd-O(4)	73.52(9)
O(1)-Nd-O(4)	140.43(10)	O(8)-Nd-O(5)	139.84(9)	O(6)-Nd-O(5)	101.20(10)
O(10)-Nd-O(5)	148.84(9)	O(3)-Nd-O(5)	75.50(9)	O(1)-Nd-O(5)	78.62(10)
O(4)-Nd-O(5)	80.98(10)	O(8)-Nd-O(7)	73.58(10)	O(6)-Nd-O(7)	73.45(9)
O(10)-Nd-O(7)	142.14(10)	O(3)-Nd-O(7)	133.86(9)	O(1)-Nd-O(7)	76.39(9)
O(4)-Nd-O(7)	125.91(9)	O(5)-Nd-O(7)	68.64(9)		

(a)

(b)

(c)

(d)

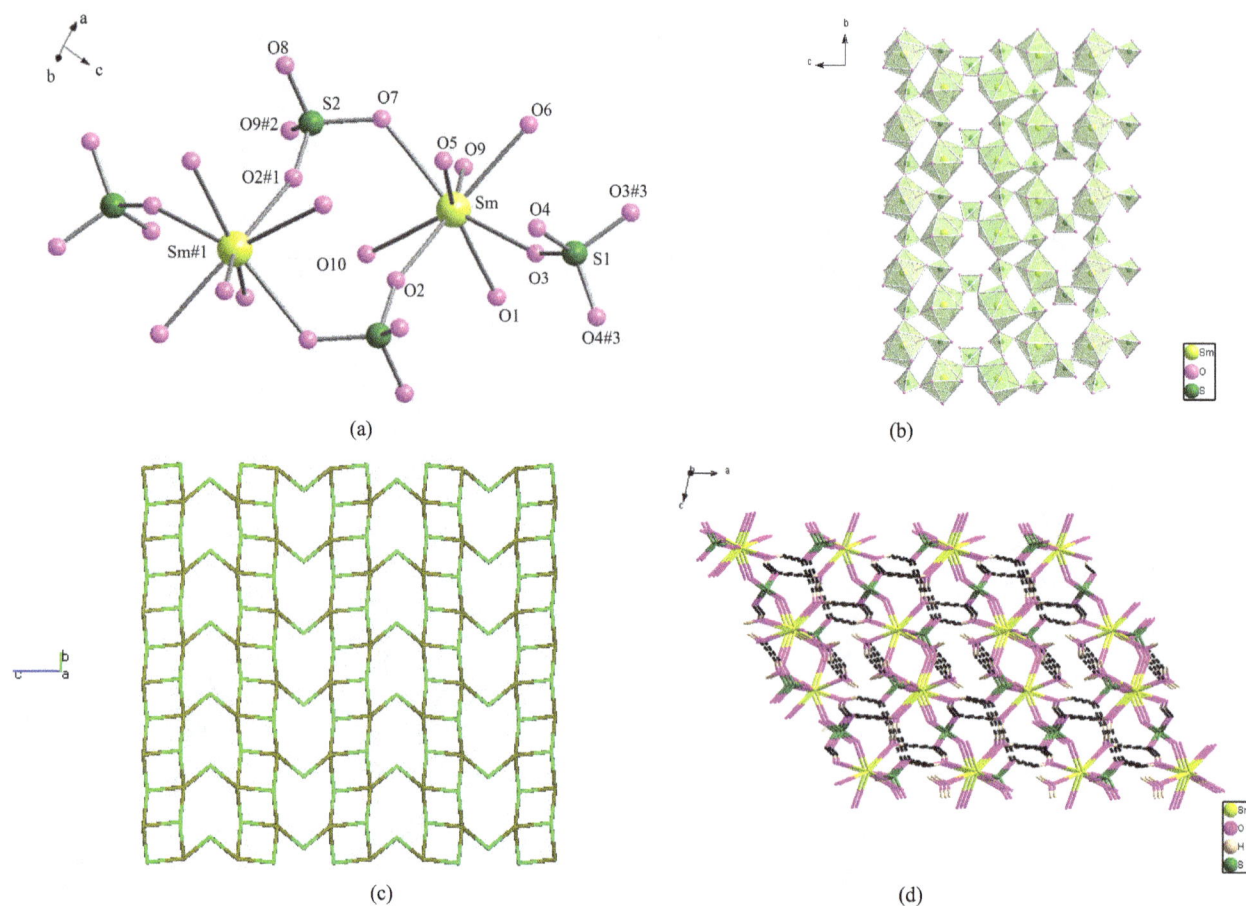

Figure S13. (a) The coordination environment of Sm in complex 4; (Symmetry codes: #1 −x + 3/2, −y + 1/2, −z + 1 #2 −x + 3/2, −y−1/2, −z + 1 #3 −x + 2, y,−z + 3/2); (b) The 2D polyhedron layer in complex 4, viewed along the a-axis; (c) Schematic representation of the 2D $(4^2.6.8^2.10)_2(4^2.6)_2(8)$ topology network in complex 4. Olive, Sm; Lime, SO_4^{2-}; (d) A view of the 3D hydrogen bonding network structure of complex 4 (black dotted lines representing hydrogen bonding).

Table S3. Selected bond distances (Å) and angles (deg) of complex 4.

Bond distances					
Sm-O(3)	2.340(2)	Sm-O(2)	2.341(3)	Sm-O(1)	2.364(3)
Sm-O(9)	2.392(2)	Sm-O(6)	2.403(3)	Sm-O(7)	2.461(2)
Sm-O(10)	2.467(3)	Sm-O(5)	2.514(3)		
Bond angles					
O(3)-Sm-O(2)	80.07(9)	O(3)-Sm-O(1)	70.86(9)	O(2)-Sm-O(1)	88.09(10)
O(3)-Sm-O(9)	126.04(8)	O(2)-Sm-O(9)	143.59(8)	O(1)-Sm-O(9)	79.49(9)
O(3)-Sm-O(6)	79.75(10)	O(2)-Sm-O(6)	147.10(9)	O(1)-Sm-O(6)	109.26(10)
O(9)-Sm-O(6)	68.79(8)	O(3)-Sm-O(7)	140.51(8)	O(2)-Sm-O(7)	100.46(8)
O(1)-Sm-O(7)	148.29(9)	O(9)-Sm-O(7)	75.61(8)	O(6)-Sm-O(7)	79.59(9)
O(3)-Sm-O(10)	134.49(10)	O(2)-Sm-O(10)	70.19(9)	O(1)-Sm-O(10)	74.41(10)
O(9)-Sm-O(10)	73.52(9)	O(6)-Sm-O(10)	140.43(10)	O(7)-Sm-O(10)	79.88(9)
O(3)-Sm-O(5)	73.94(9)	O(2)-Sm-O(5)	73.27(9)	O(1)-Sm-O(5)	142.46(10)
O(9)-Sm-O(5)	133.60(9)	O(6)-Sm-O(5)	76.33(9)	O(7)-Sm-O(5)	68.66(9)
O(10)-Sm-O(5)	125.56(9)				

(a)

(b)

(c)

(d)

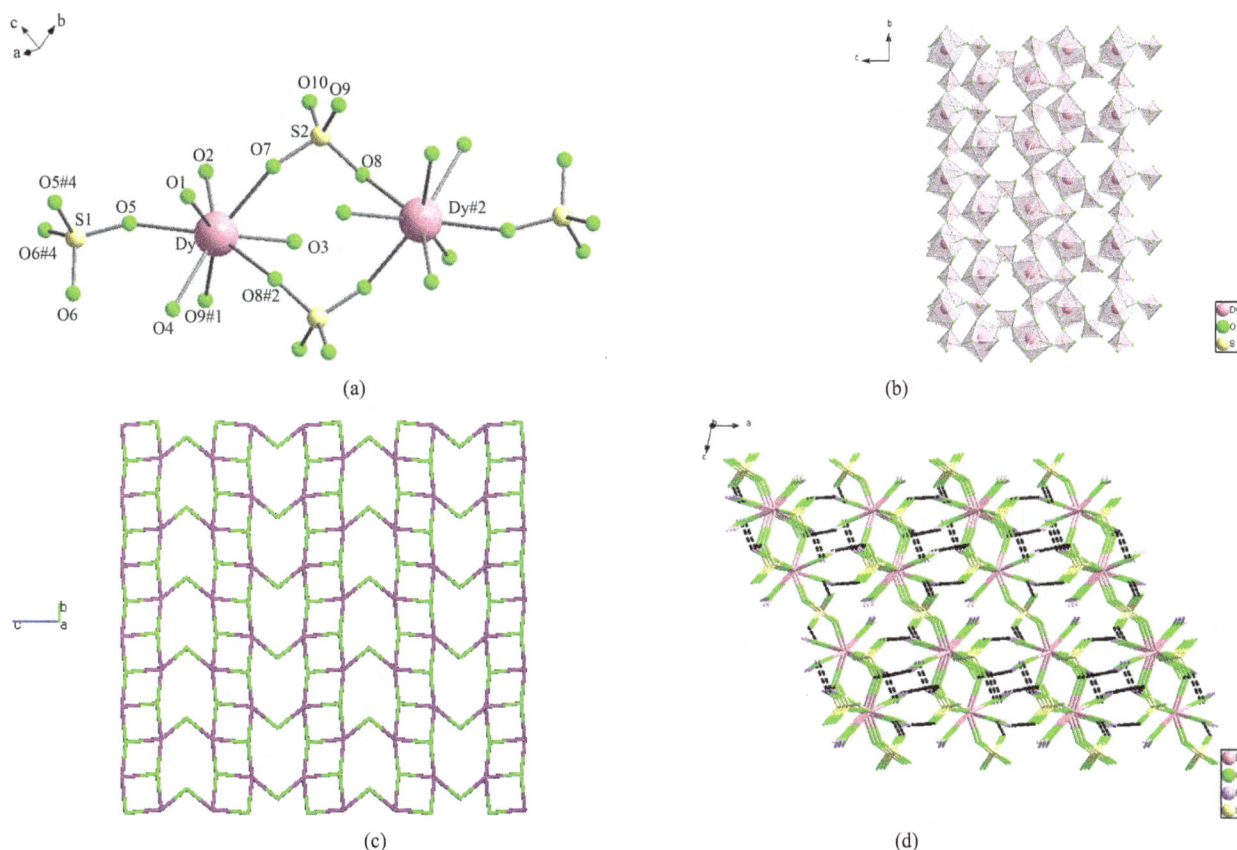

Figure S14. (a) The coordination environment of Dy in complex 5; (Symmetry codes: #1 $-x + 3/2$, $-y - 1/2$, $-z + 1$ #2 $-x + 3/2$, $-y + 1/2$, $-z + 1$ #3 $-x + 2$, y, $-z + 3/2$ #4 $1.5 - x$, $-0.5 - y$, $1 - z$); (b) The 2D polyhedron layer in complex 5, viewed along the a-axis; (c) Schematic representation of the 2D $(4^2.6.8^2.10)_2(4^2.6)_2(8)$ topology network in complex 5. Purple, Dy; Lime, SO_4^{2-} ; (d) A view of the 3D hydrogen bonding network structure of complex 5 (black dotted lines representing hydrogen bonding).

Table S4. Selected bond distances (Å) and angles (deg) of complex 5.

Bond distances					
Dy-O(9)#1	2.296(5)	Dy-O(2)	2.308(6)	Dy-O(5)	2.312(4)
Dy-O(7)	2.354(5)	Dy-O(1)	2.355(5)	Dy-O(8)#2	2.411(5)
Dy-O(3)	2.436(6)	Dy-O(4)	2.480(5)	O(8)-Dy#2	2.411(5)
O(9)-Dy#3	2.296(5)				
Bond angles					
O(9)#1-Dy-O(2)	88.7(2)	O(9)#1-Dy-O(5)	80.05(19)	O(2)-Dy-O(5)	70.50(19)
O(9)#1-Dy-O(7)	144.23(16)	O(2)-Dy-O(7)	79.61(18)	O(5)-Dy-O(7)	125.79(18)
O(9)#1-Dy-O(1)	146.58(19)	O(2)-Dy-O(1)	109.0(2)	O(5)-Dy-O(1)	79.7(2)
O(7)-Dy-O(1)	68.61(17)	O(9)#1-Dy-O(8)#2	99.55(19)	O(2)-Dy-O(8)#2	147.6(2)
O(5)-Dy-O(8)#2	141.66(17)	O(7)-Dy-O(8)#2	75.47(19)	O(1)-Dy-O(8)#2	80.80(19)
O(9)#1-Dy-O(3)	70.29(18)	O(2)-Dy-O(3)	74.6(2)	O(5)-Dy-O(3)	134.18(19)
O(7)-Dy-O(3)	74.02(16)	O(1)-Dy-O(3)	140.80(18)	O(8)#2-Dy-O(3)	78.82(19)
O(9)#1-Dy-O(4)	73.05(18)	O(2)-Dy-O(4)	143.05(18)	O(5)-Dy-O(4)	74.78(15)
O(7)-Dy-O(4)	132.84(16)	O(1)-Dy-O(4)	76.16(18)	O(8)#2-Dy-O(4)	68.67(17)
O(3)-Dy-O(4)	125.17(17)				

*Symmetry transformations used to generate equivalent atoms: #1 x, y − 1, z #2 −x + 1/2, −y + 3/2, −z #3 x, y + 1, z.

(a)

(b)

(c)

(d)

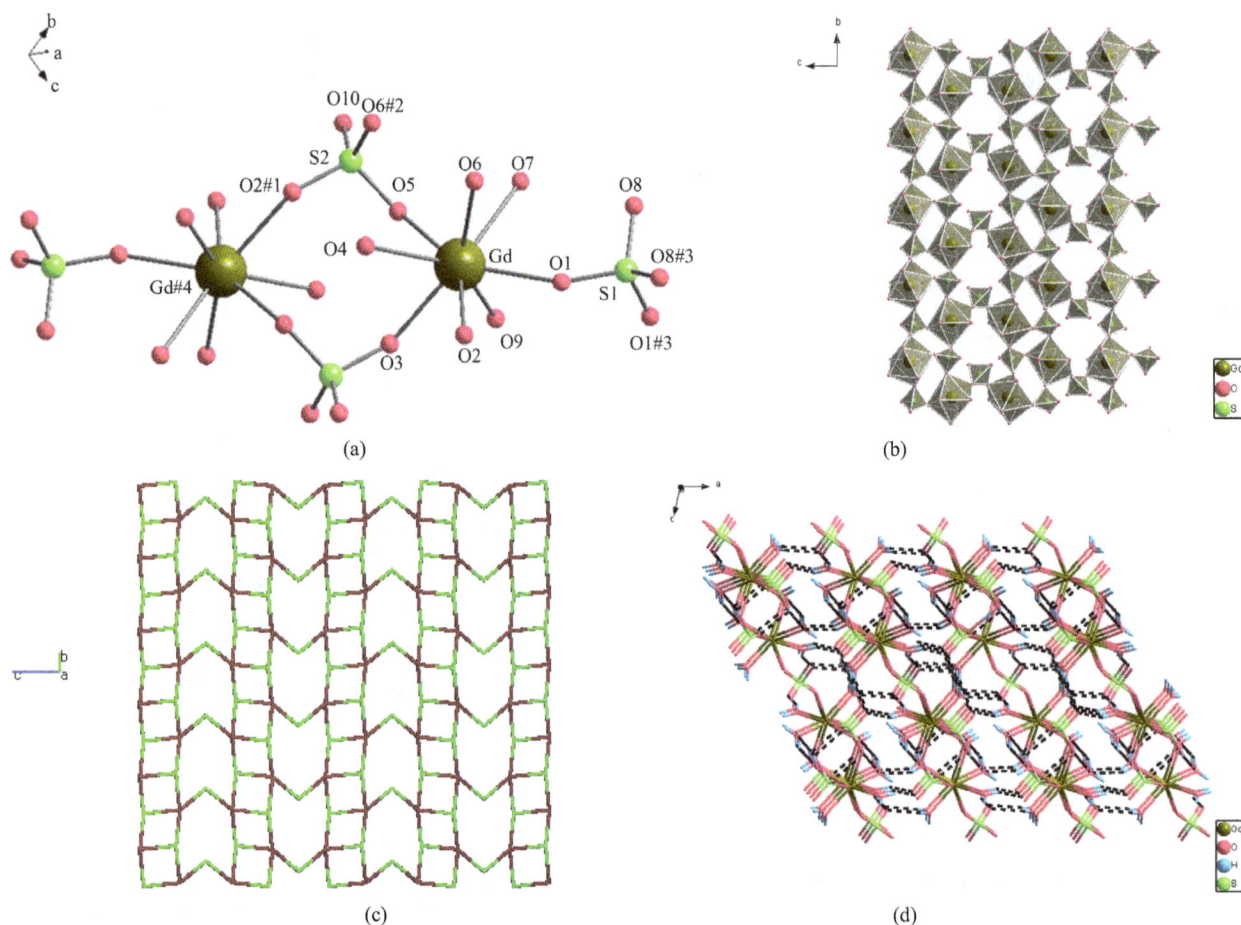

Figure S15. (a) The coordination environment of Gd in complex 7; (Symmetry codes: #1 −x + 3/2, −y − 1/2, −z + 1#2 −x + 3/2,−y + 1/2, −z + 1 #3 −x + 2, y, −z + 3/2 #4 1.5 − x, −0.5 − y, 1 − z); (b) The 2D polyhedron layer in complex 7, viewed along the a-axis; (c) Schematic representation of the 2D $(4^2.6.8^2.10)_2(4^2.6)_2(8)$ topology network in complex 7. Maroon, Gd; Lime, SO_4^{2-} ; (d) A view of the 3D hydrogen bonding network structure of complex 7 (black dotted lines representing hydrogen bonding).

Table S5. Selected bond distances (Å) and angles (deg) of complex 7.

Bond distances					
Gd-O(6)	2.331(4)	Gd-O(1)	2.336(4)	Gd-O(2)	2.355(6)
Gd-O(3)	2.385(4)	Gd-O(9)	2.389(4)	Gd-O(5)	2.455(4)
Gd-O(4)	2.470(6)	Gd-O(7)	2.513(4)		
Bond angles					
O(6)-Gd-O(1)	80.00(16)	O(6)-Gd-O(2)	88.33(17)	O(1)-Gd-O(2)	70.74(17)
O(6)-Gd-O(3)	143.93(14)	O(1)-Gd-O(3)	125.97(16)	O(2)-Gd-O(3)	79.59(15)
O(6)-Gd-O(9)	146.81(16)	O(1)-Gd-O(9)	79.30(17)	O(2)-Gd-O(9)	108.75(19)
O(3)-Gd-O(9)	68.78(15)	O(6)-Gd-O(5)	100.05(16)	O(1)-Gd-O(5)	141.04(15)
O(2)-Gd-O(5)	147.93(17)	O(3)-Gd-O(5)	75.51(15)	O(9)-Gd-O(5)	80.66(16)
O(6)-Gd-O(4)	70.33(14)	O(1)-Gd-O(4)	134.41(17)	O(2)-Gd-O(4)	74.5(2)
O(3)-Gd-O(4)	73.70(14)	O(9)-Gd-O(4)	140.78(15)	O(5)-Gd-O(4)	79.33(17)
O(6)-Gd-O(7)	73.22(15)	O(1)-Gd-O(7)	74.02(14)	O(2)-Gd-O(7)	142.52(16)
O(3)-Gd-O(7)	133.30(15)	O(9)-Gd-O(7)	76.29(16)	O(5)-Gd-O(7)	68.92(15)
O(4)-Gd-O(7)	125.65(15)				

(a)

(b)

(c)

(d)

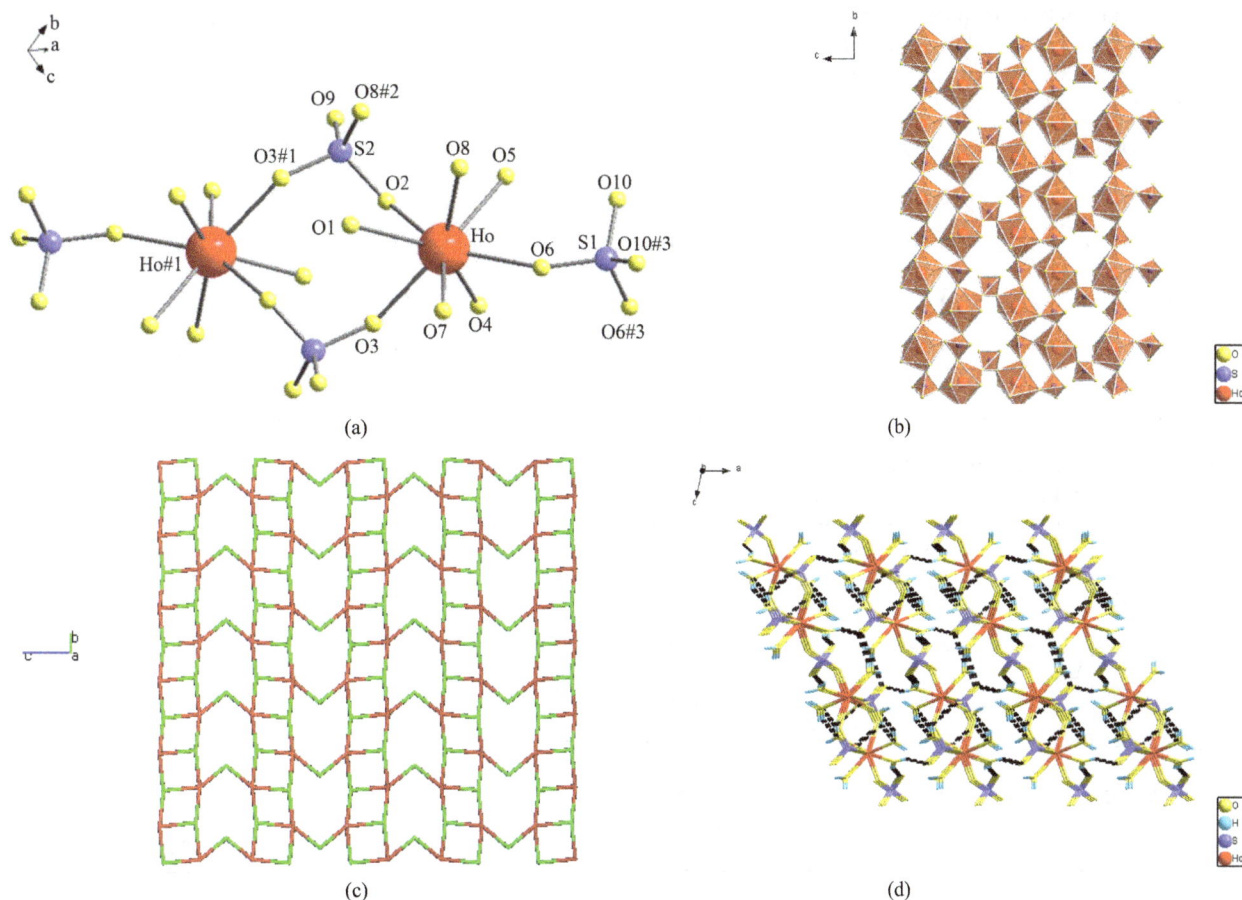

Figure S16. (a) The coordination environment of Ho in complex 8; (Symmetry codes: #1 − x + 3/2, −y − 1/2, −z #2 −x + 3/2, −y + 1/2, −z #3 −x + 2, y, −z + 1/2); (b) The 2D polyhedron layer in complex 8, viewed along the a-axis; (c) Schematic representation of the 2D $(4^2.6.8^2.10)_2(4^2.6)_2(8)$ topology network in complex 8. Red, Ho; Lime, SO_4^{2-} ; (d) A view of the 3D hydrogen bonding network structure of complex 8 (black dotted lines representing hydrogen bonding).

Table S6. Selected bond distances (Å) and angles (deg) of complex 8.

Bond distances					
O(1)-Ho	2.429(6)	O(2)-Ho	2.392(5)	O(3)-Ho	2.350(5)
O(4)-Ho	2.339(5)	O(5)-Ho	2.461(5)	O(6)-Ho	2.299(4)
O(7)-Ho	2.308(6)	O(8)-Ho	2.285(5)		
Bond angles					
O(8)-Ho-O(6)	79.99(19)	O(8)-Ho-O(7)	88.7(2)	O(6)-Ho-O(7)	70.89(19)
O(8)-Ho-O(4)	146.68(19)	O(6)-Ho-O(4)	79.7(2)	O(7)-Ho-O(4)	109.1(2)
O(8)-Ho-O(3)	144.47(16)	O(6)-Ho-O(3)	125.62(18)	O(7)-Ho-O(3)	79.39(18)
O(4)-Ho-O(3)	68.30(18)	O(8)-Ho-O(2)	99.48(18)	O(6)-Ho-O(2)	141.66(17)
O(7)-Ho-O(2)	147.2(2)	O(4)-Ho-O(2)	80.90(19)	O(3)-Ho-O(2)	75.64(18)
O(8)-Ho-O(1)	70.27(17)	O(6)-Ho-O(1)	134.10(19)	O(7)-Ho-O(1)	74.2(2)
O(4)-Ho-O(1)	140.77(18)	O(3)-Ho-O(1)	74.28(16)	O(2)-Ho-O(1)	78.83(19)
O(8)-Ho-O(5)	72.90(17)	O(6)-Ho-O(5)	74.73(15)	O(7)-Ho-O(5)	143.30(18)
O(4)-Ho-O(5)	76.41(18)	O(3)-Ho-O(5)	132.91(16)	O(2)-Ho-O(5)	68.70(17)
O(1)-Ho-O(5)	125.10(17)				

Table S7. Hydrogen bonds of complex **1**/Å and symmetry codes: j: $1/2 + x$, $1/2 + y$, z; k: $1/2 - x$, $1/2 + y$, $1/2 - z$; e: $1/2 - x$, $3/2 - y$, $1 - z$; d: $1/2 - x$, $1/2 - y$, $1-z$; b: x, $1 + y$, z; m: $-1/2 + x$, $1/2 + y$, z.

D-H···A	D-H	H···A	D···A	<D-H···A
O1-H1A···O9^j	0.9300	2.0172	2.8090	141.99
O1-H1B···O10^k	0.9300	1.9459	2.7623	145.38
O4-H4A···O5^e	0.8566	2.0407	2.7847	144.79
O4-H4B···O2^d	0.7855	2.1451	2.9182	168.11
O6-H6A···O9^b	0.8999	1.9888	2.8320	155.40
O6-H6B···O10^m	0.9000	1.8867	2.7371	156.84

Table S8. Hydrogen bonds of complex **2**/Å and symmetry codes: g: $1/2 + x$, $1/2 + y$, z; a: x, $-1 + y$, z; e: $1/2 - x$, $3/2 - y$, $-z$; d: $1/2 - x$, $1/2 - y$, $-z$; h: $1/2 - x$, $1/2 + y$, $1/2 - z$; i: $-1/2 + x$, $-1/2 + y$, z.

D-H···A	D-H	H···A	D···A	< D-H···A
O1-H1A···O2^g	0.8999	1.8819	2.7288	156.03
O1-H1B···O9^a	0.9000	1.9757	2.8210	155.82
O4-H4A···O7^e	0.9001	2.0859	2.9144	152.57
O4-H4B···O5^d	0.7370	2.1825	2.7968	141.45
O10-H10A···O2^h	0.9001	2.1579	2.7649	124.11
O10-H10B···O9^i	0.9000	2.2344	2.8173	122.05

Table S9. Hydrogen bonds of complex **4**/Å and symmetry codes: k: $1/2 - x$, $1/2 + y$, $1/2 - z$; j: $-1/2 + x$, $1/2 + y$, z; m: $1/2 + x$, $1/2 + y$, z; b: x, $1 + y$, z; d: $1/2 - x$, $1/2 - y$, $-z$; e: $1/2 - x$, $3/2 - y$, $-z$.

D-H···A	D-H	H···A	D···A	< D-H···A
O1-H1A···O8^k	0.8999	2.1331	2.7463	124.64
O1-H1B···O4^j	0.8998	2.2262	2.8012	121.30
O6-H6A···O8^m	0.8999	1.8712	2.7129	154.84
O6-H6B···O4^b	0.9000	1.9653	2.8127	156.28
O10-H10A···O5^d	0.7541	2.1869	2.9253	166.44
O10-H10B···O7^e	0.7326	2.1383	2.7803	146.79

Table S10. Hydrogen bonds of complex **5**/Å and symmetry codes: a: x, $-1 + y$, z; f: $1/2 + x$, $-1/2 + y$, z; d: $1/2 - x$, $3/2 - y$, $-z$; e: $1/2 - x$, $1/2 - y$, $-z$; j: $1 - x$, $1 - y$, $-z$.

D-H···A	D-H	H···A	D···A	<D-H···A
O1-H1A···O6^a	0.9000	2.0531	2.8321	144.17
O1-H1B···O10^f	0.9000	1.8650	2.7259	159.41
O3-H3A···O8^d	0.9000	1.9752	2.7901	149.82
O3-H3B···O4^e	0.9000	2.1834	2.9650	144.85
O4-H4A···O8^j	0.8999	2.1893	3.0411	157.70

Table S11. Hydrogen bonds of complex 7/Å and symmetry codes: h: $1/2 - x$, $-1/2 + y$, $1/2 - z$; i: $-1/2 + x$, $1/2 + y$, z; e: $1/2 - x$, $1/2 - y$, $-z$; d: $1/2 - x$, $3/2 - y$, $-z$; g: $1/2 + x$, $-1/2 + y$, z; a: x, $-1 + y$, z.

D-H···A	D-H	H···A	D···A	< D-H···A
O2-H2A···O10[h]	0.9000	2.1245	2.7551	126.37
O2-H2B···O8[i]	0.9000	2.2537	2.8251	121.04
O4-H4A···O7[e]	0.8999	2.1307	2.9604	152.91
O4-H4B···O5[d]	0.9784	2.0107	2.7882	134.83
O7-H7A···O4[e]	0.9000	2.1508	2.9604	149.26
O9-H9A···O10[g]	0.9001	1.8892	2.7313	154.94
O9-H9B···O8[a]	0.9000	1.9902	2.8360	155.98

Table S12. Hydrogen bonds of complex 8/Å and symmetry codes: c: $1/2-x$, $3/2-y$, $-z$; h: $1-x$, $1-y$, $-z$; k: x, $-1 + y$, z; l: x, $1-y$, $1/2 + z$.

D-H···A	D-H	H···A	D···A	D-H···A
O1-H1A···O5[c]	0.9001	2.1185	2.9499	153.20
O4-H4A···O9[h]	0.9000	1.8863	2.7242	154.06
O4-H4B···O10[k]	0.9000	1.9792	2.8250	155.97
O5-H5A···O1[c]	0.9000	2.1384	2.9499	149.58
O7-H7A···O9[l]	0.9000	2.1145	2.7522	127.05

8

Synthesis, crystal structure, and catalytic properties of 2,2'-bipyridyl-dicyano-palladium(II)

Xiaoli Tang, Longguan Zhu[*]

Department of Chemistry, Zhejiang University, Hangzhou, China
Email: [*]chezlg@zju.edu.cn

ABSTRACT

In the solvothermal condition, a Pd(II) complex with a *in situ* synthesis ligand, $[Pd(2,2'-bipy)_2(CN)_2]$ (1) was obtained and characterized by elemental analysis, IR, UV, TG, and powder X-ray analysis. The single crystal X-ray analysis showed that the title complex is different from the reported 2,2'-bipyridyl-dicyano-palladium(II) (2) in crystal system and structural parameters. The catalytic investigation for the reactions of the disproportionation of hydrogen peroxide and oxidation of sulfide showed that the complex 1 is an active homogeneous catalyst in the presence of imidazole and 2-sulfobenzoic acid, respectively.

Keywords: Synthesis; Thioanisol; Pd Complex; Oxidation of Sulfide; Hydrogen Peroxide; Disproportionation

1. INTRODUCTION

In situ ligand synthesis under hydro/solvothermal conditions has already yielded a large number of novel coordination complexes [1]. Under specific conditions, the solvents used in the solvothermal reaction may make some changes to produce new species [2]. Acetonitrile is a common organic solvent in the synthesis of coordination complexes, while this solvent can result in the C-C bond cleavage under hydro/solvothermal conditions through the catalysis of metal ions, such as Cu, Mn, Pb, Ir, Ru, and Cd [3]. Herein we report the first C-C bond cleavage of acetonitrile in the presence of Pd(II) under solvothermal condition.

2. EXPERIMENT

2.1. Reagents and Physical Measurements

All chemicals were directly purchased from commercial resource with reagent grade and were used as received.

Elemental analysis for C, H, and N were measured on a Perkin-Elmer analyzer model 1110. The infrared spectra were recorded on a Nicolet Nexus 470 infrared spectrophotometer in KBr pellets in the 400 - 4000 cm^{-1} region. Thermogravimetric analyses (TGA) were carried out using a Delta Series TA-SDT Q600 in nitrogen flow at a heating rate of 10°C/min with Al_2O_3 crucibles. The UV-vis spectra were performed on a SPECORD 2000 UV-vis spectrophotometer in CH_3OH at room temperature. The powder X-ray diffraction was measured by Rigaku D/MaX 2550PC with Cu-Kα radiation. The GC data were recorded on a Fuli Gas Chromatography equipped with a DB-5 capillary column. All the standard substance used in GC were purchased from Alfa Aesar.

2.2. Synthesis of Complex $[Pd(2,2'-bipy)_2(CN)_2]$ (1)

A mixture of $Pd(CH_3COO)_2$ (0.014 g, 0.062 mmol), 2-sulfobenzoic acid (0.050 g, 0.25 mmol), and 2,2'-bipyridine (0.020 g 0.13 mmol) was dissolved in methanol/acetonitrile (8 ml/8 ml) solution. The resulting mixture was sealed in a 25 ml Teflon-lined autoclave and heated to 120°C for two days. Clear solution was set aside about two weeks at room temperature, then yellow plate-like crystals were obtained. No precipitation was observed in the absence of 2,2'-bipyridine or 2-sulfobenzoic acid. No complex was formed if $PdCl_2$ was used in place of $Pd(CH_3COO)_2$. Anal. Calcd (%) for $C_{12}H_8N_2Pd$: C, 45.80; H, 2.54; N, 17.81. Found: C, 45.86; H, 2.92; N, 17.92. IR (KBr, cm^{-1}): ν = 3451(s), 3115(m), 3082 (m), 3053 (m), 2133 (m), 1606 (s), 1450 (s), 1319(w), 1168 (w), 1038 (w), 773 (s), 727 (w), 415 (w).

The complex **1** can dissolve in methanol, acetonitrile, DMF, and DMSO, while it can not dissolve in H_2O, ethanol, and CH_2Cl_2.

2.3. X-Ray Structure Determination

The single crystal with suitable size for X-ray diffraction

was selected for data collection by a Bruker Smart CCD area detector with graphite-monochromatized Mo-Kα radiation (λ = 0.71073 Å) at room temperature. The crystal was mounted on a glass fiber in a random orientation. Lorenz-polarization effect and secondary extinction were corrected [4]. The structure was solved by the heavy-atom method and successive Fourier syntheses. Full-matrix least-squares refinements on F^2 were carried out using the SHELXL-97 package [5]. All non-H atoms were anisotropically refined. The structure analysis and drawing were helped by WinGX software [6]. The detailed crystallographic data and refinement parameters for complex 1 are listed in Table 1.

Table 1. Crystallographic data and refinement parameters for complex 1.

Empirical formula	$C_{12}H_8N_4Pd$
Mr	314.62
Crystal system	Triclinic
Space group	P-1
Size/mm^3	0.05 × 0.06 × 0.08
a/Å	6.9162(6)
b/Å	8.8442(7)
c/Å	9.4753(6)
α/°	86.578(6)
β/°	80.378(6)
γ/°	74.574(7)
V/Å^3	550.78(8)
Z	2
D_c/Mg·m^{-3}	1.897
μ/mm^{-1}	1.664
θ range/°	3.1 - 25.1
Reflections collected	3409
Unique reflections	1958
Observed reflections	1751
R_{int}	0.034
parameters	152
F(000)	308
Temperature/K	295(2)
R_1, wR_2[I > 2σ(I]	0.033, 0.064
R_1, wR_2[all data]	0.041, 0.071
GOF	0.860
largest difference peak and hole/e·Å^{-3}	0.431, −0.565

2.4. Reaction of Sulfide Oxidation

A 25 mL round-bottom flask was charged with a mixture of the catalyst [0.015 mmol of complex 1 or 0.015 mmol of complex 1 and 0.01 mmol of 2-sulfobenzoic acid (2-H$_2$sb)] in CH$_3$CN (5 mL) and the substrate methyl phenyl sulfide (MPS, 0.5 mmol). The mixture was stirred at room temperature for 10 minutes. Reaction was started by the addition of the corresponding amount of H$_2$O$_2$ solution (30%, 1.5 mmol) under stir. The first sample was taken after 30 minutes and the later samples were taken at specified time with a fixed interval of 30 min. Pellucid solution was taken, and diluted by 10 times immediately and measured by gas chromatography. Each catalyzed reaction was kept for 270 minutes.

2.5. Hydrogen Peroxide Disproportionation Study

A mixture of 1 mL aqueous solution of imidazole with the concentration of 0.1 mol/L, 5 ml of water, and 1 mL of DMSO containing the synthesized complex with the concentration of 10^{-3} mol/L in a three-necked flask of 25 mL was stirred in a thermostated cell at room temperature. The flask was connected with a pump and a pressure gauge (APM-2D). With a vacuum pump, an average 5.0 kPa pressure above the solution was obtained and kept. Then zero setting was done. At this moment, hydrogen peroxide (1 ml, H$_2$O$_2$ 30% w/w) was introduced into the system by a syringe and immediately the time and the evolved oxygen pressure were recorded.

3. RESULTS AND DISCUSSION

3.1. Powder X-Ray Analysis

The purity of the synthesized complex 1 was confirmed by powder X-ray analysis. The measured XRD pattern is shown in Figure 1(a) and the calculated pattern derived from the single crystal data by Mercury [7] is depicted in Figure 1(b). These two patterns are nearly same, indicating that the bulk sample is pure.

3.2. Single Crystal Structure

The crystal structure of [Pd(2,2'-bipy)(CN)$_2$] (2) has been reported by Che and his coworkers in the orthogonal crystal system [8]. Complex 2 was directly synthesized by the reaction of Pd(CN)$_2$ and 2,2'-bipyridine in dimethylformamide. The title complex crystallizes in triclinic system, indicating the different synthetic route led to the different packing arrangement. The Pd(II) center adopts a square-planar geometry completed by two N donors from two 2,2'-bipyridine and two C atoms from two cyano anions (Figure 2 and Table 2). The Pd-N bond lengths in 1 are significantly longer than those in 2, while the Pd-C bond lengths are remarkably shorter than

(a)

h, k, l = 1, 1, -1

(b)

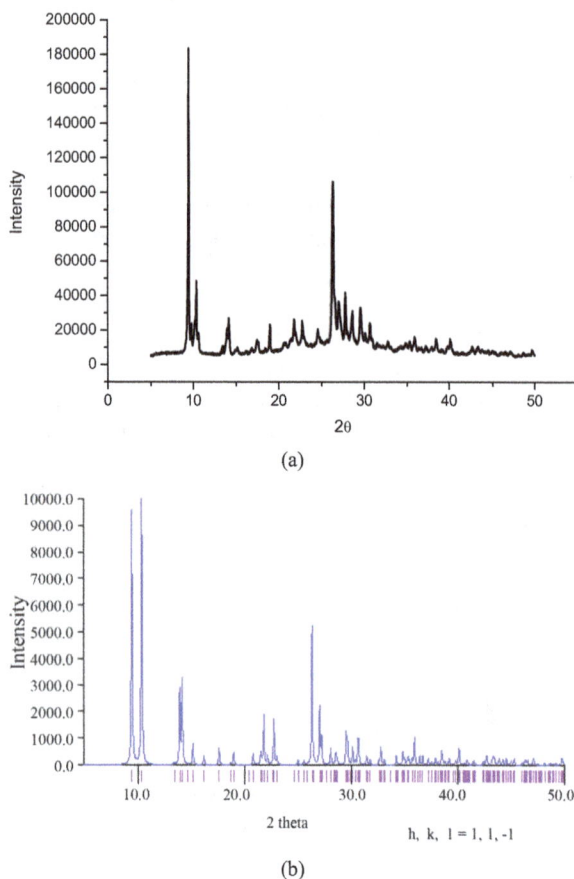

Figure 1. (a) The powder X-ray pattern of complex **1**; (b) The simulated pattern of complex **1** derived from single crystal data.

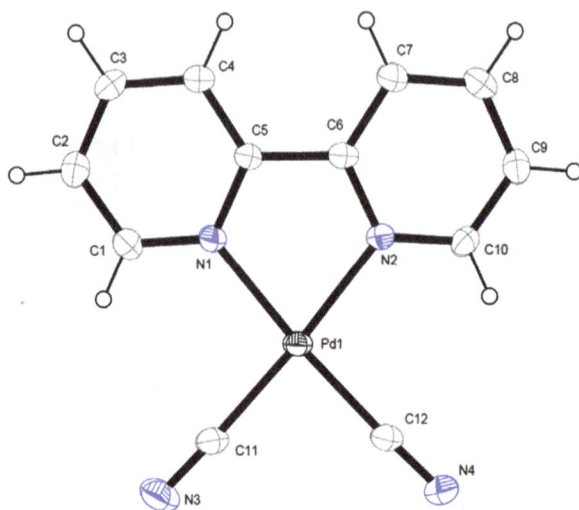

Figure 2. The view of molecular structure of complex **1** with atom labels and 40% probability displacement ellipsoids for non-H atoms.

those in **2**. The distance of Pd...Pd in neibouring molecules in **1** is 3.5579(7) Å which is slightly longer than that of **2** (3.3541(4) Å. Complex **1** has no classic hydro-

gen bond and aromatic π-π interactions, while it has non-classic hydrogen bonds of C-H...N and short contact interactions of C-H...H-C, forming a 2D layer (**Figure 3**).

3.3. TG Analysis

The TG curve (**Figure 4**) was recorded from room temperature to 700°C in the nitrogen atmosphere. The complex **1** has no solvent and it can be stable up to 200°C, then the complex starts to decompose.

3.4. UV Spectrum

The UV spectrum of complex **1** in methanol with the concentration of 4.675×10^{-5} mol/L at room temperature was shown in **Figure 5**. The spectrum shows two absorption peaks at 307 nm ($\varepsilon = 8342$ dm^3·mol^{-1}·cm^{-1}) and 320 nm ($\varepsilon = 8449$ dm^3·mol^{-1}·cm^{-1}). These bands can be assigned to characteristic π-π* transitions [9]. The absorption peak of complex **1** is red-shifted compared with the absorption of 2,2'-bipy in CH$_3$CN (282 nm), indicating that the coordination leads some change of the absorption.

3.5. Catalytic Properties

The disproportionation of hydrogen peroxide [10,11] and oxidation of MPS [12,13] were selected for the study of the catalytic properties of the complex **1**. The disproportionation of hydrogen peroxide was investigated by monitoring oxygen pressure. As is revealed from **Figure 6**, not only complex **1** but also imidazole has accelerated the reaction to the similar extent. Furthermore, existence of both **1** and imidazole (mz) have greatly accelerated the

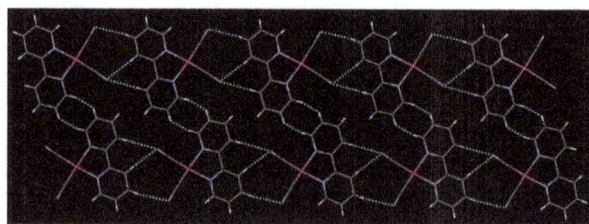

Figure 3. The view of the 2D layer formed by C-H...N and short contact interactions in complex **1**.

Table 2. Selected bond lengths (Å) and angles (°) for complex **1**.

Pd1-N1	2.056(3)	Pd1-N2	2.055(3)
Pd1-Cl1	1.958(4)	Pd1-Cl2	1.956(4)
N1-Pd1-N2	79.54(12)	N1-Pd1-Cl1	96.58(15)
N1-Pd1-Cl2	175.22(15)	N2-Pd1-Cl1	175.78(15)
N2-Pd1-Cl2	95.92(15)	Cl1-Pd1-Cl2	87.91(17)

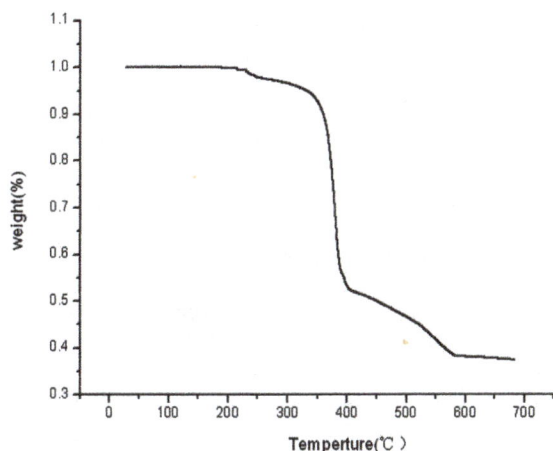

Figure 4. The TG curve of complex **1**.

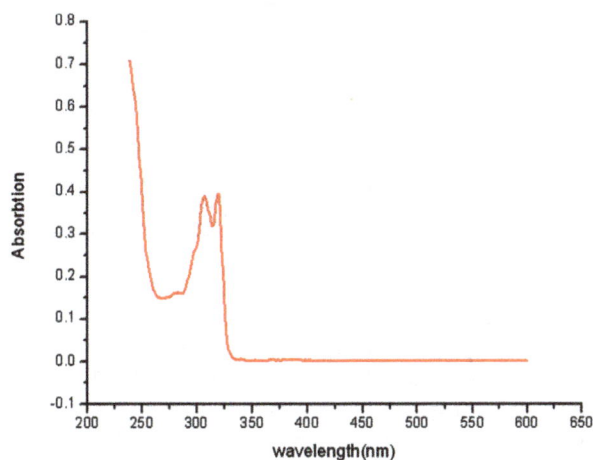

Figure 5. The UV spectrum of complex **1** in the methanol at room temperature.

reaction. The essential roles of bases have been mentioned in similar reactions [14]. We confirmed, and believe, that the base helps the subtraction of protons in H_2O_2 molecules, which may help the vicinity of HO_2^- as nucleophilic reagents to the active centers of Pd^{II} catalyst [15].

The efficiency of activating MPS (methyl phenyl sulfide) oxygenation utilizing simple oxidant H_2O_2 in 3 equiv. was studied at ambient temperature ($23°C \pm 2°C$) (**Figure 7**). Complex **1** exhibited the conversion of 45.8% and 56.8% selectivity for sulfoxide after 270 min, and the addition of 2-H_2sb in the presence of complex **1** showed significantly conversion of 89.1% and 96.4% selectivity for sulfoxide (**Figure 8**), indicating that the complex **1** is an active catalyst in the reaction of oxidation of MPS and can promote the catalytic activity through the cooperation in the presence of 2-H_2sb [16].

4. CONCLUSION

In conclusion, we synthesized a Pd(II) complex with the

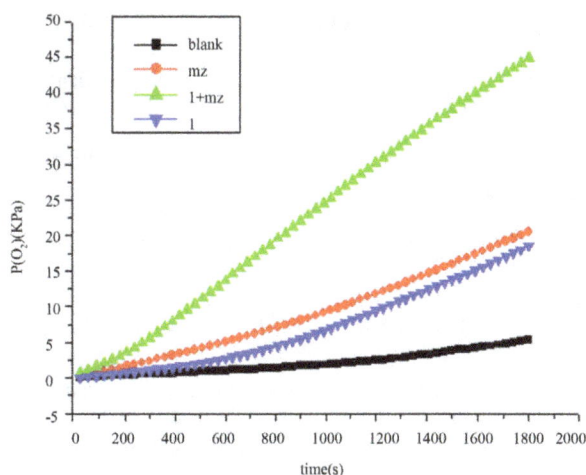

Figure 6. The disproportionation of hydrogen peroxide catalyzed by complex **1** and base.

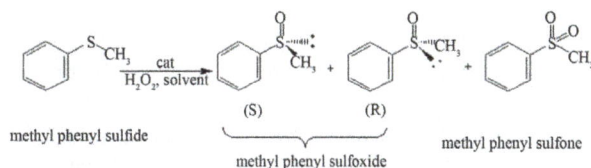

Figure 7. The scheme of the oxidation of methyl phenyl sulfide.

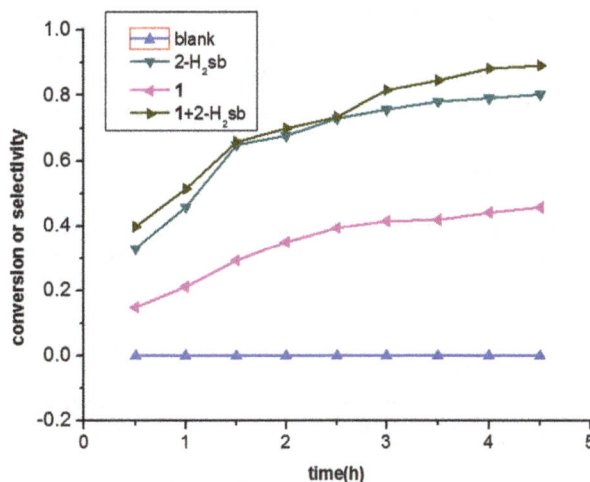

Figure 8. The oxidation of methyl phenyl sulfide by catalysts.

different crystal system from which is reported in the reference and characterized by the elemental analysis, IR, UV, TG, and powder X-ray analysis. The single crystal structure analysis showed that the complex **1** has some different structural characters from the complex **2**. The catalytic investigation for the reactions of oxidation of sulfide and the disproportionation of hydrogen peroxide showed the complex **1** is an active homogeneous catalyst, especially combined with 2-H_2sb and imidazole, respectively.

5. ACKNOWLEDGEMENTS

The authors thank the National Natural Science Foundation of China (grant No. 21073157).

REFERENCES

[1] Chen, X.M. and Tong, M.L. (2007) Solvothermal *in situ* metal/ligand reactions: a new bridge between coordination chemistry and organic synthetic chemistry. *Accounts of Chemical Research*, **40**, 162-170. http://dx.doi.org/10.1021/ar068084p

[2] Zhang, X.M. (2005) Hydro(solvo)thermal *in situ* ligand syntheses. *Coordination Chemistry Reviews*, **249**, 1201-1219. http://dx.doi.org/10.1016/j.ccr.2005.01.004

[3] Zhu, Y.L., Qu, L.L., Zhang, J., Ge, G.W., Li, Y.Z., Du, H.B. and You, X.Z. (2011) *In situ* formation of an unprecedented 3D microporous CuCN coordination polymer based on a semirigid tetrahedral linker. *Inorganic Chemistry Communications*, **14**, 1644-1647.

[4] Sheldrick, G.M. (1997) SADABS, Program for bruker area detector absorption correction. University of Göttingen, Göttingen.

[5] Sheldrick, G.M. (1997) SHELXL-97, program for crystal structure refinement. University of Göttingen, Göttingen.

[6] Farrugia, L.J. (1999) WinGX suite for small-molecule single-crystal crystallography. *Journal of Applied Crystallography*, **32**, 837-838. http://dx.doi.org/10.1107/S0021889899006020

[7] Macrae, C.F., Edgington, P.R., McCabe, P., Pidcock, E., Shields, G.P., Taylor, R., Towler, M. and van de Streek, J. (2006) Mercury: Visualization and analysis of crystal structures. *Journal of Applied Crystallography*, **39**, 453-457. http://dx.doi.org/10.1107/S002188980600731X

[8] Che, C.C., He, L.Y., Poon, C.K. and Mak, T.C.W. (1989) Solid-state emission of dicyanoplatinum(II) and palladium(II) complexes of substituted 2,2'-bipyridines and isomorphous M(bpy)(CN)$_2$ (bpy=2,2'-bipyridine, M=Pt, Pd). *Inorganic Chemistry*, **28**, 3081-3083.

[9] Yang, X.J., Janiak, C., Heinze, J., Drepper, F., Mayer, P., Piotrowski, H. and Klufers, P. (2001) Heteroleptic 5,5'-disubstituted-2,2'-bipyridine complexes of ruthenium(II): spectral, electrochemical, and structural investigations. *Inorganica Chimica Acta*, **318**, 103-116. http://dx.doi.org/10.1016/S0020-1693(01)00414-5

[10] Miao, X.H. and Zhu, L.G. (2010) Supramolecular assembly under the control of the chelating ligand for the Mn^{II}/bridging ligands/3-sulfobenzoate system and catalytic properties for the disproportionation of hydrogen peroxide. *New Journal of Chemistry*, **34**, 2403-2414. http://dx.doi.org/10.1039/b9nj00428a

[11] Dubois, L., Caspar, R., Jacquamet, L., Petit, P. E., Charlot, M.F., Baffert, C., Collomb, M.N., Deronzier, A. and Latour, J.M. (2003) Binuclear manganese compounds of potential biological significance. Part 2. Mechanistic study of hydrogen peroxide disproportionation by dimanganese complexes: the two oxygen atoms of the peroxide end up in a dioxo intermediate. *Inorganic Chemistry*, **42**, 4817-4827. http://dx.doi.org/10.1021/ic020646n

[12] Villalobos, L., Cao, Z., Fanwick, P.E. And Ren, T. (2012) Diruthenium(II,III) tetramidates as a new class of oxygenation catalysts. *Dalton Transactions*, **41**, 644-650. http://dx.doi.org/10.1039/c1dt11530h

[13] Barker, J.E. and Ren, T. (2008) Diruthenium(II,III) bis(tetramethyl-1,3-benzenedipropionate) as a novel catalyst for *tert*-butyl hydroperoxide oxygenation. *Inorganic Chemistry*, **47**, 2264-2266.

[14] Devereux, M., McCann, M., Leon, V., McKee, V. and Ball, R.J. (2002) Synthesis and catalytic activity of manganese(II) complexes of heterocyclic carboxylic acids: X-ray crystal structures of [Mn(pyr)$_2$]$_n$, [Mn(dipic)(bipy)$_2$]·/4.5H$_2$O and [Mn(chedam)(bipy)]·/H$_2$O(pyr=/2-pyrazinecarboxylic acid; dipic =/pyridine-2,6-dicarboxylic acid; chedam_/chelidamic acid (4-hydroxypyridine-2,6-dicarboxylic acid); bipy =/2,2-bipyridine). *Polyhedron*, **21**, 1063-1071. http://dx.doi.org/10.1016/S0277-5387(02)00842-2

[15] Viossat, V., Lemoine, P., Dayan, E., Dung, N.H. and Viossat, B. (2003) Synthesis, crystal structure and IR spectroscopy of Mn^{II}(2-IC)$_2$(NC)(DMSO) and [Mn^{II}(2-IC)$_2$(phen)(H$_2$O)]·DMA; (2-HIC, indole-2-carboxylic acid; phen, 1,10-phenanthroline; NC, 2,9-dimethyl-1,10-phenanthroline; DMSO, dimethyl sulfoxide; DMA, dimethyl acetamide); catalysts for the disproportionation of hydrogen. *Polyhedron*, 22, 1461-1470. http://dx.doi.org/10.1016/S0277-5387(03)00126-8

[16] Hu W.T. and Zhu, L.G. (2013) Study on the synthesis, crystal structure and catalytic activity for thioanisol of a ruthenium complex with 3-sulfobenzoate and 2,2'-bipyridine ligands. *Chinese Journal of Inorganic Chemistry*, **29**, 1109-1114.

9

Synthesis and characterization of a novel schiff base metal complexes and their application in determination of iron in different types of natural water

Mostafa M. H. Khalil[1*], Eman H. Ismail[1], Gehad G. Mohamed[2], Ehab M. Zayed[2], Ahmed Badr[1]

[1]Chemistry Department, Faculty of Science, Ain Shams University, Cairo, Egypt
[2]Chemistry Department, Faculty of Science, Cairo University, Cairo, Egypt
Email: *khalil62@yahoo.com

ABSTRACT

A novel and simple approach to the synthesis of macrocyclic Schiff base ligand resulted from the condensation of bisaldehyde and ethylenediamine was prepared (7, 8, 15, 16, 17, 18-hexahydrodibenzo (a, g) (14) annulene) (L) and its complexes were synthesized and characterized using different physicochemical studies as elemental analysis, FT-IR, ^{1}H NMR, conductivity, magnetic properties, thermal analysis, and their biological activities. The spectroscopic data of the complexes suggest their 1:1 complexe structures which are investigated by elemental analysis, FT-IR, ^{1}H NMR, conductivity, magnetic properties, thermal analysis, and their biological activities. The spectroscopic studies suggested the octahedral structure for the all complexes. The spectroscopic data of the complexes suggest their structure in which (N_2O_2) group act as a tetradentate ligand and two chlorides as monodentate ligands. Also electronic spectra and magnetic susceptibility measurements indicate octahedral structure of these complexes. The synthesized Schiff base and its metal complexes also were screened for their antibacterial and antifungal activity. Here we report the effect of a neutral chelating ligand on the complexation with iron to determine it in different types of natural water using recovery test. The activity data show that the metal complexes to be more potent/antibacterial than the parent Schiff base ligand against one or more bacterial species.

Keywords: Novel Schiff Base; Transition Metal Complexes and Natural Water

1. INTRODUCTION

A large number of Schiff bases and their complexes have been investigated for their interesting and important properties, such as their ability to reversibly bind oxygen [1], catalytic activity in the hydrogenation of olefines [2], photochromic properties [3,4] and complexing ability towards some toxic metals [5]. Schiff bases are a special class of ligands with a variety of donor atoms exhibiting interesting coordination modes towards various metals [6-8]. The azomethine linkage in Schiff bases is responsible for the biological activities such as antitumor, antibacterial, antifungal and herbicidal activities [9]. Designing a suitable polydentate Schiff base ligand to combine with a metal ion along with pseudohalide anion has opened a new area of synthesizing metal complexes of particular choice [10]. Such complexes are readily assembled from diamines and various salicylaldehyde derivatives and are amenable to combinatorial syntheses [11]. Metal Schiff base complexes have been well known for their easy synthesis, stability and wide application [12-14]. A large number of Schiff base (N_2O_2) complexes have been reported so far, and their catalytic and biological properties have been studied intensively [15-21]. Numerous techniques were available in iron determination have been reported [22-24]. However, these techniques were tedious need time and rigid conditions. Also due to the very low concentrations of iron, lead and chromium in natural water samples, their determinations demand very sensitive analytical techniques including ET-AAS, ICP-MS, ICP-AES and XRF. However, relative to flame AAS, these techniques have some disadvantages such as its high cost, slowness and greater proneness to matrix interferences as well as their high sensitive advantages [25]. The present study was under taken to throw more light on the chelation behavior of Schiff base (L) towards some transition elements, which may help in more understanding of the mode of chelation of the new Schiff base towards metals and tried to used in the recovery test with a purity of more than higher than 96% for the metal ion.

*Corresponding author.

2. EXPERIMENTAL

2.1. Materials and Solutions

The chemicals used included metal (II) and (III) chlorides. Solvents of absolute ethyl alcohol, diethylether, dimethyl-formamide and ethylenediamine were spectroscopic pure and purchased from BDH. A 10^{-4} M solution of the metal salts ($CrCl_3 \cdot 6H_2O$ (0.341 g/L), $CoCl_2 \cdot 6H_2O$ (0.119 g/L), $NiCl_2 \cdot 6H_2O$ (0.119 g/L) (BDH), $FeCl_3 \cdot 6H_2O$ (0.117 g/L) (Prolabo), $CuCl_2 \cdot 2H_2O$ (0.085 g/L) (Sigma) and $ZnCl_2 \cdot 2H_2O$ (0.086 g/L) (Ubichem)) were prepared by dissolving the accurately weighed amounts of the metal salts in the appropriate volume of de-ionized water. The metal salt solutions were standardized by the recommended procedures [26].

2.2. Instrumentation

Elemental microanalyses of the separated solid chelates for C, H, N and Cl were performed at the Microanalytical center, Cairo University. The molar conductance of the complexes in DMF was measured using Sybron-Barnstead conductometer (Meter-PM.6, E = 3406). Infrared spectra were recorded on a Perkin-Elmer FT-IR type 1650 spectrophotometer. The spectra were recorded as KBr pellets. The UV-vis absorption spectra were measured on a Shimadzu 3101pc spectrophotometer. The molar magnetic susceptibility was measured on powdered samples using the Faraday method. The diamagnetic corrections were made by Pascal's constant and $Hg[Co(SCN)_4]$ was used as a calibrant. The mass spectra were recorded by the EI technique at 70 eV using MS-5988 GS-MS Hewlett-Packard instrument. The ^{HI}NMR spectra were recorded using 300 MHz Varian-Oxford Mercury. The thermal analyses were carried out in dynamic nitrogen atmosphere (20 mL·min^{-1}) with a heating rate of 10°C·min^{-1} using Shimadzu TG-60H thermal analyzer. XPD analyses were carried out by using Philips Analytical X-Ray BV, diffractometer type PW 1840. Radiation was provided by copper target (Cu anode 2000 W) high intensity X-ray tube operated at 40 KV and 25 mA. All spectrophotometric measurements were preformed using a Unicam UV-2 spectrophotometer using 10 mm quartz cell and a blank solution as a reference. The pH measurements were made using HANNA pH meter.

2.3. Synthesis of the Compounds

2.3.1. Synthesis of Bisaldehyde

Salicylaldehyde (20 mmol) was dissolved in hot ethanolic KOH (prepared by dissolving 1.12 g (20 mmol) of KOH in 20 ml of absolute ethanol), and the solvent was then removed in vacuo. The remaining material was dissolved in DMF (15 ml) and the appropriate 1,2-dibromoethane (10 mmol) was added. The reaction mixture was refluxed

for 5 min during which KCl was separated. The solvent was then removed in vacuo and the remaining materials was washed with water and purified by crystallization.

2.3.2. Synthesis of Schiff Base

Hot solution (60°C) of ethylenediamine (0.399 g, 6.67 mmol) in 50 mL ethanol/DMF mixture as 1:1 mole ratio mixed with hot solution (60°C) of bisaldehyde (1.6 g, 6.67 mmol) in the same solvent and the reaction mixture was left under heating for 30 min. A solid mass separated was collected and washed with diethylether. Crystallization was done with ethanol and then dried over $CaCl_2$. The pale yellow ligand; L, (yield, 82%) was collected.

2.3.3. Synthesis of Metal Complexes

All the new complexes were prepared by adding hot ethanolic/DMF solution in a 1:1 molar ratio (60°C) of metal (II)/(III) (1 mmol) to hot solution (60°C) of ligand in the same solvent. The solution was stirred with heating for one hour whereupon the complexes precipitated then filtered and left for drying. A solid residue was separated and washed by diethylether. Crystallization was done with methanol and the complexes dried over anhydrous $CaCl_2$ and the metal contents were determined compleximetrically.

2.4. Biological Activity

The standard disc-agar diffusion method [27] was followed to determine the antibacterial and antifungal activeity of the synthesized compounds. The tested compounds were dissolved in DMF (which have no inhibittion activity), to get concentrations of 100 μg/mL. Uniform size filter paper disks (3 disks per compound) were impregnated by equal volume (0.1 mL) from the specific concentration of dissolved tested compounds and carefully placed on incubated agar surface. After incubation for 48 h at 37°C, inhibition of the organisms which evidenced by clear zone surround each disk was measured and used to calculate mean of inhibition zones.

2.5. The Recovery Procedure

Water samples were collected according to the recommended standard methods [28]. Water samples were spiked with a definite concentration of Fe(III) soluation and 1 ml of 10^{-4} M of the Schiff base was added. After adjusting the pH with HNO_3 and/or NaOH, the solution was shaked well and transferred to the measurement cell to ensure complete complexation. The concentration of metal ion determined spectrophotometrically and obtained the recovery percentage.

3. RESULTS AND DISCUSSION

The results of elemental analyses of the ligand and its

complexes are in good agreement with those required by the proposed formulae giving in **Table 1**. The molecular modeling of Schiff base ligand (using Chem3D ultra 8.0 program) shows that the bond lengths of all bonds in the left and right hand sides are typical due to the similarity of moieties on the two sides. Mass spectrum (**Figure 1**) of the new Schiff base showed a signal with m/z = 301 which is close to the calculated formula m/z = 294, supporting the structures of the Schiff base. The fragments observed were in good agreement with the proposed formula (**Scheme 1**).

3.1. Molar Conductance Measurements

The metal (II)/(III) complexes were dissolved in DMSO and the molar conductivities of 10^{-3} M of their solutions were measured at room temperature. The conductance values of the M(III) complexes support their electrolytic nature (1:1) types while M(II) complexes are non ionic in nature and non electrolytes. The data are listed in **Table 1**.

3.2. IR Spectral Studies

IR spectra in the region 4000 - 400 cm^{-1} have been recorded for the ligand and its complexes and the assignments of the observed frequencies have been made to specific group vibrations by comparison with the spectra of related complexes. The IR spectrum of the ligand showed a shoulder at 1597 cm^{-1} assigned to $v(C = N)$ of the azomethine stretching vibration, beside a band at 1049 cm^{-1} due to C-O-C stretching frequency. These two bands are shifted to higher or lower wave numbers in the complexes indicating the participation of the azomethine nitrogen in coordination [29] and the oxygen of the ether group, **Table 2**. On the other hand, the IR spectra of the complexes exhibited new non-ligand bands in the range 432 - 467 cm^{-1} and in the range 492 - 594 cm^{-1} assigned as M-O and M-N stretching vibrations, respectively. Therefore, it can be concluded that (L) ligand binds to the metal ions through azomethine N and the ether O and the ligand behaves as neutral tetradentate ligand.

Table 1. Analytical and physical data of L ligand and its metal complexes.

Compound	Colour (%yield)	M.P. (°C)	% Found (calcd.)					$\mu_{eff.}$ (B.M)	Λ_m $\Omega^{-1} \cdot mol^{-1} \cdot cm_2$
			C	H	N	M	Cl		
L = ($C_{18}H_{18}N_2O_2$)	Pale Yellow (82)	154	73.75 (73.47)	6.02 (6.12)	9.82 (9.52)	-----	-----	-----	-----
[Cr(L)Cl$_2$]Cl	Pale Yellow (88)	122	47.91 (47.75)	3.41 (3.98)	6.57 (6.19)	11.65 (11.47)	23.48 (23.54)	3.64	103.9
[Fe(L)Cl$_2$]Cl	Brown (77)	148	47.16 (47.34)	3.76 (3.94)	6.49 (6.14)	12.58 (12.22)	23.39 (23.34)	5.44	104.5
[Co(L)Cl$_2$]	Green (83)	114	50.51 (50.96)	4.45 (4.25)	6.25 (6.60)	13.42 (13.89)	16.48 (16.75)	4.49	32.51
[Ni(L)Cl$_2$]	Yellowish Green (89)	120	50.96 (50.98)	4.31 (4.25)	6.88 (6.61)	13.50 (13.85)	16.49 (16.76)	3.21	31.25
[Cu(L)Cl$_2$]	Green (74)	130	50.74 (50.41)	4.90 (4.20)	6.84 (6.53)	14.63 (14.82)	16.66 (16.57)	1.99	35.14
[Zn(L)Cl$_2$]	Pale Yellow (81)	260	50.34 (50.19)	4.52 (4.18)	6.08 (6.51)	15.04 (15.19)	16.79 (16.50)	-----	17.92

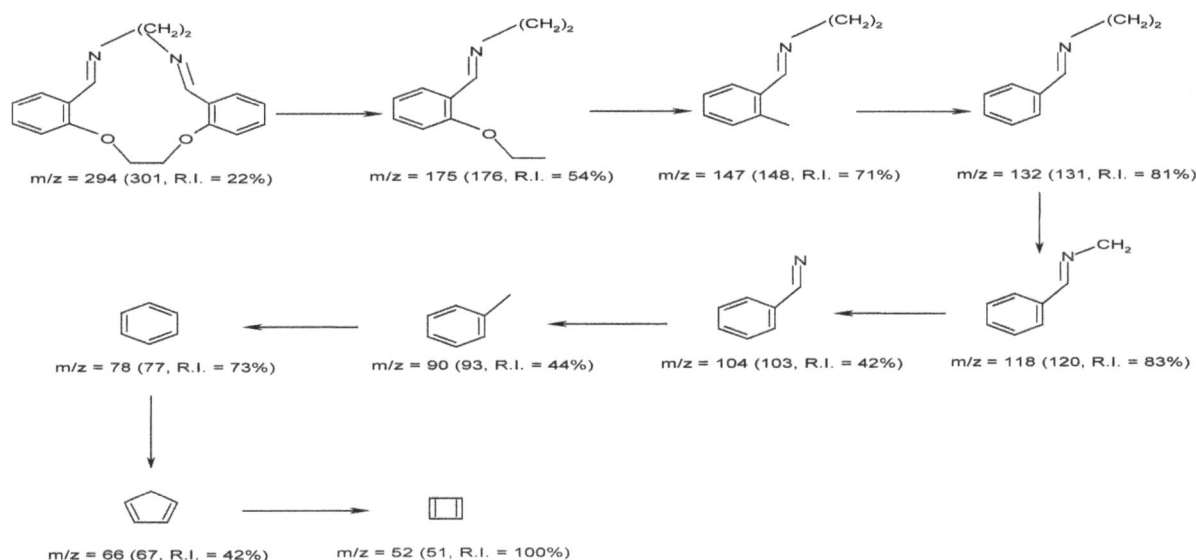

m/z = 294 (301, R.I. = 22%)   m/z = 175 (176, R.I. = 54%)   m/z = 147 (148, R.I. = 71%)   m/z = 132 (131, R.I. = 81%)

m/z = 78 (77, R.I. = 73%)   m/z = 90 (93, R.I. = 44%)   m/z = 104 (103, R.I. = 42%)   m/z = 118 (120, R.I. = 83%)

m/z = 66 (67, R.I. = 42%)   m/z = 52 (51, R.I. = 100%)

Scheme 1. Mass fragmentation pattern of L ligand.

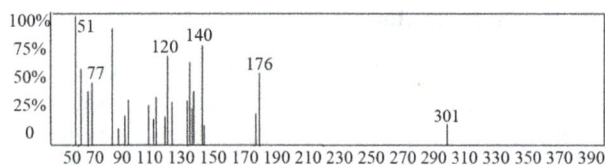

Figure 1. Mass spectra of Schiff base ligand (L).

Table 2. IR data (4000 - 400 cm^{-1}) of L ligand and its metal complexes.

Compound	υ(HC=N)$_{azomethine}$	(C-O-C)$_{ether}$	υ(M-N)$_{azo}$	υ(M-O)$_{ether}$
L	1597sh	1049sh	------	-----
[Cr(L)Cl$_2$]Cl	1611sh	1061sh	502s	440s
[Fe(L)Cl$_2$]Cl	1590sh	1038sh	498m	464m
[Co(L)Cl$_2$]	1593m	1057sh	505w	432s
[Ni(L)Cl$_2$]	1600sh	1038m	521s	436s
[Cu(L)Cl$_2$]	1589sh	1053m	502w	436m
[Zn(L)Cl$_2$]	1601m	1049m	594w	467s

Sh = sharp; m = medium; br = broad; s = small; w = week.

3.3. H^{1}NMR Spectra

The H^{1}NMR spectra of Schiff base ligand (L) is recorded in d$_6$-dimethylsulfoxide (DMSO) solution using tetramethylsilane (TMS) as internal standard. The chemical shifts of the different types of protons found in the ^{H1NMR}spectra of the Schiff base ligand is compared with its diamagnetic Zn(II) complex and shown in **Table 3**. It is found that while the ^{H1NMR}signal of (HC=N) is shifted due to complexation, the signals due to the ether CH$_2$-O does not show a significant shift in the chelation mode.

3.4. Electronic Absorption Spectra and Magnetic Moments of the Metal Complexes

The Uv-vis spectrum of the ligand showed two bands at 330 and 370 nm which are assigned to π-π* and n-π*, transition, respectively. The Uv-vis spectra of 10^{-4} M of the metal complexes, **Figure 2**, display similar absorption spectra of the ligand which are shifted to lower and higher wavelengths beside a decrease or disappearance of the peak dye to n-π* transition which confirm the coordination through azomethine nitrogen. On the other hand, π-π* transition showed shoulder at 337 nm with new band at 320 nm due to complexation. Also the d-d transition in this type of complexes may appear above 500 nm but does not appear due to the low intensity of the d-d transition. The data listed in **Table 1** summarized the measured magnetic moment values. These data support the octahedral geometry of the complexes.

Table 3. ^{H1NMR}spectral data of the organic ligand and its metal chelates.

Compound	Chemical shift, (δ) ppm	Assignment
L	4.434	(m, 2H, -OCH$_2$)
	2.725 - 2.888	(m, 2H, -C=N-CH$_2$)
	6.885 - 7.950	(m, 4H, ArH)
	8.479 - 8.493	(m, 1H, azomethineH)
[Zn(L)Cl$_2$]	4.435	(m, 2H, -OCH$_2$)
	2.729 - 2.887	(m, 2H, -C=N-CH$_2$)
	6.943 - 7.821	(m, 4H, ArH)
	8.546	(s, 1H, azomethineH)

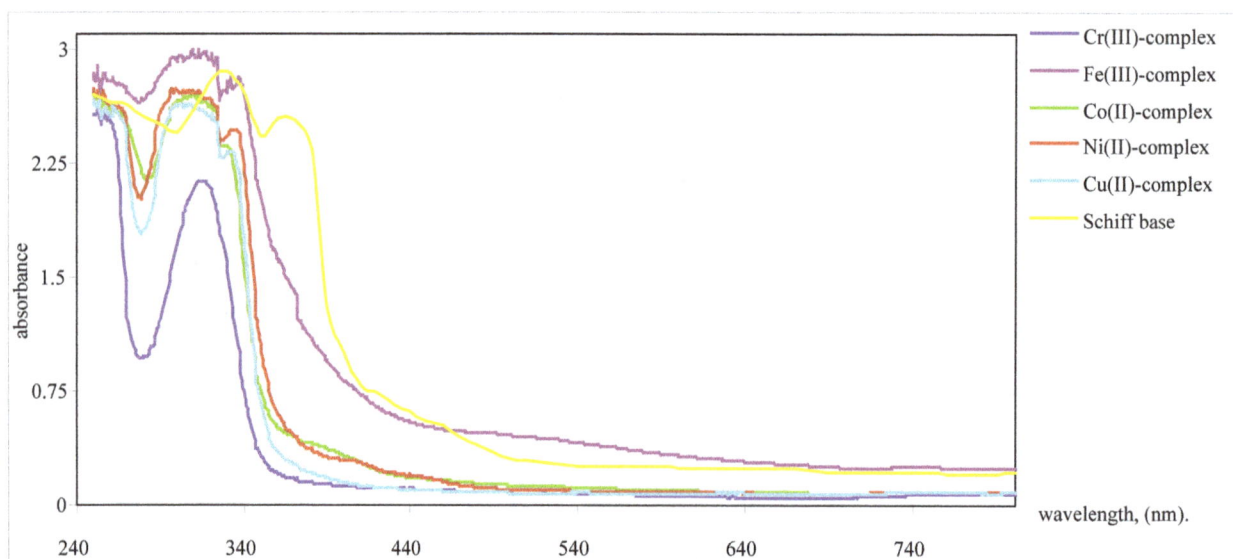

Figure 2. Uv-vis spectrum of the Schiff base and its metal complexes.

3.5. Thermal Analyses (TG and DTG) Studies

The Schiff base ligand exhibits a first estimated mass loss of 13.33% (calcd. 12.93%) at 25°C - 140°C, which may be attributed to the liberation of $2NH_3$ and $2H_2$ molecules as gases. In the following stages within the temperature range from 140°C - 750°C, the organic part $C_{17}H_8$ with CO_2 gas are lost with an estimated mass loss of 86.67% (calcd. 87.07%). The thermogram of Fe(III)-L chelate show three decomposition steps within the temperature range from 50°C - 790°C. The first step of decomposition within the temperature range from 50°C - 245°C corresponds to the loss of 2HCl gases with an estimated mass loss of 16.18% (calcd. 16.00%). While the second step occurs within the temperature range from 245°C - 520°C and corresponds to the removal of HCl,

CO and $2NH_3$ as gases with a mass loss of 21.91% (calcd. 21.58%). The final step (520°C - 790°C) corresponds to the removal of the organic part of the ligand leaving metal oxide as a residue. The overall weight loss amounts to 84.37% (calcd. 84.27%). The TG curve of the cobalt complex shows five stages of decomposition within the temperature range of 25°C - 705°C. The first three stages at 25°C - 270°C correspond to the loss of 2HCl as gases in the first step and NO and NH_3 as gases in the next steps, while the final stages involve the loss of organic molecule leaving the metal oxide as residue. The overall weight loss amounts to 82.90% (calcd. 82.33%). The Cu(II) complex decomposed in four steps. The overall weight loss amounts to 82.22% (calcd. 81.45%), **Table 4** and **Figure 3**.

Table 4. Thermoanalytical results of Schiff base and its metal complexes.

Compound	TG Range (°C)	DTG_{max} (°C)	n^*	Mass Loss (Estim) %	Total Mass Loss Calcd	Assignment	Metallic Residue
L		62					
	25 - 140	318,399	1	12.93 (13.33)		-Loss of $2NH_3$ and $2H_2$	------
	140 - 750	501,612	4	87.07 (86.67)	100 (100)	-Loss of CO_2 and $C_{17}H_8$	
[Fe(L)Cl$_2$]Cl	50 - 245	187	1	16.00 (16.18)		-Loss of 2HCl	FeO
	245 - 520	384	1	21.59 (21.91)		-Loss of HCl, CO and $2NH_3$	15.74
	520 - 790	681	1	46.68 (46.28)	84.27 (84.37)	-Loss of $C_{17}H_9$	(15.63)
[Co(L)Cl$_2$]	25 - 270	30,205	2	17.22 (17.03)		-Loss of 2HCl	CoO
	270 - 415	362	1	11.09 (11.58)		-Loss of NO and NH_3	17.67
	415 - 705	625,937	2	54.02 (54.29)	82.33 (82.90)	-Loss of $C_{18}H_{13}$	(17.10)
[Cu(L)Cl$_2$]						-Loss of 2HCl	CuO
	35 - 220	125,229	2	17.04 (17.84)		-Loss of $2NH_3$, $\frac{1}{2}O_2$ and	18.55
	220 - 775	345,593	2	64.41 (64.38)	81.45 (82.22)	$C_{18}H_{10}$	(17.78)

n^* = number of decomposition steps.

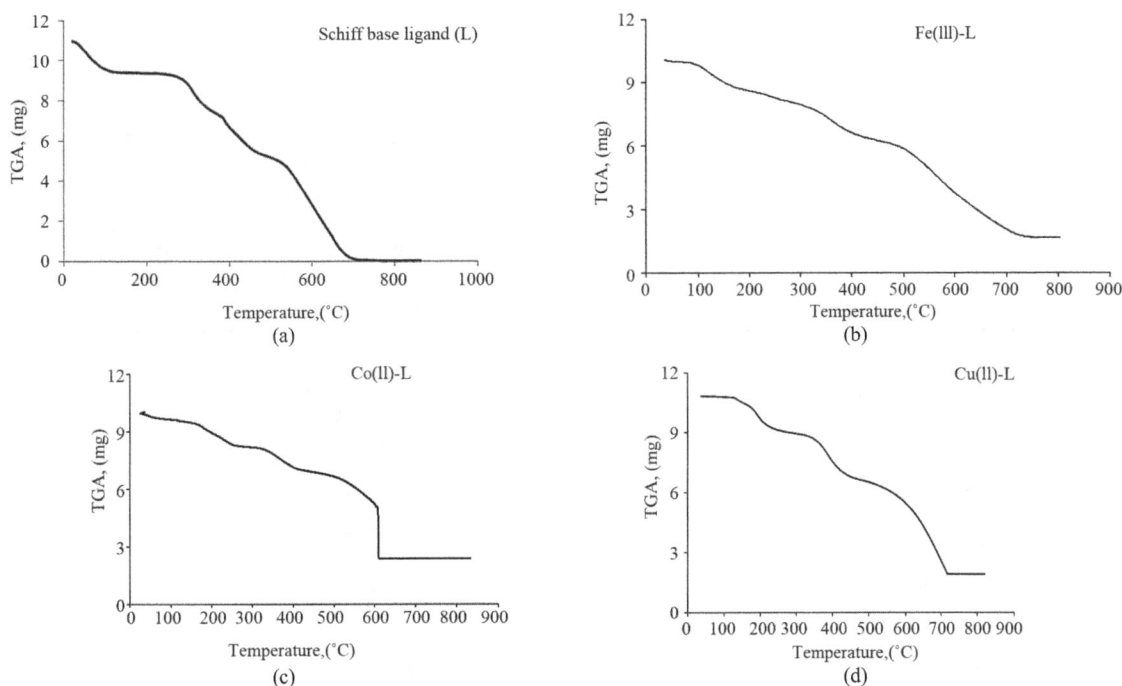

Figure 3. Thermal analyses of (a) Schiff base ligand (L), (b) Fe(III)-L, (c) Co(II)-L and (d) Cu(II)-L. complexes.

3.6. Calculation of Activation Thermodynamic Parameters

The thermodynamic activation parameters of decomposition processes of the complexes are evaluated graphically by employing the Coats-Redfern relation [30].

$$\log\left[\log\left\{W_f/(W_f - W)\right\}/T^2\right] = \log\left[AR/q\theta E^*\left(1 - 2RT/E^*\right)\right] - E^*/2.303RT$$

where W_f is the mass loss at the completion of the reaction, W is the mass loss up to temperature T; R is the gas constant, E^* is the activation energy in KJ/mol and θ is the heating rate. Since, $1\text{-}2RT/E^* \cong 1$, a plot of the left hand side of the equation against $1/T$ was drawn and the data obtained are given in **Table 5**. The high values of the activation energies reflect the thermal stability of the complexes. The entropy of activation is found to have negative values in all the complexes which indicate that the decomposition reactions proceed with a lower rate than the normal ones.

3.7. X-Ray Diffraction

The trend of the curves decreases from maximum to minimum intensity indicating that all the metal complexes are amorphous in nature in the present metal-ligand formation except the Cr(III), Ni(II), Cu(II) and Zn(II) chelates they have crystalline phase.

3.8. Structures Interpretation

On the basis of the above observations, octahedral geometry is suggested for the investigated complexes. The structures of the complexes are shown in **Figure 4**.

Table 5. Thermodynamic data of the thermal decomposition of the Schiff base ligand and its metal complexes.

Complex	Decomp. Temp. °C	A s^{-1}	E* kJ·mol^{-1}	ΔS* J·mol^{-1}	ΔH* kJ·mol^{-1}	ΔG* kJ·mol^{-1}
L	25 - 140	1.34×10^5	72.60	−33.00	72.10	80.40
	140 - 220	8.53×10^6	26.20	−52.00	93.80	133.4
	220 - 445	7.78×10^5	82.20	−104.0	128.8	152.0
	445 - 630	6.75×10^{10}	95.30	−141.0	191.1	212.0
	630 - 750	2.91×10^{11}	90.80	−181.0	239.9	262.0
[Fe(L)Cl$_2$]Cl	35 - 185	6.15×10^4	43.60	−46.00	72.60	90.30
	185 - 220	2.69×10^{11}	79.90	−83.00	106.9	115.0
	220 - 470	1.90×10^7	42.00	−106.0	139.1	160.0
	470 - 775	2.91×10^6	82.40	−146.0	177.4	202.0
[Co(L)Cl$_2$]	25 - 160	2.27×10^9	141.0	−46.00	63.80	82.00
	160 - 270	6.28×10^{10}	172.0	−75.00	95.50	122.7
	270 - 415	2.69×10^8	156.0	−125.0	152.6	163.4
	415 - 650	2.26×10^8	114.0	−191.0	210.9	266.0
	650 - 705	2.09×10^5	180.0	−252.0	280.2	318.3
[Cu(L)Cl$_2$]	50 - 245	3.19×10^9	68.30	−59.0	82.70	106.3
	245 - 520	1.31×10^{11}	106.0	−134.0	174.5	205.0
	520 - 790	1.74×10^5	116.0	−181.0	211.0	243.0

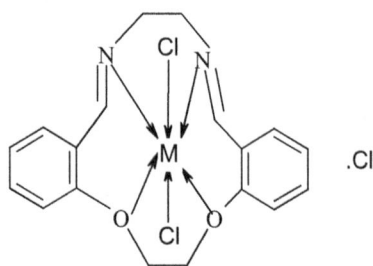

M = Ni(II), Co(II), Cu(II) and Zn(II).

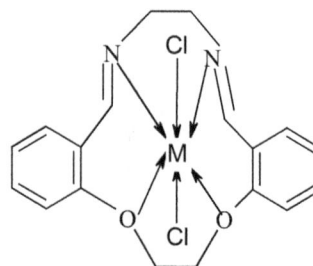

M = Cr(III) and Fe(III).

Figure 4. Structural formula of the metal complexes.

3.9. Biological Activity

The main aim of the production and synthesis of any antimicrobial compound is to inhibit the causal microbe without any side effects on the patients. In addition, it is worthy to stress here on the basic idea of applying any chemotherapeutic agent which depends essentially on the specific control of only one biological function and not multiple ones. The antibacterial activity of the parent Schiff base and its metal complexes against *Aspergillus flavus* and *Canidia albicans* fungi and, *Staphylococcus* (G^+) and *E. coli* (G^-) bacteria was tested in order to asses their potential antimicrobial agents. The biological activeity of the Schiff base ligand and its metal complexes were also compared with tetracycline (Antibacterial agent) and Amphotericin B (Antifungal agent). The data are listed in **Table 6** and **Figure 5**. According to the data it can be seen that the ligand did not show any biological activity, while all the complexes showed activity against *Staphylococcus* (G^+) and *E. coli* (G^-) bacteria and *Canidia albicans* fungi. On the other hand, no activity is observed for *Aspergillus flavus* fungi by the ligand or the complexes [31]. This means that the activity of the newly prepared Schiff base against different microorganisms is enhanced with chelation with different biological active metals.

4. APPLICATION

A New Spectrophotometric Determination to Fe(III) Ions in Natural Water

The UV-vis spectrum of the Fe(III) solution in the presence of the Schiff base ligand was measured from 250 to 900 nm at an interval of 1 nm and an average measuring time of 1 s per wavelength and the maximum wavelength is found at 313 nm. Several factors influence on the recovery method are studied for [Fe(III)-L] system in order to achieve maximum recovery. The optimum factors are pH = 3, 25°C, 1:1 (M:L) metal to ligand ratio and 30 min. In order to assess the applicability of the proposed method to recover Fe(III) from spiked water samples, the effect of some foreign ions was also investigated. These foreign ions were selected on the basis that they are normally present in natural water samples. Solutions containing various amounts of foreign ions, Fe(III), and the ligand were subjected to the described spectrophotometric procedure. The tolerable amounts of each ion, given a maximum error up to 1.23% in the recovery, are summarized in **Table 7**. As can be seen, all the investigated foreign ions with a relatively high concentration have no adverse effect on the ferric ions determinate.

Recovery tests were carried out on different types of water samples in order to assess its accuracy. The selection of these samples was done in a way to provide a wide variety of sample matrices characterized by different types of interferents. Solutions of water samples containing Fe(III) with varying a concentration were recovered with the newly Schiff base synthesized under the recommended conditions and the data listed in **Table 8** showed satisfactory recovery of Fe(III). The standard deviation and relative standard deviation of the method were also calculated and their values indicate that, this method has a greater accuracy where the small values for the standard deviation attest to the procedure's fairly high degree of precision. Therefore, the synthesis of these complexes might be established a new line for search to new ligand for determination of ferric ion in natural water.

Table 6. Biological activity of L and its metal complexes.

	Sample	Inhibition Zone Diameter (mm)			
		Escherichia coli (G^-) Mean ± RSD	*Staphylococcus aureus* (G^+) Mean ± RSD	*Aspergillus favus* (Fungus) Mean ± RSD	*Candida albicans* (Fungus) Mean ± RSD
	Control: DMSO	0.0	0.0	0.0	0.0
Standard	Tetracycline (Antibacterial agent)	33	30	----	----
	Amphotericin B (Antifungal agent)	----	----	18	19
	L	0.00	0.00	0.0	0.00
	[Cr(L)Cl$_2$]Cl	14.33 ± 0.011	14.43 ± 0.041	0.0	12.433 ± 0.191
	[Fe(L)Cl$_2$]Cl	13.23 ± 0.019	13.067 ± 0.069	0.0	11.500 ± 0.157
	[Co(L)Cl$_2$]	21.16 ± 0.007	21.667 ± 0.071	0.0	14.000 ± 0.286
	[Ni(L)Cl$_2$]	16.40 ± 0.022	22.333 ± 0.068	0.0	16.000 ± 0.250
	[Cu(L)Cl$_2$]	13.13 ± 0.012	14.000 ± 0.143	0.0	13.000 ± 0.428
	[Zn(L)Cl$_2$]	20.23 ± 0.010	25.667 ± 0.098	0.0	12.333 ± 0.528

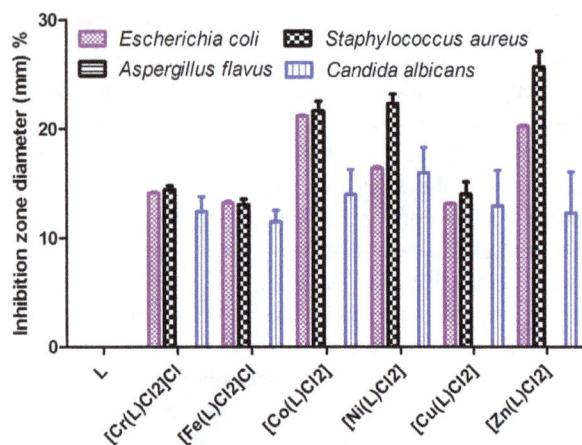

Figure 5. Percent of inhibition zone diameter (mm) relative to used standard vs. the Schiff base ligand (L) and its metal complexes.

Table 7. Effect of some foreign ions on the recovery of 5×10^{-4} mol·L^{-1} Fe(III) using 5×10^{-4} mol·L^{-1} of L ligand at pH = 3, t = 30 min, T = 25°C and λ_{max} = 313 nm.

Foreign Ion Added	[Foreign Ion]/[Fe(III)]	R(%)
Ag(I)	2000	98.77
Cd(II)	0.200	99.45
NH(I)	20.00	99.62
Sn(IV)	2000	98.89
Al(III)	200.0	99.50
K(I)	20.00	99.43
Hg(I)	20.00	99.49
Ca(II)	20.00	99.47
Na(I)	200.0	99.50
Cu(II)	20.00	99.44
Co(II)	0.200	99.68
Ni(II)	20.00	99.38
Zn(II)	200.0	99.50
Cl$^-$	200.0	99.10
Mg(II)	2000	99.77

Table 8. Recovery percent (%) of Fe(III) in spiked natural water samples using 5×10^{-4} mol·L^{-1} at pH = 3, t = 30 min, T = 25°C and λ_{max} = 313 nm.

Water Samples (location)	Fe(III)$_{spiked}$ (mg·L^{-1})	Fe(III)$_{recovered}$ (mg·L^{-1})	Recovery Percent (%)	SD $\times 10^{-3}$	RSD (%)
Distilled water	4.65	4.641	99.81	1	14.14
	8.42	8.409	99.87		
Tap water (our Lab.)	4.65	4.628	99.53	1	6.734
	8.42	8.400	99.76		
Nile water (Egypt)	4.65	4.631	99.59	4	16.44
	8.42	8.396	99.71		
Sea water (Gamasah)	4.65	4.638	99.74	1	10.88
	8.42	8.406	99.83		
Underground water (Belqas)	4.65	4.629	99.55	3	14.89
	8.42	8.403	99.80		
Well water (6th October City)	4.65	4.620	99.35	5	18.68
	8.42	8.397	99.73		

SD = standard deviation and RSD = relative standard deviation.

5. CONCLUSION

Synthesis and characterization of complexes containing N_2O_2-donor tetradentate Schiff base ligand have been described in this paper. The spectroscopic data of the complexes give good evidence of proposed structure. The results of the antibacterial screening of the test compounds indicate mild to moderate bactericidal activities and *Canidia albicans* fungi. Spectrometry procedure used to determine of iron (III) in different types of natural water within recovery test including with stable solution of the newly synthetic Schiff base as selective reagent by highly sensitive, selective and simple method than others.

REFERENCES

[1]   Chen, D. and Martell, A.E. (1987) Dioxygen affinities of synthetic cobalt Schiff base complexes. *Inorganic Chemistry*, **26**, 1026-1987. doi:10.1021/ic00254a013

[2]   Collman, J. and Hegedus, L.S. (1980) Principles and application of organotransition metal chemistry. University Science Book, Sausalito.

[3]   Zhao, J., Zhao, B., Liu, J., Xu, W. and Wang, Z. (2001) Spectroscopy study on the photochromism of Schiff bases N,N'-bis(salicylidene)-1,2-diaminoethane and N,N'-bis-(Salicylidene)-1,6-exanediamine. *Spectrochimica Acta Part A: Molecular and Biomolecular Spectroscopy*, **57**, 149-154. doi:10.1016/S1386-1425(00)00353-X

[4]   Zgierski, M.Z. and Grabowska, A. (2000) Theoretical approach to photochromism of aromatic Schiff bases: A minimal chromophore salicylidene methylamine. *Journal of Chemical Physics*, **113**, 7845. doi:10.1063/1.1316038

[5]   Sawodny, W.J. and Riederer, M. (1977) Addition compounds with polymeric chromium(II)-schiff base complexes. *Agnewandate Chemie*, **16**, 859-860. doi:10.1002/anie.197708591

[6]   Panda, B.K. and Chakravorty, A. (2005) Spectroscopic properties of inorganic and organometallic compounds. *Journal of Organometallic Chemistry*, **690**, 3169-3175. doi:10.1016/j.jorganchem.2005.04.012

[7]   Krishnapriya, K.R. and Kandaswamy, M. (2005) Coordination properties of a dicompartmental ligand with tetra- and hexadentate coordination sites towards copper(II) and nickel(II) ions Polyhedron. *Polyhedron*, **24**, 113-120. doi:10.1016/j.poly.2004.10.010

[8]   Kumar, K.N. and Ramesh, R. (1999) Synthesis of lanthanide(III) complexes of chloro- and bromo-substituted 18-memberedtetraazamacrocycles. *Polyhedron*, **18**, 1561-1568.

[9]   Rekha, S. and Nagasundara, K.R. (2006) Complexes of the Schiff base derived from 4-amino-phenyl benzimidazole and 2,2'-dehydropyrollidene-N-aldehyde with Zn(II), Cd(II) and Hg(II) halides. *Indian Journal of Chemistry*, **45**, 2421-2425.

[10] Cotton, F.A., Wilkinson, G., Murillo, C.A. and Bochmann, M. (1999) Advanced inorganic chemistry. 6th Edition, Wiley, New York.

[11] Shearer, J.M. and Rokita, S.E. (1999) Diamine preparation for synthesis of a water soluble Ni(II) salen complex. *Bioorganic & Medicinal Chemistry Letters*, **9**, 501-504. doi:10.1016/S0960-894X(99)00020-7

[12] Singh, K., Barwa, M. S. and Tyagi, P. (2006) Synthesis, characterization and biological studies of Co(II), Ni(II), Cu(II) and Zn(II) complexes with bidentate Schiff bases derived by heterocyclic ketone. *European Journal of Medicinal Chemistry*, **41**, 147-153. doi:10.1016/j.ejmech.2005.06.006

[13] Majumder, A., Rosair, G.M., Mallick, A., Chattopadhyay, N. and Mitra, S. (2006) Synthesis, structures and fluorescence of nickel, zinc and cadmium complexes with the N,N,O-tridentate Schiff base N-2-pyridylmethylidene-2-hydroxy-phenylamine. *Polyhedron*, **25**, 1753-1762. doi:10.1016/j.ica.2005.12.055

[14] Freiria, A., Bastida, R., Valencia, L., Macias, A. and Lodeiro, C. (2006) Metal complexes with two tri-aza, tri-oxa pendant-armed macrocyclic ligands: Synthesis, characterization, crystal structures and fluorescence studies. *Inorganica Chimica Acta*, **359**, 2383-2394. doi:10.1016/j.ica.2005.12.055

[15] Faniran, J. A., Patel, K. S. and Bailar, J. C. (1974) Infrared spectra of N,N'-bis(salicylidene)-1,1-(dimethyl)-ethylenediamine and its metal complexes. *Journal of Inorganic and Nuclear Chemistry*, **36**, 1547-1551. doi:10.1016/0022-1902(74)80621-4

[16] Yano, S., Takizawa, S., Sugita, H., Takahashi, T., Shioi, H., Tsubomura, T. and Yoshikawa, S. (1985) Reactions of metal complexes with carbohydrates: Synthesis and characterization of novel nickel(II) complexes containing glycosylamines derived from a monosaccharide and a diamine. An X-ray crystallographic study of (ethylenediamine) {N-(2-aminoethyl)-d-fructopyranosylamine} nickel(II)-Cl$_2$-CH$_3$OH. *Carbohydrate Research*, **142**, 179-193. doi:10.1016/0008-6215(85)85021-7

[17] Bian, H.D., Xu, J.Y., Gu, W., Yan, S.P., Liao, D.Z., Jiang, Z.H. and Cheng, P. (2003) Synthesis, structure and properties of terephthalate-bridged copper(II) polymeric complex with zigzag chain. *Inorganic Chemistry Communications*, **6**, 573-576.

[18] Mohamed, G.G., Omar, M.M. and Hindy, A.M. (2006) Metal complexes of Schiff basess, preparation, characterization and biological activity. *Turkish Journal of Chemistry*, **30**, 361-382.

[19] El-Ayaan, U., El-Metwally, N.M., Youssef, M.M. and El Bialy, S.A.A. (2007) Perchlorate mixed-ligand copper(II) complexes of β-diketone and ethylene diamine derivatives: Thermal, spectroscopic and biochemical studies. *Spectrochimica Acta Part A: Molecular and Biomolecular Spectroscopy*, **68**, 1278-1286.

[20] Surati, K.R. and Thake, B.T. (2010) Synthesis, spectral, crystallography and thermal investigations of novel Schiff base complexes of manganese(III) derived from heterocyclic β-diketone with aromatic and aliphatic diamine. *Spectrochimica Acta A: Molecular and Biomolecular Spectroscopy*, **75**, 235-242. doi:10.1016/j.saa.2009.10.018

[21] Bharty, M.K., Srivastava, A.K., Dulwere, R., Butcher, R.J. and Singh, N.K. (2011) Synthesis, spectral and X-ray structural studies of Ni(II) complexes of N'-acylhydrazine carbodithioic acid esters containing ethylenediamine or o-phenanthroline as coligands. *Polyhedron*, **30**, 990-996. doi:10.1016/j.poly.2010.12.043

[22] Ghazy, S.E., Kabil, M.A., Shallaby, A.M. and Ammar, N.S. (2001) Flotation-separation of the pollutant species of chromium, cadmium and lead from aqueous solutions and natural waters. *Indian Journal of Chemical Technology*, **8**, 211-218. doi:10.1021/ac00232a002

[23] Leyden, D.E. and Wegschider, W. (1981) Preconcentration for trace element determination in aqueous samples. *Analytical Chemistry*, **53**, 1059-1065.

[24] Zouboulis, A.I., Lazaridis, N.K. and Zamboulis, D. (1994) Powdered activated carbon separation from water by foam floation, *Separation Science and Technology*, **29**, 385-400. doi:10.1080/01496399408002490

[25] Lajunen, L.H.J. (1991) Spectrochemical analysis by atomic absorption and emission. *Royal Society of Chemistry*, Cambridge.

[26] Vogel, A.I. (1962) Quantitative inorganic analysis includeing elemental instrumental analysis. 2nd Edition, Longmans, London.

[27] Sari, N., Arslan, S., Logoglu, E. and Sakiyan, I. (2003) Antibacterial activities of some amino acid Schiff bases. *Journal of Animal Science*, **16**, 283.

[28] Batley, G.E. and Gardner, D. (1977) Sampling and storage of natural waters for trace metal analysis. *Water Research*, **11**, 745-756. doi:10.1016/0043-1354(77)90042-2

[29] Masoud, M.S., Hindawy, A.M. and Soayed, A.S. (1991) Structural chemistry of some new azo complexes. *Transition Metal Chemistry*, **16**, 372-376. doi:10.1007/BF01024086

[30] Coats, A.W. and Redfern, J.P. (1964) Kinetic parameters from thermogravimetric data. *Nature*, **201**, 68-79. doi:10.1038/201068a0

[31] Duraiswamy, S, Michael, R.E., Matthias, Z. and Karuppannan, N. (2007) Hydrolytic cleavage of Schiff bases by [RuCl$_2$(DMSO)$_4$]. *Polydedron*, **26**, 4314-4320.

10

Preparation mechanism and luminescence of Sr_2SiO_4:Eu phosphor from $(Sr,Eu)CO_3$@SiO_2 core-shell precursor

Yunsheng Hu[1], Weidong Zhuang[1*], Jianhua Hao[2*], Xiaowei Huang[1], Huaqiang He[1]

[1]National Engineering Research Center for Rare Earth Materials, General Research Institute for Nonferrous Metals, and Grirem Advcanced Materials Co., Ltd., Beijing, China
[2]Department of Applied Physics, The Hong Kong Polytechnic University, Hong Kong, China
Email: *wdzhuang@126.com; *apjhhao@polyu.edu.hk

ABSTRACT

Sr_2SiO_4:Eu phosphor for white light emitting diodes (LEDs) was synthesized by employing an as-prepared $(Sr,Eu)CO_3$@SiO_2 core-shell precursor as starting materials, and the effect of the core-shell precursor was also discussed on the crystal structure, particle morphology and luminescent properties of the resultant phosphor. The results showed that the hybrid β- and α'-Sr_2SiO_4: Eu phosphor with fine particle size and narrow distribution could be obtained at a lower firing temperature than that in conventional solid-state reaction method, and its formation mechanism was deduced to be $(Sr,Eu)CO_3$ diffusion controlled reaction process. Responded to its hybrid crystal structure, this phosphor exhibited the combined luminescence of β- and α'-Sr_2SiO_4:Eu.

Keywords: Phosphor; Precursor; Core-Shell Structure; Preparation Mechanism; Luminescence

1. INTRODUCTION

Since Eu^{2+}-activated alkaline earth orthsilicate phosphor was firstly reported on its fluorescence by Barry in 1968 [1], Me_2SiO_4:Eu (Me = Ca, Sr, Ba) had attracted little attention until the advent of white light emitting diodes (LEDs). Due to its high light conversion efficiency for near ultraviolet (NUV) and blue light, Me_2SiO_4:Eu has been an excellent commercial phosphor for white LEDs. Compared with the most popular $(Y,Gd)_3(Al,Ga)_5O_{12}$:Ce (YAG:Ce) phosphor for white LEDs, Me_2SiO_4:Eu is not only suitable for blue LED but also for NUV LED; moreover, Me_2SiO_4:Eu can produce more colorful emission to satisfy the demands of the white LEDs with lower color temperature and higher color rendering index [2-6].

Currently, the commercial Me_2SiO_4:Eu phosphor is produced by high temperature solid-state reaction method [5-10]. Such a method can achieve high light conversion efficiency of phosphor; however, it usually requires high firing temperature and introduction of flux to promote crystallization, which results in big particle size (>10 μm), broad particle distribution and irregular morphology, even flux contamination for phosphor. Taking the application properties into account, the phosphors for LEDs should have suitable particle size (<10 μm) and narrow distribution besides high brightness and desirable color coordinates. So some efforts have made to improve the particle properties of Me_2SiO_4:Eu phosphor by soft-chemistry methods [11-14]. For instance, Chang and co-authors synthesized nanometer Sr_2SiO_4 by employing $SrCO_3$@SiO_2 core-shell precursor as the starting materials; however, the luminescent center Eu was not considered and doped into the Sr_2SiO_4 host in their work [14]. In the previous work, we also developed a homogenous $(Sr,Eu)CO_3$@SiO_2 core-shell precursor, in which $(Sr,Eu)CO_3$ represents the homogenous mixture of Sr^{2+} and Eu^{3+} carbonates [15]. In this study, we employed this as-prepared core-shell precursor as starting materials to synthesize Sr_2SiO_4:Eu phosphor, and investigated its effect on the crystal structure, morphology and luminescent properties of the phosphor. Based on these results, the formation mechanism and the luminescence of Sr_2SiO_4:Eu was also discussed.

2. EXPERIMENTAL

The as-prepared $(Sr,Eu)CO_3$@SiO_2 core-shell precursor in our previous work [15] was directly employed as the starting materials to synthesize Sr_2SiO_4:Eu phosphor in the present work, wherein the molar ratio of Sr, Eu and Si was 1.9:0.1:1. To be specific, the core-shell precursor was put into an alumina crucible and directly fired in a horizontal tube furnace at 1000°C for 2 h under reducing atmosphere (95% N_2 + 5% H_2). And then the obtained phosphor powder was cooled to room temperature in the furnace for characterization.

FTIR measurements were performed on a Nicolet Magna-IR 760 Fourier transform infrared spectrometer using the standard KBr pellets technique, in the frequency interval 4000 - 400 cm^{-1}. X-ray diffraction (XRD) identification was determined by Burker D8 Advance X-ray powder diffractometer running Cu Kα radiation at 40 kV and 40 mA, and the XRD patterns were collected in the range of $15° \leq 2\theta \leq 65°$. The microstructure and morphology were detected by a JEOL JSM-6335F field emission scanning electronic microscope. The emission and excitation spectra of the phosphor were acquired by using Edinburgh FLS920P fluorescence spectrometer equipped with a 450W xenon lamp as an excitation source. All the measures were carried out at room temperature.

3. RESULTS AND DISCUSSION

Figure 1 shows the FTIR spectrum of the obtained phosphor sample (curve b). For comparison, the FTIR spectrum of (Sr,Eu)CO$_3$@SiO$_2$ precursor is also presented as curve (a). Obviously, the FTIR spectrum of the precursor mainly exhibits the characteristic vibrations of SiO$_2$ (1093, 796 and 471 cm^{-1}), CO$_3^{2-}$ group (692, 856, 1456 and 1749 cm^{-1}), CTAB (1630 and 2823 cm^{-1}) and H$_2$O (3435 cm^{-1}). After the precursor was fired into phosphor, the obtained sample shows different FTIR spectrum. As displayed in curve (b), the characteristic vibrations of CO$_3^{2-}$ group, CTAB and H$_2$O disappear, and the characteristic vibration bands of SiO$_2$ are replaced by multiple bands at 1000 - 800 cm^{-1} and 600 - 500 cm^{-1}, which could be attributed to the stretching and bending vibrations of Si-O bonds in SiO$_4$ tetrahedra, respectively [16]. It means that the vibration bands of the phosphor sample are shifted to lower frequency compared with that of pure SiO$_2$. As O/Si ratio increases from 2 (SiO$_2$) to 4 (orthosilicate), the Si-O bond length increases from 0.161 nm to 0.163 nm due to the presence of modifier oxides in the silica network, and the polymerization degree of SiO$_4$ tetrahedral gets lower, which generally corresponds to a lower vibration frequency in FTIR spectrum [17,18]. Meanwhile, it is noticed that the stretching vibrations of the phosphor sample cover a broader range (1300 - 600 cm^{-1}, the region between both dot lines in curve (b)) than that of pure SiO$_2$ (1300 - 880 cm^{-1}, the region between both dot lines in curve (a)) and split into three groups: 962, 908, and 839 cm^{-1}, which is probably arisen by the di- fferent NBO/Si ratio (NBO/Si: non-bridging oxygen per silicon) in SiO$_4$ tetrahedra [19].

To further investigate and clearly identify the phase structure, **Figure 2** depicts the XRD pattern of the phosphor sample obtained by directly firing (Sr,Eu)CO$_3$@SiO$_2$ precursor at 1000°C for 2 h. This pattern agrees well with the standard diffraction data of α'-Sr$_2$SiO$_4$ (JCPDS No. 39-1256) and β-Sr$_2$SiO$_4$ (JCPDS No. 38-0271), so it can be deduced that this phosphor sample is a mixture of

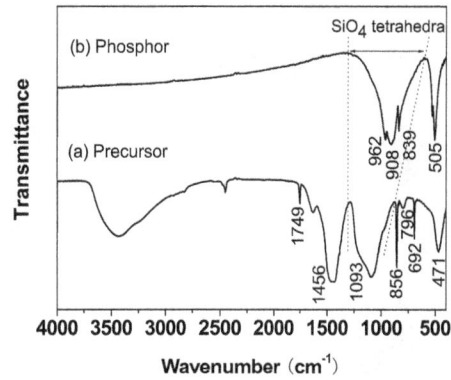

Figure 1. FTIR spectra of (Sr,Eu)CO$_3$@SiO$_2$ precursor (a) and the obtained Sr$_2$SiO$_4$:Eu phosphor sample (b).

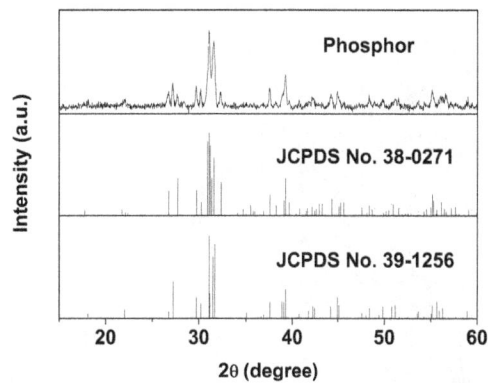

Figure 2. XRD pattern of Sr$_2$SiO$_4$:Eu phosphor sample prepared by directly firing (Sr,Eu)CO$_3$@SiO$_2$ precursor at 1000°C for 2 h.

orthorhombic α'-Sr$_2$SiO$_4$:Eu and monoclinic β-Sr$_2$SiO$_4$:Eu. The α' and β forms are the two modifications of Sr$_2$SiO$_4$, and the phase transition between low temperature β phase and high temperature α' phase occurs at about 358 K [20,21], whereas α' phase can also be stabilized at room temperature by substituting more Eu ($\geq$0.1) or small amounts of Ba^{2+} for Sr^{2+} [22,23]. In this work, the concentration of Eu activator is 0.1, so α'-Sr$_2$SiO$_4$:Eu is stably crystallized as well as β-Sr$_2$SiO$_4$:Eu as expected. Approximately estimated from the intensity of the diffraction peaks, α'-Sr$_2$SiO$_4$:Eu has much more content percentage than β-Sr$_2$SiO$_4$:Eu in this mixture. It is noted that not only is the present synthesized temperature (1000°C, no flux) lower than that in the conventional flux-assisted solid-state reaction method (usually 1300°C), but also the contamination of flux can be avoided due to the absence of flux.

To investigate the effect of core-shell precursor on the morphology of the resultant Sr$_2$SiO$_4$:Eu phosphor, **Figure 3** and **Figure 4** demonstrate the SEM images of the precursor and phosphor sample, respectively. In this precursor (**Figure 3**), most of SiO$_2$ are induced to coat on

(a)

(b)

Figure 3. SEM images of $(Sr,Eu)CO_3@SiO_2$ core-shell precursor observed at different magnifications: (a) × 5000; (b) × 30,000.

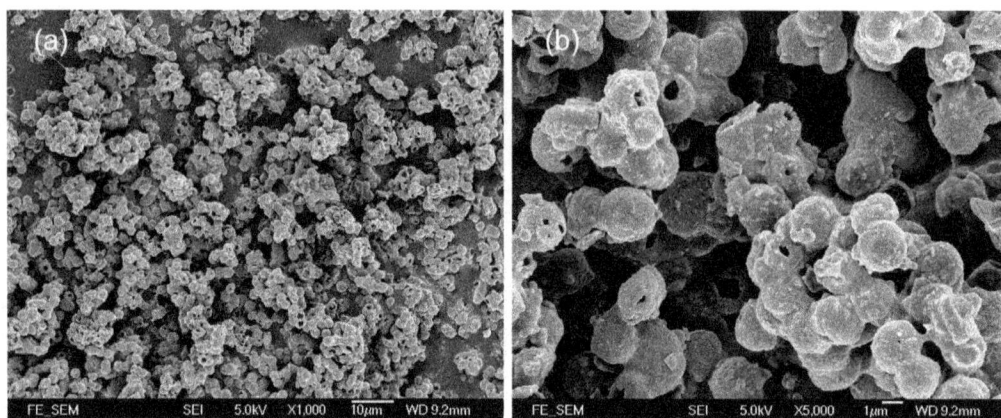

Figure 4. SEM images of Sr_2SiO_4:Eu phosphor sample observed at different magnifications: (a) ×1000; (b) × 5000.

the surface of $(Sr,Eu)CO_3$ core to form $(Sr,Eu)CO_3@SiO_2$ core-shell structure with SiO_2 shell layer about $100 \sim 200$ nm thickness; however, a few nucleate directly into SiO_2 nano-particles ($100 \sim 200$ nm) and locate onto the surface of core-shell structure. The formation mechanism and the composition of this precursor have been disclosed in our previous work [15]. It can be noted that this precursor has imperfect core-shell structure, and there appears an apparent gap without coated by SiO_2 on every core-shell particle. However, the existence of the gap accesses us to clearly understand the reaction mechanism between core and shell by morphology observation and comparison.

After the precursor was fired at 1000°C for 2 h, the obtained Sr_2SiO_4:Eu phosphor sample seems to inherit the external contour of the precursor. Overall, the particles of the phosphor still appear uniform and near-spherical morphology, as shown in **Figure 4(a)**, and the particle size is about 2 μm, slightly less than that of the precursor, which results from the condensation reaction between $(Sr,Eu)CO_3$ core and amorphous SiO_2 shell at

the high temperature. Observed from the enlargement figure (**Figure 4(b)**), the condensation reaction also leads to slight aggregation and adhesion among particles, but the profile of the single particle can be clearly recognized, which indicates that every $(Sr,Eu)CO_3@SiO_2$ particle in the core-shell precursor would act as a separate reaction unit, and it is of the suitable stoichiometrical ratio to situ produce an isolated Sr_2SiO_4:Eu micrometer particle. More noticeably, all the particles of the Sr_2SiO_4:Eu phosphor exhibit hollow morphology with an opening, and the wall thickness is slightly thicker than that of SiO_2 shell layer of the precursor. Combined with the results of XRD analysis, it seems that the shell of the precursor has transformed into thicker Sr_2SiO_4:Eu layer, while the core has disappeared past the chemical reaction at the high temperature. That is to say, the fact that the phosphor particles still maintain near single and hollow spherical shape suggests that $(Sr,Eu)CO_3$ would diffuse into SiO_2 shell and react into hollow Sr_2SiO_4:Eu phosphor until the exhaust of $(Sr,Eu)CO_3$ core. It can be deduced intuitively that this reaction mechanism can be regarded as $(Sr,Eu)CO_3$

diffusion controlled process, which is consistent with the theoretical inference in Lu and Wu' work [24].

The luminescent properties of α'-Sr$_2$SiO$_4$:Eu or β-Sr$_2$SiO$_4$:Eu have been researched widely [25-28]; however, the hybrid luminescence of the co-existent phase of α'- and β-Sr$_2$SiO$_4$:Eu has seldom been separately distinguished. In order to systematically understand the dependence of luminescent properties on the crystal structure of Sr$_2$SiO$_4$:Eu phosphor, we further investigated and analyzed the photoluminescence of this hybrid phosphor sample.

Figures 5(a) and **(b)** show the emission and excitation spectra of the phosphor sample, respectively. As illustrated in **Figure 5(a)**, the peak of the dominant emission band moves from 555 to 568 nm with the increase of the excitation wavelength from 320 to 450 nm. Whichever the excitation wavelength is, a turn at 532 nm is observed in the emission spectra as clearly shown in the inset of **Figure 5(a)**, which implies that the peak at 532 should be classified as a different emission band from the dominant one in the emission spectra. Both emission bands are the response to the co-existence of α'-Sr$_2$SiO$_4$:Eu and β-Sr$_2$SiO$_4$:Eu. The orientations of β and α' lattices are related by a simple rotation of x axis, and β-Sr$_2$SiO$_4$:Eu with stronger Si-O bond valence produces stronger crystal field and then yields shorter wavelength emission compared with α'-Sr$_2$SiO$_4$:Eu, so it can be deduced that the emission band centered at 532 nm originates from β-Sr$_2$SiO$_4$:Eu, and its emission peak position is independent on the excitation wavelength; while the dominant emission band at longer wavelength should be ascribed to α'-Sr$_2$SiO$_4$:Eu, and this emission can occur blue- or red-shift with the change of the excitation wavelength. It can be noted that α'-Sr$_2$SiO$_4$:Eu shows stronger emission intensity than β-Sr$_2$SiO$_4$:Eu in our sample, which is contrary to the results confirmed by many work [10,22,26]. The reason is that α'-Sr$_2$SiO$_4$:Eu has much more mass percentage than β-Sr$_2$SiO$_4$:Eu in this phosphor sample. Both of the emission band are produced by the 5d-4f transition of the activator Eu^{2+} occupied the ten oxygen coordinated Sr^{2+}(I). As well known, there is another Sr^{2+} site, Sr^{2+}(II), surrounded by nine oxygen in β- and α'-Sr$_2$SiO$_4$:Eu. When Eu^{2+} replaces Sr^{2+}(II), a shorter blue light emission band will be produced with peak at about 470 nm. As shown in **Figure 5(a)**, the 470 nm blue emission is obviously resolved as gibbous shoulder when the excitation wavelength is as short as 320 nm, while it disappears when the excitation wavelength is set at 400 or 450 nm. On the other hand, the phosphor sample has the strongest emission peak intensity when excited by 400 nm (NUV, near ultraviolet), and the intensity still maintains about 80% under 450nm blue light excitation. That is to say, the Sr$_2$SiO$_4$:Eu phosphor can applied to combine with NUV-LED or blue

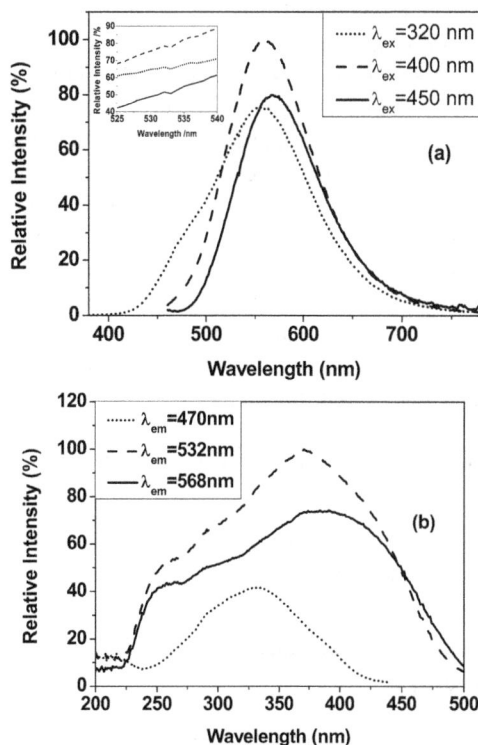

Figure 5. Emission (a) and excitation (b) spectra of the obtained Sr$_2$SiO$_4$:Eu phosphor sample.

LED.

The applicability of this Sr$_2$SiO$_4$:Eu phosphor for white LED can also be certificated by the excitation spectrum. Three excitation spectra were illustrated in **Figure 5(b)**, with monitoring emission wavelength at 568, 532 and 470 nm, respectively. Obviously, there is a very broad excitation band in 300 - 500 nm spectral regions when monitoring wavelength at 532 and 568 nm, which assures that this phosphor has high efficient light conversion for NUV and blue light. As for 470 nm blue emission, it has much lower efficiency, and its suitable excitation region is narrower (250 - 400 nm). That is the reason that 470 nm emission is nearly unresolved when this phosphor is excited by long-wavelength light.

4. CONCLUSIONS

The hybrid α'- and β-Sr$_2$SiO$_4$:Eu phosphor was successfully developed by firing (Sr,Eu)CO$_3$@SiO$_2$ core-shell precursor directly at 1000°C, and it appeared uniformly hollow near-spherical morphology with particle size about 2 μm. The morphology was resulted from the core-shell structure of the precursor and the reaction mechanism between (Sr,Eu)CO$_3$ core and SiO$_2$ shell. The mechanism was acquired visually to be (Sr,Eu)CO$_3$ diffusion controlled reaction process. Responded to its hybrid crystal structure, the phosphor exhibited the combined luminescence of α'- and β-Sr$_2$SiO$_4$:Eu. α'-Sr$_2$SiO$_4$:Eu has the longer emission wavelength, and its emission peak can blue- or red-shift with the change of the excitation wavelength; while the 532 nm emission of β-Sr_2SiO_4:Eu is independent on the excitation wavelength. Compared with the conventional high temperature solid-state reaction method, this method requires lower firing temperature and no flux contamination, and produces finer particles with narrow size distribution. More importantly, the reaction mechanism will provide some ideas to improve the particle performance of silicate phosphors, for instance, the spherical and solid Sr_2SiO_4:Eu phosphor could be obtained by employing a spherical SiO_2@ $(Sr,Eu)CO_3$ core-shell precursor as starting materials.

5. ACKNOWLEDGEMENTS

The authors would like to acknowledge the support from the National Hi-Tech. R&D Program of China (863 Program, 2010AA03A404, 2011AA03A101) and the Hong Kong Polytechnic University Grant (Grant No. J-BB9R).

REFERENCES

[1]   Barry, T.L. (1968) Fluorescence of Eu^{2+}-activated phases in binary alkaline earth orthosilicate systems. *Journal of the Electrochemical Society*, **115**, 1181-1184. doi:10.1149/1.2410935

[2]   Stefan, T., Peter, P., Gundula, R., Walter, T., Wolfgang, K. and Detlef, S. (2001) Light source comprising a light-emitting element. US Patent No. 6809347.

[3]   Park, J.K., Lim, M.A., Kim, C.H., Park, J.T. and Choi, S.Y. (2003) White light-emitting diodes of GaN-based Sr_2SiO_4:Eu and the luminescent properties. *Applied Physics Letters*, **82**, 683-685. doi:10.1063/1.1544055

[4]   Chen, L., Lin, C.C., Yeh, C.W. and Liu, R.S. (2010) Light converting inorganic phosphors for white light-emitting diodes. *Materials*, **3**, 2172-2195. doi:10.3390/ma3032172

[5]   Kim, J.S., Park, Y.H., Choi, J.C. and Park, H.L. (2005) Optical and structural properties of Eu^{2+}-doped $(Sr_{1-x}Ba_x)_2SiO_4$ phosphors. *Journal of the Electrochemical Society*, **152**, H135-H137. dio:10.1149/1.1971065

[6]   He, H., Fu, R.L., Zhang, X.L., Song, X.F., Zhao, X.R. and Pan, Z.W. (2009) Photoluminescence spectra tuning of Eu^{2+} activated orthosilicate phosphors used for white light emitting diodes. *Journal of Materials Science: Materials in Electronics*, **20**, 433-438. doi:10.1007/s10854-008-9747-5

[7]   Lee, J.H. and Kim, Y.J. (2008) Photoluminescent properties of Sr_2SiO_4:Eu^{2+} phosphors prepared by solid-state reaction method. *Materials Science and Engineering B*, **146**, 99-102. doi:10.1016/j.mseb.2007.07.052

[8]   Hsu, C., Jagannathan, R. and Lu, C. (2010) Luminescent enhancement with tunable emission in Sr_2SiO_4:Eu^{2+} phosphors for white LEDs. *Materials Science and Engineering: B*, **167**, 137-141. doi:10.1016/j.mseb.2010.01.045

[9]   Zhang, X.G., Tang, X.P., Zhang, J.L. and Gong, M.L. (2010) An efficient and stable green phosphor $SrBaSiO_4$:Eu^{2+} for light-emitting diodes. *Journal of Luminescence*, **130**, 2288-2292. doi:10.1016/j.jlumin.2010.07.006

[10] Guo, H., Wang, X.F., Zhang, X.B., Tang, Y.F., Chen, L.X. and Ma, C.G. (2010) Effect of NH_4F flux on structural and luminescent properties of Sr_2SiO_4:Eu^{2+} phosphors prepared by solid-state reaction method. *Journal of the Electrochemical Society*, **157**, J310-J314. doi:10.1149/1.3454723

[11] Kang, H.S., Hong, S.K., Kang, Y.C., Jung, K.Y., Shul, Y.G. and Park, S.B. (2005) The enhancement of photoluminescence characteristics of Eu-doped barium strontium silicate phosphor particles by co-doping materials. *Journal of Alloys and Compounds*, **402**, 246-250. doi:10.1016/j.jallcom.2005.04.143

[12] Hsu, W., Sheng, M. and Tsai, M. (2009) Preparation of Eu-activated strontium orthosilicate $(Sr_{1.95}SiO_4$:$Eu_{0.05})$ phosphor by a sol-gel method and its luminescent properties. *Journal of Alloys and Compounds*, **467**, 491-495. doi:10.1016/j.jallcom.2007.12.014

[13] Lei, B.F., Machida, K., Horikawa, T. and Hanzawa, H. (2010) Facile combustion route for low-temperature preparation of Sr_2SiO_4:Eu^{2+} phosphor and its photoluminescence properties. *Japanese Journal of Applied Physics*, **49**, 095001-095006. doi:10.1143/JJAP.49.095001

[14] Chang, Y.L., Hsiang, H., Lan, F.T., Mei, L.T. and Yen, F.S. (2010) Synthesis of Sr_2SiO_4 nanometer particles from the core-shell precursor of $SrCO_3/SiO_2$. *Journal of Alloys and Compounds*, **500**, 108-112. doi:10.1016/j.jallcom.2010.04.002

[15] Hu, Y.S., Hao, J.H., Zhuang, W.D., Huang, X.W. and He, H.Q. (2011) Synthesis of $(Sr,Eu)CO_3$@SiO_2 core-shell-like precursor for alkali earth silicate phosphors. *Journal of Rare Earths*, **29**, 911-914. doi:10.1016/S1002-0721(10)60567-4

[16] Chrysafi, R., Perraki, T. and Kakali, G. (2007) Sol-gel preparation of $2CaO·SiO_2$. *Journal of the European Ceramic Society*, **27**, 1707-1710. doi:10.1016/j.jeurceramsoc.2006.05.004

[17] Smith, J.V. and Bailey, S.W. (1963) Second review of Al-O and Si-O tetrahedral distances. *Acta Crystallographica*, **16**, 801-811. doi:10.1107/S0365110X63002061

[18] Ye, D.N., Li, Z. and He, W. (2001) Variation of the grand mean value of Si-O distances in metamorphic reactions. *Chinese Science Bulletin*, **46**, 702-704. doi:10.1007/BF03182841

[19] Park, J.H., Min, D.J. and Song, H.S. (2002) FT-IR Spectroscopic study on structure of $CaO-SiO_2$ and $CaO-SiO_2$-CaF_2 Slags. *ISIJ International*, **42**, 344-351. doi:10.2355/isijinternational.42.344

[20] Pieper, G., Eysel, W. and Hahn, T. (1972) Solid solubility and polymorphism in the system Sr_2SiO_4-Sr_2GeO_4-Ba_2GeO_4-Ba_2SiO_4. *Journal of American Ceramic Society*, **55**, 619-622. doi:10.1111/j.1151-2916.1972.tb13455.x

[21] Catti, M. and Gazzoni, G. (1983) The $\beta \rightleftarrows \alpha'$ phase transi-

tion of Sr_2SiO_4. II: X-ray and optical study, and ferroelasticity of the β form. *Acta crystallographica. Section B*, **39**, 679-684.
doi:10.1107/S0108768183003225

[22] Sun, X.Y., Zhang, J.H., Zhang, X., Luo, Y.S. and Wang, X.J. (2008) A green-yellow emitting β-Sr_2SiO_4:Eu^{2+} phosphor for near ultraviolet chip white-light-emitting diode. *Journal of Rare Earths*, **26**, 421-424.
doi:10.1016/S1002-0721(08)60109-X

[23] Nishioka, H., Watari, T., Eguchi, T. and Yada, M. (2011) Synthesis and luminescent properties of Sr_2SiO_4 phosphors. *IOP Conference Series: Materials Science and Engineering*, **18**, 102008.
doi:10.1088/1757-899X/18/10/102008

[24] Lu, C.H. and Wu, P.C. (2008) Reaction mechanism and kinetic analysis of the formation of Sr_2SiO_4 via solid-state reaction. *Journal of Alloys and Compounds*, **466**, 457-462.
doi:10.1016/j.jallcom.2007.11.066

[25] Wang, Z.J., Yang, Z.P., Guo, Q.L., Li, P.L. and Fu, G.S. (2009) Luminescence characteristics of Eu^{2+} activated Ca_2SiO_4, Sr_2SiO_4 and Ba_2SiO_4 phosphors for white LEDs. *Chinese Physics B*, **18**, 2068-2071.
doi:10.1088/1674-1056/18/5/057

[26] Lee, S.H., Koo, H.Y. and Kang, Y.C. (2010) Characteristics of α'- and β-Sr_2SiO_4:Eu^{2+} phosphor powders prepared by spray pyrolysis. *Ceramics International*, **36**, 1233-1238. doi:10.1016/j.ceramint.2010.01.007

[27] Won, Y.S. and Park, S.S. (2010) Density functional theory study on two-peak emission of Eu^{2+} activators in Sr_2SiO_4. *Journal of Physics and Chemistry of Solids*, **71**, 1742-1745. doi:10.1016/j.jpcs.2010.09.008

[28] Nguyen, H., Yeo, I. and Mho, S. (2010) Identification of the two luminescence sites of Sr_2SiO_4:Eu^{2+} and $(Sr,Ba)_2SiO_4$:Eu^{2+} phosphors. *ECS Transactions*, **28**, 167-173.
doi:10.1149/1.3367223.

11

Synthesis and characterization of a ladder-like structure compound formed by cadmium (II) and anionic nitronyl nitroxide

Jing Chen[1,2*], You-Juan Zhang[1], Kun-Tao Huang[2], Qiang Huang[2]

[1]College of Chemistry and Chemical Engineering, Anyang Normal University, Anyang, China
[2]School of Chemical Engineering and Energy, Zhengzhou University, Zhengzhou, China
Email: *chenjinghao2008@163.com

ABSTRACT

A ladder-like structure compound formed by cadmium (II) and anionic nitronyl nitroxide, [Cd$_2$(NITpBA)$_4$(H$_2$O)$_4$] (where NITpBA = 2-(4-carboxyphenyl)4,4,5,5-tetramethyl-4,5-dihydro-imidazol-1-oxyl-3-oxide), has been synthesized and characterized. X-ray analysis reveals that the compound 1 crystallizes in the monoclinic $P2_1/c$ (No. 14) space group, and [Cd$_2$(NITpBA)$_4$(H$_2$O)$_4$] units are linked into infinite chains by radical bridging ligands.

Keywords: Nitronyl Nitroxide; Anionic Radical; Ladder-Like Structure; Cd (II) Complex

1. INTRODUCTION

The field of molecule-based materials has attracted intense interest for decades [1]. The design and synthesis molecular-based magnetic materials by using metal-radical approach have become one of the very promising strategies [2]. The organic radicals can act as not only spin carriers but also building blocks. In this field, the increasing attention has been focused on the coordination complexes of paramagnetic metal ions with organic radical ligands in order to exploit new molecular magnetic materials and diverse topological structures [3,4]. The possibility to play with anionic nitronyl nitroxides increases the chances for designing magnetic molecular materials and diverse structures [5,6]. The stable anionic radical NITpBAH, named as 2-(4-carboxyphenyl)-4,4, 5,5-teramethyl-4,5-dihydro1H-imidazol-1-oxyl-3-oxide, has two different types of potentially metal-coordinating site. One is the carboxylate group; the other consists of the two oxygen atoms belonging to the nitronyl nitroxide

part. Recently, due to the increased recognition of its role in biological organisms, there has been an increasing interest in the coordination chemistry of cadmium [7].

With the purpose to explore a new polynuclear system derived from nitroxide radicals, we attempted to synthesize transition metal complexes incorporating dianion bridges. In this paper, we report the synthesis and the structural characterization of the first cadmium(II) complex with NITpBA$^-$, of formula [Cd$_2$ (NITpBA)$_4$(H$_2$O)$_4$] (NITpBA = 2-(4-carboxy-phenyl)-4,4,5,5-dihydro-imidazol-1-oxyl-3-oxide), in which a ladder-like one dimensional structure formed by cadmium (II) and anionic nitronyl nitroxide ligand.

2. EXPERIMENTAL

2.1. Materials and Measurements

All chemicals used were of reagent grade and used without further purification. The NITPBA was prepared by the literature method [8]. Elemental analysis for carbon, hydrogen and nitrogen were carried out on a Perkin-Elmer elemental analyzer model 240. The IR spectrum was taken on a Shimadzu 408 IR spectrophotometer in the 4000 - 400 cm^{-1} region with KBr pellets.

2.2. Synthesis of [Cd$_2$(NITpBA)$_4$(H$_2$O)$_4$]

To a stirred solution (10 mL) of Cd (CH$_3$COO)$_2$·2H$_2$O (0.2 mmol) in methanol was added a methanol solution (10 mL) of NITpBAH (0.4 mmol). The mixture was stirred continually for 1 h and then filtered. The filtrate was kept at room temperature for slow evaporation. After a few days, dark blue single crystals suitable for X-ray analysis appeared. Yield: 38%. Anal. Calcd. for C28 H36 Cd N4 O10 **1**: C, 47.93; H, 5.18; N, 7.99%. Found: C, 47.89; H, 5.13; N, 8.03%. IR (KBr disk): 1630 cm^{-1}

*Corresponding author.

$(v_{as}COO^-)$, 1380 cm^{-1} (v_sCOO^-), 1365 cm^{-1} (vNO).

A dark blue crystal with dimensions 0.20 × 0.20 × 0.20 mm^3 was mounted on a glass fiber in a random orientation. The determination of the unit cell and data collection was performed on a computer-controlled Bruker APEX-II CCD diffractometer. 3635 independent reflections [R(int) = 0.0203] in the range 1.63° $\leq \theta \leq$ 28.35° with $-18 \leq$ h $\leq$ 18, $-12 \leq$ k $\leq$ 11, $-12 \leq$ l $\leq$ 16 were collected by using an $\varphi - \omega$ scan technique at 293(2) K using Mo-Kα radiation with a graphite monochromator (λ = 0.71073A). The X-ray structure analysis revealed that the compound belongs to the monoclinic system, space group P2$_1$/c. Absorption corrections were applied using SADABS program. The structure of the complex was solved using direct methods by SHELXS-97 program [9]. Cadmium atom was located from an E-MAP. The other non-hydrogen atoms were determined with successive difference Fourier syntheses. The final refinement used full-matrix least-squares method by the SHELXL-97 [10]. Crystallographic data and refinement parameters are listed in **Table 1**. Atomic coordinates and selected bond distances and bond angles are given in **Tables 2** and **3**, respectively.

Table 1. Crystal data and structure refinement for the compound **1**.

Empirical formula	C28 H36 Cd N4 O10
Formula weight	701.01
Temperature	293(2) K
Wavelength	0.71073 A
Crystal system, space group	Monoclinic, P2$_{(1)}$/c
Unit cell dimensions	a = 13.732(3) A
	b = 9.2637(19) A β = 114.61(3)°.
	c = 12.641(3) A
Volume	1461.9(5) A^3
Z, Calculated density	2, 1.593 Mg/m^3
Absorption coefficient	0.811 mm^{-1}
F(000)	720
Crystal size	0.20× 0.20 × 0.20 mm
Theta range for data collection	1.63 to 28.35°.
Limiting indices	$-18 \leq$ h $\leq$ 18, $-12 \leq$ k $\leq$ 11, $-12 \leq$ l $\leq$ 16
Reflections collected/unique	12765/3635 [R(int) = 0.0203]
Completeness to theta = 28.35	99.5 %
Absorption correction	NONE
Refinement method	Full matrix least squares on F^2
Data/restraints/parameters	3635/0/197
Goodness of fit on F^2	1.030
Final R indices [I > 2 sigma(I)]	R$_1$ = 0.0235, wR$_2$ = 0.0688
R indices (all data)	R$_1$ = 0.0280, wR$_2$ = 0.0714
Extinction coefficient	0.0039(6)
Largest diff. peak and hole	0.452 and −0.317 e. A^3

3. RESULTS AND DISCUSSION

Description of the Crystal Structure

Single-crystal X-ray diffraction analysis reveals that complex **1** and the reported compound Mn$_2$(NITpBA)$_4$(H$_2$O)$_4$ [8] are isostructural. The title complex **1** is the first antimagnetic metal compound involving anionic nitronyl nitroxide ligand. An ORTEP drawing of the asymmetric unit of Cd$_2$(NITpBA)$_4$(H$_2$O)$_4$ is shown in **Figure 1**. The crystal structure of the complex **1** is shown in **Figure 2**, which reveals a ladder-like one di-

Table 2. Atomic coordinates (×10^4) and equivalent isotropic displacement parameters (A^2 × 10^3) for **1**. U(eq) is defined as one third of the trace of the orthogonalized Uij tensor.

	x	y	z	U(eq)
Cd(1)	5000	5000	5000	25(1)
O(1)	3258(1)	14443(1)	4936(1)	29(1)
O(1W)	4406(1)	4574(2)	3024(1)	35(1)
O(2)	804(1)	13117(2)	1262(1)	45(1)
O(3)	4408(1)	7205(1)	4535(1)	34(1)
O(4)	4608(1)	7533(1)	2887(1)	43(1)
N(1)	2420(1)	14427(2)	3956(1)	24(1)
N(2)	1297(1)	13830(2)	2196(1)	28(1)
C(1)	4309(1)	7925(2)	3649(2)	28(1)
C(2)	3765(1)	9375(2)	3525(1)	25(1)
C(3)	3384(2)	10095(2)	2464(2)	30(1)
C(4)	2867(1)	11413(2)	2330(1)	30(1)
C(5)	2737(1)	12030(2)	3266(2)	24(1)
C(6)	3123(1)	11310(2)	4336(1)	29(1)
C(7)	3622(2)	9983(2)	4458(2)	28(1)
C(8)	2181(1)	13402(2)	3137(1)	24(1)
C(9)	1001(1)	15364(2)	2337(2)	30(1)
C(10)	1515(1)	15485(2)	3681(2)	28(1)
C(11)	1537(2)	16317(2)	1746(2)	45(1)
C(12)	−206(2)	15554(2)	1767(2)	48(1)
C(13)	1927(2)	16963(2)	4172(2)	40(1)
C(14)	811(2)	14884(2)	4251(2)	43(1)

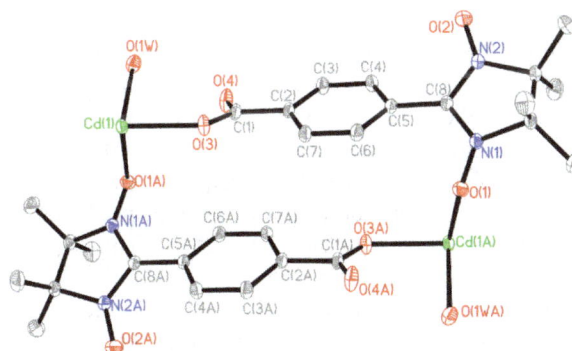

Figure 1. ORTEP of Cd$_2$(NITpBA)$_4$(H$_2$O)$_4$ drawn at the 30% probability level. All hydrogen atoms and disordered fluorine atoms are omitted for clarity.

Table 3. Bond lengths (A) and angles (°) for **1**.

Bond lengths (A)			
Cd(1)-O(3)	2.1862 (12)	O(3)-C(1)	1.261 (2)
Cd(1)-O(1W)	2.3159 (14)	O(4)-C(1)	1.248 (2)
Cd(1)-O(1)#2	2.4154 (13)	N(1)-C(8)	1.340 (2)
O(1)-N(1)	1.2933 (17)	N(2)-C(8)	1.357 (2)
O(2)-N(2)	1.2742 (19)		
Bond angles (°)			
O(3)#1-Cd(1)-O(3)	180.0	O(1W)#1-Cd(1)-O(1)#2	91.64 (5)
O(3)#1-Cd(1)-O(1W)	93.17 (5)	O(3)#1-Cd(1)-O(1)#3	93.52 (5)
O(3)-Cd(1)-O(1W)	86.83 (5)	O(1W)-Cd(1)-O(1W)#1	180.0
O(3)#1-Cd(1)-O(1)#2	86.48 (5)	N(1)-O(1)-Cd(1)#4	120.55 (10)
O(3)-Cd(1)-O(1)#2	93.52 (5)	O(1)#2-Cd(1)-O(1)#3	180.0
O(1W)-Cd(1)-O(1)#2	88.36 (5)		

Symmetry transformations used to generate equivalent atoms: #1 −x + 1, −y + 1, −z + 1; #2 −x + 1, −y + 2, −z + 1; #3 x,y − 1,z; #4 x,y + 1,z.

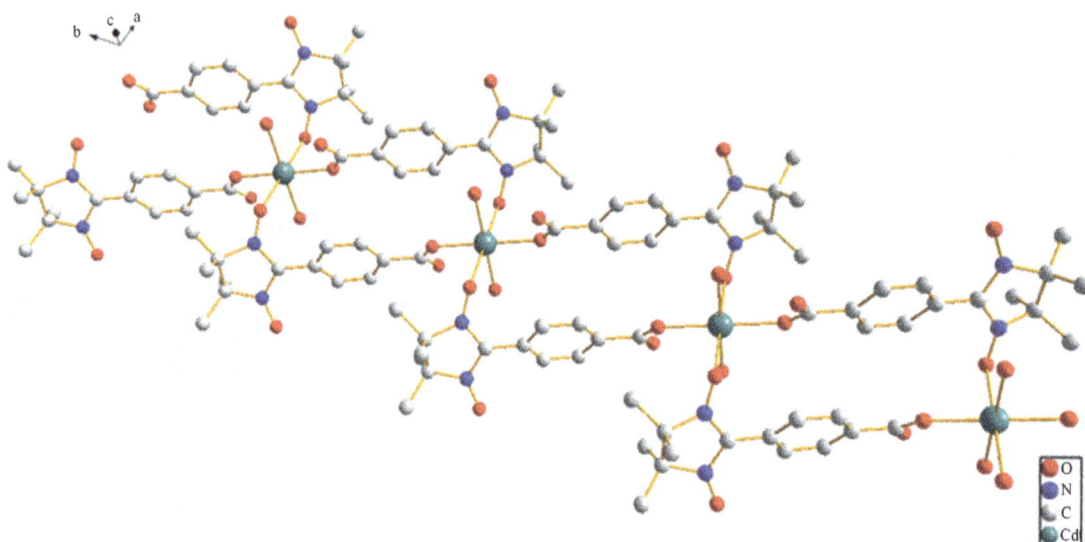

Figure 2. Draw of the ladder-like structure of the title compound **1**.

mensional chain of repeating $Cd_2(NITpBA)_4(H_2O)_4$ units.

The Cd (II) atom exhibits a slightly distorted octahedral coordination sphere and lies on an inversion center. The Cd (II) ion is coordinated by two pairs radical which *trans* to each other. One pair binds through one oxygen atom of the carboxylic anion, while the other coordinates through one of the NO group. Four oxygen atoms from nitronyl nitroxide ligand form the equatorial plane. Each nitronyl nitroxide coordinats the adjacent Cd (II) atom in the opposite fashion thus forming the chains. The Cd (II) ion is coordinatively saturated by two H_2O molecules in axial positions, thus completing the CdO_6 core. The Cd-O distances involving both types of radical oxygen atoms are markedly different to each other. The Cd-O3 (O from carboxy group) is 2.1862 A, which is shorter

than that Cd-O1 (O from nitroxide, 2.4154 A). The axial Cd-O 1W bond is equal to 2.3159 A. The N-O distances are slightly different from each other, and they are shorter than those found in NITpBAH, which are in the range of the typical distances of this kind of nitronyl nitroxide derivatives [10-12]. The band distance of the coordinated N-O in the radical is 1.2933 A, slightly longer than that of the uncoordinated N-O (1.2742 A). The bond of coordinated C1-O3 of the carboxyl group is 1.261 A, longer than that of the uncoordinated C1-O4 (1.2482 A). The fragment O1-N1-C8-N2-O2 is nearly planar, and forms a dihedral angle of 39.5(1)° with the plane of the phenyl ring. The dihedral angle between phenyl ring and the carboxylic group is 14.6(3)°. In the linear chain, the distance between successive Cd (II) ions is 9.2637 A. Interchain contacts are found between the water molecule oxygen O and the non-coordinated carboxylic oxygen O separated by 2.768 (2) A, indicative of hydrogen bonding.

In summary, we have successfully obtained the first ladder-like structure compound formed by cadmium (II) and anionic nitronyl nitroxide ligand, of formula [Cd_2 (NITpBA)$_4$(H_2O)$_4$] (NITpBA = 2-(4-carboxy-phenyl)-4,4,5,5-tetramethyl-4,5-dihydro-imidazol-1-oxyl-3-oxide anion).

5. ACKNOWLEDGEMENTS

The work was supported by the National Natural Science Foundation of China (No. 21071006), the Natural Science Foundation of Henan Province (No. 102102210457) and the Natural Science Foundation of the Henan Higher Education Institutions of China (No. 2010B150001).

REFERENCES

[1]   Kahn, O. (1993) Molecular magnetism. Vancouver Coastal Health, New York.

[2]   Caneschi, A. Gatteschi, D. Sessoli, R. and Rey, P. (1989) Toward molecular magnets: The metal-radical approach. *Accounts of Chemical Research*, **21**, 392-398. doi:10.1021/ar00167a004

[3]   Numata, Y. Inoue, K. Baranov, N. Kurmoo, M. and Kikuchi, K. (2007) Field-induced ferrimagnetic state in a molecule-based magnet consisting of a Co (II) ion and a chiral triplet bis(nitroxide) radical. *Journal of American Chemical Society*, **129**, 9902-9909. doi:10.1021/ja064828i

[4]   Maspoch, D. Domingo, N. Ruiz-Molina, D. Wurst, K. Vaughan, G. Tejada, J. Rovira, C. and Veciana, J. (2004) A robust purely organic nanoporous magnet. *Angewandte Chemie International Edition*, **43**, 1828-1832.

[5]   Fegy, K. Sanz, N. Luneau, D. Belorizky, E. and Rey, P. (1998) Proximate nitroxide ligands in the coordination spheres of manganese (II) and nickel (II) ions. Precursors for high-dimensional molecular magnetic materials. *Inorganic Chemistry*, **37**, 4518-4523. doi:10.1021/ic971618l

[6]   Laget, V. Hornick, C. Rabu, P. Drillon, M. and Ziessel, R. (1998) Molecular magnets hybrid organic-inorganic layered compounds with very long-range ferromagnetism. *Coordination Chemistry Reviews*, **178-180**, 1533-1553. doi:10.1016/S0010-8545(98)00166-0

[7]   Huang, C.F. Wei, H.H. Lee, G.H. and Wang, Y. (1998) Structure and magnetic properties of a novel chloro-bridged polymeric cadmium (II) complex with pyridyl-substituted nitrony nitroxide. *Inorganica Chimica Acta*, **279**, 233-237. doi:10.1016/S0020-1693(98)00105-4

[8]   Schiødt, N.C. Biani, F.F. Caneschi, A. and Gatteschi, D. (1996) Structure andmagnetism of nickel (II) and manganese (II) complexes of a nitronyl nitroxide carboxylic acid. *Inorganica Chimica Acta*, **248**, 139-146. doi:10.1016/0020-1693(95)04996-7

[9]   Sheldrick, G.M. (1990) SHELXS-97, program for X-ray crystal structure solution. University of Gottingen, Germany.

[10]   Sheldrick, G.M. (1997) SHELXL-97, program for X-ray crystal structure refinement. University of Gottingen, Germany.

[11]   Hirel, C. Li, L.C. Brough, P. Vostrikova, K. Pécaut, J. Mehdaoui, B. Bernard, M. Turek, P. and Rey, P. (2007) New spin-transition-like-nitroxide species. *Inorganic Chemistry*, **46**, 7545-7552 doi:10.1021/ic700851x

[12]   Chen, Y.W. Wei, H.X. Wang, X.G. Gao, D.Z. Sun, Y.Q. Zhang G.Y. and Xu, Y.Y. (2012) Syntheses, crystal structures, and magnetic properties of three 1D chain metal—Nitroxide complexes bridged by flexible dicerrboxylate anions. *Zeitschrift für anorganische und allgemeine Chemie*, **638**, 1328-1334 doi:10.1002/zaac.201200056

[13]   Chen, J. Liao, D.Z. Jiang, Z.H. and Yan, S.P. (2005) A new heterospin complex from oxamide-bridged Cu (II) binuclear units and carboxy-phenyl-substituted nitronyl nitroxide. *Inorganic Chemistry Communications*, **8**, 564-567. doi:10.1016/j.inoche.2005.02.015

12

Electrophysical properties of solid solutions of Zn$_{2-x}$(Ti$_a$Zr$_b$)$_{1-x}$Fe$_{2x}$O$_4$

Rudik Hamazaspi Grigoryan[1], **Levoh Hamazaspi Grigoryan**[2]

[1]Institute of Structural Macrokinetics and Materials Science, Russian Academy of Sciences, Moscow, Russia
[2]Institute of Problems of Chemical Physics, Russian Academy of Sciences, Moscow, Russia
Academician Semenova av. 1, Chernogolovka, Moscow, Russia
Email: rud@ism.ac.ru, hrant31@icp.ac.ru

ABSTRACT

Electrophysical properties of multicomponent system Zn$_2$TiO$_4$-Zn$_2$ZrO$_4$-ZnFe$_2$O$_4$ were investigated. The electrical conductivity, band gap, dielectric permittivity, and molar polarizability of formed solid solutions— Zn$_{2-x}$(Ti$_a$Zr$_b$)$_{1-x}$Fe$_{2x}$O$_4$ ($x = 0$ - 1.0 are defined; $\Delta x = 0.05$; a + b = 1; a:b = 1:5; 1:4; 1:3; 1:2; 1:1; 2:1; 3:1; 4:1) were defined. Electrophysical properties of samples are in linear dependence on their composition. All the synthesized samples are semiconductors with high electrical resistivity. Was confirmed the formation of two phases of variable composition with a wide homogeneity range.

Keywords: Spinel; Solid Solutions; Electrical Conductivity; Band Gap; Dielectric Permittivity; Polarizability

1. INTRODUCTION

Quickly developing electronic industry puts before chemical science problem searching for of the new compounds, possessing valuable electrophysical characteristics and their deepened physic-chemical study. Such are, in particular, hard melting complex oxides. The Special interest present the complex oxides and formed by them hard solutions, containing d-elements, differing broad area of homogeneity, expressed dielectric, semiconductor, superconductive, optical and the other characteristic.

The electrical properties of solid solutions with broad homogeneity ranges vary systematically with composition, which enables the fabrication of materials with tailored properties. There is considerable research interest in refractory mixed transition metal oxides and their solid solutions, which have broad homogeneity ranges and attractive dielectric, semiconducting, superconducting, optical and other properties [1,2]. Divalent metal orthotitanates are widely used in semiconductor technology as bulk resistors and microwave ceramics [2]. ZnO enriched

zinc orthotitanate was proposed as a candidate material for nonlinear and bulk resistors [3].

In a recent x-ray diffraction study [4], two phases (α and β) with broad homogeneity regions were identified in the Zn$_2$TiO$_4$-Zn$_2$ZrO$_4$-ZnFe$_2$O$_4$ system at 1170 K. The α-phase crystallizes in a cubic structure and the b-phase in a tetragonal inverse spinel structure. Their stability limits and the structural parameters of Zn$_{2-x}$(Ti$_a$Zr$_b$)$_{1-x}$Fe$_{2x}$O$_4$ samples were determined, and the cation distribution over the tetrahedral and octahedral interstices in the oxygen sublattice of the spinel structure was assessed. Electrophysical parameters of formed hard solution were not studied earlier.

In this paper, we report the electrical properties of solid solutions in the Zn$_2$TiO$_4$-Zn$_2$ZrO$_4$-ZnFe$_2$O$_4$ system.

2. METHODS OF THE EXPERIMENT

The Syntheses of Zn$_{2-x}$(Ti$_a$Zr$_b$)$_{1-x}$Fe$_{2x}$O$_4$ (where $x = 0$, 0 - 1.0; $x = 0.05$; a + b = 1) were realized in two ways—in low temperature plasma of hydrogen-oxygen flame [5] and well-known ceramic technology. As source material were a used especially purec reagents: ZnO, TiO$_2$, Fe$_2$O$_3$ and ZrO$_2$. For the synthesis were prepared by a mixture of $(2 - x)$ZnO + a$(1 - x)$TiO$_2$ + b$(1 - x)$ZrO$_2$ + xFe$_2$O$_3$, where a:b = 1:5; 1:4; 1:3; 1:2; 1:1; 2:1; 3:1; 4:1; $x = 0$ - 1.0, $x = 0.05$. Samples for measurement were prepared by pressing (200 MPa) divided powder (65 microns) synthesis of solid solutions in the form of tablet (d = 20 mm, l = 2 mm). The tablets were subjected to heat treatment at 1170 K for 2 h, followed by rapid cooling on a copper substrate. Conductivity of the samples was measured by a standard four-probe potentiometric method, with electrical contacts made by firing silver paste. Temperature dependence of electrical conductivity was measured in the temperature range 293 - 900 K. Dielectric permittivity was measured by the flat condenser method. Molar polarizability was evaluated using the Clauzius-Mossotti equation. Dielectric permittivity measured the method of flat condenser.

3. RESULTS AND DISCUSSION

It was established the identity of the investigated physical properties of samples of identical compositions synthesized in low-temperature plasma and ceramic technology. The following are data from the study of the samples synthesized at low temperature plasma. All of the synthesized samples are dielectrics with semiconducting behavior of conductivity with specific resistivity from 10^7 to 10^{12} ohm·cm. Comparison of the results obtained values of electrical parameters and the constructed composition—property diagram confirms biphasic of system Zn_2TiO_4-Zn_2ZrO_4-$ZnFe_2O_4$ and (1170 K) and the boundaries of homogeneous phases (**Table 1**) previously established by X-ray diffraction method [5]. In **Table 2** shows the values of the density of single-phase samples, certain pycnometric method at room temperature. As can be seen from these data, a significant change in the density of solid solutions in the whole concentration range is observed.

α-**phase**. The results of measurements of electrical parameters of the α-phase are shown in **Table 2**. As can be seen from these data, the electrical properties of solid solutions are most sensitive to iron content.

Changes in the relative content of titanium and zirconium, with a constant iron content does not lead to significant changes in the physical properties (**Table 3**). For example, a value of $x = 0.7$ (1.4 mol of iron content) change in the ratio a:b 1:5 to 4:1 leads to an increase in conductivity of one order (2.34 × 10^{-9} to 5.01 × 10^{-8} ohm^{-1}·cm^{-1}). At the same time slightly decreases the band gap (1.31 eV to 1.08 eV) and the dielectric constant and polarization parameters are changed slightly. While the increase in iron content in the α-phase leads to an increase in the specific conductivity of solid solutions of 2 - 3 orders of magnitude, depending on the relative content of titanium and zirconium.

Depending on the electrophysical parameters of the samples from the α-phase composition are almost linear for all tested values of a:b. In **Figure 1** shows the graphical dependence of the electrical conductivity and dielectric constant of solid solutions of α-phase system for cutting a:b = 3:1. As can be seen from the data **Table 2**, the polarization characteristics of the samples at the same time did not significantly change.

β-**phase.** The results of measurements of electrical parameters of samples are given in **Table 3**. Samples of β-phases are more low electrical conductivity. As in the case of α-phase, increasing iron content leads to a decrease in the band gap and hence to an increase in electrical conductivity of the samples. The lowest conductivity observed in the solid solution containing no iron, and contains the maximum number of zircon ($x = 0$, a:b = 1:5).

Table 1. The boundaries of homogeneous phases in the system $Zn_{2-x}(Ti_aZr_b)_{1-x}Fe_{2x}O_4$.

[Ti]:[Zr]	x	
	α-phase	β-phase
1:5	0.70 to 1.0	0.0 to 0.47
1:4	0.69 to 1.0	0.0 to 0.46
1:3	0.69 to 1.0	0.0 to 0.45
1:2	0.66 to 1.0	0.0 to 0.40
1:1	0.58 to 1.0	0.0 to 0.25
2:1	0.36 to 1.0	-
3:1	0 to 1.0	-
4:1	0 to 1,0	-

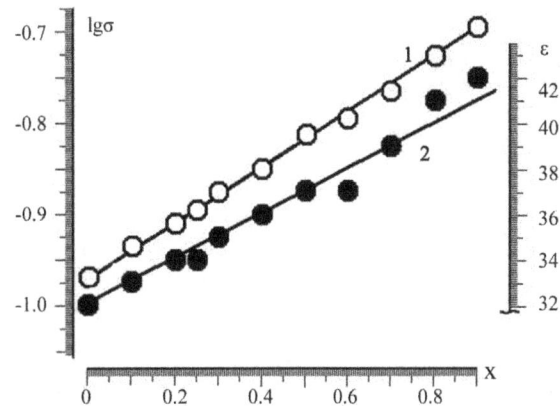

Figure 1. The dependence of the conductivity (1) and the dielectric constant (2) of solid solutions α-phase from the composition in the $Zn_{2-x}(Ti_aZr_b)_{1-x}Fe_{2x}O_4$ (a:b = 3:1).

Among the synthesized solid solutions of our best meets these criteria, the sample does not contain iron, with a maximum content of zirconium ($Zn_2Ti_{0.167}Zr_{0.833}O_4$, $\sigma = 5.68 \times 10 - 12$ ohm^{-1}·cm^{-1}). In the sample with the highest iron content ($Zn_{1.55}Ti_{0.1}Zr_{0.5}Fe_{0.9}O_4$), electrical conductivity higher by nearly three orders of magnitude ($\sigma = 2.51 \times 10^{-9}$ ohm^{-1}·cm^{-1}).

Depending on the electrophysical parameters of the samples as the α- and β-phases of the composition are almost linear for all tested sections. In **Figure 2** shows the dependence of these cuts for a:b = 1:3 and a:b = 1:5.

An exponential dependence of the electrical conduction of samples as the α- and β-phases of the temperature (**Figure 3**). Raising the temperature from 293 to 900 K leads to an increase in conductivity up to 4 orders of magnitude. Significant changes in the densities of the samples throughout the concentration range is not observed (**Table 4**).

Table 2. Depending on the electrical conductivity (σ, ohm^{-1}·cm^{-1}), the band gap (ΔE^*, eV), the dielectric permittivity (ε^{**}), molar polarization (P^{***}, cm^3) of the α-phase compositions of $Zn_{2-x}(Ti_aZr_b)_{1-x}Fe_{2x}O_4$.

x	σ	ΔE	ε	P	σ	ΔE	ε	P
	$a{:}b = 1{:}5$				$a{:}b = 1{:}4$			
0.7	2.34×10^{-9}	1.31	39	42.31	2.37×10^{-9}	1.26	39	42×23
0.8	2.14×10^{-8}	1.05	41	42.88	3.31×10^{-8}	1.18	41	42×33
0.9	2.00×10^{-7}	1.00	42	42.32	2.05×10^{-7}	1.00	42	42.34
	$a{:}b = 1{:}3$				$a{:}b = 1{:}2$			
0.7	2.37×10^{-9}	1.26	39	42.23	2.69×10^{-9}	1.29	36	41.90
0.8	3.31×10^{-8}	1.18	41	42.33	2.34×10^{-8}	1.17	39	42.13
0.9	2.05×10^{-7}	1.00	42	42.34	2.24×10^{-7}	1.04	42	42.28
	$a{:}b = 1{:}1$				$a{:}b = 2{:}1$			
0.4	-	-	-	-	4.68×10^{-9}	1.14	38	42.09
0.5	-	-	-	-	6.31×10^{-9}	1.11	40	42.28
0.6	1.00×10^{-9}	1.08	39	42.18	1.59×10^{-8}	1.08	40	42.26
0.7	3.39×10^{-9}	1.05	40	42.28	3.39×10^{-8}	1.05	42	42.39
0.8	2.00×10^{-8}	1.02	42	42.35	6.17×10^{-8}	1.02	42	42.33
0.9	1.12×10^{-7}	0.99	43	42.35	1.26×10^{-7}	0.99	43	42.34
	$a{:}b = 3{:}1$				$a{:}b = 4{:}1$			
0.0	2.01×10^{-10}	1.19	32	42.64	1.81×10^{-9}	1.18	33	43.10
0.1	4.47×10^{-10}	1.17	33	42.64	2.82×10^{-9}	1.16	34	42.75
0.2	7.94×10^{-10}	1.14	34	42.61	4.47×10^{-9}	1.14	34	41.08
0.3	1.74×10^{-9}	1.11	35	42.53	7.41×10^{-9}	1.11	35	42.58
0.4	3.16×10^{-9}	1.09	36	42.74	1.18×10^{-8}	1.09	36	42.53
0.5	7.08×10^{-9}	1.11	37	42.74	1.51×10^{-8}	1.07	37	42.50
0.6	1.12×10^{-8}	1.10	37	42.35	3.02×10^{-8}	1.05	38	42.44
0.7	2.82×10^{-8}	1.08	39	42.34	5.01×10^{-8}	1.02	38	42.32
0.8	4.47×10^{-8}	1.03	41	42.47	7.94×10^{-8}	1.00	40	42.36
0.9	1.12×10^{-7}	1.01	42	42.42	1.38×10^{-7}	0.98	42	42.44
1.0	-	-	-	-	6.61×10^{-7}	0.96	43	42.73

$^*\pm 0.02$; $^{**}\pm 1$; $^{***}\pm 0.05$.

Table 3. Depending on the electrical conductivity (σ, ohm^{-1}·cm^{-1}). The band gap (ΔE^*, eV). the dielectric permittivity (ε^{**}) and molar polarization (P^{***}, cm) of β-phase compositions of $Zn_{2-x}(Ti_aZr_b)_{1-x}Fe_{2x}O_4$.

x	σ	ΔE	ε	P	σ	ΔE	ε	P
	$a{:}b = 1{:}5$				$a{:}b = 1{:}4$			
0.0	5.68×10^{-12}	1.78	25	47.62	6.30×10^{-12}	1.81	25	45.44
0.1	3.16×10^{-11}	1.61	28	45.79	2.51×10^{-11}	1.56	28	45.40
0.2	5.24×10^{-11}	1.50	31	47.22	7.94×10^{-11}	1.48	31	45.42
0.25	3.16×10^{-10}	1.47	32	45.02	2.18×10^{-10}	1.42	33	45.28
0.3	5.01×10^{-10}	-	33	-	4.17×10^{-10}	1.41	34	45.01
0.4	1.03×10^{-9}	1.42	35	47.49	1.51×10^{-9}	1.41	35	44.69
0.45	2.51×10^{-9}	1.39	36	47.64	3.72×10^{-9}	1.38	37	44.60
	$a{:}b = 1{:}3$				$a{:}b = 1{:}2$			
0.0	1.09×10^{-11}	1.69	25	44.88	1.86×10^{-11}	1.71	27	44.92
0.1	3.31×10^{-11}	1.58	29	45.24	5.25×10^{-11}	1.60	30	45.13
0.2	1.10×10^{-10}	1.52	32	45.00	1.82×10^{-10}	1.54	33	45.21
0.3	3.02×10^{-10}	1.47	35	45.01	4.786×10^{-10}	1.45	34	44.70
0.4	1.03×10^{-10}	1.36	36	44.73	1.445×10^{-9}	1.36	35	44.57
0.45	1.74×10^{-9}	1.33	38	44.80	2.51×10^{-9}	1.30	36	44.61
				$a{:}b = 1{:}1$				
x	σ	ΔE	ε	P	x	σ	ΔE	ε	P
0.0	4.89×10^{-11}	1.48	30	44.27	0.2	3.89×10^{-10}	1.29	36	44.25
0.1	1.59×10^{-10}	1.39	33	44.27	0.25	5.13×10^{-10}	1.23	37	43.94

Table 4. Densitys ($d_{pickn.}$, g/cm^3) of solid solutions α- and β-phases compositions of $Zn_{2-x}(Ti_aZr_b)_{1-x}Fe_{2x}O_4$.

	d ± 0,01							
	4:1	3:1	2:1	1:1	1:2	1:3	1:4	1:5
0	5.27	5.28	-	5.34	5.34	5.34	5.36	5.36
0.1	5.22	5.30	-	5.32	5.32	5.32	5.30	5.29
0.2	5.20	5.30	-	-	5.30	5.30	5.28	5.25
0.25	5.16	5.29	-	-	5,28	5.28	5.28	5.25
0.3	5.16	5.31	-	-	5.29	5.29	5.28	5.27
0.4	5.16	5.30	5.55	-	5.26	5.26	5.22	5.24
0.5	5.19	5.28	5.30	-	-	-	-	-
0.6	5.18	5.27	5.24	5.36	-	-	-	-
0.7	5.15	5.25	5.55	5.30	5.52	5.52	5.42	-
0.8	5.12	5.24	5.30	5.26	5.33	5.33	5.33	-
0.9	-	5.23	5.24	5.22	5.33	5.33	5.25	-

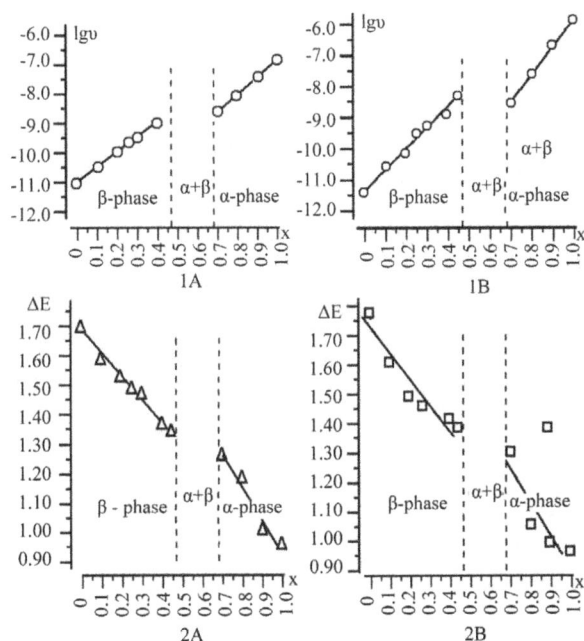

Figure 2. The Dependency to conductivity (1) and width of the forbidden zone (2) of solid solutions $Zn_{2-x}(Ti_aZr_b)_{1-x}Fe_{2x}O_4$. (A) for a:b = 1:3 and (B) for a:b = 1:5.

The proportional dependence of the electrical conductivity of samples of the content of Fe 3 and semiconductor conductivity indicate the donor-acceptor mechanism of the electrical conductivity of the samples can be explained to the valence exchange (electron hopping) between the iron ions: $Fe^{3+} \rightarrow Fe^{2+} + o^+$ and $Fe^{2+} + o^+ \rightarrow Fe^{3+}$. Other possible similar schemes ($Ti^{4+} \rightarrow Ti^{3+} + o^+$, or $Zr^{3+} + o^+ \rightarrow Zr^{4+}$), are unlikely. Increasing the tita-

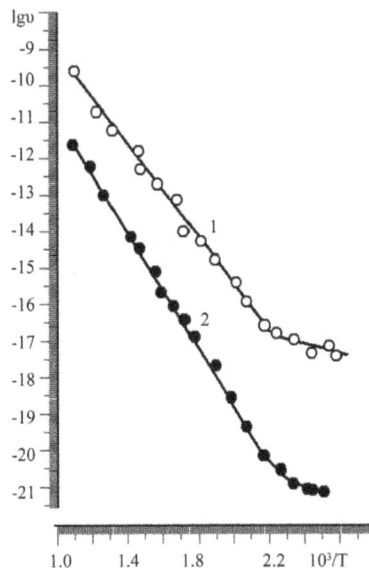

Figure 3. The temperature dependence of conductivity: $\lg \sigma = f(1/T)$ of solid solutions $Zn_{2-x}(Ti_aZr_b)_{1-x}Fe_{2x}O_4$; (1) α-phase; $x = 0.5$; a:b = 3:1; (2) β-phase, $x = 0.3$; a:b = 4:1.

nium and zircon content, accompanied by a decrease in iron content, do not lead to an increase, but on the contrary, a significant reduction in the electrical conductivity of the samples. A similar pattern is also observed when replacing ions Sn^{4+} by Fe^{3+} [6].

4. CONCLUSION

Were synthesized and investigated the electrical properties of samples of a multicomponent system Zn_2ZrO_4-Zn_2TiO_4-$ZnFe_2O_4$. In the system formed two phases with wide homogeneity ranges. It was established that all the synthesized samples are semiconductors with high electrical resistivity. The values of electrical conductivity, band gap, dielectric constant and molar polarizability of the solid solution compositions $Zn_{2-x}(Ti_aZr_b)_{1-x}Fe_{2x}O_4$ were determined. All the studed properties vary linearly with the composition of the solid solutions. Increased iron content leads to a significant increase in conductivity of the samples (up to 3 orders of magnitude), simultaneously the width of the band gap and polarizability of the samples decreased.

REFERENCES

[1] Pasynkov, V.V. and Sorokin, V.S. (1986) Materialy elektronnoi tekhniki (electronic materials). Vysshaya Shkola, Moscow.

[2] Yanagida, H.M. (1986) Tonkaya tekhnicheskaya keramika (fine technical ceramics). Metallurgiya, Moscow.

[3] Loshkarev, B.A. and Semirikov, I.S. (1967) Properties

and Structure of ZnO-TiO$_2$ Ceramics. *Neorganicheskie. Materiali*, **3**, 1467-1473.

[4] Grigoryan, R.A. and Grigoryan, L.A. (2004) Synthesis and properties of Zn$_{2-x}$(Ti$_a$Zr$_b$)$_{1-x}$Fe$_{2x}$O$_4$ solutions. *Neorganicheskie. Materiali*, **40**, 351-356.

[5] Grigoryan, R.A., Grigoryan, L.A. and Babayan, G.G.

(2001) Synthesis of ZnFeO-Zn ZrO solid solutions. *Neorganicheskie. Materiali*, **37**, 367-370.

[6] Grigoryan, R.A. and Grigoryan, L.A. (2004) Synthesis and properties of Zn$_2$TiO$_4$-Zn$_2$SnO$_4$-ZnFe$_2$O$_4$ solid solutions. *Inorganic Materials*, **40**, 295-299.

13

Effect of PVP, PVA and POLE surfactants on the size of iridium nanoparticles

Anjali Goel[*], Neetu Rani

Department of Chemistry, Kanya Gurukul Mahavidyalaya, Gurukul Kangri University, Hardwar, India
Email: [*]anjaligoel10@gmail.com

ABSTRACT

Commonly transition metal nano particle are synthesized by physical, chemical or electrochemical methods. In the present work colloidal iridium nanoparticles were synthesized by chemical oxidation method with different surfactants like poly vinyl pyrrolidone (PVP), poly vinyl alcohol (PVA) and poly oxyethylene lauryl ether (POLE). It was found that shape and size of Ir-nano particles resulted were related to kind of capping agent (surfactant) used. The characterization of the synthesized nano particle has been carried out by UV-vis, X-ray diffraction (XRD), FT-IR, scanning electron microscopy (SEM) and transmission electron microscopic (TEM) techniques. UV-vis and FT-IR confirm the oxidation of $IrCl_3$ into IrO_2 while XRD confirms the amorphous nature of the iridium nanoparticles synthesized. The morphology and size of the particle were confirmed by TEM. The average particle size determined by Scherrer equation was about 4.12 nm to 4.23 nm with PVP, 2.74 to 3.36 nm with PVA and 20.41 to 42.25 nm with POLE. Poly oxyethylene lauryl ether particles were not further analyzed because of their large size and less stability. Further particle size was confirmed with TEM, which was 4.5 nm with PVP and 7.0 nm with PVA. The particles are spherical with no agglomeration tendency.

Keywords: Nano-Materials; Surfactant; Agglomeration; PVP; PVA; POLE

1. INTRODUCTION

Transition metal nanoparticles are of continuing interest because of their fascinating catalytic, electronic, and optical properties. These ultrafine particles often agglomerate to form either lumps or secondary particles in order to minimize the total surface or the interfacial energy of the system Therefore, it is very important to stabilize the particles against adverse agglomeration at both the syn-

thesis and the usage stages. There have been numerous types of stabilizers that have been used as capping agent for stabilizing nanoparticles. It is well known that type of stabilizer that is used to cap the nanoparticles affects their stability and in turn affects its catalytic activity [1-3]. Iridium is particularly an interesting transition-metal catalyst. There have been reports of unusually high catalytic activity from molecular iridium clusters [4]. A few reports of iridium nanoparticles synthesis exist in the literature [5,6]. In the present work iridium nanoparticles are successfully prepared by refluxing the alcohol water solution of iridium trichloride in the presence of polymer such as poly vinyl pyrrolidone, polyvinyl alcohol and poly oxyethylene lauryl ether (abbreviated as PVP, PVA and POLE respectively). The water soluble alcohol like methanol, ethanol and propanol are used as oxidant for the synthesis of iridium nano clusters.

2. EXPERIMENT

2.1. Synthesis

In order to prepare stable metal cluster with a defined particle size below 10 nm and narrow size distribution, suitable solvent (oxidant) and an effective method of controlling the metal particle size has been used. In the present work iridium nanoparticles (abbreviated as Ir-nano) are prepared by refluxing the alcohol (methanol, ethanol and propanol) [7] water solution of iridium trichloride, and one of the protective agent such as polyvinyl pyrrolidone (PVP) [8] poly vinyl alcohol (PVA) or poly oxyethylene lauryl ether (POLE) [9,10]. Poly (N-vinyl-2-pyrrolidone) (PVP), poly vinyl alcohol (PVA), poly oxyethylene lauryl ether (POLE) were obtained from Merck and used as such. All the other reagents were also of analytical grade.

2.2. Characterization

UV-vis spectrophotometer—117 was used for spectral measurements. X-Ray-diffraction (XRD) measurements were performed on the dry powders using Bruker Axs

---

[*]Corresponding author.

D-8 Advance diffractometer with a scan rate $1°·min^{-1}$ and CuK_{α} ($\lambda = 0.154$ nm) radiation. Transmission electron microscopy (TEM) was carried out with a JEOL, JEM-1011 electron microscope.

3. RESULTS AND DISCUSSION

Size controlled synthesis was carried out by varying the amount of iridium precursor, NaOH, stabilizer and alcohol as described in **Tables 1, 2 & 3** respectively. For monitoring the progress of the metal colloid formation UV-vis spectroscopy is used. As can be seen from **Figure 1(c)** initially reaction mixture shows an absorption peak at 206 nm which starts disappearing with the passage of time [11]. After about 25 minutes refluxing this absorption peak disappeared completely and a new peak appeared at around 230 nm indicating the formation of IrO_2 nanoparticles [12]. No further change was observed. The colour of solution changes from yellow to black-brown. The same type of spectra was obtained with PVP and PVA as shown in **Figures 1(a)** and **1(b)** respectively.

The amorphous nature of the particles was determined by XRD. XRD was performed on the dry powders obtained after evaporation of solvent. **Figures 2, 3 & 4** show the XRD patterns of colloidal Ir-nano prepared by method a, b, c, d & e as represented in the **Tables 1, 2 & 3** respectively. The XRD spectra of Ir-nano show mainly two broad peaks at 2-theta about 28° & 40° at reflection planes $IrO_2(110)$ and $IrO_2(200)$ respectively which are characterstics of isolated IrO_2 nanoparticles. The diffraction peaks also show the amorphous nature of the sample [13]. Analysis of peaks using Scherrer equation gives an estimate of particle diameter. The diameter of particles calculated by Scherrer equation are given in **Tables 1-3**. The particles synthesized using POEL are large in size and less stable so further analysis of these particles was not carried out.

FT-IR Spectroscopy measurements were performed with the purpose of learning about the role of the solvent and stabilizer in the formation of iridium nanoparticles. The peaks were identified by literature value [14,15].

Table 1. Synthesis conditions and particle size of Ir-nano with PVP.

S.No.	Method	(PVP)/IrCl$_3$·xH$_2$O	Solvent	Solvent (ml)	[NaoH] Ml	Approx. particle size (nm)	
						XRD	TEM
1.	A	$8.69 \times 10^{-5}/7.07 \times 10^{-4}$	Methanol	35 ml	1 ml	3.16, 5.67	-
2.	**B**	$8.69 \times 10^{-5}/7.07 \times 10^{-4}$	**Methanol**	**25 ml**	**1 ml**	**4.12, 4.23**	**4.5 nm**
3.	C	$8.69 \times 10^{-5}/7.07 \times 10^{-4}$	Methanol	25 ml	2 ml	3.91, 4.46	4.2 nm
4.	D	$8.69 \times 10^{-5}/7.07 \times 10^{-4}$	Ethanol	25 ml	1 ml	43.6, 42.1, 48.4	65.16 nm
5.	E	$8.69 \times 10^{-5}/7.07 \times 10^{-4}$	n-propanol	25 ml	1 ml	41.2, 42.0, 43.4	38.6 nm

Table 2. Synthesis, conditions and particle size of Ir-nano with PVA.

S.No.	Method	PVA/Ircl$_3$·xH$_2$O	Solvent	Solvent (ml)	[NaoH] Ml	Approx. particle size (nm)	
						XRD	TEM
1	A	$4.34 \times 10^{-5}/2.23 \times 10^{-4}$	**Methanol**	**25 ml**	**1ml**	**2.74, 3.36**	**7.00 nm**
2.	B	$2.17 \times 10^{-5}/1.12 \times 10^{-4}$	Methanol	25 ml	1ml	2.92, 4.70	-
3.	C	$4.34 \times 10^{-5}/2.23 \times 10^{-4}$	Ethanol	25 ml	1ml	2.65, 4.46	-
4.	D	$2.17 \times 10^{-5}/1.12 \times 10^{-4}$	Ethanol	25 ml	1ml	6.83, 8.32, 3.38	-
5.	E	$4.34 \times 10^{-5}/2.23 \times 10^{-4}$	2-propanol	25 ml	1ml	8.19, 7.44, 7.09, 11.95	-
6.	F	$2.17 \times 10^{-5}/1.12 \times 10^{-4}$	2-propanol	25 ml	1ml	3.15, 4.03	-

Table 3. Synthesis conditions and particle size of Ir-nano with POLE.

S.No.	Method	[Iridium tri chloried]	Solvent	Solvent (ml)	[NaoH] ml	Approx.particle size (nm)	
						XRD	TEM
1.	**a.**	$4.34 \times 10^{-5}/2.23 \times 10^{-4}$	**Methanol**	**25 ml**	**1 ml**	**20.41, 41.77, 42.25**	68.19 nm
2.	b.	$4.34 \times 10^{-5}/2.23 \times 10^{-4}$	Methanol	35 ml	1 ml	27.22, 41.82, 42.25	
3.	c.	$2.17 \times 10^{-5}/1.12 \times 10^{-4}$	Methanol	25 ml	1 ml	16.38, 27.29, 20.90 28.45	-
4.	d.	$4.34 \times 10^{-5}/2.23 \times 10^{-4}$	Ethanol	25 ml	1 ml	26.92, 27.88	-
5.	e.	$4.34 \times 10^{-5}/2.23 \times 10^{-4}$	2-propanol	25 ml	1 ml	4.03, 4.11	-

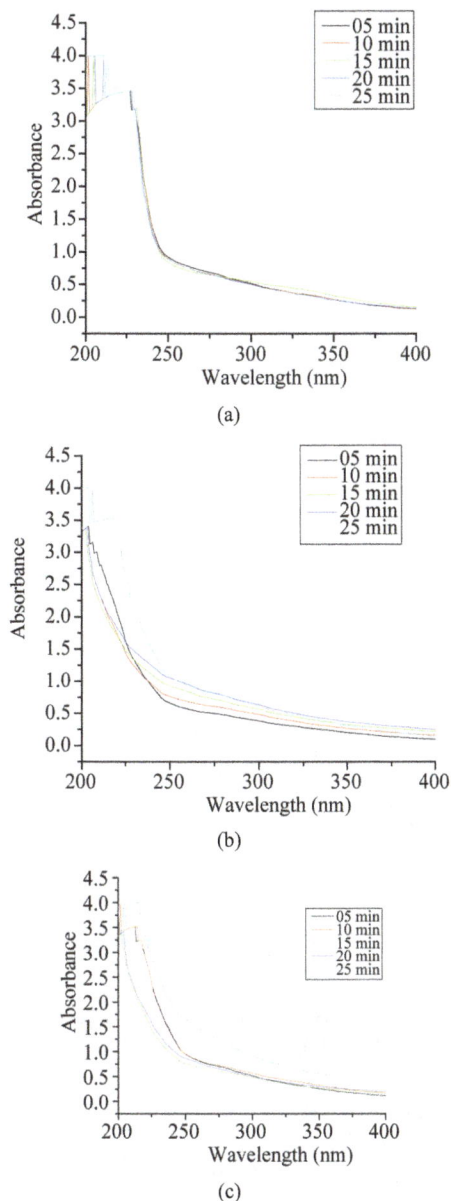

Figure 1. UV-vis spectra during the formation of Ir-nano at different time with (a) PVP-IrCl₃-methanol-water system; (b) PVA-IrCl₃-methanolwater system; (c) POLE-IrCl₃-methanol-water system.

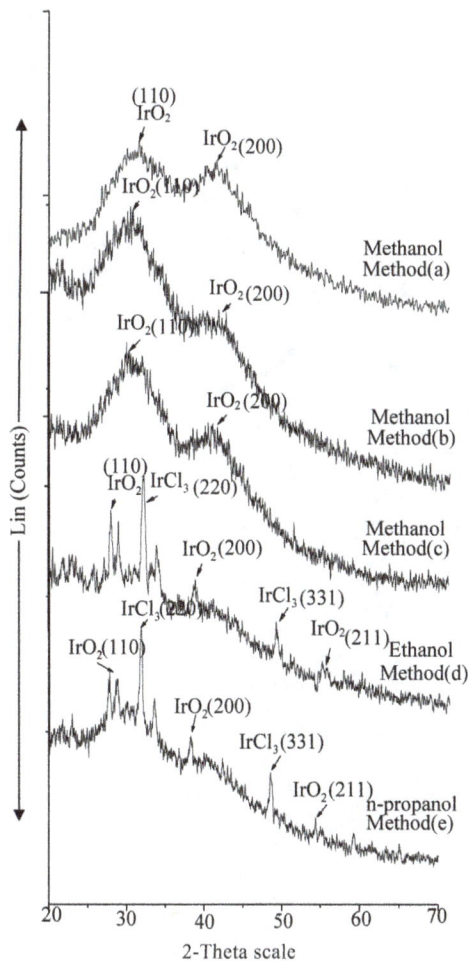

Figure 2. XRD spectra of Ir-nano prepared with methanol, ethanol and propanol using PVP as stabilizer.

The peaks under the region 2665 cm^{-1} and 2773 cm^{-1} may be due to Ir(III) but a new peak near the frequency 2015 cm^{-1} indicates the formation of iridium oxide. The red shift of resonance peak of pure stabilizer at 1700 cm^{-1} to 1664 cm^{-1} indicates the interaction of >C=O group of stabilizer with Ir-nano. The presence of same bands at about 1500 cm^{-1}, 1461 cm^{-1} (aromatic C-C str, N-H bending) show that these groups do not interact with Ir-nano.

TEM is one of the most important required methods for the characterization of nano materials. It has the unique ability to prove the size and shape of the particles.

Figures 5(a)-(c) show the transmission electron micrographs and the size distribution of the colloidal Ir-nano protected by PVP, PVA and POLE respectively. The average diameter of the metal nanoparticle was 4.2 to 4.5 nm with PVP, 7.0 nm with PVA and 68.16 nm with POLE respectively with methanol as oxidizing agent, while with ethanol and propanol particle size is larger as compared to methanol.

Thus from the above results it can be revealed that Ir-nano particles of size less than 10 nm can be synthesized using method "b" (**Table 1**), method "a" (**Table 2**) and method "e" (**Table 3**) when PVP, PVA and POLE respectively are the stabilizers.

PVP stabilized Ir-nano: A comparison of the data presented in **Table 1** show that as the boiling point of oxidant increases size of particle increases. The diameter ranges between 3.16 nm to 48.4 nm. It is also clear from the data that increase in NaOH concentration (method-c) does not show any remarkable effect on particle size but the decrease in oxidant concentration (method-b) in-

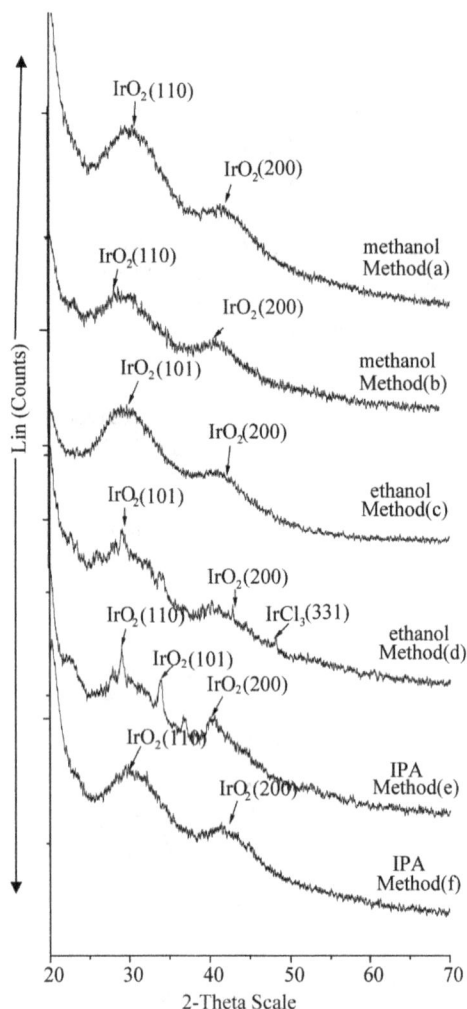

Figure 3. XRD spectra of Ir-nano with methanol, ethanol and IPA using PVA as stabilizer.

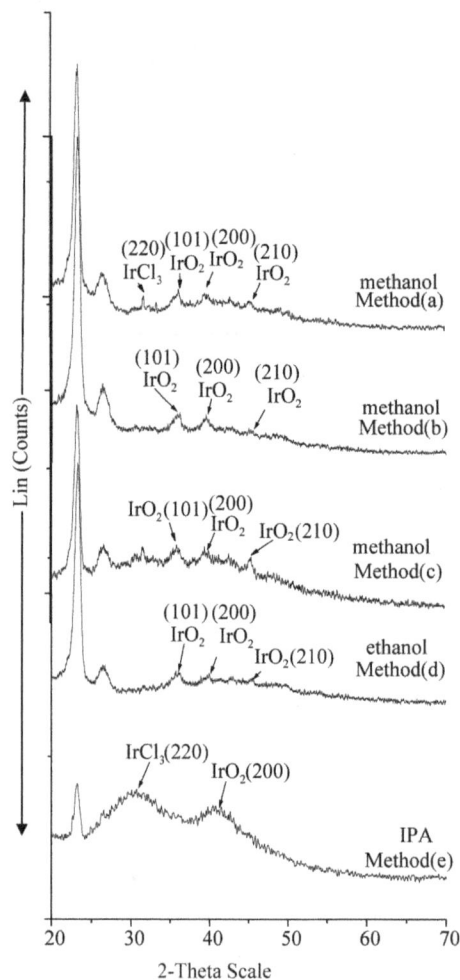

Figure 4. XRD spectra of Ir-nano with methanol, ethanol, and IPA using POLE as stabilizer.

creases the size distribution *i.e.* with 35 ml of methanol particle diameter is 3.16 to 5.67 nm which becomes less disperse from 4.12 to 4.23 with 25 ml of methanol (a & b) respectively. It may be due to the fact that the size of colloidal metal particle depends on the speed of oxidation of the metal precursor. Data also reveal that particle size is large when ethanol and propanol are the oxidant.

PVA stabilized Ir-nano: The results in **Table 2** show that Ir-nano clusters synthesized using PVA stabilizer are smallest as compared to PVP and POLE. The particles are small with size distribution between 2.65 to 11.95 nm. The data also reveal that particles synthesized by method a & e are small with size distribution of 2.74 to 3.36 and 3.15 to 4.03 nm respectively. It also demonstrates the amorphous nature of Ir-nano synthesized by method a, b & c. With methanol and ethanol particle size increases by decreasing PVA concentration, while in case of 2-propanol particle size decreases with the decreases of PVA conc.

POLE stabilized Ir-nano: From **Table 3**, it can be seen that size of Ir-nano synthesized using POLE stabilizer are large ranging from 4.03 to 42.25 nm. It is smallest (4.03 to 4.11 nm) (method-e) when 2-propanol is used as oxidant, while with methanol and ethanol the size of particle is large ranging from 20.41 to 42.25 nm. It can also be seen that with the increase of oxidant concentration (method-b) the range of particle size distribution decreases from 27.22 to 42.25 nm to 20.41 to 42.25 nm. With the decrease in stabilizer POLE concentration (method-c) the particle size decreases to 16.38 from 28.45 nm. It can be revealed that Ir-nano using 2-propanol are more amorphous than with methanol and ethanol.

Role of stabilizer: The main effect of stabilizers on particle size may be due to the steric effect and chemical bonding [16]. The stabilization of colloidal metal particles with polymers [17-19] in water is often discussed by the adsorption of the polymer on the colloidal particles. These large adsorbates provide a steric barrier which

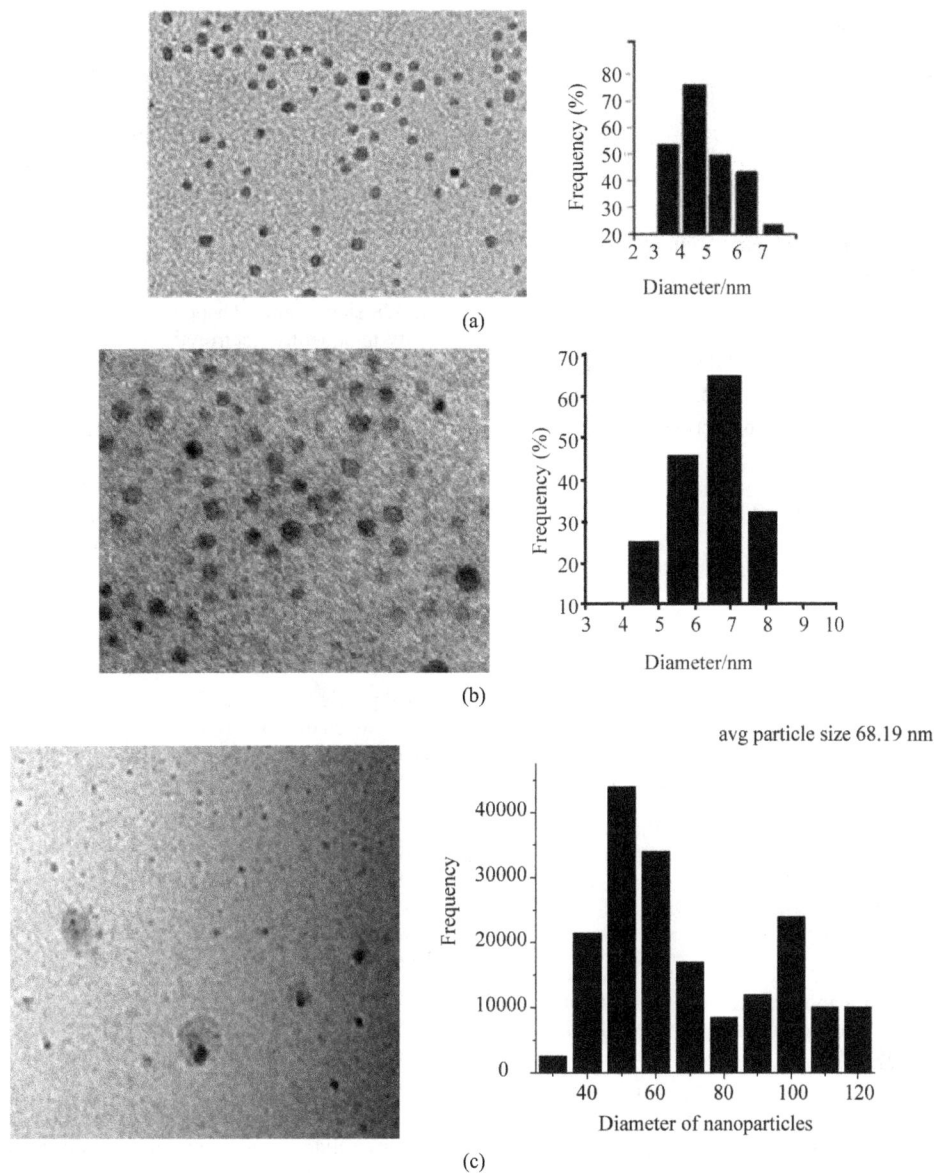

Figure 5. (a) TEM pictures of the iridium nanoparticles prepared by reduction method (b) using PVP as protective agent; (b) TEM pictures of the iridium nanoparticles prepared by reduction method (a) using PVA as protective agent; (c) TEM pictures of the iridium nanoparticles prepared by reduction method (a) using POLE as protective agent.

prevents close contact of metal nanoclusters to each other as demonstrated in (**Scheme 1**) [20].

The interaction between the surface of the metal particles and the polymers is considered to be hydrophobic. However coordination of the polymer to metal particles has been proposed by the shift of the >C=O stretching in the IR spectra of the PVP and PVA surrounding nanoparticles. It is reported that carbonyl group of PVP partly co-ordinate to the surface Pt and Pd atoms of Pt and Pd nanoparticles respectively [21-23]. PVP/PVA may play the same role for Au and other noble metal nanoparticles (Ir-nano).The large particle size and less stability of

POLE capped Ir-nano may be due to less interaction of train part of the POLE and nanoparticle.

4. CONCLUSION

The iridium nanoclusters using PVP, PVA and POLE as protecting agent have been successfully synthesized and characterized by different methods. The study show that particles which were stabilized by PVP and PVA are small and remain stabilized even after six months but POLE stabilized particles are large in size and less stable. Comparison of the data show that as the boiling point of oxidant increases size of particle increases. It is also

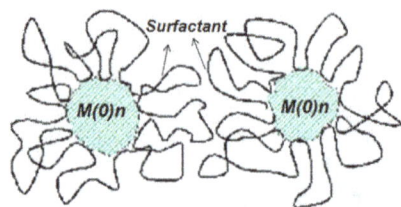

Scheme 1. The schematic representation of the steric stabilization of transition metal-nanoclusters.

clear from the data that increase in NaOH concentration does not show any remarkable effect on particle size but the decrease in oxidant concentration increases the size distribution.

5. ACKNOWLEDGEMENTS

The authors are gratefully thankful to the support of UGC for providing financial assistance.

REFERENCES

[1]   Roucoux, A., Schulz, J. and Partin, H. (2002) Reduced transition metal colloids: A novel family of reusable catalyst. *American Chemical Society*, **102**, 3757-3778.

[2]   Kishore, P.S., Viswanathan, B. and Varadarajan, T.K. (2008) Synthesis and characterization of metal nanoparticle embedded conducting polymer-polyoxometalate composites. *Nanoscale Research Letters*, **3**, 14-20. doi:10.1007/s11671-007-9107-z

[3]   Miyazaki, A., Balint, I. and Nakano, Y. (2003) Morphology control of platinum nanoparticles and their catalytic properties. *Journal of Nanoparticles Research*, **5**, 69-80. doi:10.1023/A:1024451600613

[4]   Gledison, S.F., Giovanna, M., Sergio, R.T., Gerhard, H.F., Jonder, M., Maria, C.M. and Jairton, D. (2006) Synthesis and characterization of catalytic iridium nanoparticles in imidazolium ionic liquids. *Journal of Colloid and Interface Science*, **301**, 193-204. doi:10.1016/j.jcis.2006.04.073

[5]   Zhao, Y.-X., Hernandez-Pagan, E.A., Vargas-Barbosa, N.M., Dysart, J.L. and Mallouk, T.E. (2011) A high yield synthesis of ligand-free iridium oxide nanoparticles with high electrocatalytic activity. *The Journal of Physical Chemistry Letters*, **2**, 402-406. doi:10.1021/jz200051c

[6]   Cheng, J.-B., Zhang, H.-M., Ma, H.-P., Zhong, H.-X. and Zou, Y. (2009) Prepration of $Ir_{0.4}Ru_{0.6}MoxOy$ for oxygen evolution by modified Adams' fusion method. *International Journal of Hydrogen Energy*, **34**, 6609-6613. doi:10.1016/j.ijhydene.2009.06.061

[7]   Yu, W., Liu, M., Liu, H. and Zheng, J. (1999) Prepration of polymer-stabilized nobel metal colloids. *Journal of Colloid and Interface Science*, **210**, 218-221. doi:10.1006/jcis.1998.5938

[8]   Bonet, F., Delmas, V., Grugeon, S., Herrera-Urbina, R.,

Silvert, P.-Y. and Tekaia-Elhsissen, K. (1999) Synthesis of monodisperse Au, Pt, Pd, Ru and Ir nanoparticle in ethylene glycol. *Nanostructured Materials*, **11**, 1277-1284. doi:10.1016/S0965-9773(99)00419-5

[9]   Tu, W. and Liu, H. (2000) Rapid synthesis of nanoscale colloidal metal clusters by microwave irradiation. *Journal of Materials Chemistry*, **10**, 2207-2211. doi:10.1039/b002232m

[10]   Chen, C. W., Tano, D. and Akashi, M. (2000) Colloidal platinum nanoparticles stabilized by vinyl polymers with Amide side chains; Disperson stability and catalytic activity in aqueous electrolytic solution. *Journal of Colloid and Interface Science*, **225**, 349-358. doi:10.1006/jcis.2000.6731

[11]   Solís, D. Vigueras-Santiago, E., Hernández-López, S., Gómez-Cortés, A., Aguilar-Franco, M. and Camacho-López, M.A. (2008) Textural, structural and electrical properties of TiO_2 nanoparticles using Brij-35 and P-123 as surfactants. *Science and Technology of Advanced Materials*, **9**, 025003.

[12]   Goel, A. and Sharma, S. (2009) Synthesis and characterization of nanocrystalline iridium oxide. *International Transactions in Applied Science*, **12**, 243-251.

[13]   Kundu, S. and Liang, H. (2011) Shape-selective formation and characterization of catalytically active iridium nanoparticle. *Journal of Colloid and Interface Science*, **354**, 597-606. doi:10.1016/j.jcis.2010.11.032

[14]   Baglio, V., Blasi, A.D., Denaro, T., Antonucci, V., Aricò, A.S., Ornelas, R., Matteucci, F., Alonso, G., Morales, L., Orozco, G. and Arriaga, L.G. (2008) Synthesis, characterization and evaluation of IrO_2-RuO_2 electrocatalytic powdersfor oxygen evolution reaction. *Journal of New Materials for Electrochemical Systems*, **11**, 105-108.

[15]   Hirai, H., Chawanya, H. and Toshima, N. (1985) Colloidal palladium protected with poly(N-vinyl-2-pyrrolidone) for selective hydrogenation of cyclopentadiene. *Reactive Polymers*, **3**, 127-141.

[16]   Tu, W.-X. (2008) Study on the interaction between polyvinylpyrrolidone and platinum metal during the formation of the colloidal metal nanoparticles. *Chinese Journal of Polymer Science*, **26**, 23-29.

[17]   Marie-Christine D. and Didier A. (2004) Nanoparticles: Assembly, supramolecular chemistry, quantum-size-related properties, and applications toward biology, catalysis, and nanotechnology. *Chemical Reviews*, **104**, 293-346. doi:10.1021/cr030698+

[18]   Ott, L.S., Hornstein, B.J. and Finke, R.G. (2006) A test of the transition-metal nanocluster formation and stabilization ability of the most common polymeric stabilizer, poly(vinylpyrrolidone), as well as four other polymeric protectants. *Langmuir*, **22**, 9357-9367. doi:10.1021/la060934m

[19]   Jia, C.-J. and Schüth, F. (2011) Colloidal metal nanoparticles as a component of designed catalyst. *Physical Chemistry Chemical Physics*, **13**, 2457-2487. doi:10.1039/c0cp02680h

[20]   Narayanan, R. and Mostafa, A.S.El. (2005) Catalysis with transition metal nanoparticles in colloidal solution: Nano-

particle shape dependence and stability. *The Journal of Physical Chemistry B, American Chemical Society*, **109**, 12663-12676.

[21] Toshima, N., Nakata, K. and Kitoh, H. (1997) Gaint platinum clusters with organic ligand: Prepration and catalysis. *Inorganic Chimica Acta*, **265**, 149-153. doi:10.1016/S0020-1693(97)05690-9

[22] Teranishi, T. and Miyake, M. (1998) Size control of palladium nanoparticles and their crystal structures. *Chemistry of Materials*, **10**, 594-600. doi:10.1021/cm9705808

[23] Tan, Y.-W., Dai, X.-H., Y.-F. Li and D.-B. Zhu (2003) Preparation of gold, platinum, palladium and silver nanoparticles by the reduction of their salts with a weak reductant—potassium bitartrate. *Journal of Materials Chemistry*, **13**, 1069-1075.

14

Metal-organic frameworks derived from divalent metals and 4-(1H-1,2,4-triazol-1-yl)benzoic acid

Jia Li, Gao Zhang, Yan-Tao Li, Xin Wang, Jian-Qi Zhu, Yun-Qi Tian[*]

Institute of Chemistry for Functionalized Materials, College of Chemistry and Chemical Engineering, Liaoning Normal University, Dalian, China
Email: [*]yqtian@lnnu.edu.cn

ABSTRACT

Two inorganic-organic coordination polymers, $[Cd(TBA)_2] \cdot 3H_2O$ (1) and $[Cu(TBA)_2] \cdot 2H_2O$ (2) have been synthesized by reaction of the metal ions of Cd(II) and Cu(II) with the ligand HTBA [HTBA = 4-(1H-1,2,4-triazol-1-yl)benzoic acid] under solvothermal condition. The compounds were characterized by single-crystal X-ray diffraction, elemental analysis, IR spectroscopy, X-ray powder diffraction (XRPD), and thermo gravimetric analysis (TGA). The structural analyses reveal the two independent polymers exhibit different structures. Complex 1 exhibits a 3D framework with 1D channels, and complex 2 shows a 2-fold interpenetrating 3D diamond net.

Keywords: Coordination Polymers; 4-(1H-1,2,4-triazol-1-yl)benzoic Acid; Crystal Structure; Cadmium; Copper

1. INTRODUCTION

The reticular design and synthesis of metal-organic frameworks (MOFs) are currently the hot topic for discovery of the materials with potential applications in magnetism [1-6], catalysis [7-10], luminescence and chemical sensing [11-15], as well as gas adsorption and/or separation [16-18]. However, any of a successful construction of the target MOFs shall practically depend on a deliberate design or a judicious selection of the organic ligand incorporating with a proper geometrical metal. For example, the MOFs with zeolitic or zeolite-like structures are constructed by using imidazole and its derivates incorporating with tetrahedral coordinated metals [19-20]. To create novel MOFs having the 4-connected zeolitic structures, the MOFs based on imidazole derivate of 4-(1H-imidaozol-1-yl)benzoic acid (HIBA) and divalent six-coordinated metals have been reported in our previous work [21,22], where the ligand presents two $((2.21)_{syn\text{-}syn}$

and $(2.11)_{syn})$ of the several probable coordination models (Scheme 1).

The ligand of 4-(1H-1,2,4-triazol-1-yl)benzoic acid (HTBA) is structurally similar to HIBA and can be facilely prepared by the following procedure: So far we know that the use of 4-(1H-1,2,4-triazol-1-yl)benzoic acid (HTBA) to construct a coordination polymer remains still unexplored until this work. Compared with the ligand of HIBA, HTBA shows only small structural difference with one carbon being replaced by a nitrogen atom. Thus, HTBA shall also form the various probable coordination modes shown in Scheme 1. By virtue of this consideration, novel MOF structures of HTBA ligand were expected. In our current work, we successfully obtained two novel 3D coordination polymers formulated as $[Cd(TBA)_2] \cdot 3H_2O$ (1), $[Cu(TBA)_2] \cdot 2H_2O$ (2), which are characterized by X-ray single crystal analysis, elemental analysis, IR spectroscopy, X-ray powder diffraction (XRPD), and thermo gravimetric analysis (TGA).

2. EXPERIMENTAL SECTION

2.1. Materials and Methods

All chemicals and solvents used in the syntheses were commercially available and used without further purification. The elemental analyses were carried out on a Perkin-Elmer 2400 elemental analyzer. The IR spectra were recorded (400 - 4000 cm^{-1}) on a FT-IR spectrometer Bruker TENSOR 27. X-ray powder diffraction (XRPD) analyses were carried out on a Bruker D8 Advance and thermal gravimetric analyses (TGA) was performed under static air atmosphere with a heating rate of 10°C/min by using a Perkin-Elmer Diamond thermo gravimetric analyzer.

2.2. X-Ray Single-Crystal Structure Determination

Suitable single crystals of two complexes 1 and 2 were mounted on glass fibers for X-ray measurement. Crys-

[*]Corresponding author.

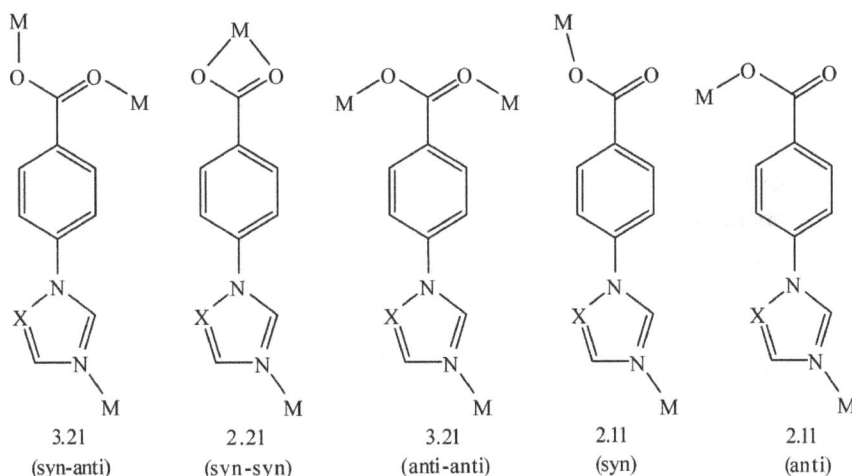

Scheme 1. The probable coordination modes of ligand (X=C or N atom).

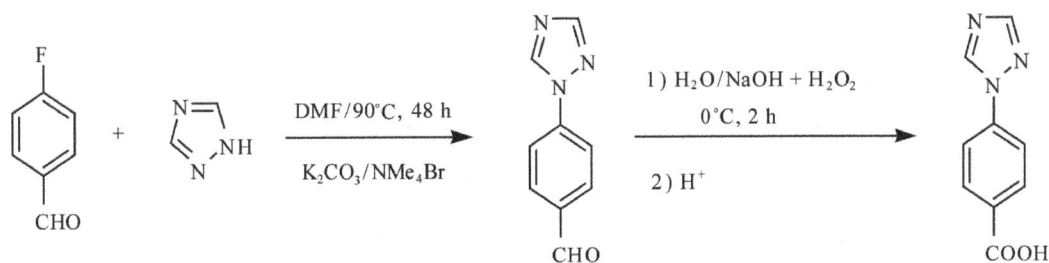

Scheme 2. Synthesis pathway of the ligand.

tallographic measurements for **1** & **2** were carried out on a Bruker SMART-Apex-II CCD diffractometer with graphite-monochromatized MoKα radiation ($\lambda = 0.71073$ Å) at 293(2) K. All absorption corrections were performed using the SADABS program [23]. Crystal structures were solved by the direct method. All non-hydrogen atoms were refined anisotropically. Hydrogen atoms of carbon atoms were fixed at calculated positions with isotropic thermal parameters, while the free lattice water molecules were located by difference Fourier maps. All calculations were performed using the SHELX-97 program [24]. However, hydrogen atoms on O1W and O2W in **1** are refined isotropically and hydrogen atom of O1W, O2W in **2** could not be located. Crystal data and details of the data collection and the structure refinement are given in **Table 1**. The selected bond lengths and bond angles of complexes **1** and **2** are listed in **Table 2**.

2.3. Synthesis of the Organic Ligands

4-(1H-1,2,4-triazol-1-yl)-benzaldehyde: a mixture of 4-fluorobenzaldehyde (5.58 g, 45 mmol), 1,2,4-triazol (2.05 g, 30 mmol), anhydrous K$_2$CO$_3$ (4.14 g, 30 mmol) and tetramethylammonium bromide (0.159 g, 0.6 mmol) in anhydrous DMF (50 mL) was stirred under nitrogen atmosphere at 90°C for 48 h. After the reaction was completed, the solvent was removed by vacuum distillation.

Then, ice water was added into the residue and the white needle product was obtained by filtration. Yield: 4.2 g (96%).

4-(1H-1,2,4-triazol-1-yl)benzoic acid (HTBA): 4-(1H-1,2,4-triazol-1-yl)-benzaldehyde (3.46 g, 20 mmol) and 0.8 g NaOH (10 mmol) were placed in a 200 mL flask, stirred at 0°C, then 30% H$_2$O$_2$ was added dropwisely. After the mixture was stirred at 0°C for 2 h and refluxed for 12 h, it was filtrated. The filtrate was then adjusted to pH 6.0 with dilute hydrochloric acid to obtain solid product. The product was washed with water and dried to obtain a white solid of 4-(1H-1,2,4-triazol-1-yl)benzoic acid (**Scheme 2**). Yield: 3.02 g (80%).

2.4. Synthesis of the Two Complexes

[Cd(TBA)$_2$]·3H$_2$O (**1**): 0.0410 g (0.2 mmol) of HTBA was dissolved in 7 mL of DMF, and 0.0308 g (0.1 mmol) of Cd(NO$_3$)$_2$·4H$_2$O was added and stirred at room temperature for 1 h. The reaction mixture was sealed in a 25 mL Teflon-lined stainless autoclave, heated at 120°C under autogenous pressure for 48 h, and then slowly cooled down to room temperature. The colorless block crystals formed were collected, washed with ethanol, and dried at room temperature; Yield, 0.0195 g; 36% (based on Cd). Formula of the compound (molecular mass): C 40.95 (calc. 41.88); H 3.45 (3.49); N 15.44 (16.29)%.

Table 1. Crystallographic data for complexes **1** and **2**.

Complex	1	2
Formula	$C_{18}H_{18}CdN_6O_7$	$C_{18}H_{16}CuN_6O_6$
M (g·mol^{-1})	515.76	475.91
Crystal system	Monoclinic	Orthorhombic
Space group	$C2/c$	$Pnna$
a (Å)	15.629(3)	13.063(2)
b (Å)	15.991(3)	9.9017(18)
c (Å)	9.2343(19)	17.472(3)
α (deg)	90	90
β (deg)	102.774(2)	90
γ (deg)	90	90
V (Å)	2250.6(8)	2260.0(7)
Z	4	4
D_{calc} (Mg/m^3)	1.602	1.399
Temperature (K)	293(2)	293(2)
Crystal size (mm)	0.53 × 0.35 × 0.21	0.56 × 0.32 × 0.20
F (000)	1088	972
μ(Mo-Kα)/mm^{-1}	1.021	1.011
Theta range for data collection (deg)	2.55 - 25.00	1.95 - 26.00
Reflections collected	5526	9407
Independent reflections	1980	2231
Parameters	151	140
$\Delta(\rho)$ (e·Å^{-3})	0.825 and −0.266	0.796 and −0.419
Goodness of fit	1.037	1.130
Final R indices [$I > 2\sigma(I)$][a]	R1 = 0.0335, wR2 = 0.1047	R1 = 0.0707, wR2 = 0.2010
R indices (all data)[b]	R1 = 0.0420, wR2 = 0.1114	R1 = 0.1269, wR2 = 0.2413

*[a] $R = \sum \|Fo| - |Fc\| / \sum |Fo|$, $wR_2 = \sum w \left(Fo^2 - Fc^2\right)^2 / \sum \left(w\left(Fo^2\right)^2\right)^{1/2}$; [$Fo > 4\sigma(Fo)$]; [b] Based on all data.

IR spectrum (KBr, cm^{-1}): 3415(s), 3117(s), 1602(s), 1535(vs), 1381(vs), 1278(w), 1148(w), 974(w), 787(m). (see **Figure S5** in supporting information).

[Cu(TBA)$_2$]·2H$_2$O (**2**): 0.0410 g (0.2 mmol) of HTBA was dissolved in 7 mL of DMF, and 0.02625 g (0.1 mmol) of Cu(ClO$_4$)$_2$·6H$_2$O was added and stirred at room temperature for 1 h. The reaction mixture was sealed in a 25 mL Teflon-lined stainless autoclave, heated at 100°C for 48 h under autogenous pressure, and slowly cooled to room temperature. The blue block crystals formed were collected, washed with ethanol, and dried at room temperature; Yield, 0.0216 g; 43.7% (based on Cu). Formula of the compound (molecular mass): C 45.64 (calc.45.39); H 3.71 (3.36); N 16.91 (17.65)%.

IR spectrum (KBr, cm^{-1}): 3428(m), 3098(m), 1673(m), 1609(s), 1552(vs), 1393(vs), 1281(m), 980(m), 850(m), 779(m), 669(w) (see **Figure S6** in supporting information).

3. RESULTS AND DISCUSSION

3.1. Synthesis of the Compounds of 1 and 2

The two compounds were prepared under solvothermal conditions in DMF. The metal ions employed in the synthesis are divalent six-coordinated transition metal cations because we hope that they can incorporate with the ligands of HTBA just as with the ligand of HIBA to form the neutral four-connected frameworks. So far, only two MOFs of **1** and **2**, representing two different structural types, have been obtained by reaction of Cd(II) and Cu(II) with the ligand of HTBA. Their crystal structures are determined and their X-ray powder diffraction patterns on the bulk samples were experimentally obtained that were in agreement with those simulated from their X-ray single-crystal data, revealing that the structures of the single crystal **1** and **2** represent those of the bulk samples.

However, the other Cd and Cu group compounds

failed to obtain, such as Zn and Ag.

3.2. Structural Description of Complexes 1 and 2

The complex **1** crystallizes in monoclinic, space group C2/c. The asymmetry unit of **1** contains half a cadmium ion, one TBA$^-$ ligand and one highly disordered lattice water molecule. The coordination environment of the cadmium ion (Cd) is depicted in **Figure 1**. In the structure of **1**, each of the metal Cd(II) ion is six-coordinated by two nitrogen atoms (N1 and N1#3) from two different TBA$^-$ ligands and four carboxylate oxygen atoms (O1#4, O1#5, O2#1, O2#2) from another four different TBA$^-$

Table 2. Selected bond distances (Å) and angles (deg) of complexes **1** and **2**.

Complex 1					
Cd-O2$^{\#1}$	2.303(3)	Cd-O2$^{\#2}$	2.303(3)	Cd-N1$^{\#3}$	2.342(3)
Cd-N1	2.342(3)	Cd-O1$^{\#4}$	2.343(2)	Cd-O1$^{\#5}$	2.343(2)
O2$^{\#1}$-Cd-O2$^{\#2}$	180.00(15)	O2$^{\#1}$-Cd-N1$^{\#3}$	89.39(9)	O2$^{\#2}$-Cd-N1$^{\#3}$	90.61(9)
O2$^{\#1}$-Cd-N1	90.61(9)	O2$^{\#2}$-Cd-N1	89.39(9)	N1$^{\#3}$-Cd-N1	180.00(16)
O2$^{\#1}$-Cd-O1$^{\#4}$	97.23(11)	O2$^{\#2}$-Cd-O1$^{\#4}$	82.77(11)	N1$^{\#3}$-Cd-O1$^{\#4}$	101.13(9)
N1-Cd-O1$^{\#4}$	78.87(9)	O2$^{\#1}$-Cd-O1$^{\#5}$	82.77(11)	O2$^{\#2}$-Cd-O1$^{\#5}$	97.23(11)
N1$^{\#3}$-Cd-O1$^{\#5}$	78.87(9)	N1-Cd-O1$^{\#5}$	101.13(9)	O1$^{\#4}$-Cd-O1$^{\#5}$	180.00(15)
Complex 2					
Cu-O1$^{\#1}$	1.975(4)	Cu-O1$^{\#2}$	1.975(4)	Cu-N1$^{\#3}$	1.994(5)
Cu-N1	1.994(5)				
O1$^{\#1}$-Cu-O1$^{\#2}$	90.6(3)	O1$^{\#1}$-Cu-N1$^{\#3}$	165.53(19)	O1$^{\#2}$-Cu-N1$^{\#3}$	89.3(2)
O1$^{\#1}$-Cu-N1	89.3(2)	O1$^{\#2}$-Cu-N1	165.53(19)	N1$^{\#3}$-Cu-N1	94.3(3)

Symmetry transformations used to generate equivalent atoms: #1 −x + 3/2, −y + 3/2, −z + 2; #2 x − 1/2, y + 1/2, z; #3 −x + 1, −y + 2, −z + 2; #4 −x + 3/2, y + 1/2, −z + 3/2, #5 x − 1/2, −y + 3/2, z + 1/2 for complex 1; #1 x − 1/2, −y + 1/2, z + 1/2 ; #2 −x, y + 1/2, z + 1/2; #3 −x − 1/2, −y + 1, z for complex 2.

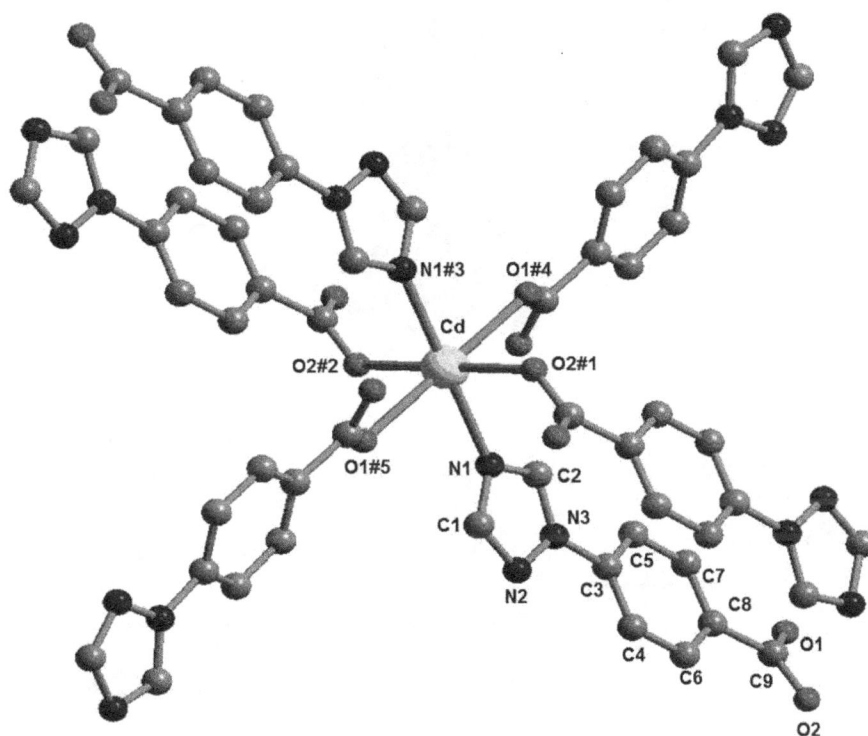

Figure 1. The coordination environment of Cd(II) in complex **1** (The lattice water molecules and hydrogen atoms have been omitted for clarity. Symmetry codes: #1 −x + 3/2, −y + 3/2, −z + 2; #2 x − 1/2, y + 1/2, z; #3 −x + 1, −y + 2, −z + 2; #4 −x + 3/2, y + 1/2, −z + 3/2; #5 x − 1/2, −y + 3/2, z + 1/2).

ligands; Each TBA⁻ ligand shows a dihedral angle of 16.8° between triazole and benzoyl rings and bridges three Cd(II) ions, adopting the $(3.21)_{syn-anti}$ coordination mode (**Scheme 1**). Therefore, all the metal ions in the structure exhibit distorted octahedral coordination geometry with the equatorial positions occupied by the carboxylate oxygens donors and the apical positions taken up by the nitrogen atoms.

The N-Cd-O angles slightly deviate from 90°, revealing the nearly regular octahedron with the equatorial Cd-O distances ranging from 2.303(3) to 2.343(2) Å and the two apical Cd-N distances are identical of 2.343(3) Å. The Cd-N and Cd-O bond distances are similar to that observed in other Cd(II) octahedral coordination polymer [25].

In order to understand the structure thoroughly, the coordinate relationship between the metals and the ligands of complex **1** are analyzed further as follows: We regard that the structure are composed of Cd(II) chains, which are bridged by carboxylate groups of TBA⁻ ligands along [001] direction (Cd⋯Cd distance of 4.617(1) Å)

(**Figure 2(b)**). And then, these chains are linked by the connected six-five-rings of TBA⁻ ligands into a 3D network with 1D channels. (Cd⋯Cd contact of 12.738(6) Å, Cd-Cd-Cd angle of 102.24(0)° and 77.76(0)° respectively) (**Figures 2(a)**, **(c)**).

The complex **2** crystallizes in orthorhombic, space group Pnna. The asymmetric unit of complex **2** consists of half a copper atom, one TBA⁻ ligand and a free water molecule. In the structure of complex 2, each of Cu(II) ions is six-coordinated by four oxygen atoms and two nitrogen atoms from four different TBA⁻ ligands. Due to Jahn-Teller effect, the Cu(II) ions display a distorted octahedral coordination geometry (**Figure 3**) with Cu-O1 bond distance of 1.975(4) Å, Cu-O2 bond distance of 2.516(0) Å and Cu-N bond distance of 1.994(5) Å, which are close to those reported in literature [21,22]. Each TBA⁻ ligand in complex **2** links two Cu(II) and shows dihedral angle of 18.7° between triazole and phenyl rings, adopting $(2.11)_{syn}$ coordination mode (**Scheme 1**).

Thus, each metal center is coordinated by four TBA⁻ ligands and each ligand in turn links two Cu(II) ions,

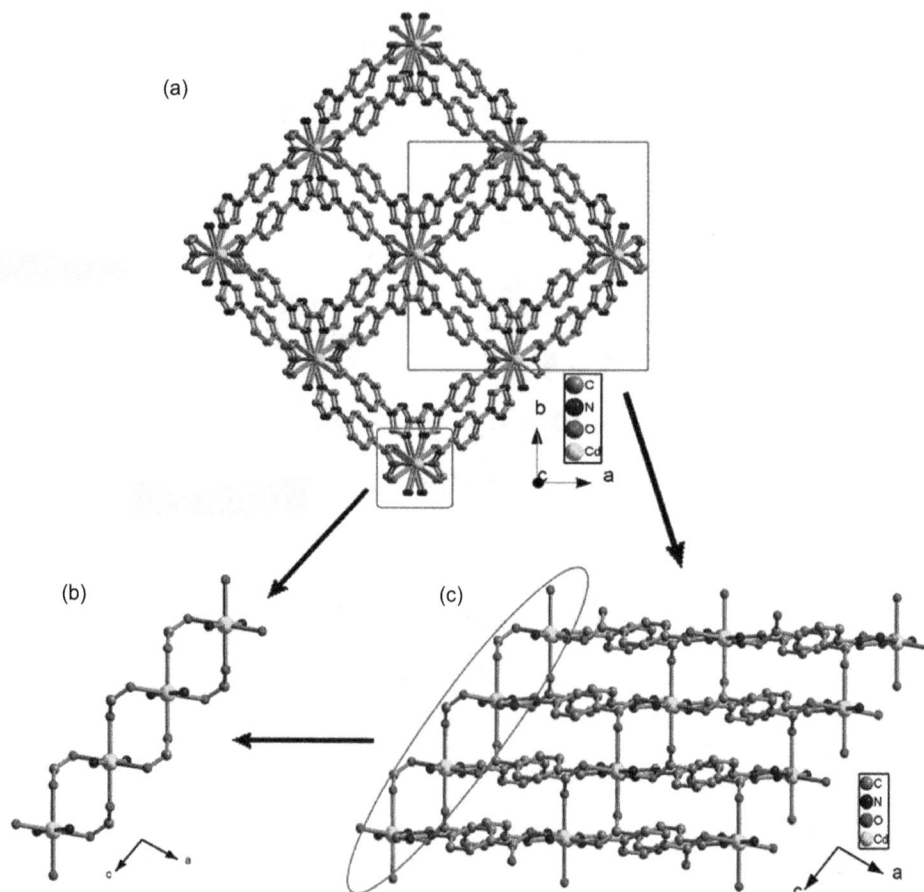

Figure 2. The framework structure of complex **1**: (a) View the whole structure along [001] to show the 1D channels; (b) The Cd(II) chains bridged by carboxylate groups oriented along [010]; (c) View the whole structure along [010] to show the connectivity of the chains with the six-five-rings of the TBA⁻ ligands.

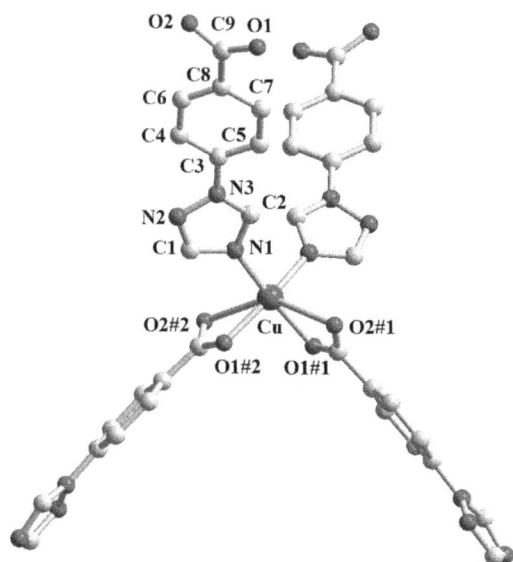

Figure 3. The coordination environment of the Cu(II) in complex **2** (The lattice water molecules and hydrogen atoms have been omitted for clarity. Symmetry codes: #1 x – 1/2, –y + 1/2, z + 1/2 ; #2 –x, y + 1/2, z + 1/2; #3 –x – 1/2, –y + 1, z for 2).

which shows a 4-connected net with diamondoid structure with the adjacent Cu···Cu distance of 11.979(1) Å and the Cu-Cu-Cu angles ranging from 86.35° to 131.18°, deviating significantly from the standard value of 109.5° for diamond network (**Figure 4(a)**). Two set of such nets interpenetrate and form the structures of compound **2** (**Figure 4(b)**), which show 1D channels of 6.0 × 3.0 Å window (Van der Waals radii was considered) (**Figure 4(c)**).

3.3. Thermogravimetric and X-Ray Powder Diffraction Analysis

Thermo gravimetric analysis (TGA) was carried out for investigating the thermal stabilities of the coordination polymers. TGA curve of complex **1** exhibits three main steps of weight losses (see **Figure S1** in supporting information). The first weight loss starts at 42°C and completes at 175°C, which corresponds to the release of three free water molecule per molecule of complex. The observed weight loss of 9.7% is close to the calculated value (9.9%). The second weight loss region occurs in the range of 175°C - 430°C. The weight loss is about 33.5% that is corresponding to the decomposition of the framework into $CdCO_3$ and TBA⁻ ligands (calculated 31.77%). The third weight loss region occurs in the range of 520°C - 630°C. The weight loss is about 33.5% that is corresponding to the complete decomposition of the TBA⁻ ligands and $CdCO_3$. The final residue is CdO because the TGA curve does not change with the temperature increase above 630°C (34.67%).

The TGA data of complex **2** indicates three steps of weight lost (See **Figure S2** in supporting information). It firstly loses one and a half of crystaline water molecules and two free water molecule per molecule of complex, the in the temperature range of 50°C - 190°C (observed 15.2%); Then, the second weight loss occurs at the range of 248°C - 300°C. The weight loss is about 14.65% that is corresponding to the decomposition of the framework 480°C. The weight loss is about 54.5% that is corresponding to the complete decomposition of the TBA⁻ ligands and $CuCO_3$. The final residue is CuO because the

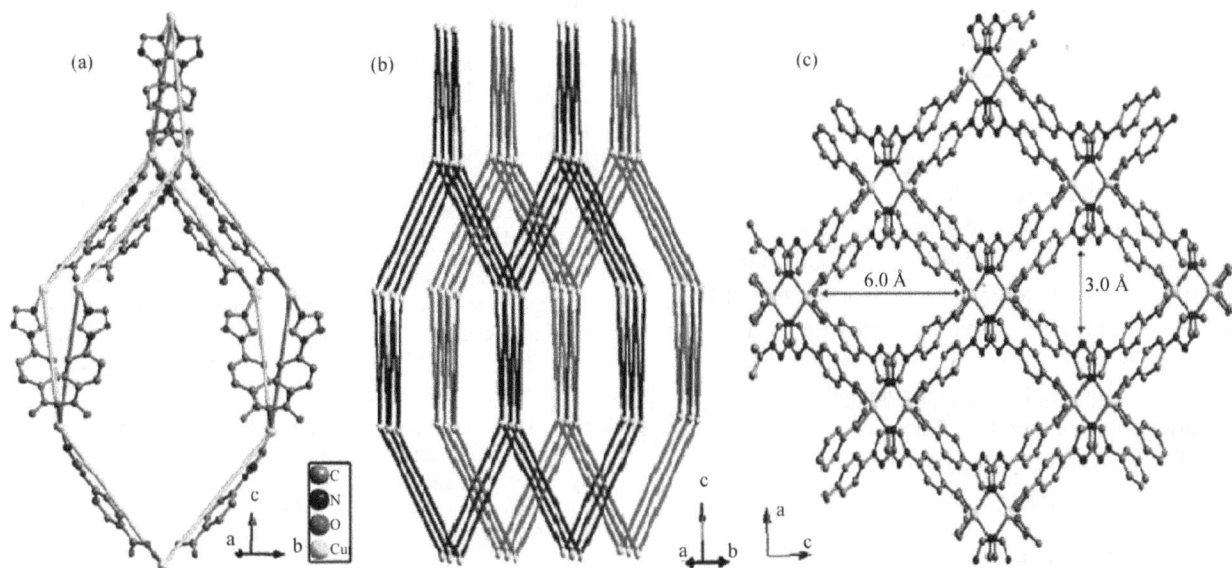

Figure 4. The structure of compound **2**: (a) A diamantane unit of the diamond network; (b) The ball-and-stick diagram to display the 2-fold interpenetrating diamondoid net (balls: metals, sticks: TBA⁻ ligands); (c) View the structure along [010] to show the 1D channels.

into $CuCO_3$ and TBA^- ligands (calculated 14.58%). The third weight loss region occurs in the range of 350°C—TGA curve does not change with the temperature increase above 480°C (56.67%).

The Powder X-ray diffraction (PXRD) was used to confirm the phase purity of the bulk samples of complexes **1** and **2** (see **Figures S3** and **S4** in the supporting information). All the peaks presented in the measured patterns closely match with the simulated patterns generated from single crystal diffraction data.

4. CONCLUSIONS

In conclusion, by use of 4-(1H-1,2,4-triazol-1-yl) benzoic acid (HTBA), two novel 3D coordination polymers of $[Cd(TBA)_2] \cdot 3H_2O$ (**1**) and $[Cu(TBA)_2] \cdot 2H_2O$ (**2**) have been successfully synthesized for the first time under solvothermal conditions. The TBA^- ligand exhibits two different coordination modes in the two compounds. Compound **1** has a tubular net consisting of metal-carboxylate chains linked by TBA^- ligands, which presents $(3.21)_{syn-anti}$ coordination mode that was not seen in the MOFs constructed by HIBA. Compound **2** demonstrates a 4-connected diamond net of 2-fold interpenetrating structure where the metals serve as the nodes and the TBA^- as linkers, which demonstrates $(2.11)_{syn}$ coordination mode that also exists in the MOF constructed by HIBA, but with a two dimensional (4, 4)-network [21,22]. Compared with the numbers of the MOFs constructed by the ligand HIBA, why only two of MOFs could have been produced for the ligand of HTBA is till an open question.

5. SUPPORTING INFORMATION AVAILABLE

TGA, XRD, IR and CIF files for compounds **1** and **2** are available on the http://www.scirp.org/joural/ojic or from the author. The atomic coordinates, isotropic thermal parameters, and complete bond distances and angles have been deposited with the Cambridge Crystallographic Data Center. CCDC Nos. 859904 (**1**) and 859905 (**2**) contain the supplementary crystallographic data for **1** - **2**. These data can be obtained free of charge via http://www.ccdc.cam.ac.uk/conts/retrieving.html, (or from The Cambridge Crystallographic Data Centre, 12, Union Road, Cambridge CB21EZ, UK; E-mail: deposit@ccdc.cam.ac.uk).

6. ACKNOWLEDGEMENTS

We are thankful for financial support from the Foundation for the Author of National Excellent Doctoral Dissertation of PR China (FANEDD) (200733).

REFERENCES

[1] Xiao, D.R., Wang, E.B., An, H.Y., Li, Y.G., Su, Z.M. and Sun, C.Y. (2006) A bridge between pillared-layer and helical structures: A series of three-dimensional pillared coordination polymers with multiform helical chains. *Chemistry—A European Journal*, **12**, 6528-6541. doi:10.1002/chem.200501308

[2] Lee, H.Y., Park, J., Lah, M.S. and Hong, J.I. (2008) One-dimensional double helical structure and 4-fold type [2 + 2] interpenetration of diamondoid networks with helical fashion. *Crystal Growth & Design*, **8**, 587-591. doi:10.1021/cg7007232

[3] Wang, X.L., Qin, C., Wang, E.B., Xu, L., Su, Z.M and Hu, C.W. (2004) Interlocked and interdigitated architectures from self-assembly of long flexible ligands and cadmium salts. *Angewandte Chemie International Edition*, **43**, 5036-5040. doi:10.1002/anie.200460758

[4] Fan, J., Yee, G.T., Wang, G. and Hanson, B.E. (2006) Syntheses, structures, and magnetic properties of inorganic-organic hybrid cobalt(II) phosphites containing bifunctional ligands. *Inorganic Chemistry*, **45**, 599-608. doi:10.1021/ic051286h

[5] Fan, J., Slebodnick, C., Angel, R. and Hanson, B.E. (2005) New zinc phosphates decorated by imidazole-containing ligands. *Inorganic Chemistry*, **44**, 552-558. doi:10.1021/ic0487528

[6] Wu, T., Yi, B.H. and Li, D. (2005) Two novel nanoporous supramo-lecular architectures based on copper(I) coordination polymers with uniform (8, 3) and $(8^2 10)$ nets: *In situ* formation of tetrazolate ligands. *Inorganic Chemistry*, **44**, 4130-4132. doi:10.1021/ic050063o

[7] Choi, H.J., Dinca, M. and Long, J.R. (2008) Broadly hysteretic H_2 adsorption in the microporous metal-organic framework Co (1, 4-benzenedipyrazolate). *Journal of the American Chemical Society*, **130**, 7848-7850. doi:10.1021/ja8024092

[8] Kaye, S.S., Dailly, A., Yaghi, O.M. and Long, J.R. (2007) Impact of preparation and handling on the hydrogen storage properties of Zn_4O (1,4-benzenedicarboxylate)$_3$ (MOF-5). *Journal of the American Chemical Society*, **129**, 14176-14177. doi:10.1021/ja076877g

[9] Dinca, M. and Long, J.R. (2007) High-enthalpy hydrogen adsorption in cation-exchanged variants of the microporous metal-organic framework $Mn_3[(Mn_4Cl)_3(BTT)_8(CH_3-OH)_{10}]_2$. *Journal of the American Chemical Society*, **129**, 11172-11176. doi:10.1021/ja072871f

[10] Han, S.S., Furukawa, H., Yaghi, O.M. and William, A.G. III (2008) Covalent organic frameworks as exceptional hydrogen storage materials. *Journal of the American Chemical Society*, **130**, 11580-11581. doi:10.1021/ja803247y

[11] Wang, Z.X., Shen, X.F., Wang, J., Zhang, P., Li, Y.Z., Nfor, E.N., *et al.* (2006) A sodalite-like framework based on octacyanomolybdate and neodymium with guest methanol molecules and neodymium octahydrate ions.

Angewandte Chemie International Edition, **45**, 3287-3291. doi:10.1002/anie.200600455

[12] Kajiwara, T., Nakano, M., Kaneko, Y., Takaishi, S., Ito, T., Yamashita, M., *et al.* (2005) A single-chain magnet formed by a twisted arrangement of ions with easy-plane magnetic anisotropy. *Journal of the American Chemical Society*, **127**, 10150-10151. doi:10.1021/ja052653r

[13] Liu, T.F., Fu, D., Gao, S., Zhang, Z.Y., Sun, H.L., Su, G. and Liu, Y.J. (2003) An azide-bridged homospin single-chain magnet: [Co(2, 2'-bithiazoline)(N$_3$)$_2$]$_n$. *Journal of the American Chemical Society*, **125**, 13976-13977. doi:10.1021/ja0380751

[14] Freeman, D.E., Jenkins, D.M., Lavarone, A.T., Freedman, D.E., Jenkins, D.M., Iavarone, A.T. and Long, J.R. (2008) A redox-switchable single-molecule magnet incorporating [Re(CN)$_7$]$^{3-}$. *Journal of the American Chemical Society*, **130**, 2884-2885. doi:10.1021/ja077527x

[15] Ohba, M., Kaneko, W., Kitagawa, S. Maeda, T. and Mito, M. (2008) Pressure response of three-dimensional cyanide-bridged bimetallic magnets. *Journal of the American Chemical Society*, **130**, 4475-4484. doi:10.1021/ja7110509

[16] Kou, H.Z., Jiang, Y.B. and Cui, A.L. (2005) Two-dimensional coordination polymers exhibiting antiferromagnetic Gd(III)-Cu(II) coupling. *Crystal Growth & Design*, **5**, 77-79. doi:10.1021/cg049897s

[17] Wang, X.F., Lv, Y., Okamura, T.-A., Kawaguchi, H., Wu, G., Sun, W.Y. and Ueyama, N. (2007) Structure variation of mercury(II) halide complexes with different imidazole-containing ligands. *Crystal Growth & Design*, **7**, 1125-1133. doi:10.1021/cg060814c

[18] Wu, G., Wang, X.F., Okamura, T., Sun, W.Y. and Ueyama, N. (2006) Syntheses, structures, and photoluminescence properties of metal(II) halide complexes with pyridine-containing flexible tripodal ligands. *Inorganic Chemistry*, **45**, 8523-8532. doi:10.1021/ic060493u

[19] Zhao, X.X., Ma, J.P., Dong, Y.B., Huang, R.Q. and Lai, T.S. (2007) Construction of metal-organic frame-works (M = Cd(II), Co(II), Zn(II), and Cu(II)) based on semirigid oxadiazole bridging ligands by solution and hydrothermal reactions. *Crystal Growth & Design*, **7**, 1058-1068. doi:10.1021/cg060583+

[20] Qin, J., Ma, J.P. and Liu, L.L. (2009) A novel two-dimensional frame-work based on unprecedented cadmium(II) chains. *Acta crystallographica*, **65**, 66-68. doi:10.1107/S0108270109000341

[21] Cui, K.H., Yao, S.Y., Li, H.Q., Li, Y.T., Zhao, H.P., Jiang, C.J. and Tian, Y.Q. (2011) Acentric and chiral four-connected metal-organic frameworks based on the racemic binaphthol-like chiral ligand of 4-(1-H(or methyl)-imidazol-1-yl) benzoic acid. *Crystal Engineering Communication*, **13**, 3432-3437. doi:10.1039/C0CE00789G

[22] Zhang, J.Z., Cao, W.R., Pan, J.X. and Chen, Q.W. (2007) A novel two-dimensional square grid cobalt complex: Synthesis, structure, luminescent and magnetic properties. *Inorganic Chemistry Communication*, **10**, 1360-1364. doi:10.1016/j.inoche.2007.08.016

[23] Sheldrick, G.M. (1996) SADABS. Program for empirical absorption correction for area detector data. University of Göttingen, Göttingen.

[24] Sheldrick, G.M. (1997) SHELXS 97. Program for crystal structure refinement. University of Göttingen, Göttingen.

[25] Seidel, R.W., Goddard, R., Zibrowius, B. and Oppel, I.M. (2011) A molecular antenna coordination polymer from cadmium(II) and 4,4'-bipyridine featuring three distinct polymer strands in the crystal. *Polymers Chemistry Journal*, **3**, 1458-1474. doi:10.3390/polym3031458

Supporting Information

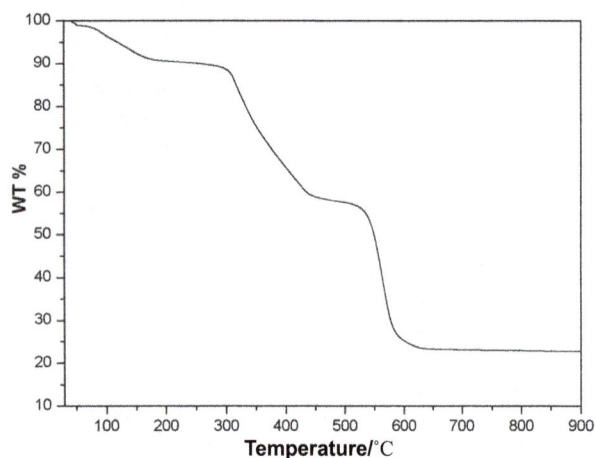

Figure S1. Thermogravimetric analysis curve of compound **1**.

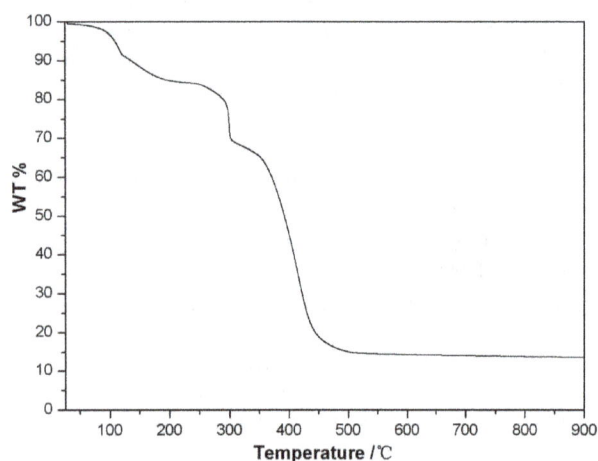

Figure S2. Thermogravimetric analysis curve of compound **2**.

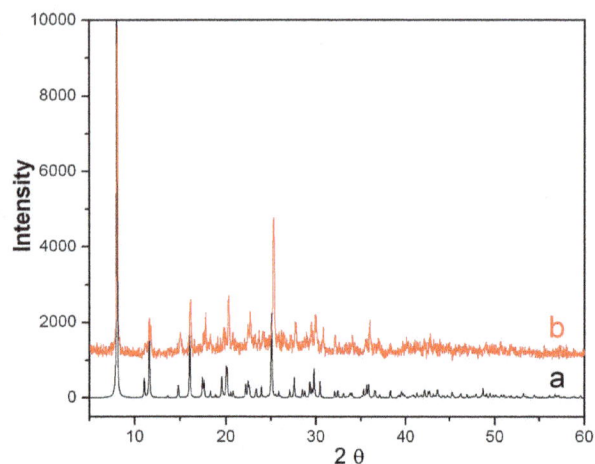

Figure S3. XRD powder patterns: (a) The simulated XPRD pattern calculated from single-crystal structure of complex **1**; (b) Experimental XPRD for complex **1**.

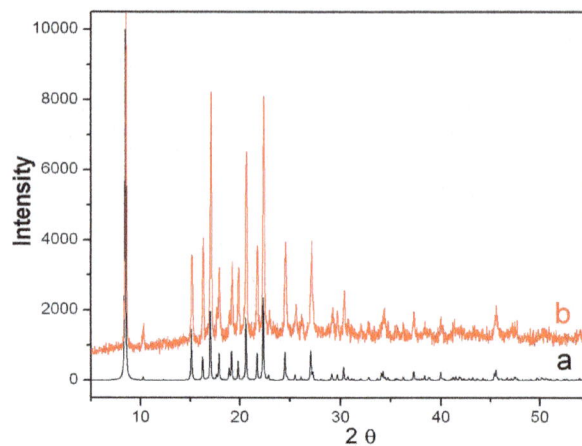

Figure S4. XRD powder patterns: (a) The simulated XPRD pattern calculated from single-crystal structure of complex **2**; (b) Experimental XPRD for complex **2**.

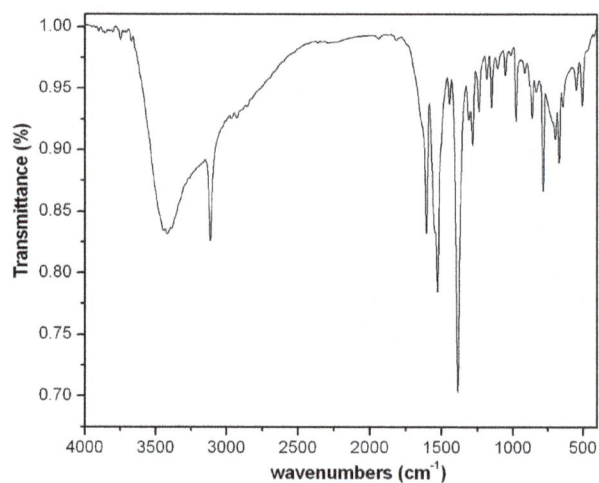

Figure S5. The infrared spectra for as-synthesized **1**.

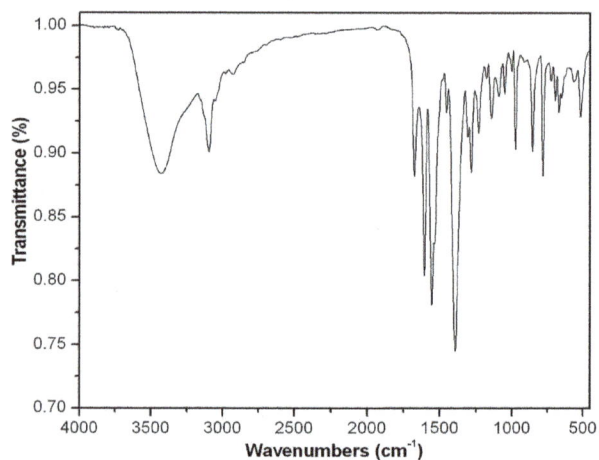

Figure S6. The infrared spectra for as-synthesized **2**.

15

Synthesis, characterization and photophysical properties of quinolin-8-olato chelated osmium(II) organometallics bearing a pendant imine-phenol motif and electrogeneration of trivalent analogue

Bikash Kumar Panda

Department of Chemistry, Jangipur College, Murshidabad, India
Email: panda_bikas@rediffmail.com, bikash.panda66@yahoo.com

ABSTRACT

The reaction of $Os(RL^1)(PPh_3)_2(CO)Br$, **1b**, with quinolin-8-ol (HQ), **2**, has furnished complexes of the type $[Os(RL^2)(PPh_3)_2(CO)(Q)]$, **3**, in excellent yield (RL^1 is C_6H_2O-2-$CHNHC_6H_4R(p)$-3-Me-5, RL^2 is C_6H_2OH-2-$CHNC_6H_4R(p)$-3-Me-5 and R is Me, OMe, Cl). In this process, quinolin-8-olato (Q) undergoes five-membered chelation, the iminium-phenolato function tautomerizing to the imine-phenol function. The trans geometry of the $Os(PPh_3)_2$ fragment is consistent with the occurrence of a single ^{31}P resonance near -6.0 ppm in **3**. In dichloromethane solution, **3** displays a quasireversible $3^+/3$ couple near 0.40 V vs. SCE (3^+ is the osmium (III) analogue of **3**). Coulometrically generated solutions of 3^+ displays a strong absorption near 340 nm, 415 nm and 500 nm and are one-electron paramagnetic (low-spin d^5, $S = 1/2$) and show rhombic EPR spectra in 1:1 dichloromethane-toluene solution at 77 K with g values near 2.44, 2.20, 1.83. Distortion parameters using the observed g values have been computed. Solutions of **3** absorb near 420 nm and emit near 510 nm at 298 K and 580 nm at 77 K. The fluorescence is believed to originate from the ^3MLCT state.

Keywords: Osmium Organometallics; Quinolin-8-Olato Chelation; Emission Properties; Trivalent Osmium

1. INTRODUCTION

The chemistry of osmium has been receiving considerable attention largely because of the variety of reactivities and due to the fascinating electron transfer properties exhibited by complexes of this metal [1-9]. Variation of the coordination environment around osmium plays a key role in modulating the redox properties of its com-

plexes. In this article we studied the reactivity of an osmium system with 8-hydroxyquinoline because complexes incorporating quinoline moiety are important in the context of biological activity [10]. It was also reported in the literature that a series of compounds derived from 8-hydroxyquinoline showed that they could possess strong antifungal, antiasthmatic and antiplatelet activity [10-13]. In particular the chemistry of osmium compounds is of current interest in the context of synthesis, reactivity and photophysical properties [1,9,14-16].

It was demonstrated that the decarbonylative metallation of 4-methyl-2,6-diformyl phenol by $M(PPh_3)_3X_2$ in the presence of primary aromatic amines ($RC_6H_4NH_2$) is known to furnish four-membered ruthenium/osmium organometallics of the type $M(RL^1)(PPh_3)_2(CO)X$ (**1**) juxtaposed to a hydrogen bonded iminium-phenolato function [17-19]. We are scrutinizing the reactivity of these compounds. It has been found that different alkynes undergoes facile insertion into the Ru-C bond of the four-membered ruthenium metallacycles leading to two carbon expansion [20-22] whereas the osmium analogue remains unreactive. Upon treatment with isonitrile ruthenium system undergoes halide substitution as well as insertion into the Ru-O bond and the resultant product then exhibits aryl migration and C-C bond formation [23, 24] while in case of osmium system isonitrile simply displaced bromide ligand furnishing the cationic complexes which are isolated as salts [25]. It is indeed found that ruthenium metallacycle is smoothly converted to the acyl system upon reaction with carbon monoxides under ambient conditions whereas the osmium analogue did not lead to insertion and afforded only an unreactive cis-dicarbonyl complex [26]. We also explore the reactivity of **1** towards bidentate monoanionic donor reagents such as acetate [27], palmitate and terephthalate [28], nitrate and nitrite [29,30] xanthate [31], pyridine-2-thiolate [32] undergo four-membered chelation and we also reported earlier that **1a** reacts with the electroneutral diimine ligands [33] such as bipyridine, phenanthroline and biimidazole [34] as well as 8-hydroxyquinoline [35] furnish five-membered ruthenium organometallics.

The above importance of osmium complexes and the richness of this reaction chemistry have now prompted us to explore the reactivity of **1b** towards quinolin-8-ol, **2**, (HQ, H stands for the dissociable phenolic hydrogen) which is suited for five-membered N, O-chelation. A facile reaction is indeed observed furnishing the novel five-membered osmium organometallics for the first time via displacement of Os-O and Os-Br bonds. Changes in the coordination mode and tautomeric state of the Schiff base ligand accompany the synthetic reaction. It may be noted that the chemistry of osmium quinolin-8-olate appears to have relatively less attention [36–39] but to the best of my knowledge organoosmium compounds incorporating 8-hydroxyquninoline are unprecedented. Another point of special interest is that such species are potentially luminescent in the visible region and are electrochemically oxidizable to the trivalent analogue.

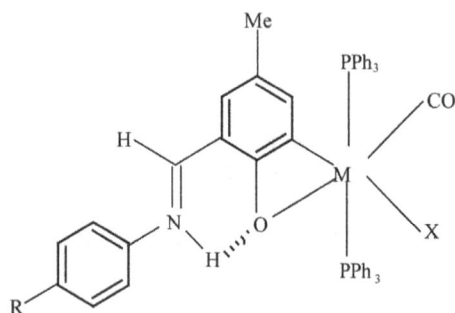

M(RL^1)(PPh$_3$)$_2$(CO)X

1a   M = Ru, X = Cl

1b   M = Os, X = Br

1

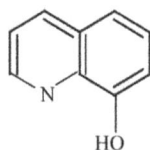

2

2. EXPERIMENTAL

2.1. Materials

The compounds Os(PPh$_3$)$_3$Br$_2$ [40] and Os(RL^1)(PPh$_3$)$_2$-(CO)Br [17] were prepared by reported methods. 8-hydroxyquinoline was obtained from Merck. The purification of dichloromethane and the preparation of tetraethyl ammonium perchlorate (TEAP) for electrochemical work were done as described in previous work [41]. All other

chemicals and solvents were of analytical grade and were used as received.

2.2. Physical Measurements

Electronic, IR and fluorescence spectra were recorded with a Shimadzu UV-1601 PC spectrophotometer, Nicolet Magna IR series II spectrometer and Perkin-Elmer model LS 55 luminescence spectrometer respectively. ^{1}H NMR spectra were obtained using a Bruker 300 MHz FT NMR spectrometer. The numbering scheme used for the ^{1}H NMR is the same as shown in drawing **3**. Spin structures are abbreviated as: s, singlet; d, doublet; t, triplet; m, multiplet. ^{31}P NMR spectra were recorded with the help of a Varian 300 MHz spectrometer using H$_3$PO$_4$ as standard. Microanalyses (C, H, N) were done by using a Perkin-Elmer 240C elemental analyzer. The magnetic behavior of the complexes was examined by a PAR 155 vibrating sample magnetometer. EPR spectra were recorded on a Varian E-109C X-band spectrometer fitted with a quartz dewar. Solution electrical conductivity was measured in acetone with a Phillips PR 9500 bridge using a platinized electrode (cell constant of 1.05). Electrochemical measurements were performed in a nitrogen atmosphere in dichloromethane solution using a CHI model 620A electrochemical analyzer. The supporting electrolyte was tetraethyl ammonium perchlorate and potentials are referenced to the saturated calomel electrode.

2.3. Synthesis of the Complexes [Os(RL^2)(PPh$_3$)$_2$(CO)(Q)] (3)

[Os(RL^2)(PPh$_3$)$_2$(CO)(Q)] **(3)** complexes were synthesized in excellent yield (~80%) by reacting Os(RL^1)-(PPh$_3$)$_2$(CO)Br (**1b**) in MeOH-CH$_2$Cl$_2$ (2:1) at room temperature with excess HQ. Details of a representative case are given below. The other compounds are prepared in an analogous manner.

2.3.1. [Os(MeL^2)(PPh$_3$)$_2$(CO)(Q)], 3(Me)

To a vigorously stirred violet solution of Os(MeL^1)-(PPh$_3$)$_2$(CO)Br (100 mg, 0.09 mmol) in a mixture of 50 ml methanol and 25 ml dichloromethane was added 65 mg (0.45 mmol) of HQ. The mixture was refluxed for 12 h when the colour of the solution changed from violet to yellow. The solvent was then removed under reduced pressure and the yellow solid thus obtained was washed with a little amount of methanol (removal of excess HQ). The residue was recrystallized from dichloromethane-hexane (1:3) mixture followed by drying in vacuo. Yield: 86 mg (81%).

Analysis: Calc. for C$_{61}$H$_{50}$N$_2$O$_3$P$_2$Os: C, 65.93; H, 4.53; N, 2.69%. Found C, 65.88; H, 4.51; N, 2.64%. ^{1}H NMR (300 MHz, CDCl$_3$, ppm, 298 K): 12.56 (s, 1H, O–H, disappeared upon shaking with D_2O), 7.80 (s, 1H, H7), 5.98 (s, 1H, H3), 1.92 (s, 3H, 4-Me), 2.32 (s, 3H, 13-Me), 8.01 (d, J_{HH} = 6.2 Hz, 1H, H16), 7.19 (d, J_{HH} = 8.5 Hz, 1H, H18), 6.11 (d, J_{HH} = 7.6 Hz, 1H, H19), 6.45 (d, J_{HH} = 8.2 Hz, 1H, H21), 6.33 - 6.38 (m, 2H, H17, H20). 6.65 - 7.05 (m, 35H, 2PPh$_3$, H5, H11, H12, H14, H15). ^{31}P NMR (300 MHz, CDCl$_3$, ppm, 298 K): –6.196 (s, 2P). IR (KBr, cm^{-1}): 1893 (v_{CO}), 1602 ($v_{C=N}$).

2.3.2. [Os(MeOL^2)(PPh$_3$)$_2$(CO)(Q)], 3(OMe)

Os(MeOL^1)(PPh$_3$)$_2$(CO)Br (100 mg, 0.089 mmol) and HQ (64 mg, 0.441 mmol) were employed. Yield: 84 mg (79%). Analysis: Calc. for C$_{61}$H$_{50}$N$_2$O$_4$P$_2$Os: C, 65.00; H, 4.47; N, 2.48%. Found C, 65.05; H, 4.41; N, 2.50%. ^{1}H NMR (300 MHz, CDCl$_3$, ppm, 298 K): 12.61 (s, 1H, O-H, disappeared upon shaking with D_2O), 7.82 (s, 1H, H7), 6.01 (s, 1H, H3), 1.90 (s, 3H, 4-Me), 3.74 (s, 3H, 13-OMe). 8.02 (d, J_{HH} = 6.5 Hz, 1H, H16), 7.20 (d, J_{HH} = 8.4 Hz, 1H, H18), 6.13 (d, J_{HH} = 7.5 Hz, 1H, H19), 6.48 (d, J_{HH} = 7.9 Hz, 1H, H21), 6.32 - 6.39 (m, 2H, H17, H20), 6.63 - 7.04 (m, 35H, 2PPh$_3$, H5, H11, H12, H14, H15). ^{31}P NMR (300 MHz, CDCl$_3$, ppm, 298 K): –6.270 (s, 2P). IR (KBr, cm^{-1}): 1896 (v_{CO}), 1606 ($v_{C=N}$).

2.3.3. [Os(ClL^2)(PPh$_3$)$_2$(CO)(Q)], 3(Cl)

Os(ClL^1)(PPh$_3$)$_2$(CO)Br (100 mg, 0.088 mmol) and HQ (64 mg, 0.441 mmol) were used. Yield: 83 mg (76%). Analysis: Calc. for C$_{60}$H$_{47}$N$_2$O$_3$P$_2$ClOs: C, 63.68; H, 4.19; N, 2.47%. Found C, 63.65; H, 4.23; N, 2.51%. ^{1}H NMR (300 MHz, CDCl$_3$, ppm, 298 K): 12.51 (s, 1H, O-H, disappeared upon shaking with D_2O), 7.79 (s, 1H, H7), 6.00 (s, 1H, H3), 1.92 (s, 3H, 4-Me). 7.99 (d, J_{HH} = 6.3 Hz, 1H, H16), 7.19 (d, J_{HH} = 8.5 Hz, 1H, H18), 6.12 (d, J_{HH} = 7.3 Hz, 1H, H19), 6.46 (d, J_{HH} = 7.8 Hz, 1H, H21), 6.33 - 6.40 (m, 2H, H17, H20), 6.62 - 7.02 (m, 35H, 2PPh$_3$, H5, H11, H12, H14, H15). ^{31}P NMR (300 MHz, CDCl$_3$, ppm, 298 K): –6.345 (s, 2P). IR (KBr, cm^{-1}): 1892 (v_{CO}), 1596 ($v_{C=N}$).

2.4. Electrochemical Generation of 3$^+$

The complexes of type 3$^+$ were generated in solution by constant potential coulometric oxidation of solutions of type 3. Details of a representative example are noted below.

A solution of 10 mg of 3(Me) in 25 ml dichloromethane solvent containing 35 mg of TEAP was oxidized coulometrically. The oxidation was performed at 0.75 V; n^e = 1.108/1.086 = 1.02, n^e = Q/Q$_1$, where Q$_1$ is the calculated coulomb count for one electron transfer and Q is the coulomb count found after exhaustive electrolysis. A part of the electrogenerated solution (5 ml) of 3(Me)$^+$ was mixed with an equal volume of toluene and the mixture was quickly frozen at 77 K, and then used

for an EPR measurement. Another part of this electro-generated orange-red solution (10 ml) is used quickly for electronic spectral study. The oxidized complex 3(Me)$^+$ was stable in solution at room temperature for few minutes only. Most of it (>90%) underwent reduction to produce 3, associated with some insoluble material of unknown composition. The oxidation of 3 also may be performed chemically using an aqueous solution of Ce^{4+}. In this case the quantity of the byproducts was more than 25%. To date, we were not successful in isolating the trivalent osmium compound 3$^+$ in its pure state. The trivalent complex 3(OMe)$^+$ and 3(Cl)$^+$ was generated in solution similarly as described above.

3. RESULTS AND DISCUSSION

3.1. Coordination with Quinolin-8-olato

In methanol-dichloromethane solution, Os(RL^1)(PPh$_3$)$_2$-(CO)Br (1) reacts smoothly with five fold excess of HQ upon stirring at room temperature, according to, Equation (1), the colour of the

$$1 + HQ \rightarrow 3 + HBr \qquad (1)$$

solution changes from violet to yellow, from which [Os-(RL^2)(PPh$_3$)$_2$(CO)(Q)] (3) is obtained in excellent yield. The R groups utilized in the present work are Me, OMe and Cl. Specific compounds are identified by putting R in parentheses: thus 3(Me) stands for [Os(MeL^2)(PPh$_3$)$_2$-(CO)(Q)]. The complex 3 has been isolated in pure form in ~ 80% yield. An isomer (because quinol-8-ol is unsymmetrical) of 3 is possible in principle but it has never been observed. Careful examination of ^{1}H NMR spectra of crude sample (before recrystallization) did not reveal the presence of any extra signals apart from those characterizing 3. We assign structure 3 to it in analogy with the stereochemistry of other Os^{II}(quinolin-8-olato) complexes [38,39] and also by comparing with acetate, thioxanthate, pyridine-2-thiolato chelated ruthenium/osmium organometallics [27,31,32].

The conversions 1→3 is attended with a prototropic shift within the salicyaldimine function. In 1 the metal is coordinated to phenolato oxygen and the Schiff base function occurs in the zwitterionic iminium-phenolato

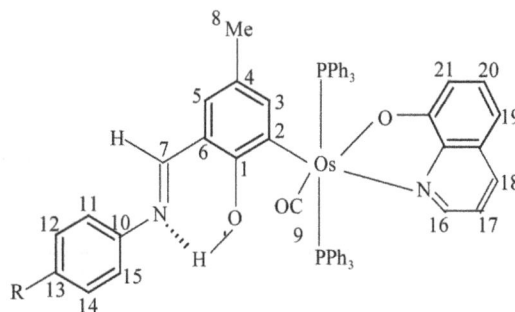

3

tautomeric form **4**. Chelation of quinolin-8-olato is attended with the cleavage of the Os-O bond and the Schiff base function becomes an imine-phenol **5**. This is fully consistent with the spectroscopic data.

Thus the C=N stretching frequency in **3** is significantly lower ($\sim$1600 cm^{-1}) than that in **1** ($\sim$1620 cm^{-1}) as expected [42]. Also the aldimine CH signal in ^{1}H NMR in **3** (**Figure 1**) occurs at lower field viz. 7.8 ppm as compared to $\sim$7.5 ppm in **1** [17]. The O-H resonance in **3** is a relatively sharp signal near 12.5 ppm having half-height width of $\sim$30 Hz. In contrast, the iminium N-H resonance in **1** is broad (width, $\sim$150 Hz) evidently due to the quadrupole moment of the nitrogen atom [17]. The prototropic transformation between **1** and **3** has certain similarities with the imine-iminium tautomerization in rhodopsins [42].

A plausible mechanism [31,32,35] for the displacement of bromide in **1** consists of *cis* attack by Q^- is shown in **6**. The anchored ligand displaces the phenolato oxygen and halide atoms achieving Q^- chelation with concomitant prototropic shift and conformational reorganization as in **7**.

In both **1** and **3**, the carbon monoxide ligand lies cis to the metallated carbon atom. It is however, located cis and trans to the phenolato oxygen in **3** and **1**, respectively which is revealed by our previous study also [32,33]. Thus the Schiff base fragment rotates by $\sim$180° around the metal-carbon bond in course of the reaction of equation 1 due to steric repulsion between the phenolic and the quinolin-8-olato function as highlighted in the hypothetical anti configuration **8** for Os(CO)(Q) fragment.

3.2. Characterization

Osmium organometallics of type **3** are non-electrolytic in

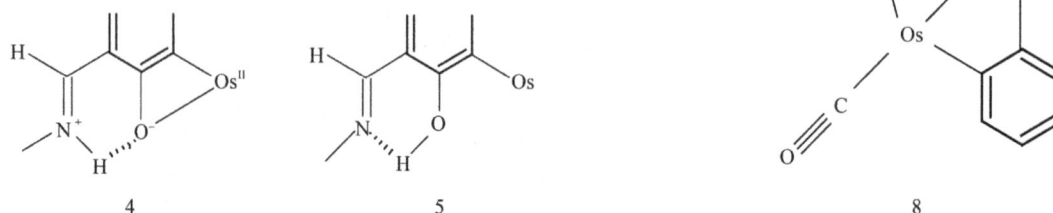

solution and are diamagnetic, consistent with a metal oxidation state of +2 having idealized t_{2g}^6 electronic con-

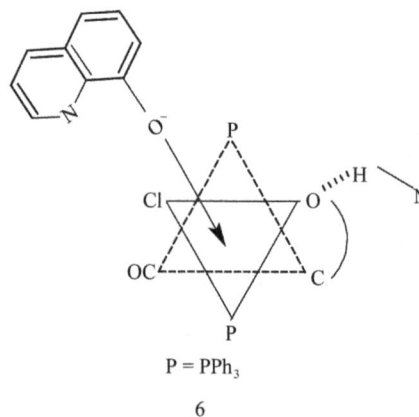

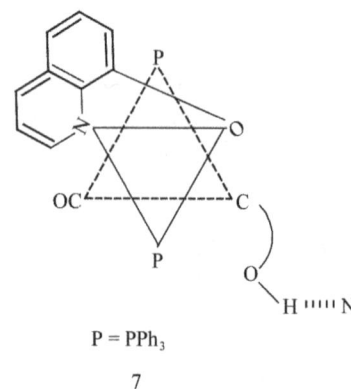

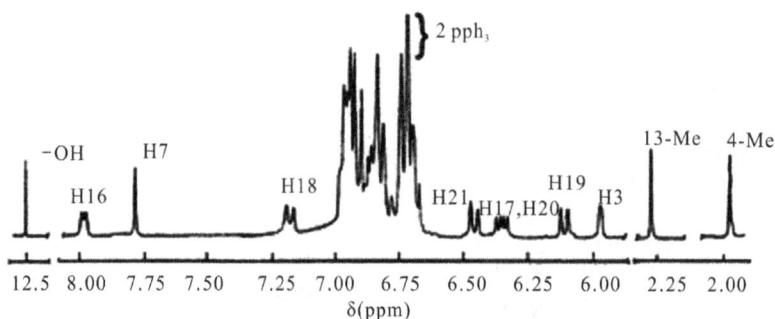

P = PPh$_3$

6

P = PPh$_3$

7

8

4

5

Figure 1. ^{1}H NMR spectrum of **3(Me)** in CDCl$_3$ solution.

figuration. In IR, the CO stretch is seen as a sharp band near 1890 cm^{-1}, the stretch is at slightly lower frequency due to the superior back bonding ability of bivalent osmium. In ^{1}H NMR (**Figure 1**) the 3-H proton of the metallated ring occur as sharp singlets near 6.0 ppm, while the C(4)-Me protons resonate near 1.9 ppm. These protons are subject to shielding by phosphine phenyl rings [17,27,43]. The PPh$_3$, Schiff base (C$_6$H$_4$R) aromatic protons form a complex multiplet in the region 6.60 - 7.05 ppm and the quinolin-8-olate aromatic protons appear in the 6.00 - 8.05 ppm region.

Detailed assignments of the ligand protons are given in the experimental section. In the previous section the chemical shifts of the imine-phenol protons were considered. The complexes uniformly display a single ^{31}P signal for the two phosphine ligands implicating *trans*-Os(PPh$_3$)$_2$ configuration. The resonance occurs near –6.0 ppm, exact data are given in the experimental section. A representative spectrum is shown in **Figure 2**. The redox and photophysical properties of the complexes are examined in later sections.

3.3. Redox Properties and EPR Spectra

In dichloromethane solution **3** displays a quasireversible one-electron cyclic voltammetric response near 0.40 V (peak-to-peak separation is ~100 mV) vs. SCE corresponding to the

$$3^+ + e \rightarrow 3 \qquad (2)$$

couple where 3^+ represents the osmium(III) analogue of **3**. A representative cyclic voltammogram is shown in **Figure 3**. As R is varied, the $E_{1/2}$ values become more positive as the electron withdrawing power of the R group increases (OMe < Me < Cl), see **Table 1**. The reduction potentials are systematically lower than those of the type **1** precursors [17] by ~200 mV, indicating better stabilization of the trivalent state in **3** compared with that in **1**. The one-electron nature of these responses has been confirmed by comparing their current heights with the standard ferrocene/ferrocenium couple under identical experimental conditions. The exhaustive coulometric oxidation at 0.75 V affords a coulomb count corresponding

to one-electron transfer (**Table 1**).

Low potentials for the Os^{II}/Os^{III} couple in the complexes of type **3** persuaded us to try and isolate the corresponding osmium(III) complexes 3^+ in their pure states. Accordingly, we tried to oxidize them both chemically and electrochemically. Unfortunately, the oxidized com-

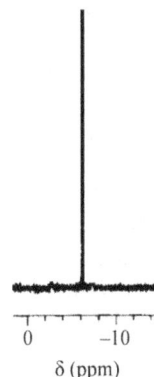

Figure 2. ^{31}P NMR spectrum of **3(OMe)** in CDCl$_3$.

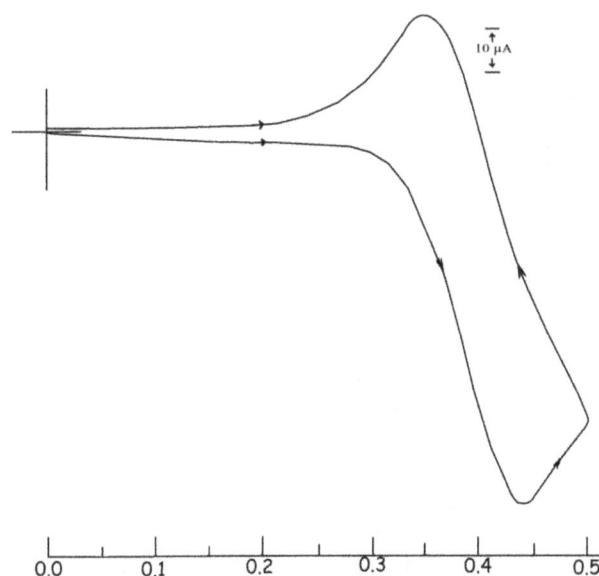

Figure 3. Cyclic voltammogram of **3(OMe)** in dichloromethane solution.

Table 1. Absorption, emission spectral and electrochemical data of 3.

Complexes	UV-Vis data[a] λ_{max}, nm (ε^b, M^{-1}·cm^{-1})		Emission data λ_{max}, nm (Φ_r)[a,c]		Electrochemical data	
			298 K	77 K	E$_{1/2}$, V (ΔE$_p$, mV)[d]	n[e]
3(Me)	346 (18690)	425 (9850)	515 (2.06 × 10^{-3})	573	0.43 (160)	1.02
3(OMe)	354 (20755)	421 (9930)	514 (1.64 × 10^{-3})	588	0.41 (124)	1.08
3(Cl)	349 (16375)	414 (9520)	509 (1.34 × 10^{-3})	582	0.46 (142)	0.97

[a]Solvent: dichloromethane; [b]molar absorption coefficient; [c]excitation at the higher wavelength absorption peak; [d]solvent: dichloromethane; ΔE_p is peak-to-peak separation; n[e] = Q/Q_1 where Q is the observed coulomb count and Q_1 is the calculated coulomb count for one-electron transfer.

plexes were not stable enough for their isolation, as they revert to the parent bivalent complexes rapidly. However, we were able to generate the cationic complexes, **3**$^+$, in solution by the controlled potential bulk electrolysis of **3**.

Trivalent organometallics **3**$^+$, generated coulometrically, have been examined in solution. Their cyclic voltammograms (initial scan cathodic) are virtually superimposible on those of **3** (initial scan anodic), showing that **3**$^+$ retains the gross structure of **3**. Orange-red solutions of **3**$^+$ are characterized by an intense band near 340 nm, 410 nm and 500 nm. A representative spectrum is shown in **Figure 4** and spectral data are given in **Table 2**. Electron paramagnetic resonance (EPR) spectra of the **3**$^+$ complexes, recorded in 1:1 dichloromethane-toluene solution at 77 K, show rhombic spectra with three distinct signals (g$_1$, g$_2$ and g$_3$, in decreasing order of magnitude). A representative spectrum is shown in **Figure 5** and the spectral data are presented in **Table 2**. The observed rhombicity of the EPR spectra is understandable in terms of the gross molecular symmetry of these complexes, containing the three non-equivalent P-Os-P, O-Os-C and N-Os-C axes. The rhombic distortion can be thought of as a combination of axial distortion (Δ, which splits t$_2$ into a and e) and rhombic distortion (V, which splits e). The splitting pattern is illustrated in **Figure 5**. Spin-orbit coupling causes further changes in the energy gaps. Thus two electronic transitions (transition energies $\Delta E1$ and $\Delta E2$; $\Delta E1 < \Delta E2$) are possible in principle within these three levels. All these energy parameters have been computed (**Table 2**) using the observed g-values, the g-tensor theory of low-spin d^5 complexes and a reported method [44,45]. The axial distortion is observed to be much stronger than the rhombic. The EPR data analysis thus shows that the **3**$^+$ complexes are significantly distorted from ideal octahedral geometry which is observed previously in the crystal structure of acetate [27] and xanthate [31] derivatives. In this connection it should be pointed out that for thioxanthate chelated osmium organometallics we indicate the better stabilization of the trivalent state but isolation of osmium(III) species via electrooxidation has not been succeeded [31] but for

quinolin-8-olato chelated osmium complexes we are able to generate the osmium(III) analogue in solution this is probably due to the five-membered chelation instead of four-membered chelation for the former.

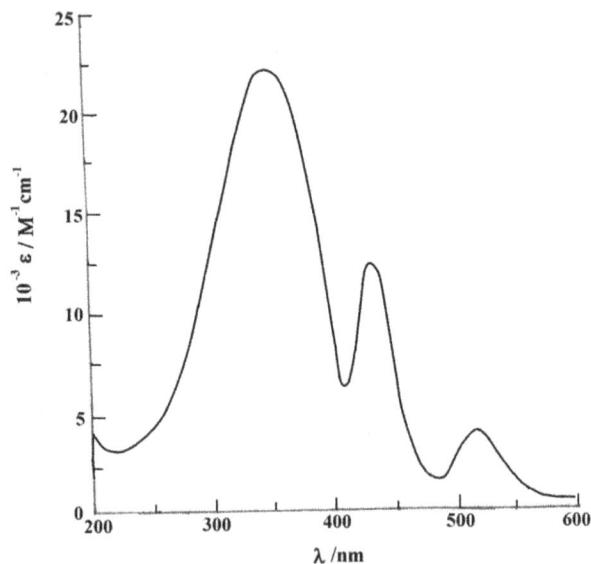

Figure 4. Electronic spectrum of **3(Cl)**$^+$ in CH$_2$Cl$_2$ solution.

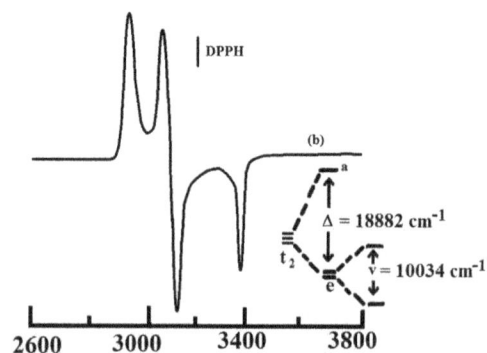

Figure 5. EPR spectra of **3(OMe)**$^+$ in frozen 1:1 dichloromethane-toluene solution at 77 K, showing the computed splitting of the t$_2$ orbitals. DPPH = diphenyl-picrylhydrazyl.

Table 2. Electronic and EPR spectral data of **3**$^+$.

Complexes	UV-Vis data[a] λ_{max}, nm (ε[b], M^{-1}·cm^{-1})			EPR g values[c] and derived parameters[d]						
				g$_1$	g$_2$	g$_3$	Δ	V	ΔE_1	ΔE_2
3(Me)$^+$	339 (23513)	409 (12748)	503 (4052)	2.469	2.161	1.815	18902	10045	14168	24498
3(OMe)$^+$	344 (24580)	414 (13544)	495 (4175)	2.443	2.185	1.843	18882	10034	14125	24455
3(Cl)$^+$	341 (22331)	419 (12418)	509 (3932)	2.442	2.264	1.832	18745	10022	14043	24322

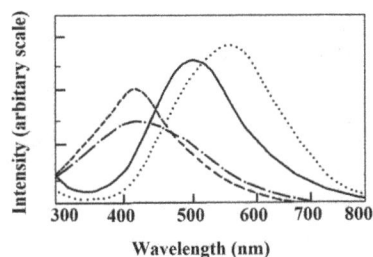

Figure 6. Absorption (– – –), excitation (– • –), and fluorescence spectra at 298 K (——) and at 77 K (• • •) of **3(Cl)** in dichloromethane solution.

3.4. Emission Properties

In dichloromethane solution complexes of type **3** display two allowed absorption bands in the regions 345 - 355 and 410 - 425 nm. The latter band is weaker in intensity and is believed to have significant $d\pi(Os) - \pi^*(Q)$ MLCT character. Selected UV-Vis spectral data are given in **Table 1**. The solutions are fluorescent at room temperature, and also at low temperature (77 K). The peaks lying in the region 505 - 515 and 570 - 590 nm at 298 K and 77 K respectively (**Figure 6**, **Table 1**) makes **3** fluorescent in the visible region. Low-temperature emission spectra are red-shifted from room temperature ones. The complexes are weak emitters, as noted from their quantum yield (Φ_r) values which are one order of magnitude lower than that of the $[Ru(bpy)_3]^{2+}$ standard [46,47] (**Table 1**). Quantum yields are calculated using Equation (3) as described previously [46,47] where Φ_{std} is 0.042 at 298 K, A is the solution absorbance at the excitation wave length, I is the relative emission intensity, η are the refractive index values of the solvent for the sample (subscript r) and the standard reference (subscript std) respectively,

$$\Phi_r = \Phi_{std}\left(A_{std}/A_r\right)\left(I_r/I_{std}\right)\left(\eta_r^2/\eta_{std}^2\right). \qquad (3)$$

Excitation spectral studies show that fluorescence is associated with the absorption band near 420 nm. A representative case is shown in **Figure 6**. Assuming that the band has MLCT character as suggested above, the possible involvement of the ^{3}MLCT state [15,16,33] incorporating $\pi^*(Q)$ contribution is implicated in the emission process. We also note that the parent organometallics of type **1** do not display any fluorescence in the visible region.

4. CONCLUSION

It is demonstrated that the metallacycles of type **1** react smoothly with HQ furnishing aryl osmium organometallics of type **3**. The conversion **1→3** is attended with cleavage of Os-O and Os-Br bonds, (N, O) chelation of quinolin-8-olate, iminium-phenolato→imine-phenol tau-

tomerization. Os^{III}/Os^{II} reduction potentials are systematically lower than those of **1** and **3** is electrooxidizable to the osmium(III) analogue 3^+ characterized by rhombic EPR spectra. The Os→Q MLCT absorption in the visible region makes **3** fluorescent with quantum yields one order of magnitude less than that of the $[Ru(bpy)_3]^{2+}$ standard.

5. ACKNOWLEDGEMENTS

The help received from Professor Animesh Chakravorty, Indian Association for the Cultivation of Science, is gratefully acknowledged. I also thank the Department of Science and Technology, and the Council of Scientific and Industrial Research, New Delhi for financial support.

REFERENCES

[1] Sauvage, J.P., Collin, J.P., Chambron, J.C., Guillerez, S., Coudret, C., Balzani, V., Barigelletti, F., Cola, L.D. and Flamigni, L. (1994) Ruthenium(II) and Osmium(II) bis(terpyridine)complexes in covalently-linked multicomponent systems: Synthesis, electrochemical behavior, absorption spectra, and photochemical and photophysical properties. *Chemical Review*, **94**, 993-1019. doi:10.1021/cr00028a006

[2] Richmond, M.G. (1995) Annual survey of ruthenium and osmium for the year 1993. *Coordination Chemistry Reviews*, **141**, 63-152.

[3] Ward, M.D. (1995) Osmium 1993. *Coordination Chemistry Reviews*, **146**, 99-113. doi:10.1016/0010-8545(96)83891-4

[4] Dyson, P.J., Johnson, B.F.G. and Martin C.M. (1996) The synthesis of ruthenium and osmium carbonyl cluster with unsaturated organic rings. *Cordination Chemistry Reviews*, **155**, 69-86. doi:10.1016/S0010-8545(96)90177-0

[5] Che, C.-M. and Huang, J.-S. (2002) Ruthenium and osmium porphyrin carbene complexes: Synthesis, structure and connection to the metal-mediated cyclopropanation of alkenes. *Coordination Chemistry Reviews*, **231**, 151-164.

[6] Jia, G. (2007) Recent progress in the chemistry of osmium carbine and metallabenzyne complexes. *Coordination Chemistry Reviews*, **251**, 2167-2187. doi:10.1016/j.ccr.2006.11.018

[7] Che, C.-M., Ho, C.-M. and Huang, J.-S. (2007) Metal-carbon multiple bonded complexes: Carbene, vinylidene and allenylidene complexes of ruthenium and osmium supported by macrocyclic ligands. *Coordination Chemistry Reviews*, **251**, 2145-2166.

[8] Esteruelas, M.A., Lopez, A.M. and Olivan, M. (2007) Osmium-carbon double bonds: Formation and reactions. *Coordination Chemistry Reviews*, **251**, 795-840. doi:10.1016/j.ccr.2006.07.08

[9] Laine, P.P., Campagna, S. and Loiseau, F. (2008) Conformationally gated photoinduced processes within photosensitizer-acceptor dyads based on ruthenium(II) and osmium(II) polypyridyl complexes with an appended pyridinium group. *Coordination Chemistry Reviews*, **252**,

2552-2571. doi:10.1016/j.ccr.2008.05.007

[10] Musiol, R., Jampilek, J., Nycz, J.E., Pesko, M., Carroll, J., Kralova, K., Vejsova, M., Mahony, J.O., Coffey, A., Mrozek, A. and Polanski, J. (2010) Investigating the activity spectrum for ring-substituted 8-hydroxyquinolines. *Molecules*, **15**, 288-304. doi:10.3390/molecules15010288

[11] Sissi, C. and Palumbo, M. (2003) The quinolone family: From antibacterial to anticancer agents. *Current Medicinal Chemistry—Anti-Cancer Agents*, **3**, 439-450. doi:10.2174/1568011033482279

[12] Musiol, R., Jampilek, J., Buchta, V., Niedbala, H., Podeszwa, B., Palka, A., Majerz-Maniecka, K., Oleksyn, B. and Polanski, J. (2006) Antifungal properties of new series of quinoline derivatives. *Bioorganic & Medicinal Chemistry*, **14**, 3592-3598. doi:10.1016/j.bmc.2006.01.016

[13] Gershon, H., Gershon, M. and Clarke, D.D. (2004) Synergistic mixtures of fungitoxic monochloro- and dichloro-8-quinolinols against five fungi. *Mycopathologia*, **158**, 131-135. doi:10.1023/B:MYCO.0000038427.42852.6a

[14] Bhaumik, C., Das, S., Maity, D. and Baitalik S. (2012) Luminescent bis-tridentate ruthenium(II) and osmium(II) complexes based on terpyridyl-imidazole ligand: Synthesis, structural characterization, photophysical, electrochemical and solvent dependence studies. *Dalton Transaction*, **41**, 2427-2438. doi:10.1039/c1dt11645b

[15] Pennington, N.S., Richter, M.M. and Carlson, B. (2010) Efficient electrogenerated chemiluminescence from osmium(II) polypyridine systems containing tetraphenylarsine or diphenylphosphine ligands. *Dalton Transaction*, **39**, 1586-1590. doi:10.1039/b912877h

[16] Hwang, K.-C., Chen, J.-L., Chi, Y., Lin, C.-W., Cheng, Y.-M., Lee, G.-H., Pi-Tai Chou, P.-T., Lin, S.-Y. and Shu, C.-F. (2008) Luminescent osmium(II) complexes with functionalized 2-phenylpyridine chelating ligands: Preparation, structural analyses, and photophysical properties. *Inorganic Chemistry*, **47**, 3307-3317. doi:10.1021/ic7023132

[17] Ghosh, P., Bag, N. and Chakravorty, A. (1996) Decarbonylative metallation of diformylphenol Schiff basses: New osmium and ruthenium organometallics incorporating the iminium-phenolato zwitterionic motif. *Organometallics*, **15**, 3042-3047. doi:10.1021/om960018b

[18] Bag, N., Choudhury, S.B., Pramanik, A., Lahiri, G.K. and Chakravorty, A. (1990) Ruthenium(II) phenolates. Synthesis and characterization of a novel four-membered metallacycle. *Inorganic Chemistry*, **29**, 5013-5015. doi:10.1021/ic00350a001

[19] Bag, N., Choudhury, S.B., Lahiri, G.K. and Chakravorty, A. (1990) A novel zwitterionic *o*-metallated Ru(II)-phenolate. *Journal of the Chemical Society, Chemical Communications*, **22**, 1626-1627.

[20] Panda, B.K. (2005) Metallacycle expansion via butadiyne/octadiyne insertion into a four-membered ruthenium organometallics. *Transition Metal Chemistry*, **30**, 712-719. doi:10.1007/s11243-005-6282-6

[21] Ghosh, K., Chattopadhyay, S., Pattanayak, S. and Chakravorty, A. (2001) Alkyne insertion into the Ru-C bond of

a four-membered metallacycle: Insertion rate and reaction pathway. *Organometallics*, **20**, 1419-1423. doi:10.1021/om000649c

[22] Ghosh, K., Pattanayak, S. and Chakravorty, A. (1998) Metallacycle expansion by alkyne insertion. Chemistry of a new family of ruthenium organometallics. *Organometallics*, **17**, 1956-1960. doi:10.1021/om970917f

[23] Panda, B.K., Chattopadhyay, S., Ghosh, K. and Chakravorty, A. (2002) Isonitrile insertion into the Ru-O bond and migratory C-C bond formation. Novel organoruthenium imidic ester and acyl species. *Organometallics*, **21**, 2773-2780. doi:10.1021/om020059

[24] Panda, B.K. (2005) Structurally characterized acylruthenium organometallics bearing a pendant aldehyde function. *Transition Metal Chemistry*, **30**, 488-495. doi:10.1007/s11243-004-6970-7

[25] Panda, B.K. and Chakravorty, A. (2005) Chemistry of a family of osmium(II) metallacycles incorporating isonitrile coordination. *Indian Journal of Chemistry*, **44A**, 1127-1132.

[26] Panda, B.K. and Chakravorty, A. (2005) Carbonylation of four-membered ruthenium and osmium metallacycles incorporating an orthometallated phenolic function: New acylruthenium and arylosmium complexes. *Journal of Organometallic Chemistry*, **690**, 3169-3175. doi:10.1016/j.jorganchem.2005.04.012

[27] Ghosh, P., Pramanik, A. and Chakravorty, A. (1996) Chemistry of a new family of carboxyl chelated ruthenium and osmium aryls incorporating the imine-phenol motif. *Organometallics*, **15**, 4147-4152. doi:10.1021/om960199h

[28] Pan, S. and Panda, B.K. (2005) Synthesis and properties of long chain carboxyl and dicarboxyl chelated ruthenium organometallics incorporating the imine-phenol motif. *Journal of Indian Chemical Society*, **82**, 16-20.

[29] Ghosh, P. and Chakravorty, A. (1997) A family of ruthenium aryls incorporating η^2-bonded nitrite or nitrate and a pendent imine-phenol function. *Inorganic Chemistry*, **36**, 64-69. doi:10.1021/ic960358p

[30] Pramanik, K., Ghosh, P. and Chakravorty, A. (1997) Synthesis and structure of osmium(II) organometallics incorporating a four-membered salicylideneiminium metallacycle and Os-η^1-NO$_2$ binding. *Journal of the Chemical Society, Dalton Transactions*, **9**, 3553-3556.

[31] Chattopadhyay, S., Panda, B.K., Ghosh, K. and Chakravorty, A. (2001) A family of thioxanthato ruthenium and osmium aryls. *Israel Journal of Chemistry*, **41**, 139-144. doi:10.1560/C80W-66CR-QHHL-GYU5

[32] Panda, B.K., Chattopadhyay, S., Ghosh, K. and Chakravorty, A. (2002) Synthesis and structure of pyridine-2-thiolato ruthenium aryls bearing a pendant imine-phenol function. *Polyhedron*, **21**, 899-904.

[33] Panda, B.K., Ghosh, K., Chattopadhyay, S. and Chakravorty, A. (2003) Chemistry of a new family of aryl ruthenium species incorporating α-diimine chelation and a pendant imine-phenol function. *Journal of Organometallic Chemistry*, **674**, 107-115. doi:10.1016/S0022-328X(03)00211-0

[34] Panda, B.K., Sengupta, S. and Chakravorty, A. (2004)

Synthesis, structure and properties of biimidazole-chelated arylruthenium complexes. *European Journal of Inorganic Chemistry*, 178-184. doi:10.1002/ejic20030029

[35] Panda, B.K. (2004) Synthesis, characterization and emission properties of quinolin-8-olato chelated ruthenium organometallics. *Journal of Chemical Science*, **116**, 245-250. doi:10.1007/BF02708274

[36] Peacock, A.F.A., Parsons, S. and Sadler, P.J. (2007) Tuning the hydrolytic aqueous chemistry of osmium arene complexes with N, O-chelating ligands to achieve cancer cell cytotoxicity. *Journal of American Chemical Society*, **129**, 3348-3357. doi:10.1021/ja068335p

[37] Kostrhunova, H., Florian, J., Novakova, O., Peacock, A.F.A., Sadler, P.J. and Brabec V. (2008) DNA interactions of monofunctional organometallic osmium(II) antitumor complexes in cell-free media. *Journal of Medicinal Chemistry*, **51**, 3635-3643. doi:10.1021/jm701538w

[38] Sinha, S., Das, P.K. and Ghosh, B.K. (1995) Synthesis and characterization of new mononitrosyl complexes of osmium-containing quinoline-8-olates. *Transition Metal Chemistry*, **20**, 59-61. doi:10.1007/BF00135403

[39] Leung, C.-F., Wong, T.-W., Lau, T.-C. and Wong, W.-T. (2005) Addition of carbenes to an Os(VI) nitride complex. *European Journal of Inorganic Chemistry*, 773-778.

[40] Hoffman, P.R. and Caulton, K.G. (1975) Solution structure and dynamics of five-coordinate d⁶ complexes. *Journal of American Chemical Society*, **97**, 4221-4228. doi:10.1021/ja00848a012

[41] Sawyer, D.T. and Roberts J.L. Jr. (1974) *Experimental Electrochemistry for Chemists*. Wiley, New York.

[42] Rhofir, C., Vocelle, D. and Sandorfy, D. (1989) Ftir study of the protonation of a retinyl schiff base in chloroform/methanol mixtures. *Research on Chemical Intermediates*, **12**, 131-139. doi:10.1163/156856789X00096

[43] Ghosh P. (1997) A four-membered C, O-Chelate: Two families of ruthenium organometallics incorporating N···O hydrogen bonding. *Polyhedron*, **16**, 1343-1349. doi:10.1016/S0277-5387(96)00390-7

[44] Griffith, J.S. (1961) *The Theory of Transition Metal Ions*. Cambridge University Press, Cambridge.

[45] Bhattacharyya, S. and Chakravorty, A. (1985) Electron spin resonance studies of distorted octahedral ruthenium(III) species. *Proceedings of the Indian Academy of Sciences—Chemical Sciences*, **95**, 159-167.

[46] Houten, J.V. and Watts, R.J. (1976) Temperature dependence of the photophysical and photochemical properties of the tris(2, 2'-bipyridyl)ruthenium(II) ion in aqueous solution. *Journal of the American Chemical Society*, **98**, 4853-4858. doi:10.1021/ja00432a028

[47] Rillema D.P., Taghdiri D.G., Jones D.S., Worl L.A., Meyaer T.J., Levy H.A. and Keller C.D. (1987) Structure and redox and photophysical properties of a series of ruthenium heterocycles based on the ligand 2,3-bis(2-pyridyl) quinoxaline. *Inorganic Chemistry*, **26**, 578-585. doi:10.1021/ic00251a018

16

Synthesis, characterization, and stability of iron (III) complex ions possessing phenanthroline-based ligands

Shawnt Tosonian*, **Charles J. Ruiz***, **Andrew Rios**, **Elma Frias**, **Jack F. Eichler**[#]

Department of Chemistry, University of California, Riverside, Riverside, USA
Email: [#]eichler@ucr.edu

ABSTRACT

It has previously been demonstrated that phenanthroline-based ligands used to make gold metallotherapuetics have the ability to exhibit cytotoxicity when not coordinated to the metal center. In an effort to help assess the mechanism by which these ligands may cause tumor cell death, iron binding and removal experiments have been considered. The close linkage between cell proliferation and intracellular iron concentrations suggest that iron deprivation strategies may be a mechanism involved in inhibiting tumor cell growth. With the creation of iron (III) phen complexes, the iron binding abilities of three polypyridal ligands [1,10-phenanthroline (phen), 2,9-dimethyl-1,10-phenanthroline (^{methyl}phen), and 2,9-di-sec-butyl-1,10-phenanthroline ($^{sec\text{-}butyl}$phen)] can be tested via a competition reaction with a known iron chelator. Therefore, iron (III) complexes possessing all three ligands were synthesized. Initial mass spectrometric and infrared absorption data indicate that iron (III) tetrachloride complex ions with protonated phen ligands (^{R}phenH^{+}) were formed: [phenH][FeCl$_4$], [^{methyl}phenH][FeCl$_4$], [$^{sec\text{-}butyl}$phenH][FeCl$_4$]. UV-Vis spectroscopy was used to monitor the stability of the complex ions, and it was found that the $^{sec\text{-}butyl}$phen-iron complex was more stable than the phen and ^{methyl}phen analogues. This was based on the observation that free ligand was observed immediately upon the addition of EDTA to the [phenH][FeCl$_4$] and [^{methyl}phenH] [FeCl$_4$] complex ions.

Keywords: Polypyridyl Ligands; Phenanthroline; Iron (III); Complex Ions

1. INTRODUCTION

Cancer is the 2nd leading cause of death in the United States. Though a wide variety of organic-based molecular compounds have been used to treat tumors, the metallotherapy cisplatin is one of the most successful anti-cancer drugs. Despite its success, cisplatin therapy has several drawbacks, including the development of resistance to the drug by tumors and the occurrence of severe side-effects during treatment [1]. It is well known that gold (III) complexes are potential alternatives to the platinum (II) based drugs [2]. Even though gold (III) complexes with polypyridyl based ligands [e.g., 1,10-phenanthroline (phen), 2,2'-bipyri-dine (bipy), and 2,2'-6',2''-terpyridine (terpy)] have shown promise as anti-cancer agents, it has also been shown that the polypyridyl ligands can have antitumor activity on their own; therefore the antiproliferative properties of the ligands must be considered when evaluating the antitumor activity of the corresponding metal-based complexes [1,3,4].

We have previously reported the synthesis of a gold (III) complex bearing a 2,9-di-sec-butyl-1,10-phenanthroline ($^{sec\text{-}butyl}$phen) ligand {[($^{sec\text{-}butyl}$phen) AuCl$_3$]} [5]. The anti-tumor activity of this gold complex compound is currently being studied, therefore it is desired to determine the potential activity of the $^{sec\text{-}butyl}$phen ligand. It has been previously hypothesized that chelating ligands can impact tumor cell death via iron depletion [6]. Additionally, Satterfield and Brodbelt reported that alkyl substituents on phenanthroline ligands have an impact on metal binding [7], and this data predicts that $^{sec\text{-}butyl}$phen should exhibit stronger metal binding than unsubstituted phen. To test whether alkyl substituted phenanthroline based ligands have different iron binding abilities, we synthesized and characterized iron complex ions with a series of phenanthroline-based ligands [phen, $^{sec\text{-}butyl}$phen, and 2,9-dimethyl-1,10-phenanthroline (^{methyl}phen)], and evaluated the relative stability of these three complex ions.

2. MATERIALS AND METHODS

2.1. General Experimental Procedures

$^{sec\text{-}butyl}$phen was synthesized according to a previously published protocol [8]; phen and ^{methyl}phen were purchased from Sigma-Aldrich and used without purification, as were all other reagents and solvents used in the

---

*These authors contributed equally to this work.
[#]Corresponding author.

course of characterizing the iron (III) complex ions. UV-Vis spectra were recorded on a Varian Cary 50 UV-Vis spectrophotometer, and IR spectra were obtained using a Perkin Elmer Spectrum One FT-IR Spectrophotometer fitted with a Universal ATR sampling accessory. Mass spectrometric analyses were performed using a Waters GCT (2008) high resolution mass spectrometer, using Electrospray Ionization (ESI) and Liquid Injection Field Desorption Ionization (LIFDI) techniques.

2.2. Synthesis of [phenH][FeCl₄]·2H₂O (1)

The iron (III) phen complex was synthesized using a previously published protocol [9]. A solution of 0.130 g $FeCl_3$ (0.802 mmole) in a 10 M aqueous HCl (10 mL) was added to a 150 mL reaction flask. To this was added 0.145 g (0.802 mmole) of 1,10-phenanthroline, which was dissolved in 10 mL of HCl. The ligand solution was added dropwise, upon which an orange color was observed. The solution was stirred overnight at room temperature, and an orange precipitate was formed. The solid was filtered and dried in vacuo, and 0.224 g of product was obtained after the solid was recrystallized from ethanol (77.3% yield). Melting point: 298 - 301°C. UV-Vis [λ_{max}/nm (ε/M^{-1}·cm^{-1})]: 230 (28,847), 270 (23,812), 320 (6623), 366 (6608). IR [v_{max}/cm^{-1}]: 3350 (NH); 3051 (CH aromatic); 2000 - 1600 (aromatic overtones); 1611; 1605; 1,580, 1517; 1345; 855. Mass spec (LIFDI): mo- lecular ion; m/z = 306; compound 1—2Cl$^-$ {[(phen)FeCl₂]+ = 306 amu}.

2.3. Synthesis of [^{methyl}phenH][FeCl₄] (2)

Compound 2 was prepared in an analogous fashion to compound 1 using 0.130 g (0.802 mmole) of $FeCl_3$ and 0.174 g (0.802 mmole) 2,9-dimethyl-1,10-phenanthroline. A yellow precipitate was isolated and recrystallized from ethanol (0.283 g, 89.1% yield). Melting point: 180 - 185°C. UV-Vis [λ_{max}/nm (ε/M^{-1}·cm^{-1})]: 225 (31,119), 280 (27,475), 317 (10,306), 364 (6318). IR [v_{max}/cm^{-1}]: 3375 (NH); 3051 (CH aromatic); 2904 (CH aliphatic); 2000 - 1600 (aromatic overtones), 1680; 1625; 1600; 1500; 1345; 865. Mass spec (LIFDI): molecular ion; m/z = 334; compound 2—2Cl$^-$ {[(^{methyl}phen)FeCl₂]+ = 334 amu}.

2.4. Synthesis of [$^{sec\text{-}butyl}$phenH][FeCl₄] (3)

Compound 3 was prepared as described for compound 1 using 0.130 g (0.802 mmole) of $FeCl_3$ and 0.235 g (0.802 mmole) 2,9-di-sec-butyl-1,10-phenanthroline. A brown-red solid was isolated and recrystallized from ethanol (0.190 g, 50.1% yield). Melting point: sample decomposition was observed over a wide temperature range above 150°C. UV-Vis [λ_{max}/nm (ε/M^{-1} cm^{-1})]: 225 (46,278), 270

(29,480), 317 (10,306), 364 (8168). IR [v_{max}/cm^{-1}]: 3200 (NH); 3068 (CH aromatic); 2962 (CH aliphatic); 2000 - 1600 (aromatic overtones); 1622; 1605; 1570, 1527; 1342; 855. Mass spec (LIFDI): molecular ion; m/z = 418; compound 3—2Cl$^-$ {[$^{sec\text{-}butyl}$phen) FeCl₂]+ = 418 amu}.

2.5. EDTA Competition Reaction

The EDTA binding reaction was modeled on a previously published protocol [10]. Stock solutions were prepared in 5 mL of acetonitrile using 10.0 mg of the phenanthroline iron complexes {2.94×10^{-5} mol [phenH] [FeCl₄], 2.51×10^{-5} mol of [^{methyl}phenH][FeCl₄], 2.12×10^{-5} mol [$^{sec\text{-}butyl}$phenH][FeCl₄]} to yield concentrations of: 0.00588 M of [phen H][FeCl₄], 0.00503 M of [^{methyl}phen H][FeCl₄], and 0.00420 M of [$^{sec\text{-}butyl}$phenH] [FeCl₄]. These stock solutions were then diluted in acetonitrile to create solutions with a final concentration of 3.43×10^{-5} M. A 0.028 M solution of EDTA in distilled water was prepared for the binding experiment. The EDTA solution was then added to the iron (III) complex ions in a 20:1 EDTA:iron molar ratio, and the UV-Vis absorption spectra were collected every minute for thirty minutes.

3. RESULTS AND DISCUSSION

3.1. Synthesis

The phen-iron complexes were synthesized by combining the phenanthroline ligand with $FeCl_3$ in a 1:1 molar ratio, in the presence of 10 M aqueous hydrochloric acid and water (**Scheme 1**). The resulting precipitates were then filtered, dried under vacuum, and recrystallized from ethanol to obtain the final complex ions. The products obtained were the complex ions comprised of the protonated phen ligand cation (^{R}phenH$^+$) and the FeCl₄$^-$ anion. It is noted that previous work describing a similar synthetic protocol reported the formation of an iron (III) phen complex where the ligand was directly coordinated to the iron (III) metal center. However, the yield of this compound was extremely low, and was likely a minor side product in the reaction [9]. Khavasi and coworkers also report that the complex ion possessing [phenH]$^+$ and [FeCl₄]$^-$ was isolated after the reaction between phen and $FeCl_3$ in hot aqueous solution [11]. Other studies do demonstrate that complexes possessing direct coord-

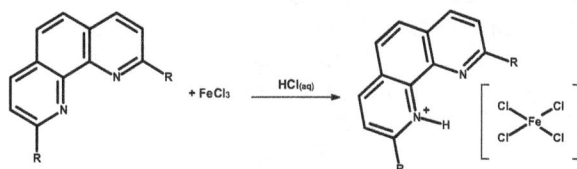

Scheme 1. Synthesis of compounds 1-3. R = H (1), -CH₃ (2), and -CH(CH₃)CH₂CH₃ (3).

ination between phen and iron (III) can be isolated, however these are obtained by reacting equimolar amounts of phen with $FeCl_3$ in non-aqueous methanol [12], or by refluxing the [phenH][$FeCl_4$] complex ion in DMSO solvent [13].

3.2. Infrared Spectroscopy

IR results for compounds 1-3 show a single resonance at 3350 cm^{-1}, 3375 cm^{-1}, and 3200 cm^{-1}, respectively, indicating the presence of a protonated secondary amine and thereby providing additional evidence that the complex ions proposed in **Scheme 1** were obtained. The presence of the ^{R}phen ligands is confirmed by the C–H aromatic resonances at 3051 cm^{-1} (1), 3051 cm^{-1} (2), and 3068 cm^{-1} (3), as well as C-N bands at 1345 cm^{-1} (1), 1345 cm^{-1} (2), and 1342 cm^{-1} (3). Though the previous reports of the [phenH][$FeCl_4$] complex ion do not provide IR data for the N-H stretch of the protonated ligand, previous reports with similar [^{R}phenH][$AuCl_4$] complex ions also indicate that the N-H stretch for the protonated ^{R}phen ligands is observed in the IR spectrum (at approximately 3180 cm^{-1}) [5].

3.3. Mass Spectrometry

In addition to the elemental analysis and IR data, mass spectrometry indicates that [^{R}phenH][$FeCl_4$] complex ions were isolated. This is evidenced by the fact that electrospray ionization (ESI) only produced the individual [^{R}phenH]$^+$ ion (observed in positive ion mode), or the [$FeCl_4$]$^-$ ion (observed in negative ion mode; see Supplemental Data **Figure 1** for representative data). In order to determine if the complex ions could be ionized without breaking the ion pair, Liquid Introduction Field Desorption Ionization (LIFDI) was subsequently used. LIFDI is known for being a soft ionization technique [14], thus the complex ions were expected to remain associated during the ionization process. Upon ionization via LIFDI positively charged molecular ions composed of [(^{R}phen)$FeCl_2$]$^+$ fragments, formed upon the loss of two chloride ligands, were indeed detected (see **Figure 1** and Supplemental Data **Figures 2** and **3**). It is also noted that compounds 2 and 3 exhibited fragment peaks in the LIFDI mode corresponding to the [^{R}phenH]$^+$ protonated ligands, further corroborating the final structural assignments for this class of compounds (compound 1 had a [(phen)$FeCl_2$]$^+$ fragment with m/z = 306; compound 2 had a [(^{methyl}phen)$FeCl_2$]$^+$ fragment with m/z = 334 and a [^{methyl}phenH]$^+$ fragment with m/z = 209; compound 3 had a [($^{sec-butyl}$phen)$FeCl_2$]$^+$ fragment with m/z = 418 and a [$^{sec-butyl}$phenH]$^+$ fragment with m/z = 293).

3.4. Ultraviolet Visible (UV-Vis) Spectroscopy

Compounds 1-3 were characterized by UV-Vis spec-

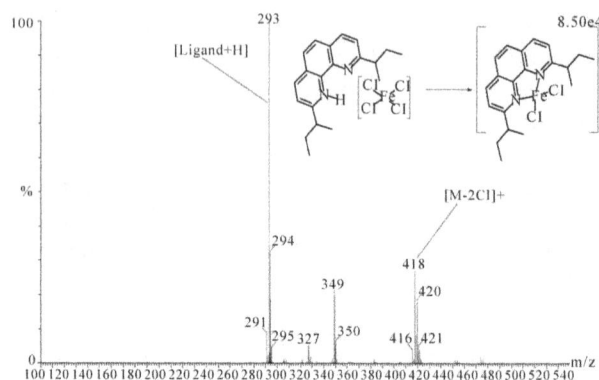

Figure 1. Representative LIFDI mass spectrometry data for the [^{R}phen H][$FeCl_4$] complex ions. The spectrum shown is for the [$^{sec-butyl}$phenH][$FeCl_4$] complex ion (compound 3). The molecular ion at m/z = 418 corresponds to the complex ion after the loss of two Cl$^-$ ligands; the fragment peak at m/z = 293 corresponds to the protonated $^{sec-butyl}$phen ligand.

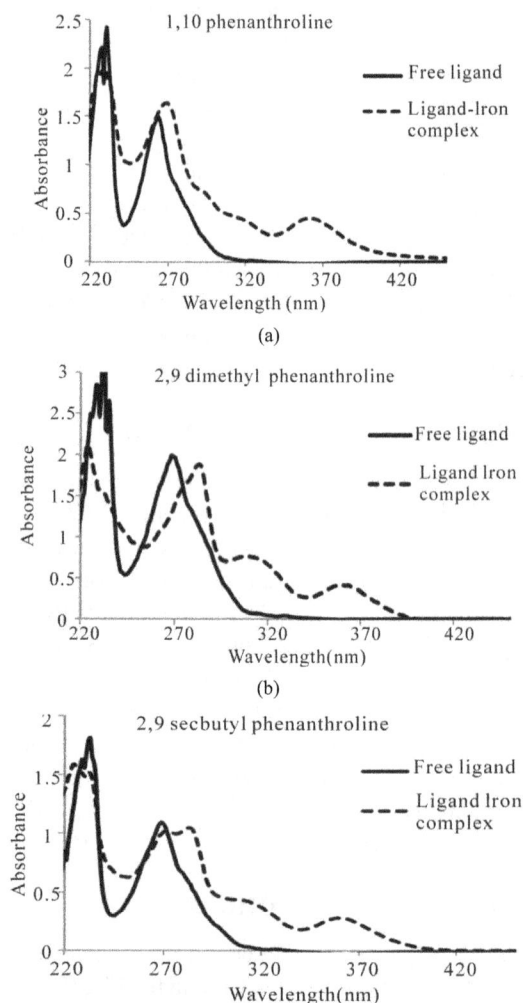

Figure 2. Uv-Vis absorbance spectra for the [^{R}phenH] [$FeCl_4$] complex ions and free ^{R}phen ligands. (a) ^{R}phen and [phenH] [$FeCl_4$] (1); (b) ^{methyl}phen and [^{methyl}phenH] [$FeCl_4$] (2); (c) $^{sec-butyl}$phen and [($^{sec-butyl}$phenH)] [$FeCl_4$] (3).

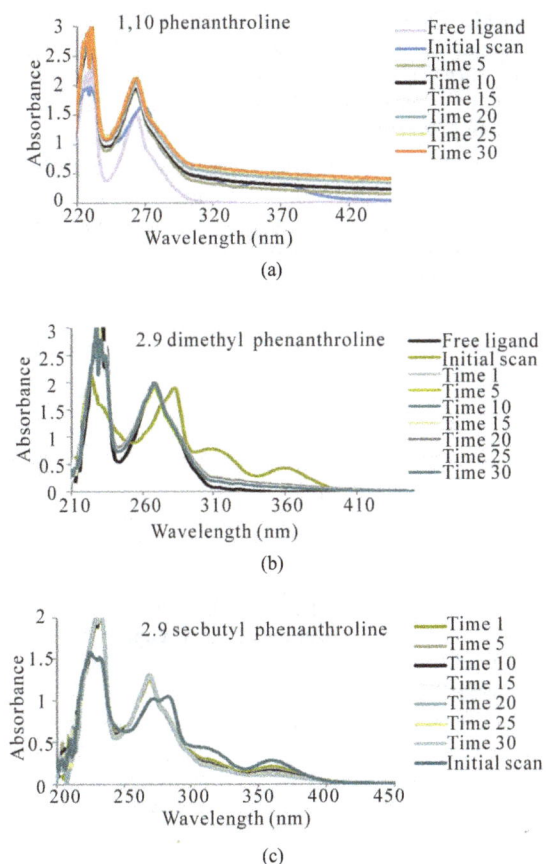

Figure 3. EDTA stability monitored by UV-Vis absorbance. Spectra are shown at 5-minute intervals after the addition of EDTA in a 20:1 molar ratio. (a) [phenH][FeCl$_4$] (1); (b) [^{methyl}phenH][FeCl$_4$] (2); (c) [($^{sec-butyl}$phenH)][FeCl$_4$] (3).

troscopy. Although direct coordination of the ligand with the iron metal center was not achieved using the synthetic protocol described herein, the complex ions have absorption spectra that confirm the presence of both the protonated ^{R}phen ligand and [FeCl$_4$]$^-$ anion. In fact, in the absence of the elemental analysis, IR, and mass spectrometry data, the absorption profiles for the [^{R}phenH] [FeCl$_4$] complex ions could be misinterpreted as evidence for direct metal binding of the phen ligand; the π $\rightarrow$ π^* charge transfer of the ligand red shifts from 260 - 270 nm for the free ligand (265 nm for phen; 270 nm for ^{methyl}phen; 270 nm for $^{sec-butyl}$phen) to 270 - 290 nm in the complex ions (275 nm for [phenH][FeCl$_4$]; 285 nm for [^{methyl}phenH][FeCl$_4$]; 285 nm for [$^{sec-butyl}$phenH] [FeCl$_4$]), and two absorption bands assigned as a ligand-to-metal charge transfer are observed at approximately 315 - 320 nm and 360 - 370 nm (see **Figure 2**). The red shift in the π $\rightarrow$ π^* charge transfer hasbeen previously reported as a result of the phen ligand coordinated to the metal iron, as have the presence of similar ligand-to-metal charge transfer bands in phenanthroline iron (III) chloride com-

plexes [11]. The similarity between UV-Vis absorption spectra for complex ions possessing protonated phen ligands and coordination complexes possessing coordinated phen ligands has been noted before, specifically for [^{R}phenH][AuCl$_4$] complex ions and [(^{R}phen)AuCl$_3$] coordination complexes [5].

3.5. EDTA Stability Studies

Even though the ^{R}phen ligands did not directly coordinate to the iron (III) metal centers, competition binding experiments were carried out to determine if the substitution on the phen ligand might impact the stability of the [FeCl$_4$]+ cation in solution. In order to test the stability of the complex ions, compounds 1-3 were treated with an excess of EDTA and the UV-Vis spectra monitored for a 30-minute period. The un-substituted [phenH][FeCl$_4$] complex ion (compound 1) and methyl substituted [^{methyl}phenH][FeCl$_4$] complex ion (compound 2) showed an immediate dissociation of the free ligand upon addition of EDTA, evidenced by the blue shift of the π $\rightarrow$ π^* charge transfer band, and the complexation of the iron (III) metal center, evidenced by the elimination of the ligand-to-metal charge transfer bands at 315 - 320 nm and 360 - 370 nm (see **Figures 3(a)** and **(b)**). The decomposition of compounds 1 and 2 in the presence of EDTA was essentially instantaneous, as these spectral changes were observed within the first minute upon adding the EDTA solution.

Conversely, the *sec*-butyl substituted [$^{sec-butyl}$phenH] [FeCl$_4$] complex ion (compound 3) displayed behavior that was significantly different from the other two compounds in the series. Although the π $\rightarrow$ π^* charge transfer band immediately shifts back to the absorbance maximum observed in the free ligand, there is a gradual decrease in the ligand-to-metal charge transfer bands at approximately 320 nm and 370 nm (see **Figure 3(c)**). These results suggest that the $^{sec-butyl}$phen ligand stabilizes the [^{R}phenH][FeCl$_4$] complex ion. This could potentially be attributed to the fact that $^{sec-butyl}$phen is a stronger base than phen and ^{methyl}phen, thereby inhibiting the deprotonation of the ligand by EDTA. However, given the fact that the ligand-to-metal charge transfer bands are eliminated more slowly over time compared to the [phenH] [FeCl$_4$] and [^{methyl}phenH][FeCl$_4$] complex ions, it appears that the $^{sec-butyl}$phen ligand may provide some protection to iron (III) metal center. Satterfield and Brodbelt have previously quantified the ability of substituted phen ligands to bind metal cations, and their studies indicate that the larger *sec*-butyl substituent on the 2,9-positions of the phen ligand enhance metal binding [7]. Thus, the fact the $^{sec-butyl}$phen ligand provides some protection of the iron (III) metal center is not completely unexpected. The LIFDI mass spectrometry data seem to provide further evidence of an interaction between the ^{R}phen ligands and iron (III) metal center in solution, given the fact that the molecular ions for compounds 1-3 are comprised of a $[(^R\text{phen})\text{FeCl}_2]+$ cation.

4. SUMMARY

It has been shown that reacting phen and 2,9-di-alkyl-substituted phen ligands with iron (III) chloride in the presence of aqueous acid results in the isolation of complex ions comprised of protonated ^{R}phen ligands and FeCl_4^- anions. Even though direct coordination of the ^{R}phen ligands was not observed, EDTA stability studies indicate that the $[^{sec\text{-butyl}}\text{phenH}][\text{FeCl}_4]$ complex ion (compound 3) is more stable than the unsubstituted phen and ^{methyl}phen congeners, and LIFDI mass spectrometry analyses provide evidence of an interaction between the ^{R}phen ligands and the iron (III) metal center in solution. Future work will be focused on developing a synthetic protocol that results in the direct coordination between $^{sec\text{-butyl}}$phen ligand and iron (III) metal center, and then comparing its binding strength to the analogous phen and ^{methyl}phen coordination complexes. Ultimately, this data will lend more insight about the potential role of $^{sec\text{-butyl}}$phen as an iron chelator in the course of tumor cell antiproliferative studies.

5. ACKNOWLEDGEMENTS

The authors would like to acknowledge the UCR MARC U STAR program (funded by the NIH) for supporting EF during the summer of 2012, and the UCR STEM Pathways for supporting CR and AR during the summer of 2012. The authors would also like to thank the UCR High Resolution Mass Spectrometry facility for their work in completing the ESI and LIFDI mass spectrometry experiments (under NSF grant CHE-0742001).

REFERENCES

[1]   Wein, A.N., Stockhausen, A.T., Hardcastle, K.I., Saadein, M.R., Peng, S., Wang, D. and Eichler, J.F. (2011) Tumor cytotoxicity of 5,6-dimethyl-1,10-phenanthroline and its corresponding gold (III) complex. *Journal of Inorganic Biochemistry*, **105**, 663-668. doi:10.1016/j.jinorgbio.2011.01.006

[2]   Sun, R. and Che, C. (2009) The anti-cancer properties of gold (III) compounds with dianionic porphyrin and tetradentate ligands. *Coordination Chemistry Reviews*, **253**, 1682-1691. doi:10.1016/j.ccr.2009.02.017

[3]   Messori, L., Abbate, F., Marcon, G., Orioli, P., Fontani, M., Mini, E., Mazzei, T., Carotti, S., O'Connell, T. and Zanello, P. (2000) Gold (III) complexes as potential antitumor agents: Solution chemistry and cytotoxic properties of some selected gold (III) compounds. *Journal of Medicinal Chemistry*, **43**, 3541-3548. doi:10.1021/jm990492u

[4]   Cinellu, M.A., Maiore, L., Manassero, M., Casini, A.,

Arca, M., Fiebig, H.-H., Kelter, G., Michelucci, E., Pieraccini, G., Gabbiani, C. and Messori, L. (2010) [Au$_2$(phen^{2Me})$_2$(μ-O)$_2$](PF$_6$)$_2$, a novel dinuclear gold (III) complex showing excellent antiproliferative properties. *ACS Medicinal Chemistry Letters*, **1**, 336-339.

[5]   Hudson, Z.D., Sanghvi, C.D., Rhine, M.A., Ng, J.J., Bunge, S.D., Hardcastle, K.I., Macbeth, C. and Eichler, J.F. (2009) Synthesis and characterization of gold (III) complexes possessing 2,9-dialkylphenanthroline ligands: To bind or not to bind? *Dalton Transactions*, **28**, 7473-7480. doi:10.1039/b823215f

[6]   Richardson D.R., Tran, E.H. and Ponka, P. (1998) The potential of iron chelators of the pyridoxal isonicotinoyl hydrazone class as effective antiproliferative agents. *Blood*, **86**, 4295-4306.

[7]   Satterfield, M. and Brodbelt, J.S. (2001) Relative binding energies of gas-phase pyridyl ligand/metal complexes by energy-variable collisionally activated dissociation in a quadrupole ion trap. *Inorganic Chemistry*, **40**, 5393-5400. doi:10.1021/ic010356r

[8]   Pallenberg, A.J., Koenig, K.S. and Barnhart, D.M., (1995) Synthesis and characterization of some copper (I) phenanthroline complexes. *Inorganic Chemistry*, **34**, 2833-2840. doi:10.1021/ic00115a009

[9]   Kulkarni, P., Padhye, S. and Sinn, E. (1998) Communication: The first well characterized Fe (phen) Cl^3 complex: Structure of aquo mono(1,10-phenanthroline) iron (III) trichloride: [Fe (phen) Cl^3 (H^{2}O)]. *Polyhedron*, **17**, 2623-2626. doi:10.1016/S0277-5387(97)00515-9

[10]   Hara, Y. and Akiyama, M. (2001) An iron reservoir model based on ferrichrome: Iron (III)-binding and metal (III)-exchange properties of tripodal monotopic and ditopic hydroxamate ligands with an l-alanyl-l-alanyl-n-hydroxy-β-alanyl sequence. *Journal of the American Chemical Society*, **123**, 7247-7256. doi:10.1021/ja003251g

[11]   Khavasi, H.R., Amani, V. and Safari, N. (2008) (2,2'-Biquinoline-2N,N')dichloridoiron(II). *Zeitschrift für Kristallographie—New Crystal Structures*, **223**, 41-42.

[12]   Eckenhoff, W.T., Biernesser, A.B. and Pintauer, T. (2012) Structural characterization and investigation of iron (III) complexes with nitrogen and phosphorus based ligands in atom transfer radical addition (ATRA). *Inorganica Chimica Acta*, **382**, 84-95. doi:10.1016/j.ica.2011.10.016

[13]   Amani, V., Nasser, S., Khavasi, H.R. and Mirzaei, P. (2007) Iron (III) mixed-ligand complexes: Synthesis, characterization and crystal structure determination of iron (III) hetero-ligand complexes containing 1,10-phenanthroline, 2,2'-bipyridine, chloride and dimethyl sulfoxide, [Fe(phen)Cl$_3$(DMSO)] and [Fe(bipy)Cl$_3$(DMSO)]. *Polyhedron*, **26**, 4908-4914. doi:10.1016/j.poly.2007.06.038

[14]   Gomer, R. and Inghram, M.G. (1955) Applications of field ionization to mass spectrometry. *Journal of the American Chemical Society*, **77**, 500. doi:10.1021/ja01607a096

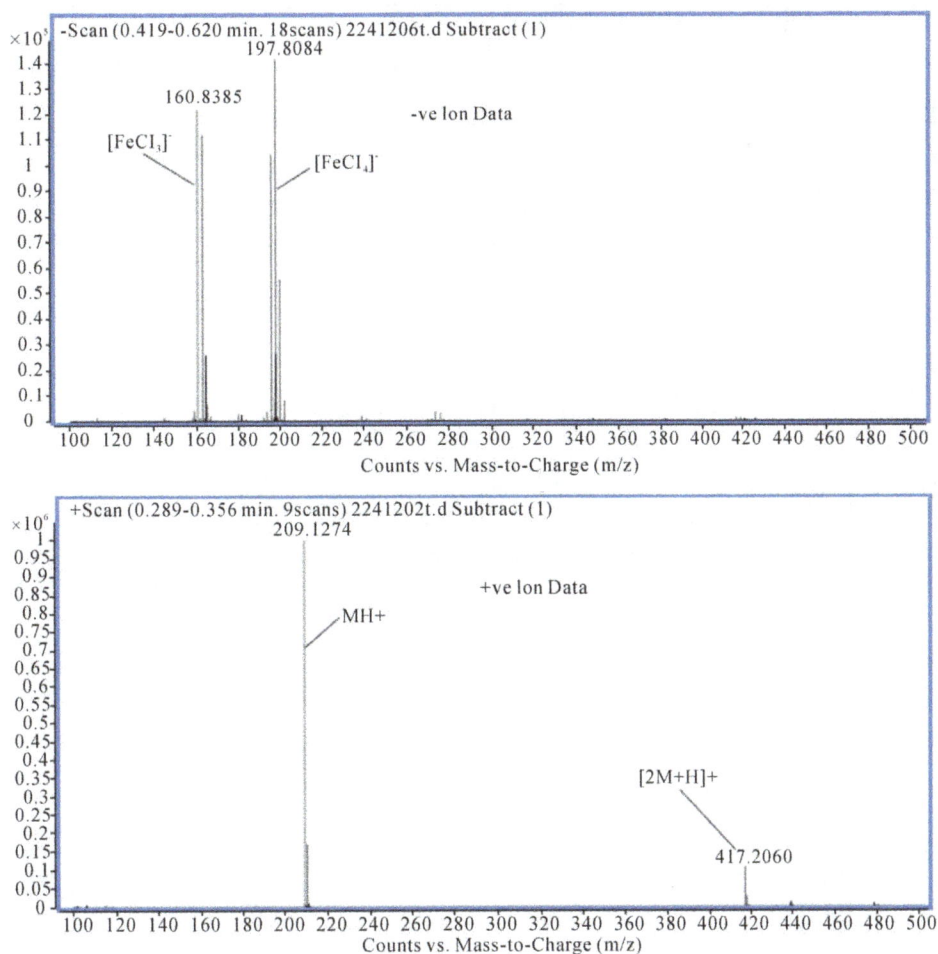

Supplemental Figure 1. Representative ESI mass spectrometry data. The ESI mass spectrum of compound 2 is shown. The (^{methyl}phenH)$^+$ molecular ion is observed at m/z = 209 (identified on the spectrum as MH$^+$) in positive ion mode, and the (FeCl$_4$)$^-$ molecular ion is observed at m/z = 198 in the negative ion mode.

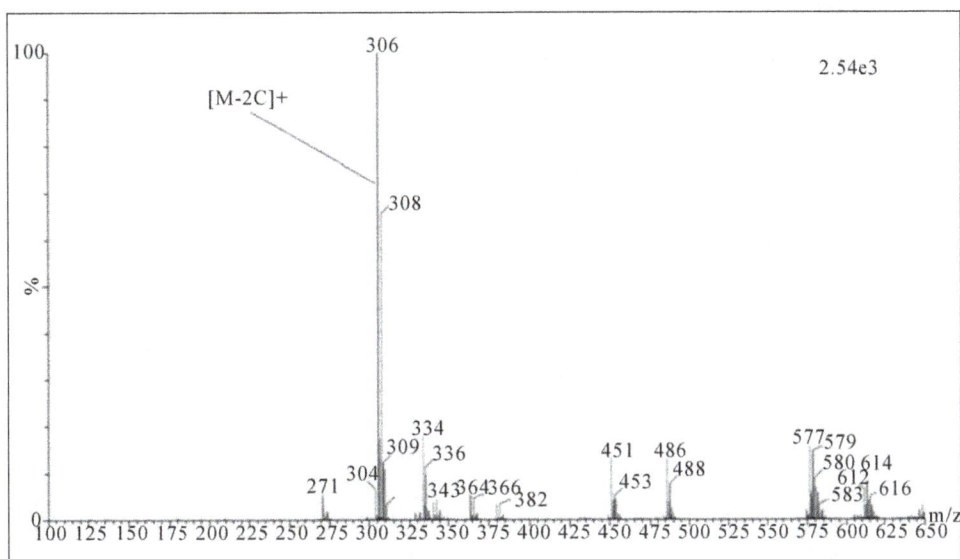

Supplemental Figure 2. LIFDI mass spectrometry analysis of compound 1. The [(phen)FeCl$_2$]$^+$ molecular ion is observed at m/z = 306 (identified on the spectrum as [M-2Cl]$^+$).

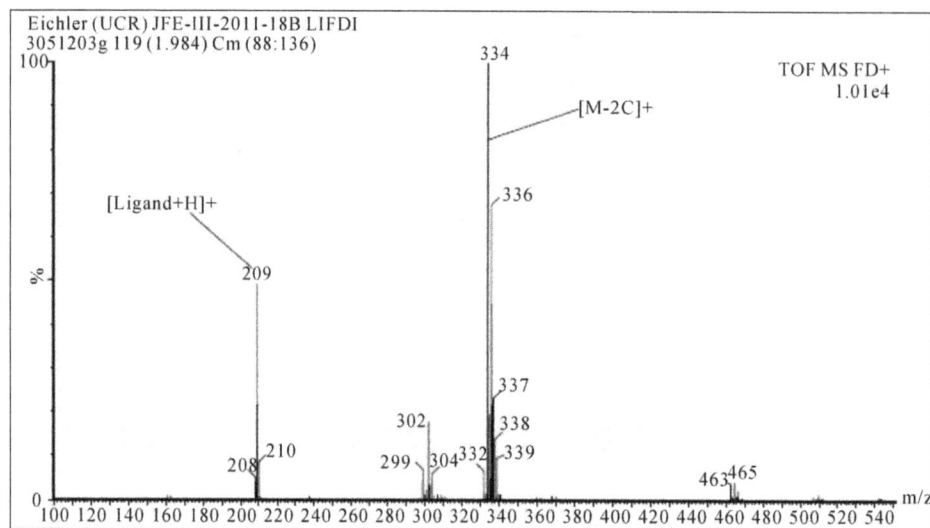

Supplemental Figure 3. LIFDI mass spectrometry analysis of compound 2. The [(^(methyl)phen) FeCl$_2$]$^+$ molecular ion is observed at m/z = 334 (identified on the spectrum as [M-2Cl]$^+$).

17

Synthesis, structures, and properties of two new binuclear complexes based on carboxyl-substituted nitronyl nitroxide: [M$_2$(NITpBA)$_4$(H$_2$O)$_2$] (M = Zn and Cu)

Jing Chen[1,2*], You-Juan Zhang[1], Kun-Tao Huang[2], Qiang Huang[2], Jun-Jie Wang[1]

[1]College of Chemistry and Chemical Engineering, Anyang Normal University, Anyang, China
[2]School of Chemical Engineering and Energy, Zhengzhou University, Zhengzhou, China
Email: *chenjinghao2008@163.com

ABSTRACT

Two new binuclear complexes, [M$_2$(μ-NITpBA)$_4$(H$_2$O)$_2$] (M = Zn 1 and Cu 2) [where NIpBA = 2-(4-carboxyl-phenyl)-4,4,5,5-tetramethyl-4,5-dihydro-1H-imidazol-1-oxyl-3-oxide], were stylized and characterized. Magnetic susceptibility measurements revealed antiferromagnetic exchange interactions in the two complexes.

Keywords: Binuclear Complex; Carboxyl-Substituted; Nitronyl Nitroxide; Antiferromagnetic Exchange

1. INTRODUCTION

The past decade has triggered fast-growing interests in nitronyl nitroxide radicals (NITR) as building blocks in the engineering of molecular-based magnet. The reasons are mainly that they are the most stable paramagnetic ligands, even in combination with metal ions where they can also act as bridging ligands [1]. Carboxylate bridges mediate exchange interactions between coordinated metal centres [2,3]. Metal complexes with bridging carboxylates as well as stable organic radical ligands are of considerable interest to the field of molecular magnetism [4]. In an effort to bring together the two areas of research, carboxyl-substituted nitronyl nitroxide, as an important anionic ligand, was deemed favorable for the realization in reactions with metal ions. To the best of our knowledge, to prepare transition metal complexes with nitronyl nitroxide-substituted carboxylate ligands, many investigations on the scope have been recently performed, but the related reports are very scarce [5,6]. The only structurally characterized transition metal compounds are 1D-chain systems in which two nitronyl

nitroxides NITpBA⁻ are coordinated with two metal centers via one NO-group and a monodentate carboxylate group [7,8]. Miller's group even reported on copper (II) compounds with NITpBA⁻ assumed to be in a bidentate-bridging mode, but did not provide structural information on the related radical-metal compounds [9]. Rentschle's group examined the reaction of NITpBAH with different metal salts, bases, and terminal capping coligands following with various methods but could not isolate any dimeric compounds. Ligand exchange reaction of NITpBAH with copper (II) acetate hydrate, however, leads to the desired system. Recrystallization from hot dimethylsulfoxide/methanol (2:1) leads to formation of small green-blue crystals Cu$_2$(μ-NITpBA)$_4$/H$_2$O/dmso [10].

With the purpose of obtaining materials with unusual molecular high nuclearity spin clusters, we were interested in the preparation of carboxylate-bridged metal complexes with pendant organic radical substituents. In this paper, we present the simple synthesis and properties of two new metal–radical binuclear complexes [M$_2$(μ-NITpBA)$_4$(H$_2$O)$_2$] (M = Zn 1 and Cu 2).

2. EXPERIMENTAL

2.1. General

All reagents used in the synthesis were of analytical grade without further purification. 2-(4-carboxyl-phenyl)-4,4,5,5-tetramethyl-4,5-dihydro-1H-imidazol-1-oxyl-3-oxide was prepared by the literature method [11]. Elemental analyses (C, H, and N) were carried out with a Perkin Elmer 240 C elemental analyzer. IR spectra were recorded from 400 to 4000 cm⁻¹ on an Avatar-360 spectrophotometer using KBr pellets. Variable-temperature magnetic susceptibilities were measured with a MPMS-7SQUID magnetometer at a magnetic field of 2000 G. Diamagnetic corrections were made with Pascal's con-

---

*Corresponding author.

stants for all constituent atoms.

2.2. Synthesis of [M$_2$(μ-NITpBA)$_4$(H$_2$O)$_2$] (M = Zn 1 and Cu 2)

2.2.1. [Zn$_2$(μ-NITpBA)$_4$(H$_2$O)$_2$] 1

NITpBAH (0.4 mmol, 0.1076 g) was dissolved in 4 mL 0.2 M NaOH. The complex 1 was synthesized by the addition of NITpBA$^-$ and ZnCl$_2$ (0.2 mmol, 0.0275 g) to 30 mL of methanol-water. The mixture was stirred for 2 h and then filtered. The resulting blue filtrate was kept at room temperature for slow evaporation. After a few days, dark blue crystals of compound 1 suitable for X-ray analysis were obtained. Anal. Found: C, 53.12; H, 5.30; N, 8.61% Cacl. for C56 H68 N8 O18 Zn2 (1): C, 52.83; H, 5.39; N, 8.81%. IR (KBr disk): 1612 cm^{-1} (v_{as}COO$^-$), 1413 cm^{-1} (v_sCOO$^-$), 1355 cm^{-1} (vNO).

2.2.2. [Cu$_2$(μ-NITpBA)$_4$(H$_2$O)$_2$] 2

The complex 2 was prepared in a procedure similar to complex 1 except that ZnCl$_2$ was replaced by CuCl$_2$ Dark blue crystals were obtained expectedly. Anal. Found: C, 52.53; H, 5.16; N, 8.74% Cacl. for C56 H68 N8 O18 Cu2 (2): C, 52.98; H, 5.40; N, 8.83%. IR (KBr disk): 1610 cm^{-1} (v_{as}COO$^-$), 1407 cm^{-1} (v_sCOO$^-$), 1350 cm^{-1} (vNO).

2.2.3. X-Ray Data Collection and Structure Determination

X-ray diffraction intensities were collected on a Bruker Smart CCD diffractometer equipped with a graphite-monochromated Mo Kα radiation (λ = 0.71073 Å) by using a ω - φ scan technique at room temperature. Absorption correction were applied using SADABS program [12]. The structures were solved by direct methods with SHELXS-97 program [13] and refined with SHELXS-97[14] by full matrix least-squares technique on F^2. All the non-hydrogen atoms were refined with anisotropic temperature para-meters. Hydrogen atoms of organic ligands were fixed in ideal positions. The relevant parameters of the crystal structures for complex 1 and 2 are listed in **Table 1**, and the selected bond lengths and bond angles are given in **Tables 2** and **3**, respectively.

Table 1. Crystal data and structure refinement for the two complexes 1 and 2.

	1	2
Formula	C56 H68 N8 O18 Zn2	C56 H68 N8 O18 Cu2
Formula weight	1271.92	1268.26
Crystal system	Monoclinic	Monoclinic
Space group	P2(1)/c	P2(1)/c
Unit cell dimensions (Å, °)		
a	10.415(2)	10.392(2)
b	12.770(3)	12.650(3)
c	22.987(5)	23.116(5)
β	98.02(3)	97.85(3)
Volume (Å^3), Z	3027.3(11), 2	3010.5(10), 2
F(000)	1328	1324
θ range for data collection (°)	1.83 to 28.40	1.84 to 28.40
Limiting indices	$-13 \leq h \leq 13$ $-15 \leq k \leq 16$ $-30 \leq l \leq 30$	$-13 \leq h \leq 13$ $-16 \leq k \leq 16$ $-30 \leq l \leq 30$
Reflections collected	51,237	52,552
Independent reflections	7419 [R(int) = 0.1174]	7505 [R(int) = 0.0545]
Completeness (%)	97.9	99.3
Goodness-of-fit on F^2	1.021	1.014
Final R indices ($I > 2 \sigma(I)$)	R_1 = 0.0771, wR_2 = 0.1852	R_1 = 0.0594, wR_2 = 0.1582
R indices (all data)	R_1 = 0.1399, wR_2 = 0.2293	R_1 = 0.0896, wR_2 = 0.1785

Table 2. Selected bond distances (Å) and angles (°) for the complex 1.

Bond distances			
Zn(1)-Zn(1)#1	2.9672(12)	O(6)-C(15)	1.254(6)
Zn(1)-O(1W)	1.964(4)	N(1)-O(3)	1.275(7)
Zn(1)-O(2)#1	2.025(3)	N(2)-O(4)	1.298(16)
Zn(1)-O(5)	2.025(4)	O(8)-N(4)	1.288(5)
Zn(1)-O(1)	2.030(4)	O(7)-N(3)	1.276(6)
Zn(1)-O(6)#1	2.035(4)	N(4)-C(22)	1.340(6)
O(1)-C(1)	1.246(7)	C(22)-N(3)	1.354(6)
O(2)-C(1)	1.249(6)	N(1)-C(8)	1.233(8)
O(5)-C(15)	1.244(6)	C(8)-N(2)	1.447(13)
Bond angles			
C(1)-O(2)-Zn(1)#1	130.0(4)	O(2)#1-Zn(1)-O(6)#1	90.03(16)
C(15)-O(6)-Zn(1)#1	132.3(3)	O(5)-Zn(1)-O(6)#1	158.99(16)
O(1W)-Zn(1)-O(2)#1	100.2(2)	O(1)-Zn(1)-O(6)#1	87.75(17)
O(1W)-Zn(1)-O(5)	104.5(2)	O(1W)-Zn(1)-Zn(1)#1	171.23(17)
O(2)#1-Zn(1)-O(5)	87.60(16)	O(2)#1-Zn(1)-Zn(1)#1	77.48(12)
O(1W)-Zn(1)-O(1)	101.2(2)	O(5)-Zn(1)-Zn(1)#1	83.94(11)
O(2)#1-Zn(1)-O(1)	158.59(17)	O(1)-Zn(1)-Zn(1)#1	81.38(13)
O(5)-Zn(1)-O(1)	86.91(16)	O(6)#1-Zn(1)-Zn(1)#1	75.16(11)
O(1W)-Zn(1)-O(6)#1	96.5(2)	C(1)-O(1)-Zn(1)	124.4(4)

Symmetry codes: #1: x + 1, y, z + 2.

Table 3. Selected bond distances (Å) and angles (°) for the complex 2.

Bond distances			
Cu(1)-Cu(1)#1	2.6457(1)	O(6)-C(15)	1.253(4)
Cu(1)-O(2)#1	1.952(2)	O(5)-C(15)	1.253(4)
Cu(1)-O(1)	1.955(3)	O(3)-N(1)	1.248(9)
Cu(1)-O(5)	1.960(2)	N(2)-O(4)	1.275(6)
Cu(1)-O(6)#1	1.971(2)	O(7)-N(3)	1.279(5)
Cu(1) -O(1W)	2.121(3)	N(4)-O(8)	1.271(4)
O(2)-C(1)	1.255(4)	N(4)-C(22)	1.340(5)
O(1)-C(1)	1.254(5)	C(22)-N(3)	1.339(5)
Bond angles			
C(1)-O(2)-Cu(1)#1	124.1(2)	O(1)-Cu(1)-O(1W)	96.67(14)
C(15)-O(6)-Cu(1)#1	124.8(2)	O(5)-Cu(1)-O(1W)	100.95(14)
O(2)#1-Cu(1)-O(1)	167.82(12)	O(6)#1-Cu(1)-O(1W)	90.97(14)
O(2)#1-Cu(1)-O(5)	88.99(12)	O(2)#1-Cu(1)-Cu(1)#1	83.00(9)
O(1)-Cu(1)-O(5)	88.35(12)	O(1)-Cu(1)-Cu(1)#1	84.96(9)
O(2)#1-Cu(1)-O(6)#1	90.66(12)	O(5)-Cu(1)-Cu(1)#1	86.15(8)
O(1)-Cu(1)-O(6)#1	89.49(12)	O(6)#1-Cu(1)-Cu(1)#1	81.96(8)
O(5)-Cu(1)-O(6)#1	168.05(11)	O(1W)-Cu(1)-Cu(1)#1	172.74(13)
O(2)#1-Cu(1)-O(1W)	95.51(14)		

Symmetry codes: #1: x, y + 2, z.

3. RESULTS AND DISCUSSION

3.1. Description of Crystal Structures

Crystal analysis show that compounds 1 and 2 are iso structural. They crystallize in the monoclinic system, $P2_1/c$ (No.14) space group, which are different from that lecture [10] (triclinic system, space group P_1). The asymmetric unit consists of one-half of a dimeric metal ion bridged by O-carboxyl from four nitronyl nitroxide. The two halves are related to each other through a center of inversion located in the middle of the M-M axis. The coordination geometry of Zn (II) and Cu (II) lies at the center of the distorted octahedral geometry formed by five oxygen atoms, which come from four equivalent nitroxide ligand, one water molecule, and the M-M distances of the neighboring metal atoms is [Zn1-Zn1(A) 2.9672 Å and Cu1-Cu1(A) 2.6457Å], which is rather short and may be regarded as the normal M-M bonding range[15,16]. Their molecular structures are depicted in **Figures 1** and **2**, respectively.

In complex 1, the C1-O1 and C1-O2 bonds from the same carboxylate group are equal to 1.246 and 1.249 Å, respectively, whereas the distance of C15-O5 and C15-O6 are of 1.244 and 1.254 Å, respectively. The Ph-COO bonds of 1.5093 Å are close to that reported in the literature [11]. The C-C bond lengths of the phenyl rings show no alternation and are located in the aromatic region ranging from 1.361 to 1.394 Å. The Zn-O$_{water}$ distance of 1.964 Å is shorter than that of Zn-O whose oxygen atoms are from Ph-COO$^-$. Four oxygen atoms from Ph-COO$^-$ of the nitroxide ligand compose the equatorial plane. The O5-Zn1-O6A angle in the equatorial plane is 158.99°, and the O1-Zn1-O2A angle is equal to 158.59°. The O1W-Zn1-Zn1A angle in the axial position is 171.23°. The Zn (II) center is displaced by 0.324 Å from the basal O4-plane toward the axial oxygen atom from H$_2$O. The fragment O4-N2-C8-N1-O3 is non-planar, and it forms a dihedral angle of 65.1° with the plane of the phenyl ring (C6-C5-C4). Conversely, the fragment O8-N4-C22-N3-O7 is nearly planar and forms a dihedral angle of 35.1° with the plane of the phenyl ring (C19-C18-C20). The IR spectrum shows the N-O stretching vibration of the NITpBA at 1355 cm^{-1} as well as the antisymmetric v_{as}COO$^-$ and the symmetric v_sCOO$^-$ stretching mode of the caroxyl group at 1612 and 1413 cm^{-1}, respectively, in accordance with a bidentate-bridging mode [10,17,18].

In complex 2, the Cu1-Cu1A distance is 2.6457 Å, which is a typical value for dimeric copper (II) carboxylate adducts [19,20]. The Cu-O$_{water}$ distance of 2.121 Å is longer than that of Cu-O whose oxygen atoms come from Ph-COO$^-$. The Cu (II) core is displaced by 0.205 Å from the basal O4-plane toward the axial oxygen atom of the solvent. Mean deviation from base O4-plan

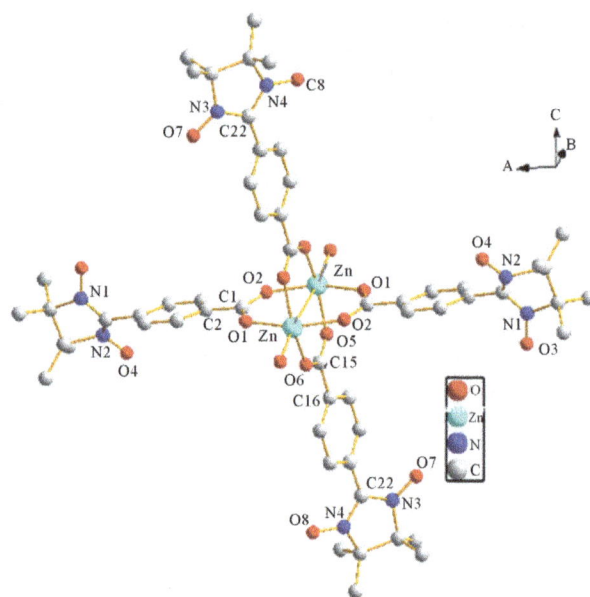

Figure 1. View of the moiety of the complex 1 [Zn$_2$(NITpBA)$_4$(H$_2$O)$_2$]. All hydrogen atoms are omitted for clarity.

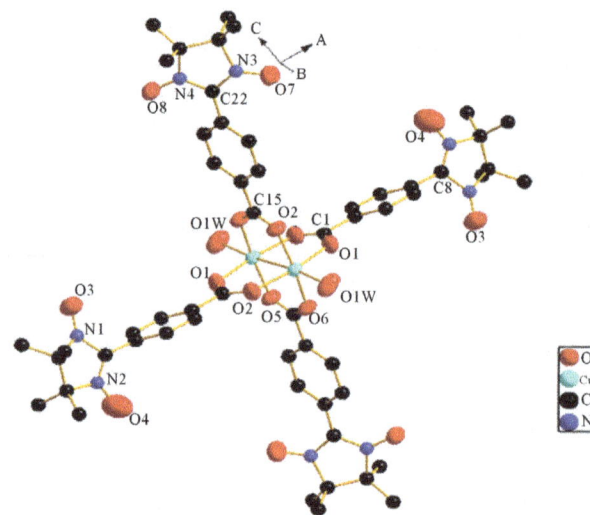

Figure 2. View of the moiety of the complex 2 [Cu$_2$(NITpBA)$_4$(H$_2$O)$_2$]. All hydrogen atoms are omitted for clarity.

is 0.0009 Å. The fragment O4-N2-C8-N1-O3 is non-planar, where the mean deviation from the plane is 0.0676 Å, and forms a dihedral angle of 86.1° with the plane of the phenyl ring (C4-C5-C6). The fragment O8-N4-C22-N3-O7 is nearly planar, and the plane for ms a dihedral angle of 36.4° with the phenyl ring (C19-C18-C20). The product shows IR absorptions at 13 50, 1407 and 1610 cm^{-1}, respectively.

3.2. Magnetic Properties

We have then examined the temperature dependences of

$\chi_M T$ and χ_M for complex 1 and 2 in order to analyze the exchange coupling in this six-spin system. Temperature dependences of the molar magnetic susceptibility for two complexes were measured in a temperature range of 2 K to 300 K at a magnetic field of 2000 G. The results are given in **Figures 3** and **4**, respectively.

For complex 1, the $\chi_M T$ at room temperature has a value of 1.535 emu·K·mol^{-1}, which is close to the expected value for uncoupled spins of S = 4/2 (1.5 emu·K·mol^{-1}) for two zinc centers and four nitronyl nitroxide radical-ligands. It was regarded as a tetra-radical system to evaluate the exchange coupling constants. The $\chi_M T$ value decreases slowly and reaches 1.43 emu·K·mol^{-1} at 100 K as the temperature is lowered, and subsequently decreases rapidly during further lowering of temperature. Based on the structural results, it undergoes a major magnetic interaction in the present system, in which the exchange interaction of NITpBA radicals through Zn (II) core exists [21]. The result indicates that a weak anti-ferromagnetic exchange interaction is predominant, which is agreement with literatures [22,23].

For complex 2, the $\chi_M T$ value at room temperature is 2.14 emu·K·mol^{-1}, which is significantly lower than that of the expected value for six uncoupled spins of S = 1/2 (2.25 emu·K·mol^{-1}) for two copper centers and four nitronyl nitroxide. The result indicates that a strong anti-ferromagnetic exchange interaction is predominant. The result is in accord with those reported [10,24].

4. CONCLUSION

Two new binuclear complexes formulae of $[M_2(\mu\text{-}NITpBA)_4(H_2O)_2]$ (M = Zn 1 and Cu 2) [where NITpBA =2-(4-carboxyl-phenyl)-4,4,5,5-tetramethyl-4,5-dihydro-1H-imidazol-1-oxyl-3-oxide)] were synthesized and structurally characterized. The simple procedure of the NITpBAH with metal chloride yielded dimeric complexes 1 and 2. The temperature dependences of the magnetic susceptibility show dominant itram-olecular anti-ferromagnetic exchange interaction in complexes 1 and 2. According to the result of the complex 1, it demonstrates that the magnitude of the intramolecular magnetic exchange has little relationship with the crystal system and space group.

5. ACKNOWLEDGEMENTS

This work was supported by the National Natural Science Foundation of China (No. 21071006), the Natural Science Foundation of Henan Province (No. 102102 210457) and the Natural Science Foundation of the Henan Higher Education Institutions of China (No. 2010 B150001).

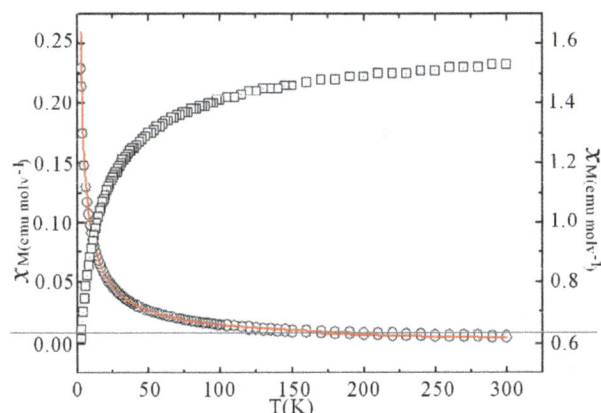

Figure 3. Temperature dependece of χM (○) and χMT (□) vs T for the complex 1

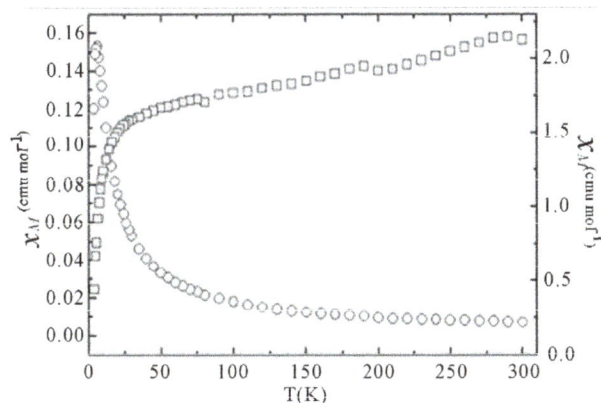

Figure 4. Temperature dependece of χM (○) and χMT (□) vs T for the complex 2

REFERENCES

[1]   Caneschi, A., Gatteschi, D., Sessoli, R. and Rey, P. (1989) Toward molecular magnets: The metal-radical approach. *Accounts of Chemical Research*, **22**, 392-398. doi:10.1021/ar00167a004

[2]   Boskovic, C., Pink, M., Huffman, J.C., Hendrickson, D.N. and Christou, G. (2001) Single-molecule magnets: Ligand-induced core distortion and multiple Jahn-Teller isomerism in [Mn12O12(O2CMe)8(O2PPh2) 8(H2O)4]. *Journal of the American Chemical Society*, **123**, 9914-9915. doi:10.1021/ja016341+

[3]   Aubin, S.M.J., Sun, Z., Eppley, H.J., Rumberger, E.M., Guzei, I.A., Folting, K., Gantzel, P.K., Rheingold, A.L., Christou, G. and Hendrickson, D.N. (2001) Single-molecule magnets: Jahn-Teller isomerism and the origin of two magnetization relaxation processes in Mn12 complexes. *Inorganic Chemistry*, **40**, 2127-2146. doi:10.1021/ic000911+

[4]   Kahn, O. (1993) Molecular magnetism. VCH, Weinheim.

[5]   Inoue, K. and Iwamura, H. (1993) Magnetic properties of the crystals of *p*-(1-oxyl-3-oxido-4,4,5,5-tetramethyl-2-imi-dazolin-2-yl)benzoic acid and its alkali metal salts.

Chemical Physics Letters, **207**, 551-554.
doi:10.1016/0009-2614(93)89046-K

[6]   Ballester, G., Coronado, E., Galán-Mascarós, J.R., Gi-
      ménez-Saiz, C., Nuez, A. and Romero, F.M. (2001) A
      new approach for the synthesis of magnetic materials bas-
      ed on nitroxide free radicals and inorganic coordination
      polymers. *Polyhedron*, **20**, 1659-1662.

[7]   Schiødt, N.C., Fabrizi de Biani, F., Caneschi, A. and
      Gatteschi, D. (1996) Structure and magnetism of nickel
      (II) and manganese (II) complexes of a nitronyl nitroxide
      carboxylic acid. *Inorganica Chimica Acta*, **248**, 139-146.
      doi:10.1016/0020-1693(95)04996-7

[8]   Zhao, Q.H., Li, L.C., Jiang, Z.H., Liao, D.Z., Yan, S.P.
      and Fang, R.B. (2004) Synthesis and structure of a novel
      one-dimensional chain compound formed by copper (II)
      and nitronyl nitroxide radical. *Journal of Chemical Crys-
      tallography*, **34**, 191-193.
      doi:10.1023/B:JOCC.0000021563.00957.85

[9]   Del Sesto, R.E., Arif, A.M. and Miller, J.S. (2000) Cop-
      per (II) benzoate nitroxide dimers and chains: Structure
      and magnetic studies. *Inorganic Chemistry*, **39**, 4894-
      4902. doi:10.1021/ic0007106

[10] Schatzschneider, U., Weyhermüller, T. and Rentschler, E.
      (2002) Copper complexes with mono- and bidentate-
      bridging nitronyl nitroxide-substituted benzoate ligands.
      Inorganica Chimica Acta, **337**, 122-130.
      doi:10.1016/S0020-1693(02)01103-9

[11] Bätz, C., Amann, P., Deiseroth, H.J. and Dulog, L. (1994)
      Die oxidation zweier neuer 1,3-dihydroxyimidazolidine-
      darstellung und Struktur von Nitronylnitroxid-Radikalen.
      Liebigs Annalen der Chemie, **7**, 739-740.
      doi:10.1002/jlac.199419940715

[12] Sheldrick, G.M. (1997) SHELXS97, program for the so-
      lution of crystal structure. University of Göttingen, Göt-
      tingen.

[13] Sheldrick, G.M. (1997) SHELXS97, program for the re-
      finement of crystal structure. University of Göttingen,
      Göttingen.

[14] Sheldrick, G.M. (1996) SADABS. University of Göttin-
      gen, Göttingen.

[15] Ohmura, T., Mori, W., Takei, T., Ikeda, T. and Maeda, A.
      (2005) Structure and magnetic behaviour of mononuclear
      and dinuclear Cu (II)/Zn (II) monocarboxylate-pyridine
      derivatives studied by crystal engineering. *Materials*

Science-Poland, **23**, 729-732.

[16] Chui, S.S.Y.S., Lo, M.F., Chammant, J.P.H., Orpen, A.G.
      and Williams, I.D. (1999) A chemically functionalizable
      nanoporous material [Cu$_3$(TMA)$_2$(H$_2$O)$_3$]$_n$. *Science*, **283**,
      1148-1150. doi:10.1126/science.283.5405.1148

[17] Deacon, G.B. and Phillips, R.J. (1980) Relationships be-
      tween the carbon-oxygen stretching frequencies of car-
      boxylato complexes and the type of carboxylate coordi-
      nation. *Coordination Chemistry Reviews*, **33**, 227-250.
      doi:10.1016/S0010-8545(00)80455-5

[18] Alcock, N.W., Tracy, V.M. and Waddington, T.C. (1976)
      Acetates and acetato-complexes. Part 2. Spectroscopic
      studies. *Journal of the Chemical Society, Dalton Trans-
      actions*, **21**, 2243-2245. doi:10.1039/dt9760002243

[19] Lewis, J., Lin, Y., Royston, L. and Thompson, R. (1965)
      The chemistry of polynuclear compounds. Part III.
      Magnetic properties of some carboxylic acid derivatives
      of copper (II). *Journal of the Chemical Society*, **17**, 6464-
      6477. doi:10.1039/jr9650006464

[20] Lippard, S.J. and Doedens, R.J. (2007) Structure and
      metal-metal interactions in copper (II) carboxylate com-
      plexes. *Progress in Inorganic Chemistry*, **21**, 209-231.

[21] Castro, I., Sletten, J., Calatayud, M.L., JuIve, M., Cano,
      J., Lloret, F. and Caneschi, A. (1995) Synthesis and mag-
      netic properties of a tetranuclear copper (II) complex with
      a mu.-1,2,3,4-squarato coordination mode. Crystal struc-
      ture of *u*-1,2,3,4-squarato)tetrakis [(tris(2-aminoethyl)-
      amine)copper (II)] perchlorate. *Inorganic Chemistry*, **34**,
      4903-4909. doi:10.1021/ic00123a027

[22] Gao, D.Z., Zhao, W.W., Chen, Y.W. and Sun, Y.Q. (2009)
      Synthesis, crystal structure and magnetic behavior of a
      new Zn (II) nitronyl nitroxide complex [Zn(NIT$_3$Py)$_2$Cl$_2$].
      Chinese Journal of Inorganic Chemistry, **25**, 738-741.

[23] Wang, S.P., Gao, D.Z., Liao, D.Z., Jiang. Z.H. and Yan,
      S.P. (2006) Two novel diamagnetic-metal coordination
      polymers of nitronyl nitroxide radical incorporating iso-
      phthalate bridges. *Transition Metal Chemistry*, **31**, 214-
      219. doi:10.1007/s11243-005-6396-x

[24] Chen, J., Liao, D.Z., Jiang, Z.H. and Yan, S.P. (2005) A
      new heterospin complex from oxamide-bridged Cu (II)
      binuclear units and carboxyl-phenyl-substituted nitronyl
      nitroxide. *Inorganic Chemistry Communications*, **8**, 564-
      567. doi:10.1016/j.inoche.2005.02.015

18

Thermally induced gelation of alumina shaping-neutron scattering and rheological measurements

**Papiya Biswas[1], Kotikalapudi Rajeswari[1], Somasani Chaitanya[1], Roy Johnson[1*],
Swapnil A. Prabhudesai[2], Veerendra K. Sharma[2], Subhankur Mitra[2], Ramaprosad Mukhopadhyay[2]**

[1]Centre for Ceramic Processing, International Advanced Research Centre for Powder Metallurgy and New Materials, Hyderabad, India
[2]Solid State Physics Division, Bhabha Atomic Research Centre, Trombay, India
Email: *royjohnson@arci.res.in

ABSTRACT

Thermally induced gelation forming based on methylcellulose is recently being explored as a simple and environmentally benign process. Alumina slurry containing 0.1 wt% methylcellulose is subjected to Quasi Elastic Neutron Scattering (QENS) and rheological measurements in gelation temperature regimes to evolve a possible mechanism of the forming process. A reduction in diffusivity of water in the slurry from 2.16 to 1.92×10^{-5} $cm^2 \cdot s^{-1}$ after exposure to 55°C is observed with QENS. This is found to be well correlated with a steep increase in viscosity from 1.2 Pa.s till 50°C to 50,000 Pa.s at 55°C. QENS studies revealed the diffusion of water occurs by jump diffusion with the jump lengths distributed randomly. Further, for the entire sample much longer residence time is found as compared to bulk water, which is due to hydrophilic interaction of water molecules with the methylcellulose in the slurry. Reduction in diffusivity of water along with the steep increase in viscosity could be understood as the strong, cross-linked polymer-solvent irreversible gel formation in presence of alumina which is responsible for the retention of a consolidated shape of the ceramic green body. Samples maintained the integrity while heat treatments achieving close to theoretical density values of 3.98 $g \cdot cm^{-3}$ at 1550°C.

Keywords: Ceramics; Polymers; Heat Treatment; Molecular Dynamics

1. INTRODUCTION

Methylcellulose is a water soluble organic compound wi-

th very good compatibility and solubility characteristics that makes it a choice of binder for various ceramic processing [1,2]. Methylcellulose is presently used widely for ceramic shaping by viscous plastic processing such as extrusion and injection molding [3,4]. Methylcellulose is also broadly used for the coating purpose in pharmaceutical, textile and in paper industry. Methylcellulose is composed of two components such as naturally occurring cellulose and methyl ($-CH_3$) side groups. Cellulose itself is water insoluble, due to the intra and intermolecular hydrogen bonds. However, when a certain number of the hydroxyl groups are substituted by methoxyl groups, some of the hydrogen bonds are broken and methylcellulose becomes water soluble. The solubility is also a strong function of the degree of substitution and generally the commercial products have an intermediate degree of substitution [5]. Though methylcellulose is water soluble at lower temperatures, at elevated temperatures beyond 50°C, hydrophobic effects manifest into gelation. Recently, this unique gelation property of methylcellulose is employed for aqueous gel casting process for slurry leading to the shaping of simple and complex shapes [6,7]. The process is proven to have advantages of defect free processing and improved densification [8]. So it is of interest to study the change in viscosity with respect to temperature and correlate with diffusion of water in alumina slurry containing the methylcellulose and after exposing the same to the gelation temperature.

Sarkar *et al.* [9] has studied the gelation of aqueous methylcellulose. An initial drop in viscosity observed during the study was primarily attributed to the gradual loss of water of hydration from methylcellulose. It was observed that as the temperature is increased beyond 50°C the dehydration also increases leading to polymer-polymer interactions resulting in a network structure leading to the drastic increase in viscosity as have been observed in the viscosity versus temperature plot. The authors thus ascribed complex viscosity as a function of temperature to study the manifestation of gelation.

The effect of gelation may influence the dynamics of water in alumina slurry. Therefore, it is of interest to study the diffusion of water molecules inside alumina slurry exposed under different conditions. Diffusion of water can be studied using several experimental techniques including PFG-NMR [10] and QENS [11]. QENS, in particular, has been employed widely due to its ability to give spatial as well as temporal information on a wide range of time (10^{-10}-10^{-13} s) and length (few angstroms) scales. It is very well suited to study the diffusion in hydrogenous material such as water confined in porous media [12-18], hydrocarbons adsorbed in zeolites [19], micellar system [20] and so forth due to its large scattering cross section with hydrogen. This technique provides quantitative as well as qualitative information about the dynamics. The broadening of peaks observed over the instrumental resolution limit can be correlated to the dynamics of the system under evaluation. The quantitative information obtained from the analysis of the data entails the information about the correlation time, length scale and activation energy while qualitative information pertains to the geometrical mechanism of the motion. In a QENS experiment, information on molecular dynamics is obtained by considering the dynamical structure factor, $S(Q,\omega)$, which is a measure of the probability that an incident neutron undergoes a scattering process with an atom by exchanging energy $\hbar\omega$ and momentum $\hbar Q$. Characteristic times and the spatial extent of the molecular dynamics can be obtained by analyzing the ω and Q dependence of $S(Q,\omega)$ respectively.

Hence, the objective of the current work is therefore to correlate the gelation process occurring in alumina slurry by complex viscosity measurements with a function of temperature and corresponding diffusion of water by QENS. Further the gelation process was also successfully employed for colloidal shaping of alumina ceramics as a simple and environmentally benign process.

2. EXPERIMENTAL PROCEDURE

2.1. Preparation of the Slurry

High purity alumina powder (AHPA 0.5, Sasol, USA) with specifications of 99.96% purity with a BET surface area of 8 m^2/g and average particle size (D$_{50}$) of 0.5 μm is used in the present study. Aqueous slurry is exposed in distilled water using Darvan 821A (R. T. Vanderbilt Co., Inc., Norwalk, CT, USA) as a dispersant and milled for 4 - 5 hrs in polypropylene bottles in a pot jar mill. Thus obtained alumina slurry with >75 wt% of solid loading was further mixed with 0.1 wt% of methylcellulose (Dow Chemical Company).

2.2. Rheological Measurements

Rheological behaviour of the slurries was measured at varying shear rates (MCR 51, Anton Paar, Austria) to determine the flow properties. The variation in viscosity as a function of temperature was also studied at low shear rates of 10 s^{-1} in the temperature range of 25°C - 70°C at the heating rate of 1°C/min.

2.3. Neutron Scattering Experiment

Neutron scattering measurements were carried out by using the QENS spectrometer at Dhruva reactor, Trombay [21] on alumina slurry samples. The alumina slurry samples, containing 0.1 wt% methylcellulose, were exposed at different temperatures of 45°C (designated as sample #1), 50°C (designated as sample #2) and 55°C (designated as sample #3). In addition sample #2 is re-heated and equilibrated to 60°C for 4 hrs and subsequent QENS data recorded at room temperature. The spectrometer is used in multi angle reflecting crystal (MARX) mode, which uses a combination of a large analyser crystal for energy analysis and a position sensitive detector for detecting the scattered neutrons. In the present configuration, this instrument has an energy resolution of 200 μeV with incident neutron energy of 5 meV as obtained from standard vanadium sample. The quasielastic data were recorded on all samples in the wave vector transfer (Q) range of 0.67 Å^{-1} - 1.8 Å^{-1} at 27°C.

2.4. Shaping of Alumina Components

Alumina slurry containing the 0.1 wt% of methylcellulose was casted into molds as per the gelation procedure described elsewhere [8]. The samples of uniform geometry were cut from various configurations after and the green density was measured using dimensional method. The samples were heat treated without cracks at the rate of 3°C/min during the binder burn out regime and could be sintered close to theoretical densities on sintering the samples at a peak temperature of 1550°C.

3. RESULTS AND DISCUSSIONS

3.1. Viscosity Measurements

The plot of apparent viscosity with shear rate for alumina slurry containing the 0.1 wt% of methylcellulose is shown in **Figure 1(a)**. It is evident that the as received slurries are pseudo plastic at lower shear rates and as the shear rate is increased beyond 200 s^{-1} the slurries exhibit close to Newtonian behavior. The change in the complex viscosity with respect to temperature is shown in **Figure 1(b)**. In the case of alumina slurry with 0.1 wt% concentration of methylcellulose the slurry had an initial viscosity of 1.2 Pa.s at room temperature. The viscosity remains

σ_{coh} (any element) $<< \sigma_{inc}$ (hydrogen), one can write,

$$\frac{d^2\sigma}{\partial\omega\partial\Omega} \propto \frac{k}{k_0}\Big[\sigma_{inc}S_{inc}(Q,\omega)\Big] \quad (2)$$

Therefore, in a neutron scattering experiment from the alumina slurry the observed dynamics mainly corresponds to the self-correlation function of the protons $S(Q,\omega) = S_{inc}(Q,\omega)$. In general, this scattering law can be written as [12]

$$S(Q,\omega) = A(Q)\delta(\omega) + \Big[1 - A(Q)\Big]L(\Gamma,\omega) \quad (3)$$

where the first term is the elastic part and the second is the quasielastic one. $L(\Gamma,\omega)$ is a Lorentzian function with a half width at half maxima (HWHM) Γ. The variation of HWHM, Γ provides information about the time scale of the motion. The contribution of the elastic scattering out of total scattered spectra is called the Elastic Incoherent Structure Factor (EISF). Therefore, $A(Q)$ in **Eq.3** is nothing but the EISF, which represents the space Fourier transform of the particle distribution, taken at infinite time and averaged over all the possible initial positions. A negligible contribution of EISF generally corresponds to the presence of pure long-range translational motion. However for localized motion e.g., rotational motion, this term is expected to have a non-zero value and the variation of EISF with Q is used to determine the information about the geometry of the molecular motion.

It may be noted that in a QENS experiment, the measured data is inherently convoluted with the resolution function of the instrument. To analyse the QENS data, it is customary to assume a theoretical scattering function (**Eq.3**), convolute with the instrumental resolution function and then obtain the dynamical parameters involved in the model scattering function by least squares fit to the experimental data.

The quasielastic spectra were recorded in the wave vector transfer (Q) range of 0.67 - 1.8 Å^{-1} at 27°C for slurry samples #1, #2 and #3 *i.e.* exposed at 45°C, 50°C and 55°C respectively. QENS measurements have been also carried out on sample #2 after exposing to 60°C. Significant quasi-elastic broadening was observed for all the samples over the instrument resolution. Thus, this broadening is related to the dynamical motion of water molecules in the alumina slurry. It is found that the sample #1 and #2 show larger quasi elastic (QE) broadening compared to sample #3, that is, the slurry exposed at 55°C. However, QE broadening in samples #1 and #2, that is, slurry exposed at 45°C and 50°C are found to be similar. This indicates qualitatively that diffusivity of water in slurry exposed at 45°C and 50°C is very similar however different than slurry exposed at 55°C.

This suggested that the water molecules in slurry exposed at 55°C experience different environment while

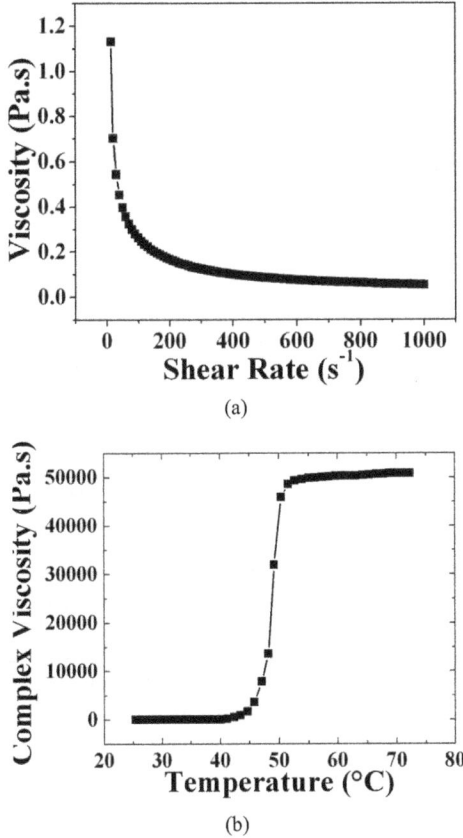

Figure 1. Plots of (a) viscosity with shear rates for the alumina slurry and (b) complex viscosity versus temperature for alumina slurry.

unchanged or with marginal decrease till the slurry reaches a temperature of 50°C which demonstrated a very prominent increase (**Figure 1(b)**) in viscosity to 50,000 Pa.s revealing the gelation.

3.2. Quasi elastic Neutron Scattering Measurements

In a neutron scattering experiment the scattered intensity is analyzed as a function of both energy and momentum transfer. The quantity measured is the double differential scattering cross section representing the probability that a neutron is scattered with energy change $dE = \hbar d\omega$ into the solid angle $d\Omega$ [12],

$$\frac{d^2\sigma}{\partial\omega\partial\Omega} \propto \frac{k}{k_0}\Big[\sigma_{coh}S_{coh}(Q,\omega) + \sigma_{inc}S_{inc}(Q,\omega)\Big] \quad (1)$$

where σ is the scattering cross-section, subscripts coh and inc denote the coherent and incoherent components and $S(Q,\omega)$ is the scattering law. k and k_0 are the final and initial wave vectors. $Q = k - k_0$ is the wave vector transfer [$Q = 4\pi\sin\theta/\lambda$, where 2θ is the scattering angle in case of elastic scattering] and $\hbar\omega = E - E_0$ is the energy transfer. For systems containing hydrogen atoms as diffusing compared to that in other samples exposed at lower temperatures. Therefore, sample #2 (slurry exposed at 50°C) is reheated to 60°C and further QENS measurement were carried out at room temperature. It is found that after heat treatment of sample #2, observed QE broadening is reduced and found very similar to the slurry exposed at 55°C. This suggests that after heating, slurry exposed at 50°C manifests the same environment to water molecules as that in sample exposed at 55°C. To extract detailed information QENS data are analysed with the model scattering law as given in **Eq.3**. The parameters, $A(Q)$ and $\Gamma(Q)$ were obtained by least squares fit. The resulting fits are shown in **Figures 2** and **3** for all the measured samples at typical Q values.

In the present case for all the samples, the elastic intensity are found to be negligible and a single Lorentzian function is good enough to describe the data, indicating that the observed dynamics correspond to the translational diffusion of the sorbed water molecules. The simplest model of translational motion is Brownian diffusion, which is described by Fick's law, in which HWHM (Γ) of the quasielastic component (Lorentzian function) varies linearly with Q^2, $\Gamma(Q) = DQ^2$, where D is the self-diffusion coefficient of fluids. This occurs when the interactions between the particles are weak. When the intermolecular interactions are significant, the diffusion mechanism deviates from the Fick's law. Here, two characteristic times are involved with the dynamical process. These are, the jump time τ_1 during which the particle diffuses and the residence time τ during which it undergoes oscillatory motions around any point r. Further, when jump time τ_1 is negligible compared to the residence time τ, the corresponding diffusion mechanism is known as jump diffusion. The molecule moves in the form of discrete jumps where the molecule remains at a given site for a specific time called the residence time before jumping instantly to another site separated from the first by a distance called the jump length. It has been shown [22,23] that for an isotropic distribution of jump length $\rho(r)$, the scattering law is a Lorentzian with a half width at half maximum (HWHM) given by

$$\Delta\omega(Q) = \frac{1}{\tau} \frac{\int_0^\infty \left[1 - \frac{\sin(Qr)}{Qr}\right]\rho(r)\mathrm{d}r}{\int_0^\infty \rho(r)\mathrm{d}r} \qquad (4)$$

In the case of a liquid, equipotential sites are not very well ordered as one might expect the jump length make a distribution which may be a Gaussian or random distribution. The corresponding jump diffusions are known as Hall and Ross [24] and Singwi-Sjölander [25] jump diffusions respectively. It may be noted that in all these jump diffusion models, the difference is that of a micro-

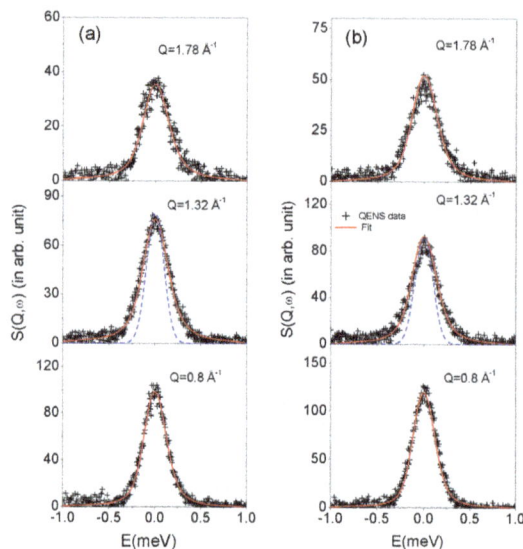

Figure 2. Fitted QENS spectra measured at room temperature from 0.1 wt% methylcellulose-alumina slurry samples exposed at 50°C (sample #2) (a) before exposing to high temperature (b) and after exposing to 60°C, at some typical Q values. Instrument resolution is shown by dashed line in the middle panel.

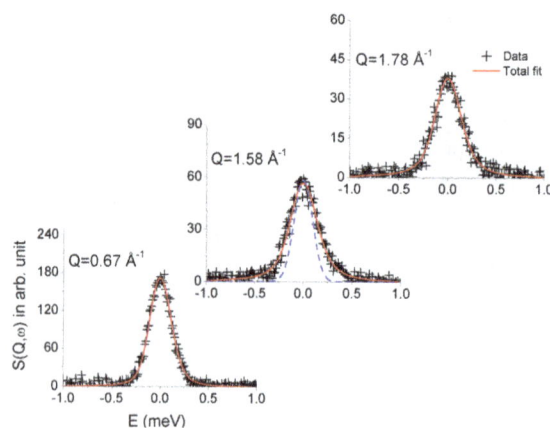

Figure 3. Fitted QENS spectra measured at room temperature from 0.1 wt% methylcellulose-alumina slurry sample exposed at 55°C (sample #3) at some typical Q values. Instrument resolution is shown by dashed line in the middle panel.

scopic detail. In case the system is studied at a larger length scales or equivalently at smaller Q values, the information about these finer details is lost and the diffusion process looks very much like the Brownian diffusion. The variation of $\Gamma(Q)$ with Q^2 is therefore linear at low Q values and the diffusion coefficient can be obtained simply from the slope of this curve in the region of small Q values. At higher Q values however, as finer details of the jump diffusion start to emerge, the variation of $\Gamma(Q)$ with Q^2 is no more linear and tend to saturate to a value, which is indicative of the residence times involved in the jump process. In the present case for all the samples, the variation of $\Gamma(Q)$ with Q^2 obtained from the experimental data were found to be described well by the Singwi-Sjölander model of jump diffusion, which employs a random distribution of jump lengths. In this model the variation of $\Gamma(Q)$ with Q^2 is given by [25]

$$\Gamma(Q) = \frac{DQ^2}{1 + DQ^2\tau} \quad (5)$$

where D is the diffusion coefficient and τ is the residence time. The variation of Γ as obtained from the fit for both samples is plotted in **Figure 4** as a function of Q^2 along with the experimental values. The deviation of the graph from the linear behavior at high Q justifies the jump diffusion model. The solid line in **Figure 4** corresponds to the least squares fit with the experimentally obtained values assuming **Eq.5**. The values of the parameters, D and τ as obtained from the fit are shown in **Table 1**.

Mean jump length $L = \left\langle r^2 \right\rangle^{\frac{1}{2}}$ for the water in the Al slurry can be obtained using Einstein relation $D = \dfrac{\left\langle r^2 \right\rangle}{6\tau}$ and also given in **Table 1**. It may be noted that the residence time obtained for sample #1 is 5.34 ps, which is much larger than that of water in bulk 1.1 ps. This is probably the result of attractive interaction of water molecules with the methylcellulose in the slurry sample, which makes the motion of water molecules hindered, thereby increasing the residence time. It is clear that diffusion of water in slurry sample exposed at 55°C is found to be hindered than the slurry exposed at lower temperatures (45°C and 50°C). Residence time of water molecule is also found to be larger for slurry exposed at 55°C. This could be interpreted in terms of restricted diffusion of water due to gelation. As the temperature increased above to 50°C, cross-linked polymer-solvent gel formation takes place, which restrict the diffusion of water, and observed diffusivity is found to be reduced. QENS results obtained from heated (60°C) slurry sample exposed at 50°C (sample #2) further confirmed our interpretation. It is found that diffusivity of water in sample #2 after heating to 60°C is found to be similar to that observed in slurry sample exposed at 55°C. Results obtained for neutron scattering, a microscopic technique, are found to be consistent with the macroscopic viscosity measurements, which suggest that temperature beyond 50°C increases polymer-polymer interactions resulting in a network structure.

3.3. Shaping of Alumina Components

The samples shaped by employing the gelation process with the green density values of 1.95 g·cm^{-3} is shown in **Figure 5(a)** along with the samples sintered to 1550°C. The sintered samples (shown in **Figure 5(b)**) exhibited a density close to theoretical density of 3.98 g·cm^{-3}, which is also complimented by microstructures depicted in **Figure 5(c)**.

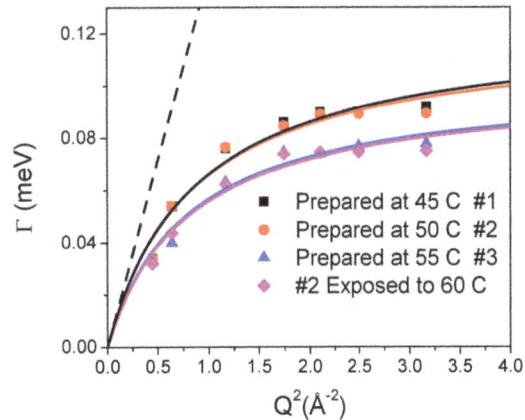

Figure 4. Variation of HWHM Γ, quasielastic component for 0.1 wt% methylcellulose-alumina slurry samples exposed at different temperatures with Q^2. The solid line corresponds to the fit assuming random jump diffusion model whereas dashed line show as expected with Fick's law.

Table 1. Diffusion coefficients and residence time are given for studied Al slurry samples at room temperature.

Sample	Diffusivity ($\times10^{-5}$) cm^2·s^{-1}	Residence time (ps)	Jump length (Å)
Al slurry (45°C) #1	2.16 (5)	5.34 (5)	2.63 (4)
Al slurry (50°C) #2	2.20 (5)	5.45 (5)	2.68 (4)
Al slurry (55°C) #3	1.92 (4)	6.47 (6)	2.71 (5)
#2 reheated to 60°C	1.89 (4)	6.53 (6)	2.72 (5)
Bulk water [26]	2.5	1.10	1.28

4. CONCLUSION

Thermally induced gelation process of methylcellulose, a derivative of naturally occurring cellulose has been successfully employed for the shaping of alumina parts from the slurry. Rheological and neutron scattering measurements were employed for tracking the gelation process. Rheological measurements of alumina slurry while insitu heating has exhibited a steep increase in viscosity in the temperature range of 50°C - 55°C. Neutron scattering studies also revealed a reduction in the diffusivity of water after exposing the sample at and above 55°C. It is found that the water molecules in alumina slurry undergo random jump diffusion. The motion is sub-diffusive and the residence time is much longer compared to bulk water. It is found that on gelation (sample #3) the diffusion is hindered and residence time goes up which is in conjunction with the viscosity data. The jump lengths are found to be more or less similar for all the samples. It is

(a)

(b)

(c)

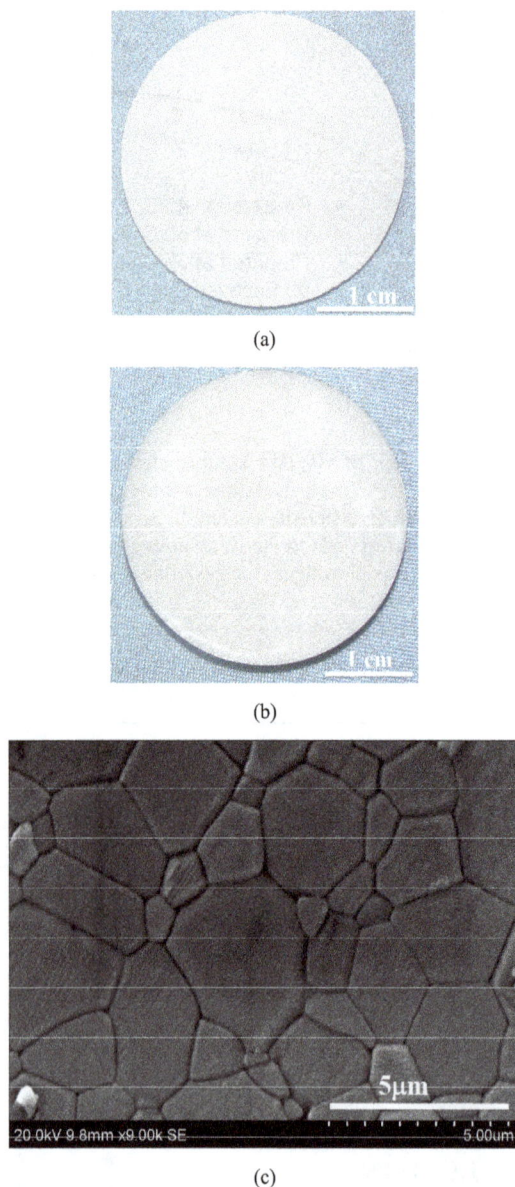

Figure 5. (a) Thermal gel cast green samples; (b) Sintered samples; and (c) Microstructure of sintered samples at 1550°C.

explained that cross-linked polymer network formation is responsible for the above observed phenomena, which can also be correlated to consolidated green alumina part on casting. Sintered samples have demonstrated the density close to theoretical value in combination with a dense micro-structure confirming the possibility of defect free shaping through methylcellulose based thermally induced gel casting process.

REFERENCES

[1]   Scheutz, J.E. (1986) Methyl cellulose polymers as bind-ers for extrusion of ceramics. *Journal of the American Ceramic Society*, **88**, 1556-1559.

[2]   Lange, F.F. (1989) Powder processing science and technology for increased reliability. *Journal of the American Ceramic Society*, **72**, 3-10. doi:10.1111/j.1151-2916.1989.tb05945.x

[3]   Rivers, R.D. (1978) Method of injection moulding powder metal parts. US Patent 4113480.

[4]   Bayer, R. and Knarr, M. (2012) Thermal precipitation or gelling behaviour of dissolved methyl cellulose (MC) derivatives—Behaviour in water and influence on the extrusion of ceramic paste. Part 1: Fundamental of MC derivatives. *Journal of European Ceramic Society*, **32**, Article ID: 10071018. doi:10.1016/j.jeurceramsoc.2011.11.025

[5]   Koda, S., Hori, T., Nomura, H. and Kawaizumi, F. (1991) Hydration of methyl cellulose. *Polymer*, **32**, 2806-2810. doi:10.1016/0032-3861(91)90112-V

[6]   Xu, X., Wen, Z., Lin, J., Li, N. and Wu, X. (2010) An aqueous gel-casting process for   -LiAlO$_2$ ceramics. *Ceramic International*, **36**, 187-191. doi:10.1016/j.ceramint.2009.07.017

[7]   Biswas, P., Swathi, M., Ramavath, P., Rajeswari, K., Suresh, M.B. and Johnson, R. (2012) Diametral deformation behaviour and machinability of methyl cellulose thermal gel cast processed alumina ceramics. *Ceramic International*, **38**, 6115-6121. doi:10.1016/j.ceramint.2012.04.059

[8]   Hareesh, U.S., Anantharaju, R., Biswas, P., Rajeswari, K. and Johnson, R. (2011) Colloidal shaping of alumina ceramics by thermally induced gelation of methyl cellulose. *Journal of the American Ceramic Society*, **94**, 749-753. doi:10.1111/j.1551-2916.2010.04188.x

[9]   Sarkar, N. and Greminger Jr., G.K. (1983) Methyl cellulose polymers as multifunctional processing aids in ceramics. *American Ceramic Bulletin*, **62**, 1280-1284.

[10]   Kärger, J. and Ruthven, D.M. (1992) Diffusion in zeolites and other microporous solids. Wiley Interscience, New York.

[11]   Bée, M. (1988) Quasielastic neutron scattering. Adam Hilger, Bristol.

[12]   Sharma, V.K., Mitra, S., Kumar, A., Yusuf, S.M., Juranyi, F. and Mukhopadhyay, R. (2011) Diffusion of water in molecular magnet Cu$_{0.75}$Mn$_{0.75}$[Fe(CN)$_6$]·7H$_2$O. *Journal Physics: Condensed Matter*, **23**, 446002-446009. doi:10.1088/0953-8984/23/44/446002

[13]   Sharma, V.K., Mitra, S., Singh, P., Jurany, F. and Mukhopadhyay, R. (2010) Diffusion of water in nano-porous polyamide membranes: Quasielastic neutron scattering study. *The European Physical Journal Special Topics*, **189**, 217-221. doi:10.1140/epjst/e2010-01325-9

[14]   Sharma, V.K., Singh, P.S., Gautam, S., Mitra, S. and Mukhopadhyay, R. (2009) Diffusion of water in nanoporous NF polyamide membrane. *Chemical Physics Letter*, **478**, 56-60. doi:10.1016/j.cplett.2009.07.045

[15]   Sharma, V.K., Singh, P.S., Gautam, S., Maheshwari, P., Dutta, D. and Mukhopadhyay, R. (2009) Dynamics of water sorbed in reverse osmosis polyamide membrane.

Journal of Membrane Science, **326**, 667-671.
doi:10.1016/j.memsci.2008.11.003

[16] Mitra, S., Mukhopadhyay, R., Tsukushi, I. and Ikeda, S. (2001) Dynamics of water in confined space (porous alumina): QENS study. *Journal Physics*: *Condensed Matter*, **13**, 8455-8465.
doi:10.1088/0953-8984/13/37/302

[17] Chakrabarty, D., Gautam, S., Mitra, S., Gil, A., Vicente, M.A. and Mukhopadhyay, R. (2006) Dynamics of absorbed water in saponite clay: Neutron scattering study. *Chemical Physics Letter*, **426**, 296-300.
doi:10.1016/j.cplett.2006.05.131

[18] Mitra, S., Pramanik, A., Chakrabarty, D., Jurányi, F., Gautam, S. and Mukhopadhyay, R. (2007) Diffusion of water adsorbed in hydrotalcite: Neutron scattering Study. *Journal of Physics*: *Conference Series*, **92**, 012167-012171. doi:10.1088/1742-6596/92/1/012167

[19] Sharma, V.K., Gautam, S., Mitra, S., Rao, M.N., Tripathi, A.K., Chaplot, S.L. and Mukhopadhyay, R. (2009) Dynamics of adsorbed hydrocarbon in nanoporous zeolite framework. *Journal of Physical Chemistry B*, **113**, 8066-8072.
doi:10.1021/jp9014405

[20] Sharma, V.K., Mitra, S., Sakai, V.G., Hassan, P.A., Embs, J.P. and Mukhopadhyay, R. (2012) The dynamical landscape in CTAB micelles. *Soft Matter*, **8**, 7151-1760.

doi:10.1039/c2sm25515d

[21] Mukhopadhyay, R., Mitra, S., Paranjpe, S.K. and Dasannacharya, B.A. (2001) Quasielastic neutron scattering facility at Dhruva reactor. *Nuclear Instrumentation Method A*, **474**, 55-66.

[22] Chudley, C.T. and Elliott, R.J. (1961) Neutron scattering from a liquid on a jump diffusion model. *Proceedings Physics Society*, **77**, 353-361.
doi:10.1088/0370-1328/77/2/319

[23] Egelstaff, P.A. (1967) An introduction to the liquid state. Academic Press, London.

[24] Hall, P.L. and Ross, D.K. (1981) Incoherent neutron scattering functions for random jump diffusion in bounded and infinite media. *Molecular Physics*, **42**, 673-682.
doi:10.1080/00268978100100521

[25] Singwi, K.S. and Sjölander, A. (1960) Diffusive motions in water and cold neutron scattering. *Physical Review*, **119**, 863-871. doi:10.1103/PhysRev.119.863

[26] Tiexiera, J., Bellissent-Funel, M.C., Chen, S.H. and Dianoux, A.J. (1985) Experimental determination of the nature of diffusive motions of water molecules at low temperatures. *Physical Review A*, **31**, 1913-1917.
doi:10.1103/PhysRevA.31.1913

19

Spectral studies of transition metal complexes with 25, 26 dioxo1,6,12,17,23,24 hexaazacyclohexacosa 1,5,12,16 tetraene macrocyclic ligand (L)

Sulekh Chandra[1]*, Poonam Pipil[2]

[1]Department of Chemistry, Zakir Husain Delhi College, University of Delhi, New Delhi, India
[2]Department of Chemistry, Rajdhani College, University of Delhi, New Delhi, India
Email: *schandra_00@yahoo.com, www.poonam.pipil@gmail.com

ABSTRACT

Mn(II), Co(II), Ni(II) and Cu(II) complexes with this ligand (L) *i.e.* 25, 26 dioxo1,6,12,17,23,24 hexaazacyclohexacosa1,5,12,16 tetraene have been prepared. All the complexes were characterized by elemental analysis, molar conductance measurements, magnetic susceptibility measurements, mass, i.r. electronic, and e.p.r. spectral studies. The molar conductance measurements in DMF indicate that the Mn(II), Co(II), and Cu(II) are non-electrolyte whereas the complexes of Ni(II) are electrolyte. Thus, these complexes may be formulated as $[M(L)X_2]$ and $[Ni(L)]X_2$ (where M = Mn(II), Co(II) and Cu(II) and X = Cl^- and NO_3^-). All the complexes are of the high spin type expected Ni(II) complexes which are diamagnetic. On the basis of spectral studies, an octahedral geometry may be assigned of Mn(II) and Co(II). Square planar is for Ni(II) complexes, and tetragonal geometry for Cu(II) complexes.

Keywords: Mass; IR; Magnetic Moment; Electronic EPR; X-Ray Diffraction Studies

1. INTRODUCTION

Schiff bases are widely studied because of increasing recognition in biological system [1] e.g. antimicrobial, antiviral, antifungal, antitumour and other biological activities particularly with first row of transition metal complexes [2-9]. Macrocyclic compounds are similar in structure and reactions to many naturally occurring compounds, Which are known to exhibit selective cation complexation [10,11]. The chemical properties of macrocyclic complexes can be turned to force metal ions to adopt unusal coordination geometry. Transition metal macrocyclic complexes have received much attention as an active part of metalloenzymes [12] as biomimic model compounds [13] due to their resemblence with natural proteins like hemerythrin and enzymes. Synthesis of transition metal complexes of new aza macrocycles with different cavity sizes donor types, ring substituents etc. under different experimental conditions have resulted in a large number and variety of compounds [14,15]. Because of the numerous areas of chemistry where aza macrocyclic complexes have found a niche, the preparation of new macrocyclic ligands with ever more elaborate structures is also a vital area of research. In view of the above applications, the synthesis and characterization of transition metal complexes with such ligand is highly desirable. In this paper we report the synthesis and characterization of transition metal complexes with 25, 26 dioxo1,6,12,17,23,24 hexaaza cyclohexaneosa 1,5,12,16 tetraene macrocyclic ligand [L]. (Figure 1).

2. CHEMISTRY

All the chemicals used were of Anala R grade, and were purchased from Aldrich. Metal salts were purchased from E. Merck and were used as received. All solvents used were of standard/spectroscopic grade.

2.1. Synthesis of Complexes

All the complexes were prepared by template method. An ethanolic solution of diethylenetriamine [10 mL] and glutaricanhydride (10 mL) were mixed. To this soluction the ehtanolic solution of metal salt (0.001 mol) was added. The resulting solution was boiled under refluxed for six hours. The solution was then concentrated upto half of its volume under reduced pressure. On cooling the solution is precipitated out. It was filtered, washed with cold ethanol and dried over P_4O_{10} under vaccum.

2.2. Physical Measurements

The C, H, and N were analyzed on a Carlo-Erba 1106 elemental analyzer. Molar conductance was measured on the ELICO (CM82T) conductivity bridge. Magnetic susceptibility was measured at room temperature on a Gouy-balance using $CuSO_4 \cdot 5H_2O$ as a callibrant Electron-impact mass spectra were recorded on JEOL, JMS, DX-303 mass spectrometer, IR spectra were recorded on a Perkin-Elmer 137 instrument in CsI pellets. EPR spectra of the complexes were recorded as polycrystalline sample and in the DMF solution, at liquid nitrogen temperature for Co(II) and at room temperature for Mn(II) and Cu(II) complexes on E4-EPR spectrometer using the DPPH as the g-marker.

3. RESULTS AND DISCUSSION

Complexes: On the basis of elemental analysis, the complexes were found to have the composition as shown in **Table 1**.

The molar conductance measurement of the complexes in DMF correspond to be nonelectrolytic nature for Mn(II), Co(II) and Cu(II) while 1:2 electrolytes for Ni(II) complexes. Thus, these complexes may be formulated as $[M(L)X_2]$ and $[Ni(L)]X_2$ (where M = Mn(II),

Co(II), and Cu(II) and X = Cl^- and NO_3^-). In the IR spectra of complexes, the shifting in lower side of a band [v(N-H)] and the band of amides II and III, suggest the coordination through nitrogen of –NH group (N_4).

3.1. i.r. Spectra of the Complexes

The presence of the obsorption bands at 1410 - 1421, 1299 - 1319 and 1010 - 1031 cm^{-1}, in the i.r. spectra **Figures 2(a)-(c)** of the Mn(II), Co(II) and Cu(II) of nitrato complexes suggest the both nitrate groups are coordinated to the central metal ion in a unidentate fashion [16-18]. The Ni(II) nitrato complex exhibits a sharp strong peak at 1387 cm^{-1}, which is in favour of an un-coordinated nitrate group [16,17].

3.1.1. Manganese(II) Complexes

The magnetic moment recorded at room temperature lies in the range 5.90 - 5.96 B.M. corresponding to five unpaired electrons. Electronic spectra of Mn(II) complexes in DMSO solution display four weak intensity adsorption bands (**Table 2**) in the 15,870 - 16,743 region (ε = 24 - 26 1·mol^{-1}·cm^{-1}), 20,356 - 21,685 (ε = 33 - 36 1·mol^{-1} cm^{-1}) 24,787 - 27,006 (ε = 61 − 64 1·mol^{-1}·cm^{-1}). These bands may be assigned to the transitions, $^6A_{1g} - {}^4A_{1g}({}^4G)$, $^6A_{1g} - {}^4E_g$, $^6A_{1g}({}^4G)$ (10B + 5C), $^6A_{1g} - {}^4E_g$ (^4D)(17B + 5C) and $^6A_{1g} - {}^4T_{1g}$ (^4P), respectively [19].

The e.p.r. spectra were recorded at room temperature

Table 1. Molar conductance and elemental analysis data of complexes.

Complex	Colour	Molar Conductance ($\Omega^1 \cdot cm^{-1} \cdot mol^{-1}$)	% Yield	Elemental Analysis Data Calculated (Found) %				M. Pt (°C)
				C	H	N	Metal	
[Mn(L)Cl$_2$] C$_{20}$H$_{18}$MnN$_6$S$_2$Cl$_2$	Brown	02	69	45.0 (45.1)	3.3 (3.4)	15.6 (15.8)	10.5 (10.3)	280
[Mn(L)(NO$_3$)$_2$] C$_{20}$H$_{18}$MnN$_8$S$_2$O$_6$	Light Brown	09	66	41.0 (40.8)	3.1 (3.0)	19.0 (19.1)	9.2 (9.4)	288
[Co(L)(NO$_3$)$_2$] C$_{20}$H$_{18}$CoN$_6$S$_2$Cl$_2$	Pink	10	65	44.6 (44.4)	3.4 (3.3)	15.6 (15.7)	11.3 (11.0)	290
[Co(L)(NO$_3$)$_2$] C$_{20}$H$_{18}$CoN$_8$S$_2$O$_6$	Pink	08	61	40.5 (40.7)	3.1 (3.2)	18.8 (19.0)	9.9 (10.0)	299
[Ni(L)Cl$_2$] C$_{20}$H$_{18}$NiN$_2$S$_2$Cl$_2$	Red	158	64	44.6 (44.2)	3.4 (3.6)	15.5 (15.7)	11.2 (10.0)	298
[Ni(L)(NO$_3$)$_2$] C$_{20}$H$_{18}$NiN$_2$S$_2$O$_6$	Orange	146	60	40.2 (40.5)	3.1 (3.5)	18.8 (19.0)	9.8 (10.0)	297
[Cu(L)Cl2] C$_{20}$H$_{18}$CuN$_6$S$_2$Cl$_2$	Blue	06	62	44.1 (44.5)	3.3 (3.2)	15.8 (15.5)	11.6 (11.8)	288
[Cu(L)(NO$_3$)$_2$] C$_{20}$H$_{18}$CuN$_8$S$_2$O$_6$	Shiny Blue	09	61	40.2 (40.4)	3.0 (2.9)	18.8 (18.9)	10.9 (10.8)	290

Table 2. Magnetic moments and electronic spectral data of the complexes.

Complex	λ Max (cm^{-1})	μeff (B.M.)	ε (L mol^{-1}·cm^{-1})
[Mn(L)Cl$_2$)]	15,872, 21,685 24,785, 38,217	5.90	24, 36, 61, 135
[Mn(L)(NO$_3$)$_2$]	16,743, 20,356, 27,006, 35,426	5.96	26, 33, 64, 128
[Co(L)Cl$_2$]	9896, 14,906, 17,341, 39,046	4.85	56, 67, 82, 157
[Co(L)(NO$_3$)$_2$]	9510, 14,285, 18,645, 38,819	4.76	55, 65, 84, 154
[Ni(L)]Cl$_2$)	1651, 22,088, 30,114,	Diamagnetic	52, 92, 132
[Ni(L)](NO$_3$)$_2$)	15,811, 20,983, 28,547,	Diamagnetic	51, 90, 128
[Cu(L)Cl$_2$]	9983, 15,991, 28,057	1.96	48, 58, 149
[Cu(L)(NO$_3$)$_2$]	10,560, 16,384, 29,514	1.90	49, 58, 151

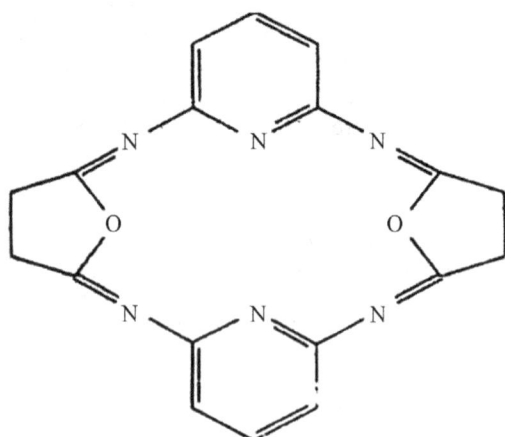

Figure 1. Structure of ligand.

as polycrystalline samples and in DMSO solution. The polycrystalline spectra give an isotropic signal centered at 2.04 - 2.16. The spectra recorded in DMSO solution gives well-resolved six line spectra due to the hyperfine interaction between the unpaired electrons with the ^{55}Mn nuclear (1 = 5/2).

3.1.2. Cobalt(II) Complexes

At room temperature the magnetic moment measurements of cobalt(II) complexes lies in the range 4.76 - 4.85 B.M corresponding to three unpaired electron (**Table 2**). The electronic spectra of all the cobalt(II) complexes **Figure 3** exhibit aborption in the region 6510 - 9896 (ε = 55 - 65 L·mol^{-1}·cm^{-1}), 14,285 - 14,906 (ε = 65 - 67 m·mol^{-1}·cm^{-1}), 17341-18645 (ε =82 - 74 L·mol^{-1}·cm^{-1})

and 39,046 - 38,819 cm^{-1} (ε = 154 - 157 L·mol·cm^{-1}). These bands may be assigned to the transitions: $^4T_{1g}$ (F) $- ^4T_{2g}$ (F) (v_1), $_4T_{1g} - ^4A_{2g}$ (v_2) and $^4T_{1g}$ (F) $- ^4T_{1g}$ (P) (v_3), respectively [20].

It is difficult to make the assignments for the fourth band and it may be considered as a charge transfer band. The position of these suggests an octahedral environment around the cobalt (II) ion.

The e.p.r. spectra of the Co(II) complexes were recorded as polycrystalline samples and in DMSO solution at liquid nitrogen temperature (LNT).

The g-values were found to be almost the same in both cases in polycrystalline samples as well as in the solution, (**Table 3**).

It indicates that the complexes have the same geometry in the solid as well as in solution. The large deviation in g-values, in the e.p.r. spectra, from the spin only value (g = 2.0023) is due to the large angular momentum contribution.

3.1.3. Nickel(II) Complexes

At room temperature these complexes show the diamagnetic behavior, indicating the square planar environment around the Ni(II) ion.

The electronic spectra of **Figure 4** the Ni(II) complexes, exhibit three absorption bands in the range 15,812 - 16,510 (ε = 51 - 53 L·mol^{-1}·cm^{-1}), 20,983 - 22,088 (ε = 90 - 92 L·mol^{-1}·cm^{-1}) and 28,547 - 30,114 cm^{-1} (ε = 128 - 132 L·mol^{-1}·cm^{-1}). An examination of these bands indicates that the complexes have square planar geometry. These bands may be assigned to the three spin allowed

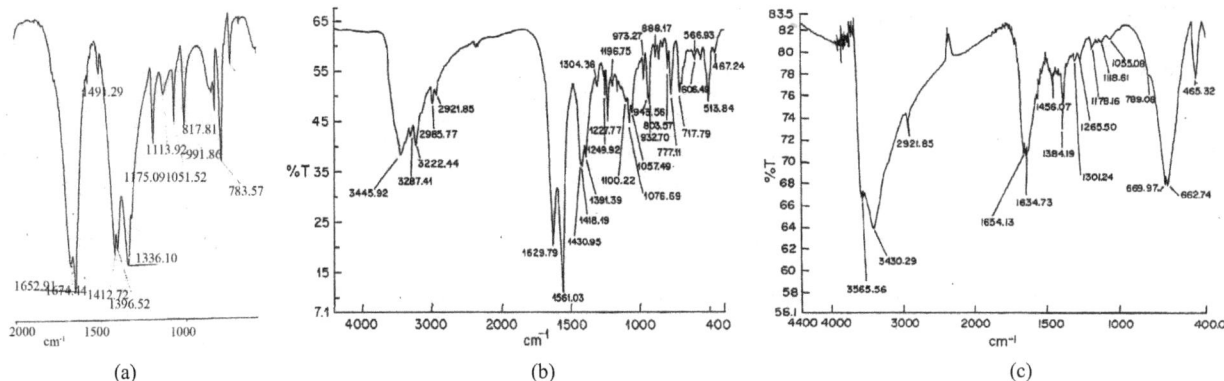

Figure 2. (a) IR Spectrum of $[Co(L^5)(NO_3)_2]$; IR Spectrum of $[Ni(L^5)(NO_3)_2]$; IR Spectrum of $[Cu(L^5)(NO_3)_2]$.

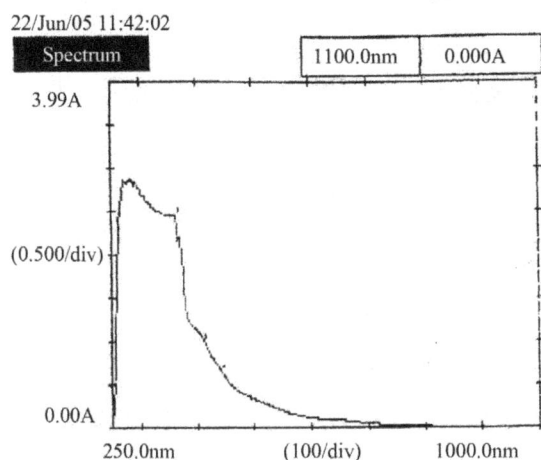

Figure 3. UV Spectrum of $[Co(L^5)(NO_3)_2]$.

transitions: $^1A_{1g}$ (D) (v_1), $^1A_{1g}$ (D) – $^1B_{2g}$ (G) (v_2) and $^1A_{1g}$ (D) – ^1E_g (G) (v_3), respectively [21].

3.1.4. Copper(II) Complexes

The magnetic moment of the Cu(II) complexes at room temperature lie in the range 1.90 - 1.96 B.M corresponding to one unpaired electron. The complexes may be considered to possess a tetragonal geometry.

The electronic spectra copper (II) complexes **Figure 5** display three bands (**Table 2**), in the 9987 - 10,560 (ε = 48 - 49 L·mol^{-1}·cm^{-1}) 15,991 - 16,384 (ε = 58 L·mol^{-1}·cm^{-1}) 28,057 - 29,514 (ε = 149 - 151 L·mol^{-1}·cm^{-1}) ranges. These bands corresponding to the transitions $B_{1g} - {}^2A_{1g}\left(d_x^2 - y^2 - d_z^2\right)v_1$, $^2B_{1g} - {}^2B_{2g}\left(d_x^2 - y^2 - d_{zy}\right)v_2$ and $^2B_{1g} - {}^2E_g\left(d_x^2 - y^2 - d_{zy}, d_{yz}\right)v_3$, respectively [22,23].

E.p.r. spectra of the Cu(II) complexes **Figure 6** were recorded, at room temperature, as polycrystalline samples and in DMSO solution, on the X-band at 9.3 GHz under the magnetic field strength 3400 G. Polycrstalline spectra show a well-resolved anisotropic broad signal [20].

The analysis of spectra give $g_{<<}$ = 2.20 - 2.26 and $g\perp$ = 2.06 - 2.10 (**Table 3**). The trend $g_{<<} > g \perp$ 2.0023, observed for the complexes, under study, indicate that the unpaired electron is localized in the $d_x^2 - y^2$ orbital of the Cu(II) ion and the spectral figures and characteristic for the axial symmetry, Tetragonally elongated geometry is thus confirmed for the aforesaid complexes [24,25].

$G = (g_{<<} - 2)/(g\perp - 2)$, which is the measurement of the exchange interaction between the metal centers in a solid sample of the complex has also been calculated. According to Hathaway and Billing [26] if G > 4, the exchange interaction is negligible but G < 4 indicates the considerable exchange interaction in the solid complexes. The complexes reported here gives the 'G' value in the range 2.43 - 2.60, which is >4, indicating exchange interaction in the solid complexes.

3.2. Ligand Field Parameters

Various ligand field parameters were calculated for the complexes and are listed in **Table 4**.

The Values of Dq in Co(II) complexes were calculated from transition energy ratio diagram using the v_3/v_2 ratio [27]. Our results are in agreement with the complexes reported earlier [28]. The Nephelauzetic parameter β was readily obtained by using the relation: β = B (Complex)/B (free ion), where B (free ion) for Mn(II) is 786 cm^{-1}, for Ni(II) is 1041 and for Co(II) is 1120 cm^{-1} [29]. The values of β lie in the range of 0.46 - 0.070 range. The values of β and indicate the appreciable covalent character of metal ligand "σ" bond. The graphical information contained in the Orgel energy leveldiagram [18,19]. Parameter B and C are a liner combination of certain Columb's exchange integral and are generally treated as empirical parameters obtained from the spectra of free ions. In Mn(II) the values of B and C are calculated from the second transition because and depend on B and C parameters. The calculated values of the ligand field parameters are given in **Table 4**. Slater condon shortly parameters F_2 and F_4 are related to the Recah

Table 3. EPR spectral data of the complexes.

Complex	Temp.	Date as polycrystalline sample				Date in DMF solution			
		g_{11}	g_1	g_{iso}	G	g_{11}	g_1	g_{iso}	G
[Mn(L)Cl₂]	RT	--	--	2.0052	--	--	--	2.0015	--
[Mn(L)(NO₃)₂]	RT	--	--	2.0094	--	--	--	2.0018	--
[Co(L)Cl₂]	LNT	2.418	2.0098	3.75787	--	2.419	2.0178	3.7642	--
[Co(L)(NO₃)₂]	LNT	2.358	2.0175	3.70267	--	2.3396	2.0060	3.6793	--
[Cu(L)Cl₂]	RT	2.0930	2.0598	3.4662	0.87797	2.1345	2.0715	3.5155	0.8309
[Cu(L)(NO₃)₂]	RT	2.119	2.668	3.89767	0.63063	2.0810	2.6896	3.87407	0.6626

22/Jun/05 12:06:38

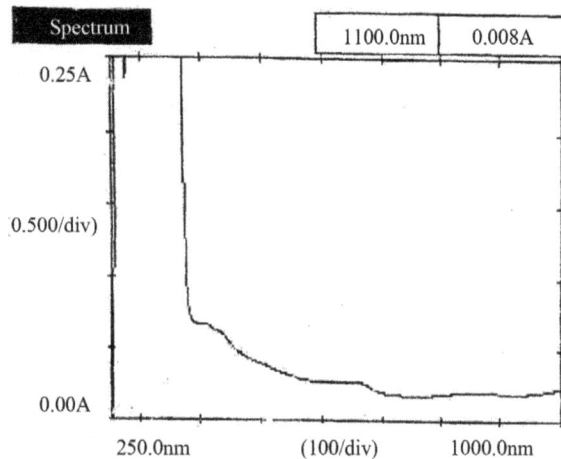

Figure 4. UV Spectrum of [Ni(L^5)(NO₃)₂].

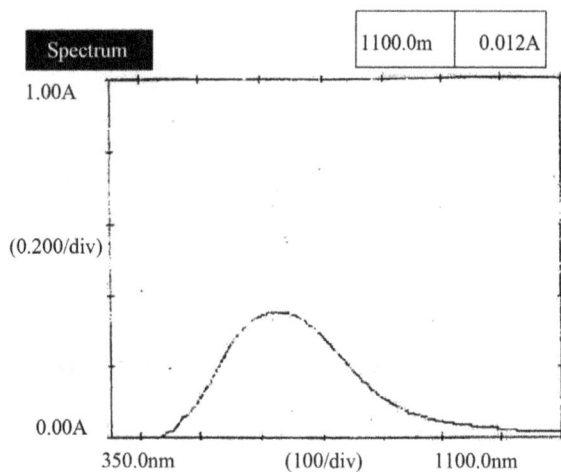

Figure 5. UV Spectrum of [Cu(L^5)(NO₃)₂].

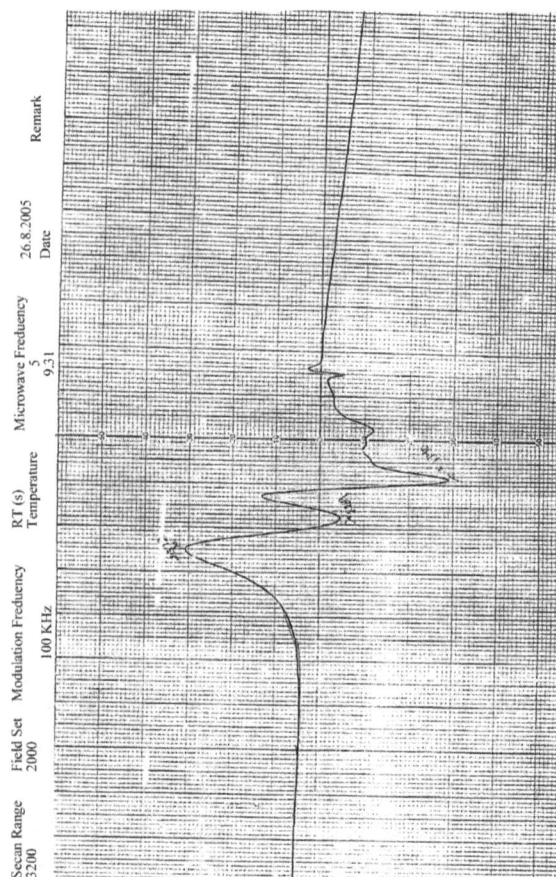

Figure 6. EPR Spectrum of [Cu(L^5)SO₄)].

parameter B and C as: $B = F_2 - 5F_4$ and $C = 35F_4$. The effect of covalence is to reduce the positive charge on the metal ion as consequence of the inductive effect to the ligand with reduced positive charge, the radial extension of the d orbital increases, this decreases the electron-electron repulsion, lowering the energy of the ^3P states.

Table 4. Ligand field parameters of the complexes.

Complex	Dq (cm⁻¹)	B (cm⁻¹)	β	LFSE (kJ·mol⁻¹)
[Mn(L)Cl₂]	1851.85	613.70	0.75	
[Mn(L)(NO₃)₂]	1891.08	518.25	0.61	
[Co(L)Cl₂]	990	501	0.53	121.50
[Co(L)(No₃)₂]	951	510	0.50	133.75
[Ni(L)Cl₂]	1650	750	0.66	231.90
[Ni(L)(No₃)₂]	1581	718	0.80	237.23

4. XRD

The XRD spectra have been recorded for all the complexes. For all the complexes the value of d is found to be same. This indicates that the complexes have same geometry. Hence on the basis elemental analyses, magnetic moment molar conductance, IR, EPR and X-ray diffraction pattern following structure of the complexes may be suggested **Figures 7(a)** and **(b)**.

5. CONCLUSION

Complexes of Mn(II), Co(II), Cu(II) & Ni(II) with tetradentate macrocyclic ligand have been prepared and characterized. The proposed study reveled six coordinated octahedral geometry for Co(II), Mn(II) and square planer geometry for Ni(II) and tetragonal geometry for Cu(II) complexes. Various ligand fields and bonding parameters have been calculated and discussed.

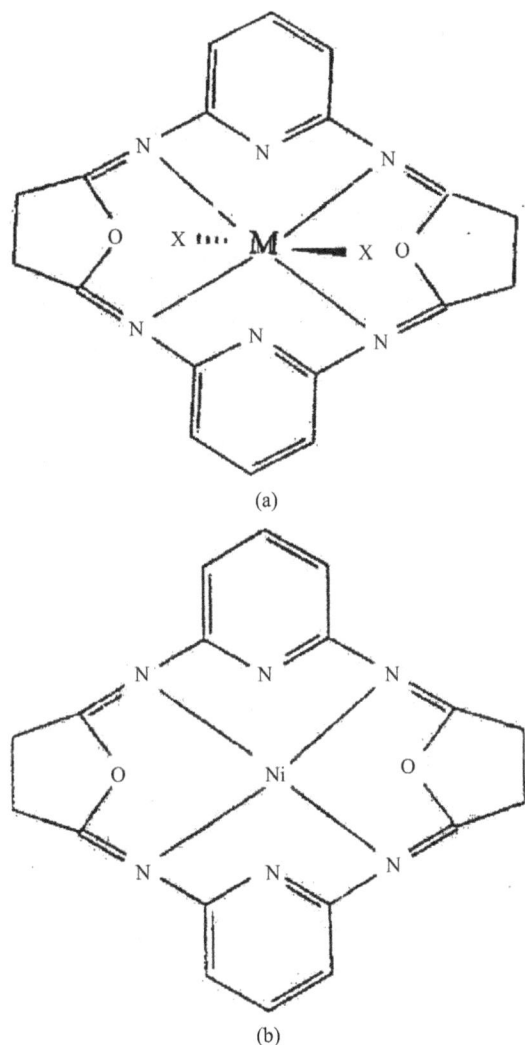

(a)

(b)

Figure 7. (a) [M(L)X$_2$]X = Cl^{-1}, NO$_3^-$, CH$_3$COO$^-$
M= Mn(II), Co(II) and Cu(II); (b) [Ni(L)]X$_2$, X=Cl^{-1}.

6. ACKNOWLEDGEMENTS

Authors are thankful to UGC New Delhi for financial assistance and I.I.T. Bombay for recording EPR spectra.

REFERENCES

[1] Mishra, D., Naskar, S.D., Drew, M.G.B. and Chatto Phadhyay, S.K. (2006) Synthesis, spectroscopic and redox properties of some ruthenium (II) thiosemi carbozone complexes: Structural description of four of these complexes. *Inorganica Chimica Acta*, **359**, 585-592. http://dx.doi.org/10.1016/j.ica.2005.11.001

[2] El-Wahab, Z.H.A., Mashaly, M.M., Salman, A.A., *et al.* (2004) Co(II), Ce(III) and Uo2 (VI) bis-salicylatothiosemicarbazide complexes: Binaryt and ternary complexes thermal studies and antimicrobial activity. *Spectrochimica Acta Part A*, **60**, 2861-2873.

[3] Kolocouris, Dimas, K., pannecouque, C., *et al.* (2002) New 2-Cl-adamantrylcarbopnye) pysidine and 1-acety-ladamantane thiosemicarbozones-thicorbonohydrazones cell growth inhibitory, antiviral and antimicrobal activity evaluation. *Bioinorganic and Medicinal Chemistry Letters*, **12**, 723-727.

[4] Argiielles, R., Touson-Touceda, M.C.P., Cao, R., *et al.* (2009) Complexes of 2-acetyl-y butysolactone and 2-furancarbaldehyde thiosemi carbazones: Antibacterial and antifungal activity. *Journal of Inorganic Biochemistry*, **103**, 35-42.

[5] Ainscough, E.W., Brodie, A.M., Denny, W.A. Finlay, G.J. and Ranford, J.D. (1998) Nitrogen, sulfar and oxygen donor adducts with copper (II) complexes of antitumer 2-formylphridinethiosemicarbazone analogs physicoche- mical and cytotoxic studies. *Journal of Inorganic Bio- chemistry*, **70**, 175-185. http://dx.doi.org/10.1016/S0162-0134(98)10011-9

[6] Kothari, R. and Sharma, B. (2011) Synthesis, characterization and in-vitro antimicrobial investigation of some transition metal complexes with the Schiff base of aromatic aldehyde. *Journal of Chemistry & Chemical Sciences*, **1**, 158-163.

[7] Chandra, S. and Kumar, A. (2013), Spectral Thermal and Morphological studies of chromium nanoparticals. *Spectrochimica Acta Part A*, **102**, 25-255.

[8] Chandra, S., Bargujar, S., Nirwal, R. and Yadav, N. (2013) Synthesis, spectral characterization and biological evaluation of copper(II) and nickel(II) complexes with thiosemicarbazones derived from a bidentate Schiff base. *Spectrochimica Acta Part A*, **106**, 91-98.

[9] Chandra, S. and Ruchi (2013) Synthesis, spectroscopic characterization molecular modeling and antimicrobial activities of Mn(II), Co(II), Ni(II), Cu(II) complexes containing the tetradentate aza Schiff base ligand. *Spectrochimica Acta Part A*, **103**, 338-348.

[10] Lindog, L.F. (1990) The chemistry of Macrocylic ligand complexes. Cambridge University Press, Cambridge.

[11] Rastogi, A., Anurag and Narayan, R. (2011) Ni(II) and Co(II) complexes of azamacrocycles drived from trichloromethane. *Journal of Coordination Chemistry*, **64**, 875-892.

http://dx.doi.org/10.1080/00958972.2011.558192

[12] Chandra, S. and Gupta, L.K. (2005) Electronic, EPR, Magnetic and mass spectral studies of mono and homo binuclear Co(II) and Cu(II) complexes with a novel macrocyclic ligand. *Spectrochimica Acta Part A*, **62**, 1102-1106.

[13] Chandra, S. and Gupta, L.K. (2005) Mass, EPR, IR, electronic studies on newly synthesized macrocyclic ligand & its trasition metal complexes. *Spectrochimica Acta Part A*, **62**, 1125-1130.

[14] Chandra, S. and Sharma, S.D. (2002) Cr(III), Mn(II), Co(II), Ni(II), Cu(II), and Pd (II) complexes of 12 memberd tetraaza (N_4) macrocyclic ligand. *Transition Metal Chemistry*, **27**, 732-735.
http://dx.doi.org/10.1023/A:1020309322470

[15] Raman, N. (2009) Synthesis, spectral charactrization, *in vitro* antimicrobial evaluation and DNA cleavage studies of few macro cyclic complexes. *Journal of the Chemical Society*, **86**, 1143-1149.

[16] Chandra, S. and Gupta, L.K. (2004) Mass, IR, electronic and EPR spectral studies on transition Metal complexes with a new tetradentate 12 Membued new Macrocyclic ligand. *Spectrochimica Acta Part A*, **60**, 3079.
http://dx.doi.org/10.1016/j.saa.2004.01.030

[17] Chandra, S. and Gupta, L.K. (2004) Spectroscopic, characterization of tetradentate Macrocyclic ligand and its transition metal complexes. *Spectrochimica Acta Part A*, **60**, 2767. http://dx.doi.org/10.1016/j.saa.2004.01.015

[18] Shakir, M., Varkey, S.P. and Hameed P.S. (1993) Synthesis characterization & antifungal studies of some tetra macrocyclic complexes. *Polyhedron*, **12**, 2775.

[19] Rao, C.R.K. and Zacharias, P.S. (1997) Synthesis & magnetic properties of dithiooxamide—Bridged Ni(II) complexes. *Polyhedron*, **16**, 1201.

[20] Chandra, S. and Kumar, U. (2004) synthesis magnetic and spectral studies on Cu(II) complexes with bidentate thiosemicarbazone. *Synthesis and Reactivity in Inorganic and Metal-Organic Chemistry*, **34**, 1417-1430.

[21] Chandra, S. and Gupta, K. (2001) synthesis & spectal studies on Cr(III) Mn(II) Fe(III), Co(II), Ni(II) and Co(III) complexes of 14 membered and 16 membered microcyclic legand. *International Journal of Quantum Chemistry*, **40A**, 775-779.

[22] Chandra, S. and Sharma, A.K. (2009) Ni(II) & Cu(II) complexes with base legand 2,6 diascetyl pyridine bis (arbohydrzaon): Synthesis and I.R. mass ^{1}HNMR electronic and EPR. *Spectrochimica Acta Part A*, **72**, 851-857.

[23] Mohammad, G.G., Hindy, M.M., *et al.* (2005) synthesis characterzation and biological activity of some transition metals with schiff base derived from 2 thiphen carboxaldehyde and amino benzoic acid. s *Spectrochimica Acta Part A*, **62**, 1140-1150.

[24] Chandra, S. and Kumar, U. (2004) spectroscopic characterization of Co(II) complexes of indoxyl N(4) methyl thiosemicarbazone. *Spectrochimica Acta Part A*, **60**, 2825-2829.

[25] Guzar, S.H. and Qin-Han, J.I.N. (2008) Synthesis characterization and spectral studies of new Co(II) and Cu(II) complexes of pyrrolyl 1-2-carboxaldehyde isonicotinoyl hydrazone. *Journal of Applied Sciences*, **8**, 2480-2485.

[26] Chandra, S., Raiza, S., Tyagi, M. and Gautam, A. (2007) Synthesis spectroscopic and antimicrobeal studies on bivalent Ni(II) and Cu(II) complexes of bis (thiosemicarbozone). *Bioinorganic Chemistry and Applications*, **7**, Article ID: 51483.

[27] Lever, A.B.P. (1968) inorganic & electronic spectroscopy. 1st Edition, Elsevier, Amesterdam, 249-360.

[28] Chandra, S. and Gupta, L.K. (2004) Spectroscopic studies on Mn(II), Co(II), Ni(II) and Cu(II) complexes with N doner tetradentate microcyclic ligand drived from ethylcinnamate moiety. *Spectrochimica Acta Part A*, **60**, 2411.

[29] Chandra and Sangeetika S. (2002) Synthesis & spectral studies on Cu(II), Co(II) complexes of macrocyclic ligand containing thiosemicarbazone moiety. *International Journal of Quantum Chemistry*, **41**, 1629.

20

An experimental study of hydrolytic behavior of thulium in basic and near-neutral solutions

S. A. Stepanchikova[1], R. P. Biteykina[1], A. A. Sava[2]

[1]Institute of Geology and Mineralogy, Siberian Division of Russian Academy of Sciences, Novosibirsk, Russia
[2]Novapharm Ltd., Sydney, Australia
Email: step@igm.nsc.ru

ABSTRACT

Hydrolytic equilibria of Tm (III) in KOH solutions were studied at 25°C. A spectrophotometry with *m*-cresol purple and 2-naphthol as pH indicators was used at an ionic strength of not more than 0.0005. The results indicate that in freshly prepared solutions at pH ranging between 6 and 10 Tm is presented as Tm^{3+}, $Tm(OH)^{2+}$, $Tm(OH)_2^+$ and $Tm(OH)_3^0$. The stepwise stability constants of hydroxide complexes calculated at zero ionic strength were obtained as coefficient of linear regression equations from the graph of optical densities of the indicators in Tm solutions at varying pH.

Keywords: Rare Earth Elements; Thulium; Stability Constants; Spectrophotometry; pH-Indicators

1. INTRODUCTION

Recently, the studies of rare earth elements (REE) complexing have been considerably intensified due to evaluating the fate of the REE compounds in the environment [1-3]. Producing and accumulating of significant quantities of REE during nuclear fission in uranium and Plutonium reactors are a potential source of their formation. The lanthanides complexes are among the most important compounds in natural waters regarding predominant anions [4]. According to the data of its abundance in the Earth's crust thulium appears to be the rarest among rare earth elements after promethium [5]. However, due to a potential risk of uptake by human bodies, animals and plants, further research in the chemistry of thulium complexes affecting the process of metabolism is needed. It has already been proven [6] that REE complexes could affect the ionic channels functioning and potentially lead to interrupting of the impulse transmission among the nerve cells.

According to the review of the published literature, thulium appears to be the least researched element amongst lanthanides due to an exclusively complex technology of its production and very high prices. The published experimental data on thulium hydrolytic behavior at 25°C are very limited, and being obtained under different experimental conditions, these are often incomparable with each other [7-10]. For example, in [7] the value of the stability constant of the $Tm(OH)^{2+}$ monohydroxocomplex, assayed by potentio metric titration in the Tm_2O_3 solutions at the ionic strength of 0.3, is reported as 5.78 log.unit The corresponding stability constant of the isotope [170]Tm, assessed by chelating organic ligands extraction together with radio-chemical labelling, is 9.6 log.unit and differs by nearly 4 log.unit [8]. The submitted data demonstrate a dispersion typical for Tm hydroxoforms stability in aqueous solutions. Hydroxocomplexes of higher order usually are evaluated at considerable ionic strengths have very often, either not been identified, or the researchers anticipate presence of different complexes in similar experimental conditions. This could be considered as one of the major explanation for a great variability of the published data.

We studied hydrolythic equilibria in Tm^{3+} solutions with a possible participation of higher order hydroxoforms besides of $Tm(OH)^{2+}$. To minimize the experimental errors caused by extrapolation to the zero ionic strength we studied hydrolytic reactions at minimal ionic strengths.

2. MATERIALS AND METHODS

Spectrophotometric pH measurement of the solutions containing variable $TmCl_3$ concentrations and constant concentrations of the acid-alkaline indicators and KOH were carried out. An increase in Tm concentration resulted in a decrease of the absorption of the deprotonized form of the indicator due to an increase in the protons quantity as per following reaction:

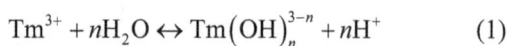

$$Tm^{3+} + nH_2O \leftrightarrow Tm(OH)_n^{3-n} + nH^+ \qquad (1)$$

The pH values were calculated from the measuring absorption densities of the indicator using tabulated values of the ionization constants [9]. The obtained values of ligand number n were used to calculate thermodynamic values of the formation constants of hydroxocomplexes. In the solutions containing m-cresol purple the measured pH values were between 6 and 8 and in 2-naphtol-containing solutions these were between 9 and 10.

3. CALCULATIONS

Reactions of hydroxocomplexes formation in solutions of trivalent metal ions

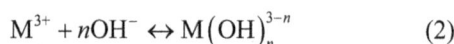

$$M^{3+} + nOH^- \leftrightarrow M(OH)_n^{3-n} \qquad (2)$$

are characterized by stepwise ^oK_n and total $^o\beta_n$ stability constants expressed at zero ionic strength by the equations:

$$^oK_n = \frac{\left[M(OH)_n^{3-n}\right]}{\left[M(OH)_{n-1}^{3-(n-1)}\right]\left[OH^-\right]} \qquad (3)$$

$$^o\beta_n = \frac{\left[M(OH)_n^{3-n}\right]}{\left[M^{3+}\right]\left[OH^-\right]^n} \qquad (4)$$

Both indicators are weak organic acids (HA) that react with a strong inorganic base KOH:

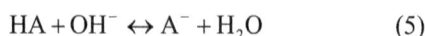

$$HA + OH^- \leftrightarrow A^- + H_2O \qquad (5)$$

where HA and A^- are the protonated and deprotonated forms of the indicator, respectively.

If the equilibrium constant K_{BHA} for reaction (5)

$$K_{BHA} = \left[A^-\right]/[HA]\left[OH^-\right] \qquad (6)$$

is known, the equilibrium concentrations [HA], [A^-] and consequently [OH$^-$] and pH of the solution under study can be evaluated from the spectra.

The reaction between potassium hydroxide and the indicator (being a weak acid) neutralizes its HA part into KA salt (5). Then the solution denoted as number 1, containing the acid and KOH without Tm, has buffering properties. Therefore, adding the increasing of protons concentration would be equivalent to titration of alkaliscent solutions by a strong acid. The protons would react

with A^- increasing the concentration of the weak acid HA. If the buffering capacity of the Tm-containing solutions is insufficient, the remainder of protons would neutralize OH$^-$-ions in the solution (i). The equation for protons created in the hydrolytic reactions can be written as:

$$\left[H^+(i)\right]_{hydr} = \left[HA(i)\right] - \left[HA(1)\right] + \left[OH^-(1)\right] - \left[OH^-(i)\right] \qquad (7)$$

If hydrolysis is treated as splitting off a proton from a water molecule in the hydrate shell of the rare earth ion, the number of produced protons would be equal to the number of hydroxide ions bound into complexes:

$$\left[H^+\right]_{hydr} = \sum_{p=1}^{p=N} p \cdot \left[Tm(OH)_p^{3-p}\right] \qquad (8)$$

and therefore the ligand number $\bar{n}$ can be calculated from Equation (9), where C_{Tm} is the analytical concentration of Tm. The index i is omitted for simplicity in the Equations (8) and (9).

The calculation algorithm was based on the stepwise approach and consisted of the following stages:

1) The activity coefficients were initially assumed to be equal to 1.

2) HA, A^- and consequently OH$^-$ equilibrium concentrations were calculated using optical densities.

3) n values were calculated using Equation (8).

4) ^oK_n values were expressed from (9) and calculated as the linear lest square method parameters.

5) Concentration of hydroxocomplexes and activity coefficients were calculated using the values obtained. Activity coefficients were evaluated by the Debye-$Hü$ ckel equation in the second approximation.

6) The program was returned to Step 2, until all the calculated values became constant according to the preset accuracy.

The complex forms Tm^{3+}, $Tm(OH)^{2+}$ and $Tm(OH)_2^+$, interlinked by the constants $^oK_1$ and $^oK_2$, were assumed to be present in solutions containing m-cresol purple. They were calculated by the regressive equations:

$$\frac{1-\bar{n}}{2-\bar{n}} = \frac{1}{^oK_1} \cdot \frac{\bar{n}}{(2-\bar{n}) \cdot [OH^-(i)]^2 \cdot \gamma_1(i) \cdot \gamma_3(i)} - {}^oK_2 \qquad (10)$$

where $\gamma_1(i)$ and $\gamma_1(i)$ are the activity coefficients for uni- and trivalent ions.

$Tm(OH)_2^+$ $Tm(OH)_3^0$ complexes were suggested in 2-naphthol-containing solutions within the measured pH

$$\bar{n} = \frac{\sum_{p=1}^{p=N} p \cdot \left[Tm(OH)_p^{3-p}\right]}{C_{Tm}} = \frac{\sum_{p=1}^{p=N} p \cdot \left[Tm(OH)_p^{3-p}\right]}{\left[Tm^{3+}\right] + \sum_{p=1}^{p=N} \left[Tm(OH)_n^{3-p}\right]} \qquad (9)$$

interval. The stability constants $^{o}K_{3}$ were evaluated as per following equation:

$$\frac{\bar{n}-2}{3-\bar{n}} = {}^{o}K_{3} \cdot \left[OH^{-}(i)\right] \cdot \gamma_{1}(i)^{2} \qquad (11)$$

4. EXPERIMENT

The absorption spectra of pH-indicators solutions containing "analytical grade" $TmCl_3$ (sourced from the Novosibirsk plant of chemical reagents) have been measured. The solutions for spectroscopy were prepared from 0.01 M $TmCl_3$ aqueous stock solution with pH 5.84. The Tm concentration was controlled spectroscopically with an arsenazo [11]. Concentration of KOH (analytical grade ALDRICH) was monitored by titration with HCl. The indicators ("Indicator grade" ACROSS ORGANICS) were monitored by comparing of their extinctions with the corresponding values obtained from the reagent purified by vacuum sublimation and multiple crystallization [12]. Twice-distilled water boiled for longer than 2 hours was used.

The spectra of solutions were measured in closed quartz cells with 5 cm optical length using UV VIS spectrophotometer Specord M40. The experimental error was evaluated using the law of errors propagation. Its value depended on the accuracy of solution preparation and photometric measurement and did not exceed 1.2% at 25°C. The accuracy of experimental evaluation of the stability constants was calculated as standard deviations of the linear regression parameters.

5. RESULTS AND DISCUSSION

5.1. Measurements in Solutions of m-Cresol Purple

The sulfonaphthalein indicator m-cresol purple (mCP) was used by Clayton and Byrne for surface and deep-water spectrophotometric pH measurement of sea water [13]. According to the results it exists in three forms— H_2I, HI^- and I^{2-}. In the visible range within the interval of the measured pH values mCP's spectrum is represented by distinct intensive bands corresponding to its protonated (HI^-, $\nu_{max} = 23,041 \text{ cm}^{-1}$) and deprotonated ($I^{2-}$, $\nu_{max} = 17,301 \text{ cm}^{-1}$, $\varepsilon_{max} = 32,163$) forms. According to our preliminary measurements, the extinction values of HI^- were ranging from 0 to 408 within the 15,400 to 17,300 cm^{-1} interval. That allowed to disregard HI^- absorption in our measuring of absorption of the I^{2-} band. This indicator was also used by the authors of [9] when studying a comparative hydrolythic behavior of REE.

The chemical equilibrium between two forms of mCP: $I^{2-} + H^+ \leftrightarrow HI^-$ is described by stepwise formation constant:

$$K_a = \frac{\left[HI^{-}\right]}{\left[H^{+}\right]\left[I^{2-}\right]} \qquad (12)$$

The value of pK_a ionization at 25°C and an ionic strength 0.7 is equal to 8.146 according to [13]. We recalculated the values of pK_a for the zero ionic strength.

Figure 1 shows as an example the spectra of mCP vs pH in KOH and $TmCl_3$ solutions. They are characterized by distinct isobestic point at 20,500 cm^{-1}. The initial analytical concentrations of KOH and mCP in this experiment were constant and equal to 2.6×10^{-4} and 3.9×10^{-5}, respectively. The concentration of $TmCl_3$ in series of 10 solutions varied from 0.0 to 5.0×10^{-5} M. The pH values were estimated as negative logarithm of hydrogen ions activity and were calculated by comparing of I^{2-} absorption in solutions under study (i) with the absorption in solution (0) with KOH concentration of 0.01 M, were all the indicator is in the I^{2-} form.

The published data confirms that mCP forms complexes with Fe (III) [14]. As for complex formation with Tm in our experiment: in this case the absorption of I^{2-} had to be decreased with increasing of metal concentration without any isobestic point. Should the Tm hydroxocomplexes be also formed, the isobestic point would be displaced vertically down. The pattern of spectra in **Figure 1** demonstrates that under conditions of our experiment the only change in pH occurs due to the change of indicator forms but not of its concentration.

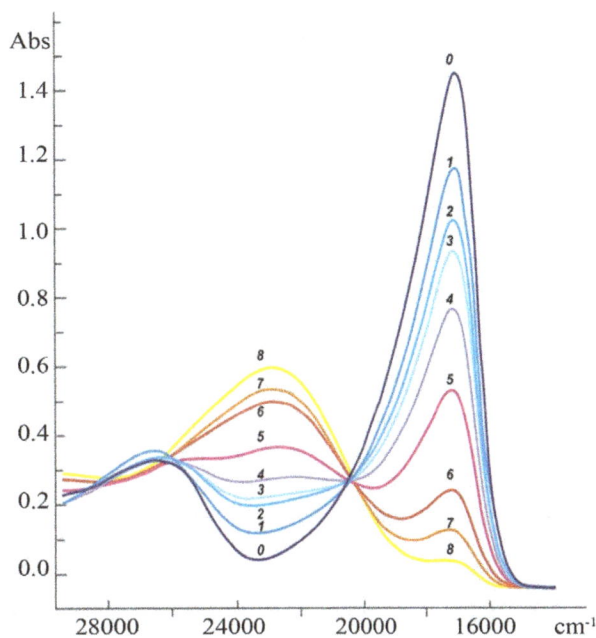

Figure 1. Effect of $TmCl_3$ concentration (M) on spectra m-cresol purple in KOH solutions; (0) 0.0; (1) 0.0; (2) 1.0×10^{-5}; (3) 1.5×10^{-5}; (4) 2.0×10^{-5}; (5) 2.5×10^{-5}; (6) 3.0×10^{-5}; (7) 3.5×10^{-5}; (8) 4.0×10^{-5}. Concentration of KOH in solution (0) is 0.01 M.

Table 1 shows the values of $\log^o K_1$ and $\log^o K_2$ averaged for 9 wave numbers within the studied 15,800 - 17,300 cm^{-1} interval, and their standard deviations. **Table 2** compares our experimental values of $\log^o K_1$ with the published data. The experimental conditions differ considerably both from ours and among themselves. These are due to the different reagent concentration, measured pH values and the proposed forms of hydroxocomplexes present in the solutions within the studied pH range. This could explain the observed data disagreement.

Unfortunately, we could not find in literature the experimental data on the values of $^o K_2$ for Tm. However these are available for erbium [15], where hydrolysis of Er by measuring of solubility of its freshly precipitated hydroxides was studied. The logarithmic values of stability constants for erbium solutions with concentrations up to 0.1 M in the 1 M $NaClO_4$ medium (under conditions excluding the presence of CO_2) are as follows: $\log^o \beta_1 = 7.7$, $\log^o \beta_2 = 13.5$, $\log^o \beta_3 = 18.9$, $\log^o \beta_4 = 19.2$. Taking into account the difference of experimental conditions one can consider that the value $\log^o \beta_2 = 15.09$ for thulium which we have obtained at the zero ionic strength satisfactorily agrees with the results of [15].

5.2. Measurements in Solutions of 2-Naphthol

The ionization constant of 2-naphthol K_a is determined by the equation:

$$K_a = \frac{[NapO^-] \cdot [H^+]}{[NapOH]} \tag{13}$$

where NapOH and $NapO^-$ are the protonated and deprotonated forms of 2-naphthol, respectively. The values of pK_a were defined spectrophotometrically up to 400°C in [16] and then used by the authors [17] for measuring equilibrium constants in solutions of sulfuric acid and ammonia, as well as boric acid [18] at elevated temperatures and pressure. According to [17], the value of $p^o K_a$ at 25°C is 9.63. **Figure 2** presents the absorption spectra of 2-naphthol in solutions KOH and $TmCl_3$ which depend upon pH. The absorption band in the interval 26,000 - 30,000 cm^{-1} belongs to $NapO^-$. **Table 3** presents data for the averaged values of $\log^o K_3$ and their standard deviations. The measurements were carried out in the range from 27,200 to 28,800 cm^{-1} in nine points of the spectra. The initial analytical concentration KOH and 2-naphthol were constant and equal to 3.0×10^{-4} and 1.0×10^{-4} M respectively. The $TmCl_3$ concentration varied from 0.0 M to 4.5×10^{-5} M in the series of 10 solutions. The β_3 values are published by Fatin-Rouge and Bünzliin [19]. The authors had determined the hydrolysis constant for a series of rare earths elements by potentiometeric titration of 0.002 M solution $Ln(ClO_4)_3$ at a constant ionic strength supported by the adding 0.1 M NaCl. However, only complex forms Tm^{3+} and $Tm(OH)_3^0$ were suggested within pH range of 8.5 - 11. The measured value of $^*\beta_3$-21.14 could be considered as agreeing with our results (19.35) if the difference in the experimental conditions is accounted for.

6. CONCLUSIONS

The number of bound hydroxide ions as well as thermodynamic values of the stability constants of the first three

Table 1. Experimental data for calculations of $\log^o K_n$ and their standard deviations (sd) in solutions of *m*-cresol purple measured at ν = 16,400 cm^{-1}.

No. solutions	C_{Tm}	Optical density	pH calc.	*n* calc.	$\log^o K_n$	
0	0.0	0.6868				
1	0.0	0.6047	7.890	0.00	$\log^o \overline{K}_1$	8.58
2	0.000010	0.5754	7.735	1.03	sd $\log^o \overline{K}_1$	0.07
3	0.000015	0.5406	7.589	1.06	$\log^o \overline{K}_2$	6.52
4	0.000020	0.5050	7.464	1.04	sd $\log^o \overline{K}_2$	0.26
5	0.000025	0.4132	7.199	1.06		
6	0.000030	0.3301	6.986	1.02		
7	0.000035	0.2222	6.698	0.96		
8	0.000040	0.1442	6.442	0.88		
9	0.000045	0.0576	5.978	0.79		
10	0.000050	0.0372	5.773	0.71		

Table 2. Comparison of the first hydrolysis constant of thulium at 25°C from literature data and those obtained in this work.

Reference	Method	I in M, max.	C_{Tm}, max	Interval of measured pH	$\log^o K_1$
[4] Frolova and Kumok, 1966	potentiometry	0.3	9×10^{-3}	~(5 - 6)	5.78
[5] Guillaumont et al., 1971	extraction; radiochemical measurements	0.1	~10^{-7}	~(2 - 8.5)	9.60
[3] Klungness and Byrne, 2000	potentiometry; spectrophotometry	0.7	0.01	~(4 - 10)	6.61
This work	spectrophotometry using pH-indicators	5×10^{-4}	5×10^{-5}	6 - 8	8.58

Table 3. Experimental data for calculations of $\log^o K_n$ and their standard deviations (sd) in solutions of 2-naphthol measured at $v = 27,600$ cm^{-1}.

No. solution	C_{Tm}	Optical density	pH calc.	n calc.	$\log^o K_n$	
0	0,0	0.7964				
1	0.0	0.5189	9.908	0.00	$\log {}^o\overline{K}_3$	4.26
2	0.000005	0.4791	9.815	4.67	sd $\log {}^o\overline{K}_3$	0.25
3	0.000010	0.4663	9.786	3.20		
4	0.000015	0.4470	9.743	2.84		
5	0.000020	0.4093	9.647	2.93		
6	0.000025	0.3868	9.597	2.73		
7	0.000030	0.3705	9.576	2.51		
8	0.000035	0.3330	9.493	2.49		
9	0.000040	0.2971	9.411	2.43		
10	0.000045	0.2811	9.374	2.27		

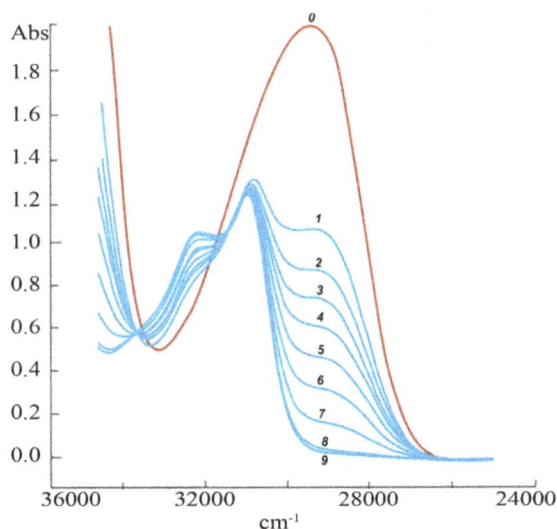

Figure 2. Effect of TmCl$_3$ concentration (M) on spectra 2-naphthol in KOH solutions: (0) 0.0; (1) 0.0; (2) 5×10^{-6}; (3) 1.0×10^{-5}; (4) 1.5×10^{-5}; (5) 2.0×10^{-5}; (6) 2.5×10^{-5}; (7) 3.0×10^{-5}; (8) 3.5×10^{-5}; (9) 4.0×10^{-5}.

Tm (III) hydroxo complexes have been evaluated by the indicator spectrophotometric method in the absence of polymer forms, side reactions and hydroxides precipitate at 25°C at minimal ionic strength. Although the research was initially intended to form a basis for estimating stability of hydroxide, carbonic and mixed hydroxocarbonate forms of REE at elevated temperatures, the employed methodology at near-room temperatures has proved to be advantageous over the direct potentiometeric measurements Our methodology utilizes tabulated values of ionization constants and extinction coefficients of pH indicators that removes the need for electrode calibration during each measurement. The employed spectroscopic technique allows carry out direct observations of the hydrolythic behavior of elements in solutions.

The work was supported by the Russian Branch for Basic Research, project No. 11-05-00662-a.

REFERENCES

[1] Powell, K.J., Brown, P.L., Byrne, R.H., Gajda, T., Hefter, G., Leuz, A.-K., Sjöberg, S. and Wanner, H. (2011) Chemical speciation of environmentally significant metals with inorganic ligands. Part 4: The Cd^{2++} OH$^-$, Cl$^-$, CO$_3^{2-}$ SO$_4^{2-}$, and PO$_4^{3-}$ systems (IUPAC Technical

Report). *Pure and Applied Chemistry*, **5**, 1163-1214. doi:10.1351/PAC-REP-10-08-09

[2] Powell, K.J., Brown, P.L., Byrne, R.H., Gajda, T., Hefter, G., Leuz, A., Sjöberg, S. and Wanner, H. (2009) Chemical speciation of environmentally significant metals with inorganic ligands. Part 4: The Pb^{2++} OH^-, Cl^-, CO_3^{2-}, SO_4^{2-}, and PO_4^{3-} systems (IUPAC Technical Report). *Pure and Applied Chemistry*, **12**, 2425-2476. doi:10.1351/PAC-REP-09-03-05

[3] Wood, S.A. (2010) Advances in the aqueous solutions chemistry of trivalent rare earth elements at temoeratures up to 250°C: Relevance to near-field radioactive waste disposal environments. *GSA Denver Annual Meeting*, Geological Society of America Abstracts with Programs, 31 October-3 November 2010, **5**, 357.

[4] Dubinin, A.V. (2006) Geochemistry of rare earth elements in the ocean. Nauka, Moscow.

[5] Riabchikov, D.I. and Riabouhin, V.A. (1966) Analytical chemistry of rare earth elements and yttrium. Nauka, Moscow.

[6] Ganjali, M.R., Norouzi. P. and Adergania, B.A. (2007) Thulium (III) ions monitoring by a novel thulium (III) microelectrode based on a S-N schiff base. *Electroanalysis*, **19**, 1145-1151. doi:10.1002/elan.200603833

[7] Frolova, U.K., Kumok, V.N. and Serebrennikov, V.V. (1966) Hydrolisis of rare earth elements and yttrium in water solutions. Izvestia Visshih Uchebnih Zavedenij SSSR. *Khimija I Khimitcheskaja Tekhnologija*, **2**, 176-179.

[8] Guillaumont, R., Desire, B. and Galin, M. (1971) Premiere constante d'hydrolise des lanthanides. *Radiochemical and Radioanalytical Letters*, **3**, 189-197.

[9] Klungness, G.D. and Byrne, R.H. (2000) Comparative hydrolysis behavior of the rare earth and yttrium: The influence of temperature and ionic strength. *Polyhedron*, **19**, 99-107. doi:10.1016/S0277-5387(99)00332-0

[10] Baes, C.F. and Mesmer, R.E. (1976) The hydrolysis of cations. John Wiley, New York.

[11] Marchenko, Z.I. (1971) Photometric definitions of elements. Mir, Moscow.

[12] Soumillion, J.Ph., Vandereecken, P., Van Der Auweraer, M. and De Schryver, F.C. (1989) Photophysical analysis of ion pairing of β-naphtholate in medium polarity solvents: Mixtures of contact and solvent-separated ion pairs. *Journal of American Chemical Society*, **111**, 2217-2225. doi:10.1021/ja00188a041

[13] Clayton, T.D. and Byrne, R.H. (1993) Spectrophotometric seawater pH measurement: Total hydrogen ion concentration scale calibration of *m*-cresol purple and at-sea results. *Deep-Sea Research*, **40**, 2115-2129. doi:10.1016/0967-0637(93)90048-8

[14] Chamjangali, M.A., Ghadamali, B. and Salek-Gilani, N. (2009) New induction period based spectrophotometric method for the determination of iron (II) in pharmaceutical products. *Acta Chimica Slovakia*, **50**, 434-440.

[15] Kragten, J. and Decnop-Weever, L.G. (1983) Hydroxide complexes of lanthanides-V*. Erbium in perchlorate medium. *Talanta*, **2**, 131-133. doi:10.1016/0039-9140(83)80033-2

[16] Xiang, T. and Johnston, K.P. (1994) Acid-base behavior of organic compounds in supercritical water. *Journal of Physical Chemistry*, **98**, 7915-7922. doi:10.1021/j100083a027

[17] Xiang, T., Johnston, K.P., Wofford, W.T. and Gloyna, E.F. (1996) Spectroscopic measurement of pH in aqueous sulfuric acid and ammonia from sub- to supercritical conditions. *Industrial Engineering Chemistry Researches*, **35**, 4788-4795. doi:10.1021/ie960368y

[18] Wofford, W.T., Gloyna, E.F. and Johnston, K.P. (1998) Boric acid equilibria in near-critical and supercritical water. *Industrial Engineering Chemistry Researches*, **37**, 2045-2051. doi:10.1021/ie9707634

[19] Fatin-Rouge, N. and Bünzli, J.-C.G. (1999) Thermodynamic and structural study of inclusion complexes between trivalent lanthanide ions and native cyclodextrins. *Inorganica Chimica Acta*, **293**, 53-60. doi:10.1016/S0020-1693(99)00227-3

Synthesis, characterization, spectroscopic and crystallographic investigation of Cobalt(III) schiff base complex with two perpendicular diamine coumarin ligands

Imen Ketata[1], Lassad Mechi[1], Taïcir Ben Ayed[1,2], Michal Dusek[3], Vaclav Petricek[3], Rached Ben Hassen[1*]

[1]Unité de Recherche de Chimie des Matériaux et de l'Environnement, Université de Tunis El Manar, Tunis, Tunisia
[2]INSAT, Université de Carthage, Tunis, Tunisia
[3]Institute of Physics, Academy of Sciences of the Czech Republic, Praha, Czech Republic
Email: [*]rached.benhassen@fss.rnu.tn

ABSTRACT

New transition metal complex of Cobalt(III) of the ligand (*E*)-3-(1-(2-aminoethylimino)ethyl)-4-hydroxy-2H-chromen-2-one, derived from condensation of ethylene diamine with 3-acetyl-4-hydroxy-chromene-2-one have been synthesized by reaction of cobalt(III) salt and the ligand, in amounts equal to metal-ligand molar ratio of 1:2. Both the Schiff base and the complex of Co(III) were characterized by IR, UV-vis, ^{1}H NMR- and ^{13}C NMR-spectroscopy techniques. Single crystal X-ray diffraction investigation, at low temperature T = 120 K, shows that the cobalt complex is triclinic P-1, a = 10.426(5) Å, b = 11.3234(2) Å, c = 15.729(5) Å, $\alpha(°)$ = 70.102(4), $\beta(°)$ = 86.049(4), $\gamma(°)$ = 82.497(4), Z = 2, and its structure consists of isolated $[Co(III)(C_{13}H_{13}N_2O_3)_2]^+$ complex cations with distorted octahedral geometry, ClO_4^- counter anions, acetone solvent and water molecules. The crystal cohesion is stabilized by hydrogen bonds between ligands and water molecules, and ionic interactions between complex cations and counter anions.

Keywords: Hydroxycoumarin Derivative; Co^{III} Complex; ^{1}H NMR; UV-Visible; Single Crystal X-Ray Diffraction

1. INTRODUCTION

Coumarins and their derivatives [1,2] were found to exhibit a variety of biological and pharmacological activeties and have attracted considerable interest because of their potential beneficial effects on human health [3,4].

A series of coumarin derivatives were recently synthesized and studied for their structure and anticoagulant activity and cytotoxic [5-10]. A number of coumarins have been investigated for complexing ability [11,12]. The mechanisms of cytotoxic action of a large number of different coordination compounds have been discussed in relation to the development of new antitumor agents.

A wide range of medicinal applications of metal complexes of coumarins was investigated. It was found that in some cases, the metal complexes obtained revealed a higher activity than their biological ligands [9-11].

In the past decade, Schiff bases have received much attention, mainly because of their extensive applications in the synthesis of coordination compounds and catalysis [13-17]. This is due to the fact that Schiff bases offer opportunities for inducing substrate chirality, tuning factors of metal centered electronic and improve the solubility and stability of homogeneous or heterogeneous catalysts, *i.e.* [18,19]. They are also increasingly important as biochemical, analytical and antimicrobial reagents [20]. Further, significant advances have also been achieved in the field of straightforward preparation of new chiral ligands, such as chiral diamines and complexes, used for asymmetric catalysis [21]. For example, complexes based on chiral diamines have been employed for chiral hydrogenations and transfer hydrogenations [22]. Both homogeneous and heterogeneous diamine complexes have proved highly effective in enantioselective catalysis [23,24].

In continuation of our previous research [25-27] on synthesis and characterization of new asymmetric Schiff base complex, we describe here the synthesis and characterization of the tridentate Schiff base (E)-3-(1-(2-aminoethylimino)ethyl)-4-hydroxy-2H-chromen-2-one (1),

---

[*]Corresponding author.

then its complexation reaction with cobalt(III) to obtain the complex numbered as (2).

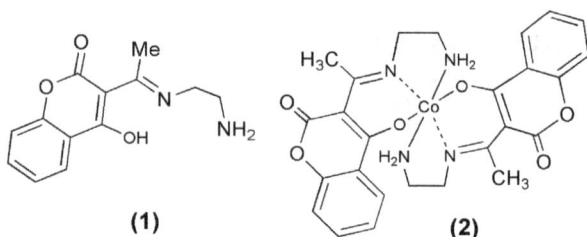

(1)   **(2)**

2. EXPERIMENTAL

2.1. Measurements

Proton nuclear magnetic resonance (^{1}HNMR) spectra were run on a Bruker AV-300 NMR spectrometer and chemical shifts expressed as d (parts per million) values with TMS as internal standard. Multiplicities of proton resonance are designated as singlet (s), doublet (d), triplet (t), quartet (q), and multiplet (m). IR spectra were recorded in KBr on a Bruker FT-IR spectrophotometer. The electronic spectra in the 200 - 900 nm range were obtained on a Varian spectrophotometer 50 scan, using solutions with the concentration of 10^{-5} M in DMSO. Single crystal was characterized by an Oxford Diffraction Gemini R Ultra CCD diffractometer. All the chemicals are commercially available and they were used without further purification. All the solvents were dried using standard methods before use.

2.2. Synthesis of the Schiff Base (1)

The reaction of 3-acetyl-4-hydroxy-*2H*-chromen-2-one with the ethylene diamine in the presence of methanol provided a high yield, as detailed below.

A solution of (0.6 mL, 5 mmol) ethylene diamine in methanol was added drop wise stirring to a methanol solution of (2 g, 5 mmol) 3-acetyl-4-hydroxy-chromene-2-one, already prepared by the method cited in [28], and then the mixture was refluxed for 1 h. The precipitated Schiff base was filtered and recrystallized from DMSO. Yield: 1.7 g (85%).

IR (selected, cm^{-1}): 3097 [ν(C-H arom)], 3207 [ν(NH$_2$)], 1637 [ν(C=O)], 1618 [ν(C=N)], 1585 [ν(C=C)], 1279 [ν(C-O)].

UV-vis λ_{max}/nm (DMSO): 235, 250, 320.

^{1}HNMR (300 MHz, DMSO): δ 13.76 (s, H, OH), 7.23 - 8.27 (m, 4H, H arom), 4.00 (s, 2H, NH$_2$), 2.69 (d, 2H, CH$_2$), 2.48 (t, 2H, CH$_2$), 2.07 (s, 3H, CH$_3$).

^{13}CNMR (75MHz, DMSO): δ 179.63 C(21), 177.20 C(10), 161.80 C(8), 152.99 C(6), 134.12 C(2), 125.63 C(4), 123.68 C(3), 120.04 C(1), 116.20 C(9), 96.47 C(5), 42.81 C(25), 34.40 C(24),18.31 C(52).

2.3. Synthesis of the Complex [Co(C$_{13}$N$_2$O$_3$H$_{13}$)$_2$]ClO$_4$, 2(C$_3$H$_6$O), H$_2$O (2)

In the synthesis of (2), one equivalent of ethylene diamine condenses with 4-hydroxy-2H-chromen-2-one to give 1:1 Schiff base (1) that is sequestered as, Co^{III} complex (2) (**Figure 1**).

A solution of (0.34 g, 1 mmol) of Co(C$_2$H$_3$O$_2$)$_2$·4H$_2$O dissolved in 50 mL methanol was added to the freshly prepared Schiff base (0.41 g, 2 mmol). Upon addition of the metal, the solution immediately turned brown. An aqueous solution of sodium perchlorate to this solution was added and then the color turned red. The mixture was stirred for 3 h and the solid was collected by gravity filtration and dried under vacuum. Yield: 70%.

IR (selected, cm^{-1}): 3414 [ν(OH)], 1602 [ν(C=N)], 1418 [ν(C-O)], 1084.19 [ν(Cl-O) (ClO$_4$)], 523 [ν(M_N)], 418 [ν(M_O)].

UV-vis λ_{max}/nm (DMSO): 552, 486, 346, 261.

^{1}H_NMR: (300MHz, DMSO): δ 7.19 - 7.8 (m, 4H, H-arom.), 3.99 (t, 2H, CH$_2$), 2.90 (s, 3H, CH$_3$), 2.49 (d, H, NH), 2.48 (d, 2H, CH$_2$).

^{13}C NMR (75 MHz, DMSO): δ 172.56 (C10), 161.85 (C21), 152.8 (C8), 133.42 (C4), 125.92 (C2), 123.63 (C4), 119.38 (C1), 115.56 (C5), 100.83 (C9), 55.54 (C25), 42.94 (C24), 22.35 (C52), 119.38 (C3).

2.4. X-Ray Data Collection and Structure Solution and Refinement

Single crystals, suitable for X-ray data collection, were obtained by slow evaporation of the solvent from a solution of (2) in CH$_3$COCH$_3$. The crystals were filtered and washed with acetone. The diffraction data were collected with an Oxford Diffraction Gemini R Ultra CCD. Semi-empirical absorption corrections from equivalents (multi-scan) were applied, using graphite-monochromated Cu Kα

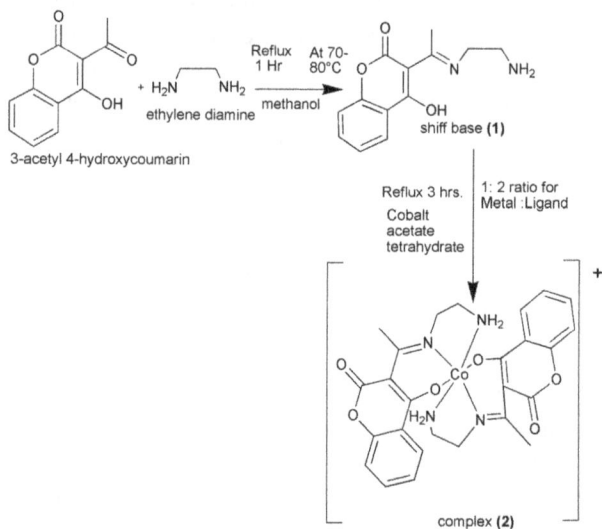

Figure 1. The synthesis steps of the title compound.

radiation at 120 K. The structure was solved with direct methods using SIR 2004 [29]. An anisotropic refinement, by full-matrix least-squares was performed with JANA2006 [30], and the residual finally converged to $R_1 = 0.08$. Non-hydrogen atoms were anisotropic and the hydrogen atom positions were included in the riding mode, except for those of water molecule, where they are introduced and refined with distance and angle restrictions. Acetone solvent molecules are not coordinated to cobalt. They just, fill voids in the crystal lattice. One of the two acetone molecules is quite involved in disorder and so, we let it to possess two positions in the same void. For this purpose, a split model in which two slightly different positions for the atoms C45, C44, O46 and C47 was used. Both refined positions had different occupancy and their sum was significantly different from 1. The disorder of this solvent molecule has negative influences on the structure because of the relatively high number of electrons that are disordered with the solvent. This, together with the very weak diffractions, meant that the R values are higher than usual, but the molecular structure is clear. Further details are given in **Table 1**. Selected bond lengths and angles are listed in **Table 2**. The structure exhibits intermolecular hydrogen bonds (**Table 3**).

Table 1. Crystal data and structure refinement for cobalt complex.

Empirical formula	C30 H46 Cl1 Co1 N4 O13
Formula weight	783.06 g·mol^{-1}
Temperature	120 (K)
Wavelength (Å)	1.5418 Å
Space group	'P - 1'
a (Å)	10.4261(6)
b (Å)	11.3234(5)
c (Å)	15.7297(7)
α (°)	70.102(4)
β (°)	86.049(4)
γ (°)	82.497(4)
Volume (Å^3)	1730.62(15)
Z	2
Dcalc (Mg/m^3)	1.503
F(000)	848
Crystal size (mm^3)	0.4 * 0.15 * 0.02
Absorption coefficient (mm^{-1})	5.222
Theta range for data collection	3.01° - 62.69°
Range of h, k, l	−12/12, −11/11, −17/17
Reflections collected number	23897
Total Reflections number	5397
[R_{int} = 0.06]	[$R \sigma(I)$ = 0.13]
Refinement method	full-matrix least-squares on F^2
Refined parameters/ Restraints/constraints	476/2/152
Final R indices [I > 3 σ(I)]	R1 = 0.08, wR1 = 0.09

Table 2. Selected bond distances (Å) and angles (°).

Bond Lengths (Å)			
Co-O(5)	1.882(4)	C(26)-C(27)	1.522(10)
Co-O(9)	1.900(5)	C(24)-C(25)	1.503(8)
Co-N(4)	1.944(5)	C(23)-C(51)	1.514(8)
CO-N(1)	1.911(5)	C(21)-C(52)	1.503(9)
CO-N(2)	1.915(5)	C(20)-O(4)	1.212(9)
CO-N(3)	1.932(6)		
N(4)-C(24)	1.483(8)		
N(1)-C(21)	1.310(7)		

Selected Angles (°)			
O(5)-CO-O(9) O1 C1 C2	92.3(2)	CO-N(1)-C(21)	127.7(4)
O(5)-CO-N(4)	177.9(2)	CO-N(1)-C(25)	110.4(4)
O(5)-CO-N(1)	93.21(19)	CO-N(2)-C(25)	126.7(5)
O(9)-CO-N(1)	86.4(2)	C(4)-O(7)-C(8)	123(5)
O(9)-CO)-N(2)	92.9(2)	N(4)-CO-N(3)	91.8(2)
O(9)-CO-N(3)	177.34(19)		
N(1)-CO-N(3) O1 C1 C2	94.1(2)		
N(2)-CO-N(3)	86.6(2)		
O(1)-O(5)-C(10)	124.9(4)		
O(1)-N(1)-C(21)	127.7(4)		

Table 3. Hydrogen bonds for the title compound (Å and °).

D-H···A	d(D···H)	d(H···A)	d(D···A)	<(DHA)
N3-H2n3...O8^b	0.87	2.14	2.963(7)	156.78
N3-H1n3...Ow^a	0.87	2.19	2.999(6)	154.19
N4-H2n4...O'1	0.87	2.23	3.05(7)	157.63
N4-H1n4...Ow^a	0.87	2.16	2.904(8)	143.35
Ow-Hw1...O4	0.96(8)	1.89(8)	2.752(7)	148(11)
Ow-Hw2...O'2^c	0.96(5)	2.31(4)	3.118(8)	142(5)
Ow-Hw2...O'4^c	0.96(5)	2.28(5)	3.018(9)	133(4)

^{a}x − 1, y, z; ^{b}−x, 1 − y, −z; ^{c}1 − x, −y, 1 − z.

3. RESULTS AND DISCUSSIONS

The Schiff base was isolated in 85% yield and was obtained cleanly, melting point 170°C, yellow in color, insoluble in common organic solvents, partially soluble in chloroform but soluble in dimethyl sulfoxide (DMSO). The reaction of the ligand and metal salt gave the corresponding metal complex in good yield, stable and non-hygroscopic in nature. The complex was prepared red in color and was found to be soluble in DMSO, CH$_3$COCH$_3$ but was only partially soluble in other organic solvents such as CHCl$_3$ and CH$_3$OH. It decomposes at about 322°C.

3.1. Spectral Characterization of the Ligand and Complex

The IR, ^{1}H NMR and ^{13}C NMR spectra of the synthesized materials confirm the formation of ligand and complex with the proposed structure of the type $[Co(C_{13}N_2O_3H_{13})_2]ClO_4, 2(C_3H_6O), H_2O$.

3.2. IR Spectrum

The IR spectrum of the Schiff base exhibits characteristic intensity band at 1618 cm^{-1} which is attributed to the v(C=N) vibration. Another band at 1637 cm^{-1} is assigned to v(C=O), lactonyl carbon of the coumarin moiety. The Schiff base has a strong absorption band at 3207 cm^{-1}, which corresponds to the NH$_2$ stretching frequency [31]. A medium intensity band at 1585 cm^{-1} is regarded as an aromatic C=C stretching vibrations.

In comparison with the spectrum of the Schiff base, the complex exhibits the band of v(C=N) in the region 1602 cm^{-1}, showing the shift of band to lower wave numbers indicating that, the azomethine nitrogen is coordinated to the metal ion. The high intensity band due to enolic C-O appeared at 1325 cm^{-1} in the Schiff base, appeared as a medium to high intensity band in the 1418 cm^{-1} region in the complex.

These observations support the formation of M-O bonds via deprotonation. The broad band of medium intensity occurring in the 3414 cm^{-1} region, corresponds to of water molecules.

The bands in the region of 410 - 375 and 524 - 480 cm^{-1} in the complex are assigned to stretching frequencies of (M-O) and (M-N) bonds respectively. The absorption bands near 523 cm^{-1} and 418 cm^{-1} are related to the coordination of Schiff-base to cobalt ion through Co-O and Co-N bands [32].

3.3. Electronic Spectrum

The electronic spectrum (**Figure 2**) of the free Schiff base ligand shows absorption bands in the ranges 215 - 260 and 270 - 367 nm. The absorption maxima at 235 nm and 250 nm are due to π-π* transitions for the benzene ring and enol ring respectively. That observed at 320 nm is reasonably assigned to π-π* transition for azomethine moiety. The shoulder at 338 nm corresponds to n-π* transitions for azomethine and NH$_2$ groups [33]. In the spectrum of the complex (**Figure 2**), the absorption bands at ca. 261, 320, 346 nm of the enol, azomethine and NH$_2$ chromophores π-π* and n-π* transitions are shifted compared to the ligand indicating that the enolic oxygen, the imine and amine nitrogen are involved in the coordination with metal ion.

The spectrum of the metal complex shows (inset of **Figure 2**) that the absorption around 400 - 600 nm is due to ligand to metal charge transfer and d-d* transition band of the metal in the complex.

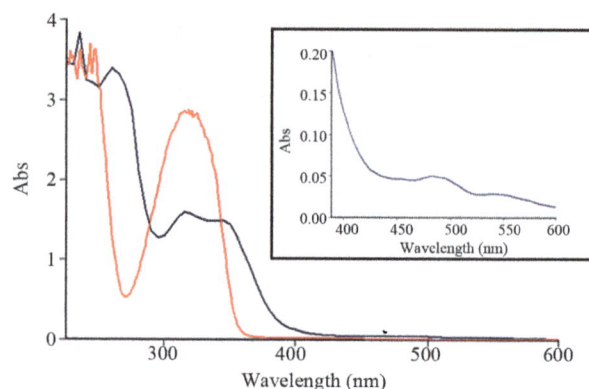

Figure 2. UV-vis absorption spectra of ligand (red spectrum) and complex (blue spectrum). The inset shows the vis. Spectrum of the complex in the region 400 - 600 nm.

3.4. NMR Spectra

The Schiff base and the complex has been characterized by ^{1}H and ^{13}C NMR spectra.

In ^{1}H NMR spectrum of the Schiff base, signal at 13.76 is ascribed to enolic-OH and the methyl protons from the acetyl fragment are identified at 2.07 ppm. The ^{1}H-NMR spectrum of the Schiff base showed the characteristic coumarin aromatic proton signals in the 7.23 - 8.27 ppm. Two additional signals at 2.48 ppm and 2.69 ppm are ascribed to CH$_2$-CH$_2$. The signal around 4.00 ppm is ascribed to NH$_2$.

The comparison of the ^{1}H-NMR spectra of the complex and ligand leads to: The resonance due to enolic-OH, around 13.76 ppm in the Schiff base disappears in the spectrum of complex; this confirms that, the hydroxyl group reacted with metal ion via deprotonation.

In ^{13}C NMR spectra of the Schiff base, the signals appeared in the region 116.20 - 152.99 ppm are assigned to aromatic carbons. The signals at 179.63, 161.80, 18.31 ppm are due to C=N, C=O and CH$_3$ respectively. Additional signals at 34.40 ppm and 42.81 ppm are ascribed to CH$_2$-CH$_2$.

The comparison of the ^{13}C NMR spectra of the complex and ligand leads to: The azomethine (C=N) carbon signal occurring, as single at 179.63 ppm in the ligand, moves upfield to 161.85 upon complexation. This provides further support for the coordination of the ligand through the azomethine nitrogens to metal ions.

3.5. X-Ray Structure Determination of Complex Cobalt

A relatively high coordination number of 6 is achieved by Co, involving two tridentate Schiff base ligands, as shown in **Figure 3** The molecular structure of the complex consists of discrete $[Co(C_{13}H_{13}N_2O_3)_2]^+$ cation, ClO_4^- counter anion, acetone solvent and water molecules.

Figure 4 Shows the packing of the structure. The cobalt(III) center has a distorted octahedral environment,

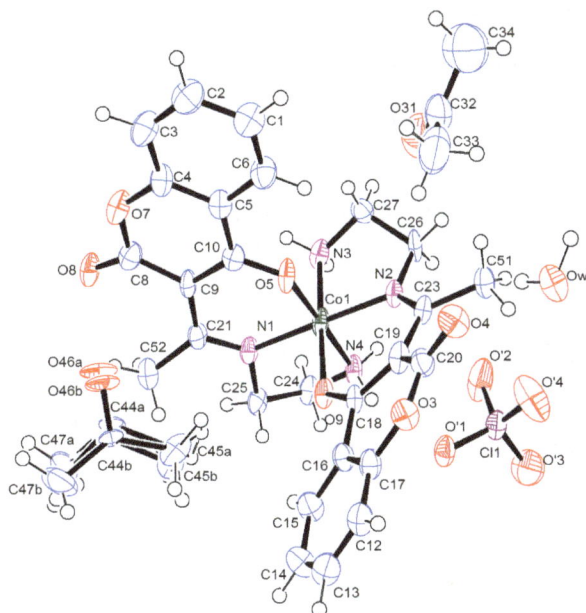

Figure 3. ORTEP stereoscopic projection of the molecular structure of [Co(L)₂]ClO₄, 2(C₃H₆O), H₂O. Thermal ellipsoids are given at 50% probability.

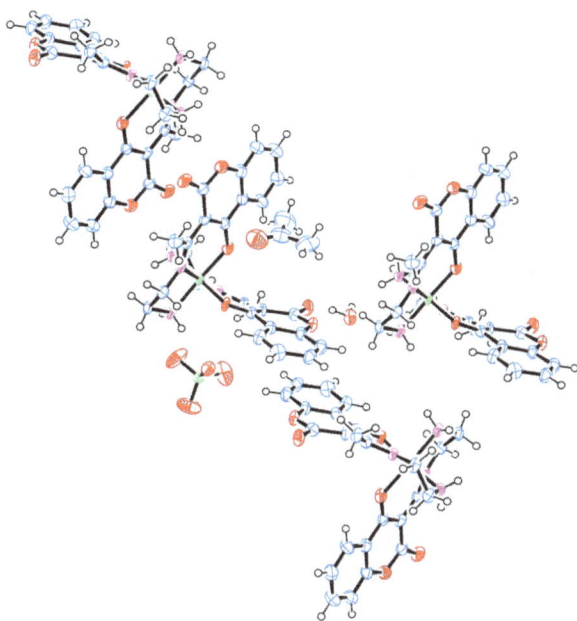

Figure 4. The packing of the structure.

with four nitrogen atoms and two oxygen atoms of the two perpendiculars tridentate ligands.

The complex coordinates through imido nitrogen and enolic oxygen with both the nitrogen donors and the enolic oxygen donors being cis. The Co-O [1.882(4) Å and 1.944(5) Å] bond lengths are in the range of those expected for Co^{III} in an octahedral environment [34]. The distortion, from regular octahedral geometry is evident

from the N(1)-Co-N(2) bond angle of 179.2(2)°, from O(5)-Co-N(3) and O(9)-Co-N(4) where the angles are 90.3(2)° and 85.6(2)° respectively. The calculated average values of the distortion indices [34,35], corresponding to the different angles and distances in the octahedron, when assuming that the Co-O and Co-N bonds are identical, [DI((O/N)Co(O/N)) = 0.003, DI(Co(O/N)) = 0.011, DI((O/N)(O/N)) = 0.055] exhibit a pronounced distortion of the Co(O/N) distances, if compared to (O/N)Co(O/N) angles and (O/N)(O/N) distances.

The diamino fragment is twisted at a torsion angle of 13.3(1)° from the fused ring plane. The dimensions in the shift base correspond closely to standard values [36] except for the bonds C9-C8, C9-C10 and C10-C5, where C9-C10, a double bond, is longer (at 1.414(9) Å) than the expected 1.340Å, C10-C5 is longer (at 1.447(10) Å) than the expected 1.440 Å, and C9-C8 is shorter (at 1.456(9) Å) than the expected 1.465 Å.

The bond angles C3-C4-O7 and C6-C5-C10 at the junction of the two rings are 117.6(6)° and 123.6(6)°, respectively. Here, as well as in other coumarin compounds [37], an important asymmetry in the O-C-O bond angles was detected; O7-C8-O8=114.8(6)°.

The crystal structure cohesion is stabilized by N---H···O hydrogen bonds between neighboring ligands, and O---H···O links between water molecules and the ligands on the one hand and counter-anions on the other, and ionic interact- tions between complex cations and counter-anions, these hydrogen bonds link the molecules into a three-dimen- sional frame. In the O---H···O hydrogen bonds, water mole-cules were proton donors. Strong H-bonding was observed when the lactone group was a potential hydro- gen acceptor (see **Table 3** for the hydrogen bond [Ow- Hw1···O4]).

4. CONCLUSION

The newly synthesized Schiff base which acts as a tridentate ligand was derived in good yield from condensation of ethylene diamine with 3-acetyl-4-hydroxy-chromene-2-one. It coordinated with the metal ion with two azomethine nitrogen and enolic oxygen atom by deprotonation. The bonding of ligand to metal ion is confirmed by spectroscopic investigation. The structure of [Co(L)₂]$^+$ was determined by single crystal X-ray diffraction and revealed that the NNO-donor sets of two ligand fragments form a distorted octahedral configuretion at the Cobalt ion. A perpendicularity in the arrangement of the two Schiff base ligands is observed. The structure of the complex revealed the existence of network in one-dimensional zigzag chain propagating along the c axis in which the perchlorate ions, the acetone solvent and water molecules are inserted between the *ab* planes.

REFERENCES

[1] Galm, U., Heller, S., Shapiro, S., Page, M., Li, S. and Heide, L. (2004) Antimicrobial and DNA gyrase-inhibitory activities of novel clorobiocin derivatives produced by mutasynthesis. *Antimicrobial Agents and Chemotherapy*, **48**, 1307-1312. doi:10.1128/AAC.48.4.1307-1312.2004

[2] Tao, J., Hu, S., Pacholec, M. and Walsh, C.T. (2003) Synthesis of proposed oxidation-cyclization-methylation intermidiates of the coumarin antibiotic biosynthetic pathway. *Organic Letters*, **5**, 3233-3236. doi:10.1021/ol035101r

[3] O'Kennedy, R. and Thornes, R.D. (1997) Coumarins: Biology, applications and mode of action. John Wiley & Sons, New York.

[4] Hoult, J.R. and Paya, M. (1996) Pharmacological and biochemical actions of simple coumarins natural products with therapeutic potential. *General Pharmacology*, **27**, 713-722. doi:10.1016/0306-3623(95)02112-4

[5] Kam, C.M., Kerrigan, J.E., Plaskon, R.R., Duffy, E.J., Lollar, P., Suddath, F.L. and Powers, J.C. (1994) Mechanism-based isocoumarin inhibitors for blood coagulation serine proteases. Effect of the 7-substituent in 7-amino-4-chloro-3-(isothioureidoalkoxy)isocoumarins on inhibitory and anticoagulant potency. *Journal of Medicinal Chemistry*, **37**, 1298-1306.

[6] Lakin, K.M., Smirnova, T.V., Vishnyakova, G.M., Lobanova, E.G., Novikova, N.V. and Sklyarenko, V.I. (1989) New water-soluble anticoagulants of the coumarin series. *Pharmaceutical Chemistry Journal*, **23**, 824-826.

[7] Finn, G.J., Creaven, B. and Egan, D.A. (2001) Study of the in vitro cytotoxic potential of natural and synthetic coumarin derivatives using human normal and neoplastic skin cell lines. *Melanoma Research*, **11**, 461-467. doi:10.1097/00008390-200110000-00004

[8] Kerr, J.S., Li, H.Y., Wexler, R.S., Robinson, A.J., Robinson, C.S., Boswell, G.A., Kranthanser, C. and Harlow, P.P. (1997) The characterization of potent novel warfarin analogs. *Thrombosis Research*, **88**, 127-136. doi:10.1016/S0049-3848(97)00224-7

[9] Ammar, H.O., Ghorab, M., El Nahhal, S.A. and Makram, T.S. (1997) Interaction of oral anticoagulants with methyl xanthines. *Pharmazie*, **52**, 946-950.

[10] Kostova, I. (2005) Antineoplastic activity of new lanthanide (cerium, lanthanum and neodymium) complex compounds. *Current Medicinal Chemistry—Anti-Cancer Agents*, **5**, 29-46. doi:10.2174/1568011053352550

[11] Mandakmare, A.U. and Navwade, M.L. (1997) Stability constants of Fe(III), Cr(III), Al(III) chelates with some substituted coumarins. *Oriental Journal of Chemistry*, **13**, 155-158.

[12] El-Ansary, A.L. and Omar, N.M.M. (1988) New lanthanide complexes of 4-methyl-7-hydroxycoumarin and their pharmacological activity. *Egyptian Journal of Chemistry*, **31**, 511-520.

[13] Goswami, N. and Eichhorn, D.M. (2000) Incorporation of thiolate donation using 2,2'-dithiodibenzaldehyde: Synthesis of Ni, Fe, and Cu complexes. Crystal structures of [M(tsalen)] (tsalen = N,N'-ethylenebis(thiosalicylidene)-imine); M = Cu, Ni) showing a tetrahedral distortion of Cu(tsalen). *Inorganica Chimica Acta*, **303**, 271-276. doi:10.1016/S0020-1693(00)00047-5

[14] Labisbal, E., Rodriguez, L., Sousa-Pedrares, A., Alonso, M., Vizoso, A., Romero, J., Garcia-Vazquez, J.A. and Sousa, A. (2006) Synthesis, characterisation and X-ray structures of diorganotin(IV) and iron(III) complexes of dianionic terdentate Schiff base ligands. *Journal of Organometallic Chemistry*, **691**, 1321-1332. doi:10.1016/j.jorganchem.2005.09.052

[15] Yu, T.Z., *et al.* (2006) Synthesis, crystal structure and electroluminescent properties of a Schiff base zinc complex. *Inorganica Chimica Acta*, **359**, 2246-2251. doi:10.1016/j.ica.2006.01.019

[16] Hoshina, G., Tsuchimoto, M., Ohba, S., Nakajima, K., Uekusa, H., Ohashi, Y., Ishina, H. and Kojima, M. (1998) *European Journal of Inorganic Chemistry*, **37**, 142-147. doi:10.1021/ic9705958

[17] Canali, L. and Sherrington, D.C. (1999) Utilisation of homogeneous and supported chiral metal (salen) complexes in asymmetric catalysis. *Chemical Society Reviews*, **28**, 85-93. doi:10.1039/a806483k

[18] Opstal, T. and Verpoort, F. (2003) Synthesis of highly active ruthenium indenylidene complexes for atom-transfer radical polymerization and ring-opening-metathesis polymerization. *Angewandte Chemie International Edition*, **42**, 2876-2879. doi:10.1002/anie.200250840

[19] Clercq, B.D., Lefebvre, F. and Verpoort, F. (2003) Immobilization of multifunctional Schiff base containing ruthenium complexes on MCM-41. *Applied Catalysis A*, **247**, 345-364. doi:10.1016/S0926-860X(03)00126-1

[20] Tumer, M., Koksal, H., Sener, M.K. and Serin, S. (1999) Antimicrobial activity studies of the binuclear metal complexes derived from tridentate Schiff base ligands. *Transition Metal Chemistry*, **24**, 414-420. doi:10.1023/A:1006973823926

[21] Desimoni, G., Faita, G. and Jorgensen, K.A. (2006) C(2)-symmetric chiral bis(oxazoline) ligands in asymmetric catalysis. *Chemical Reviews*, **106**, 3561-3651. doi:10.1021/cr0505324

[22] Fache, F., Schulz, E., Tommasino, M.L. and Lemaire, M. (2000) Nitrogencontaining ligands for asymmetric homogeneous and heterogeneous catalysis. *Chemical Reviews*, **100**, 2159-2232. doi:10.1021/cr9902897

[23] Thomas, J.M. and Raja, R.J. (2004) Catalytic significance of organometallic compounds immobilized on mesoporous silica: Economically and environmentally important examples. *Journal of Organometallic Chemistry*, **689**, 4110-4124. doi:10.1016/j.jorganchem.2004.07.052

[24] Jones, M.D. and Mahon, M.F. (2008) Synthesis of Rh(I) diamine complexes and their exploitation for asymmetric hydrogen transfer processes. *Journal of Organometallic Chemistry*, **693**, 2377-2382. doi:10.1016/j.jorganchem.2008.04.020

[25] Dreos, R., Mechi, L., Randaccio, L., Siega, P., Zangrand, E. and Ben Hassen, R. (2006) Synthesis and X-ray crystal structure of a new μ-hydroxo dinuclear cobalt complex containing one cis-β folded and one planar salen moiety. *Journal of Organometallic Chemistry*, **691**, 3305-3309. doi:10.1016/j.jorganchem.2006.04.006

[26] Dreos, R., Mechi, L., Nardin, G., Randaccio, L. and Siega, P. (2005) Alternative cocrystallization of "almost" enantiomers and true enantiomers in some cis-β-organocobalt salen-type complexes with α-amino acids. *Journal of Organometallic Chemistry*, **690**, 3815-3821. doi:10.1016/j.jorganchem.2005.05.017

[27] Mechi, L., Chtiba, S., Hamdi, N. and Ben Hassen, R. (2009) 4-Hydr-oxy-3-[(2E)-3-(3,4,5-trimethoxy-phen-yl)-prop-2-eno-yl]-2H-chromen-2-one. *Acta Crystallographica Section E*, **65**, 1652-1653. doi:10.1107/S1600536809022569

[28] Dholakia, V.N., Parekh, M.G. and Trivedi, K.N. (1968) Studies in 4-hydroxy coumarins. *Australian Journal of Chemistry*, **21**, 2345-2347. doi:10.1071/CH9682345

[29] Burla, M.C., Caliandro, R., Camalli, M., Carrozzini, B., Cascarano, G.L., De Caro, L., Giacovazzo, C., Polidori, G. and Spagna, R. (2005) SIR2004: An improved tool for crystal structure determination and refinement. *Journal of Applied Crystallography*, **38**, 381-388. doi:10.1107/S002188980403225X

[30] Petricek, V., Dusek, M. and Palatinus, L. (2006) Jana 2006. Institute of Physics, Praha.

[31] Bellamy, L.J. (1975) The infra-red spectra of complex molecules. 3rd Edition, Chapman and Hall, London.

[32] Bhattacharjee, S. and Anderson, J. (2006) Comparison of the epoxidation of cyclohexene, dicyclopentadiene and 1,5-cyclooctadiene over LDH hosted Fe and Mn sulfonato-salen complexes. *Journal of Molecular Catalysis A*, **249**, 103-110. doi:10.1016/j.molcata.2005.12.042

[33] Zhang, Y., Zhao, J., He, L., Zhao, D. and Zhang, S. (2006) Manganese(III) salen complex anchored onto MCM-41 as catalyst for the aerobic epoxidation of olefins. *Microporous and Mesoporous Materials*, **94**, 159-165. doi:10.1016/j.micromeso.2006.03.040

[34] Ertl, A., Huches, J.M., Pertlik, F., Foit, F., Wright, S.E., Brandstatter, F. and Marler, B. (2002) Polyhedrom distrtions in tourmaline. *The Canadian Mineralogist*, **40**, 153-163. doi:10.2113/gscanmin.40.1.153

[35] Baur, W.H. (1974) The geometry of polyhedral distortions predictive relationships for the phosphate group. *Acta Crystallographica*, **30**, 1195-1215.

[36] Gupta, S.K., Hitchcock, P.B. and Kushwah, Y.S. (2002) The crystal structure of 4-methyl-2,6-dibenzoylphenol and its conversion into a mononuclear cobalt(III) complex by treatment with cobalt(II) chloride and propane-1,3-diamine. *Polyhedron*, **21**, 1787-1793. doi:10.1016/S0277-5387(02)01044-6

[37] Traven, V.F., Manaev, A.V., Safronova, O.B., Chibisova, T.A., Lyssenko, K.A. and Antipin, M.Y. (2000) Electronic structure of π systems: XVIII.[1] photoelectron spectrum and crystal structure of 3-acetyl-4-hydroxycoumarin. *Russian Journal of General Chemistry*, **70**, 798-808.

22

Solid state transformation of *cis* and *trans* methylcyclopentadienyl molybdenumdicarbonyltriphenylphosphineiodide on pelleting utilizing different diluents

Olalere G. Adeyemi[*], Umaru Salami

Department of Chemical Sciences, Redeemer's University, Redemption City, Nigeria
Email: [*]drlereadeyemi@yahoo.com, adeyemio@run.edu.ng

ABSTRACT

$[(\eta^5\text{-}C_5H_4Me)Mo(CO)_2PPh_3I]$ undergoes solid state transformation on the formation of a good pellet for FT IR measurement. There was a formation of the products mixture on pelleting using different diluents of group I metal salts on either the *cis* or the *trans* isomer of the $[(\eta^5\text{-}C_5H_4Me)Mo(CO)_2PPh_3I]$ complex. The *cis* or the *trans* isomer gave the same IR spectra *i.e.* a mixture of *cis* and *trans* isomer of the complex. It does not matter the isomer started with in the course of solid state transformation reaction, an equilibrium ratio of 30/70 (*trans/cis*) will still be achieved. The solid state IR spectra show very strong peaks at v_{co} 1957, 1947 and strong peaks at 1867, 1853 cm^{-1}. The individual IR *cis/trans* isomer will therefore show at 1947 and 1853/1957 and 1867 cm^{-1}. The solution IR spectra gave, *cis* = 1961, 1875 and *trans* = 1963, 1882 cm^{-1} in dry CHCl$_3$. Hence, most of the solid state IR measurement of the organometallic complex of the type $(\eta^5\text{-}C_5H_4Me)Mo(CO)_2(PPh_3)I$ on pelleting will give isomer mixture.

Keywords: Solid State; *cis/trans*-$[(\eta^5\text{-}C_5H_4Me)Mo(CO)_2PPh_3I]$; Pelleting; Diluents; FT IR

1. INTRODUCTION

Inorganic diluents or dispersants that are insoluble in the melt have been known as additives to control the rate of solid-solid reactions. Addition of a high melting point inert solid to a melt can be used to control the rate of a solvent-free reaction just as the rates of some reactions are controlled by solvents. A good example is in a solventless substitution reaction of $Mn(CO)_4(PPh_3)Br$ with PPh_3, a range of dispersants or diluent (KBr, Al_2O_3, Na_2SO_4, $NaNO_2$, SiO_2, Na_2CO_3, $NaC_2H_3O_2$, $NaNO_3$, sucrose and TiO_2) are ground and filtered only to provide a comparable particle size. The reaction was investigated and monitored by in situ DRIFTS. The result of the above study showed that the chemical nature of the diluent matrix influenced the rate of the solventless reaction. Therefore, when the reaction was carried out in sucrose and SiO_2, the rate of the reaction was fastest. When Al_2O_3 and TiO_2 were used, the reaction was slower [1]. Also the role of the nature of the solid dispersant or diluent has been investigated in solventless reactions involving thallium salts of tris (pyrazolylborate). The grinding of thallium salts of tris (pyrazolylborate), Tp with Mn(II), Co(II) and Ni(II) salts in an agate mortar has been reported to yield Mn, Co and Ni tris (pyrazolylborate) metal complexes of the type TpMCl via a substitution type of reaction [2]. Supramolecular complexes of the formula $[Co^{III}(\eta^5\text{-}C_5H_4COOH)(\eta^5\text{-}C_5H_4COO)]_2\cdot M^+X^-$ were formed when organometallic zwitterion $[Co^{III}(\eta^5\text{-}C_5H_4COOH)(\eta^5\text{-}C_5H_4COO)]$ reacts quantitatively as a solid polycrystalline phase with a number of crystalline alkali salts MX (M = K$^+$, Rb$^+$, Cs$^+$, NH$_4^+$; X = Cl$^-$, Br$^-$, I$^-$, PF$_6^-$) [1].

In the same vein, manual grinding of the organometallic complex $[Fe(\eta^5\text{-}C_5H_4COOH)_2]$ with a number of solid bases, namely 1,4-diazabicyclo[2.2.2]octane, $C_6H_{12}N_2$, 1,4-phenylenediamine, $p\text{-}(NH_2)_2C_6H_4$, piperazine, $HN(C_2H_4)_2NH$, *trans*-1,4-cyclohexanediamine, $p\text{-}(NH_2)_2C_6H_{10}$, and guanidinium carbonate $[(NH_2)_3C]_2[CO_3]$, generates quantitatively the corresponding adducts,

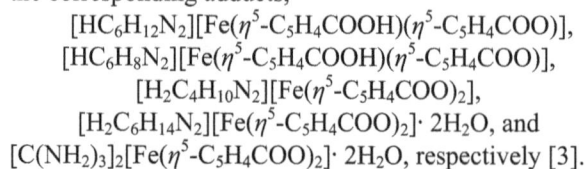

$[HC_6H_{12}N_2][Fe(\eta^5\text{-}C_5H_4COOH)(\eta^5\text{-}C_5H_4COO)]$,
$[HC_6H_8N_2][Fe(\eta^5\text{-}C_5H_4COOH)(\eta^5\text{-}C_5H_4COO)]$,
$[H_2C_4H_{10}N_2][Fe(\eta^5\text{-}C_5H_4COO)_2]$,
$[H_2C_6H_{14}N_2][Fe(\eta^5\text{-}C_5H_4COO)_2]\cdot 2H_2O$, and
$[C(NH_2)_3]_2[Fe(\eta^5\text{-}C_5H_4COO)_2]\cdot 2H_2O$, respectively [3].

Therefore, environmental concerns in synthetic chemistry have led to a reconsideration of reaction methodologies. This has resulted in investigations into atom economy, the use of supercritical CO_2, ionic liquids, and

---

[*]Corresponding author.

other procedures to reduce the disposal problems associated with most chemical reactions. One obvious route to reduce waste entails generation of chemicals from reagents in the *absence* of solvents [4]. Little is known about reactivity patterns of organometallic complexes in the solid state, until it was discovered that organometallic complexes, of the type CpML4, undergo *cis-trans* ligand isomerization reactions in the solid state [5-8].

Hence, in this study we report on the effect of pelleting on organometallic compounds, *trans-* and *cis-*[(η^5-C$_5$H$_4$Me) Mo(CO)$_2$PPh$_3$I]. Attempt were made on the effect of diluents on iron organometallic complexes [(η^5-C$_5$H$_5$) Fe(CO)$_2$ PPh$_3^+$][PF$_6^-$], as its influenced the products formed after pellet formation. Hence these diluents KBr, KCl, NaCl, CaCl$_2$, NaNO$_3$, BaSO$_4$, CaCO$_3$ and Al$_2$O$_3$ were screened for either good pellet formation or set out reactions with the complexes. It was observed that BaSO$_4$, CaCO$_3$ and Al$_2$O$_3$ could not form a good pellet for FT IR measurement of the complexes. A simple process of forming a good pellet for FT IR measurement could generate a reaction because of high amount of energy involved. In general, when two solids are ground together, the heat generated in the grinding process may be sufficient to either create a melt at the surface or completely melt the solid reagents. This could arise from the generation of a "hot spot" (an exotherm) [1,9] that could lead to a self-sustaining reaction.

2. EXPERIMENTAL

2.1. General

(η^5-C$_5$H$_4$Me)Mo(CO)$_3$I were prepared by the standard procedures used to synthesise other ring-substituted analogues [10,11]. TrimethylamineN-oxide dihydrate (Aldrich) was used as received. All reactions were carried out using standard Schlenk techniques under nitrogen.

2.2. Preparation of a Mixture of *trans-* and *cis-*[(η^5-C$_5$H$_4$Me)Mo(CO)$_2$PPh$_3$I]

A mixture of the *cis* and *trans* isomers were prepared in good yield by following a standard procedure in the literature [11].

2.3. Preparation of [(η^5-C$_5$H$_5$)Fe(CO)$_2$ PPh$_3^+$][PF$_6^-$]

The procedures are the same for the preparation of [(η^5-C$_5$H$_5$)Fe(CO)$_2$ PPh$_3^+$][PF$_6^-$] in the literature [12].

2.4. Refractometric Measurement to Ascertain the Purity of the Salts

Abbe Refractometer (Optic Ivymen) was used to determine the purity of the samples.

2.5. Pellets Formation

The diluents were dried for 12 hrs in an oven to ensure complete removal of moisture. About 10 mg of pure diluents (e.g. NaCl) were crushed to fine powder using an agate mortar and pestle. About 2 mg of the solid organometallic complex was added and gently ground together with diluents until fully mixed. The die set was assembled and the mixture was added into the die and goes between two stainless-steel discs and pressed to form a good, thin and transparent pellet. Opaque pellets gave poor spectra and white spots in the pellets.

3. RESULTS AND DISCUSSIONS

3.1. Preparation of the Complexes *trans-* and *cis-*[(η^5-C$_5$H$_4$Me)Mo(CO)$_2$PPh$_3$I]

The method developed by Blumer *et al.* [13] for the synthesis of the related unsubstituted compounds was adopted for the synthesis of *trans-* and *cis-*[(η^5-C$_5$H$_4$Me)Mo(CO)$_2$ PPh$_3$I]. Isomer separation was achieved by dissolving the crude material in CH$_2$Cl$_2$ followed by mixing with a small quantity of silica gel. The yellow powder left after removing the CH$_2$Cl$_2$ was chromatographed on a silica gel column (60 cm) with a 1:10 CH$_2$Cl$_2$/hexane mixture to afford the desired complexes [14].

3.2. Preparation of [(η^5-C$_5$H$_5$)Fe(CO)$_2$ PPh$_3^+$][PF$_6^-$]

The preparation of [(η^5-C$_5$H$_5$)Fe(CO)$_2$ PPh$_3^+$][PF$_6^-$], follows the ETC PPh$_3$ ligand replacement for I$^-$ on (η^5-C$_5$H$_5$) Fe(CO)$_2$I to obtain the complexes [12]. The yield and spectroscopic information were not at variant with the established results in the literature.

3.3. Purity of the Salts

Before the formation of a good pellet suitable for FTIR measurement, the diluents were carefully dried at 105°C until a constant weight obtained. This is to remove moisture and other volatile component in the diluents. The refractive index of the salts at the specified temperature, 33.4°C depicts the state of high purity of the salts when compared with reference refractive indexes.

3.4. Effect of Pelleting on v_{co} in Methylcyclopentadienyl Molybdenum Dicarbonyl Triphenylphosphine Iodide Using Different Diluents

We reported earlier a thermal transformation of *cis-* or *trans-*(η^5-C$_5$H$_4$Me)Mo(CO)$_2$(PPh$_3$)I in the solid state where the color and shape of the starting materials showed no visible change and no obvious decomposition during the heating process [11]. The *cis* and *trans* products was achieved by means of TLC, and solution IR and NMR spectroscopy on the products after the reaction, revealed the formation of the isomer materials. The isomerisation reactions for the new complexes were studied and the general trend was for the isomerisation reaction to occur from the *trans* to the *cis* isomer dependent on electronic factors associated with ligand orientation effects. An extension of the work was on the formation of the products mixture on pelleting either the *cis* or the *trans* isomer of the complex. **Figure 1** shows the isomerisation reaction during the formation of a suitable pellet for FT IR spectroscopy. While **Figure 2** shows the FT IR mixture of the *cis* or the *trans* isomer of the complex.

The *cis* or the *trans* isomer gave the same IR spectra *i.e* a mixture of *cis* and *trans* isomer of the complex.

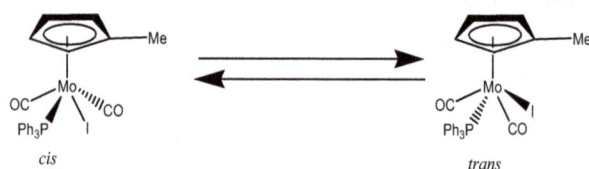

Figure 1. Solid state *cis/trans* isomerisation reaction of $(\eta^5\text{-}C_5H_4Me)Mo(CO)_2(PPh_3)I$ during pelleting.

Therefore, it does not matter the isomer started with in the course of solid state transformation reaction, an equilibrium ratio of 30/70 (*trans/cis*) will still be achieved. This result shows that the solid state isomerisation of $(\eta^5\text{-}C_5H_4Me)Mo(CO)_2(PPh_3)I$ still maintain a bidirectional reaction process. The IR spectra show very strong peaks at v_{co} 1957, 1947 and strong peaks at 1867, 1853 cm^{-1}. As shown in **Figures 2** and **3**.

The individual IR *cis/trans* isomer will therefore show at 1947 and 1853/1957 and 1867 cm^{-1}. The solution IR spectra gave, *cis* = 1961, 1875 and *trans* = 1963, 1882 cm^{-1} in dry $CHCl_3$ [11]. It should be noted that the same isomerisation products were obtained for the halide salts of group I investigated, namely, KCl, KBr and NaCl, **Figure 3**.

It is therefore possible to draw out some information on pelleting in the IR measurement of this complex:

1) Most of the solid state IR measurement of the organometallic complex of the type $(\eta^5\text{-}C_5H_4Me)Mo(CO)_2$ $(PPh_3)I$ on pelleting will give isomer mixture.

2) It can results in facile synthesis of *cis/trans*-$(\eta^5\text{-}C_5H_4Me)Mo(CO)_2(PPh_3)I$.

3) It can leads to isomerisation reaction as in this process.

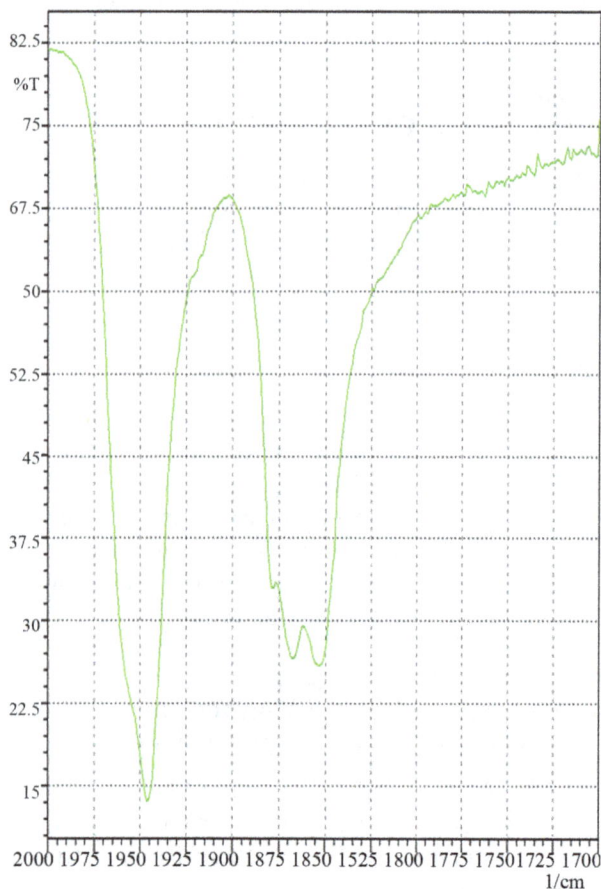

Figure 2. shows the FT IR spectrum of the isomer mixture in the solid state in NaCl.

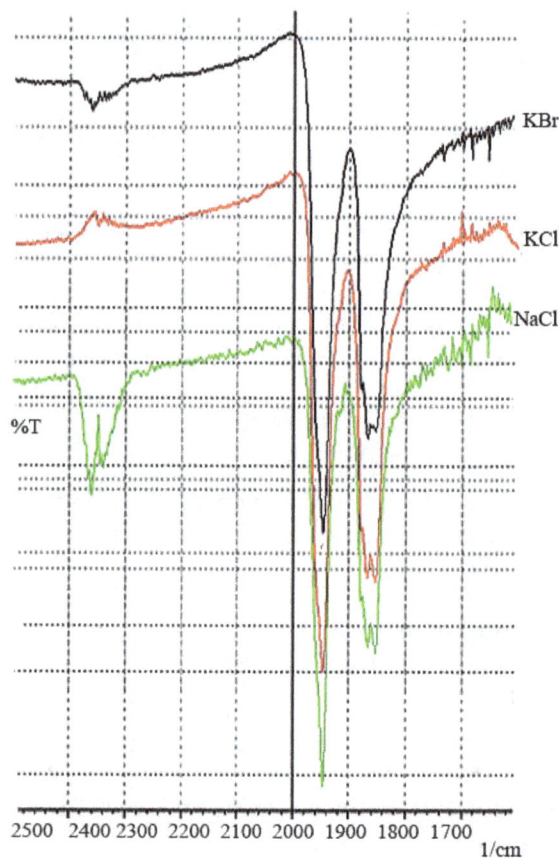

Figure 3. shows the FT IR spectra of the isomer mixture in the solid state in different diluents.

4) It is a fast reaction.

5) It is a neat and greener reaction.

Hence, when two solids are ground or pelleting together, the heat generated in the process set out a reaction to generate an exotherm that could then lead to a self-sustaining reaction [1,9]. It is not unlikely that no other product(s) formed during the creation of a suitable pellet for IR measurement. It is also not unlikely that there is halides exchange, a possibility of Iodide in the complex exchange for Chloride or Bromide in the diluents during the formation of a good pellet for FT IR measurement. Supporting evidence in **Figure 3** revealed that there is a change in the absorption intensity especially when pelleting the complex with KBr, while the intensities of NaCl and KCl remain constant.

A non isomeric organometallic product of the type $[\eta^5\text{-}(C_5H_5)Fe(CO)_2PPh_3]^+[PF_6]^-$ was investigated to determine the effect of the diluents through FT IR measurement and as a control for whether the solid state transformation of the $(\eta^5\text{-}C_5H_4Me)Mo(CO)_2(PPh_3)I$ complex actually occurred. **Figure 4** shows the spectrum of the $[\eta^5\text{-}(C_5H_5)Fe(CO)_2PPh_3]^+[PF_6]^-$ complex in NaCl diluent after pelleting.

It was observed that very strong peaks at ν_{co} 2008 and 2049 cm^{-1} correspond to the absorption of CO attached to the iron centre. A lower wave number of this kind reflects the degree of back bonding to the carbon monoxide ligand. The peaks position in the solid state is not significantly different from the peaks position in solution IR that occurred at 2014 and 2062 cm^{-1} in dry $CHCl_3$ [12] **Figure 5** depicts the behaviour of complex in different diluents.

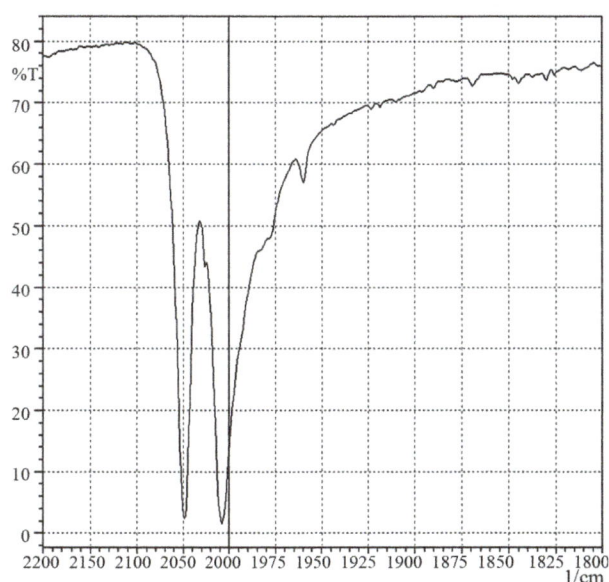

Figure 5. FTIR Spectra of $[\eta^5\text{-}(C_5H_5)Fe(CO)_2PPh_3]^+[PF_6]^-$ in different diluents.

The same pattern was therefore observed in the diluents used. This could informed that these diluents are suitable for the formation of good pellet for IR without decomposition. It is therefore possible that the anion exchange reaction, (where Cl^- or Br^- exchange for PF_6^- did not occur because from our experience the halide counter anion are not very stable as the PF_6^- anion. It could therefore infered that isomerization actually took place on pelleting $cis/trans\text{-}(\eta^5\text{-}C_5H_4Me)Mo(CO)_2(PPh_3)I$ and no isomerisation on $[\eta^5\text{-}(C_5H_5)Fe(CO)_2PPh_3]^+[PF_6]^-$ during pelleting.

4. CONCLUSION

It was established that pelleting can generate enough heat to set up a chemical reaction. Therefore on forming a suitable pellet using different diluents of group I metal salts leads to the isomerisation reactions of transforming *cis* to *trans* and the reverse, until isomeric mixture of *cis/trans* (30/70) is achieved.

5. ACKNOWLEDGEMENTS

We wish to thank the Redeemer's university for providing enabling environment and facilities for this research work.

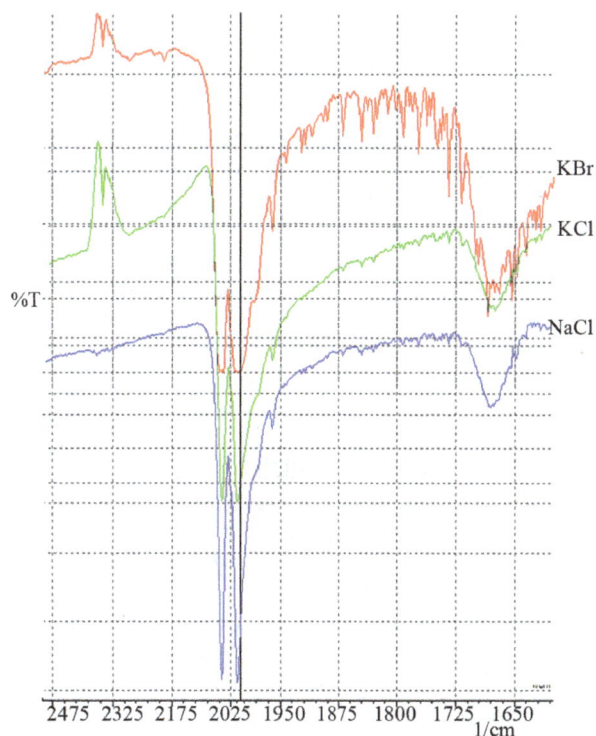

Figure 4. FTIR Spectrum of $[\eta^5\text{-}(C_5H_5)Fe(CO)_2PPh_3]^+[PF_6]^-$ in NaCl.

REFERENCES

[1]   Bala, M.D. and Coville, N.J. (2007) Organometallic chemistry in the melt phase. *Journal of Organometallic Che-*

mistry, **692**, 709-730.
doi:10.1016/j.jorganchem.2006.10.046

[2]   Kolotilov, S.V., Addison, A.W., Trofimenko, S., Dougherty, W. and Pavlishchuk, V.V. (2004) Efficient mechanochemical synthesis of tris (pyrazolylborate) complexes of manganese(II), cobalt(II) and nickel(II). *Inorganic Chemistry Communications*, **7**, 485.
doi:10.1016/j.inoche.2004.01.009

[3]   Braga, D., Maini, L., Polito, M., Mirolo, L. and Grepioni, F. (2003) Assembly of hybrid organic-organometallic materials through mechanochemical acid-base reactions. *Chemistry: A European Journal*, **9**, 4362-4370.
doi:10.1002/chem.200305017

[4]   Adeyemi, O.G. and Coville, N.J. (2003) Solvent-free organometallic migratory insertion reactions. *Organometallics*, **22**, 2284-2290. doi:10.1021/om0301738

[5]   Cheng, L. and Coville, N.J. (1996) Phase-dependent diagonal and lateral isomerism of the rhenium complexes $(\eta^5\text{-}C_5H_4Me)Re(CO)_2Br_2$. *Organometallics*, **15**, 867.
doi:10.1021/om950548s

[6]   Coville, N.J. and Cheng, L. (1998) Organometallic chemistry in the solid state. *Journal of Organometallic Chemistry*, **571**, 149. doi:10.1016/S0022-328X(98)00914-0

[7]   Coville, N.J. and Levendis, D.C. (2002) Organometallic chemistry: Structural isomerization reactions in confined environments. *European Journal of Inorganic Chemistry*, 3067.
doi:10.1002/1099-0682(200212)2002:12<3067::AID-EJIC3067>3.0.CO;2-4

[8]   Bogadi, R., Levendis, D.C. and Coville, N.J. (2002) Solid-state reaction study of the cis-trans isomerisation of $(\text{eta}^5\text{-}C_5H_4Me)Re(CO)[P(OPh)_3]Br_2$: A new mechanism for the isomerisation reaction. *Journal of the American Chemical Soceity*, **124**, 1104.
doi:10.1021/ja010695j

[9]   Cave, G.W.V., Raston, C.L. and Scott, J.L. (2001) Recent advances in solventless organic reactions: Towards benign synthesis with remarkable versatility. *Chemical Communications*, **21**, 2159-2169. doi:10.1039/b106677n

[10]   King, R.B. (1955) Organometallic synthesis. Academic Press, New York.

[11]   Adeyemi, O.G., Eke, U.B., Cheng, L., Cook, L.M., Billing, D.G., Mamba, B.B., Levendis, D.C. and Coville, N.J. (2004) Solid-state isomerisation reactions of $(\eta^5\text{-}C_5H_4R)$-$M(CO)_2(PR_3')I$ (M = W, Mo; R = ^{t}Bu, Me; R = Ph, O^{i}Pr$_3$). *Journal of Organometallic Chemistry*, **689**, 2207-2215.
doi:10.1016/j.jorganchem.2004.04.009

[12]   Adeyemi, O.G. and Liu, L.-K. (2007) Electron-transfer chain catalysis in phosphine replacement reaction: Determination of relative donor capability of arylpyridylphosphines. *Inorganica Chimica Acta*, **360**, 2464-2470.
doi:10.1016/j.ica.2006.12.051

[13]   Blumer, D.J., Barnett, K.W. and Brown, T.L. (1979) Trimethylamine *N*-oxide promoted reactions of manganese, molybdenum and tungsten carbonyl complexes. *Journal of Organometallic Chemistry*, **173**, 71.
doi:10.1016/S0022-328X(00)91236-1

[14]   Adeyemi, O.G., Fernandes, M.A., Cheng, L., Eke, U.B., Levendis, D.C., and Coville, N.J. (2002) The solid-state isomerization of *cis*- and *trans*-$(\eta^5\text{-}C_5H_4Me)Mo(CO)_2$-$(P(O^iPr)_3)I$. *Comptes Rendus Chimie*, **5**, 387-394.
doi:10.1016/S1631-0748(02)01386-3

23

Molecular and electronic structure of several 2,3-dithienylquinoxalines and their 2:1 complexes with silver(I) nitrate

Guy Crundwell[1*], Stefanie Cantalupo[1], Paul D. C. Foss[1], Brian McBurney[1], Kristen Kopp[1], Barry L. Westcott[1], James Updegraff III[2], Matthias Zeller[2], Allen D. Hunter[2]

[1]Department of Chemistry & Biochemistry, Central Connecticut State University, New Britain, USA
[2]YSU Structure Center, Department of Chemistry, Youngstown State University, Youngstown, USA
Email: [*]crundwellg@ccsu.edu

ABSTRACT

We have synthesized three bis (thienyl) quinoxaline-silver(I) complexes; however, unlike analogous silver(I) complexes of pyridylquinoxaline that utilize (N, N) bidentate behavior from the quinoxaline and pyridyl ring nitrogens, the bis(thienyl) quinoxaline ligands did not utilize the bonding potential of the thienyl rings to give (N, S) bonding modes. PES spectra modeling of these ligands indicates that the preferential metal bonding via only the quinoxaline nitrogen atoms is due to the N-rich, but S-poor, characters of the frontier orbitals.

KEYWORDS

X-Ray Diffraction; Silver(I) Complexes; Thiophene; Quinoxalines; Photoelectron Spectroscopy

1. INTRODUCTION

Complexes of quinoxalines and similar N-containing, aromatic heterocycles to silver(I) centers have been used to construct a variety of 2- and 3-D solid state molecular networks [1-5]. Molecular solids having $Ag(L)^+$ and $Ag(L)_2^+$ structural units have shown a wide range of structural variation that is guided by a few basic principles. For example, mono- or bis-quinoxaline silver complexes tend to have linear geometries about Ag [6,7]; such complexes with heterocyclic aromatic ligands, tend to interact via π-π stacking interactions building higher dimensional structural networks [8,9]; their counteranions tend to direct extended structures through weak associations with the silver in the complex (e.g. PF_6^{-1}, BF_4^{-1}) [6,10] or through direct coordination (e.g. $CF_3CO_2^-$, NO_3^-) [11-14]; and solvent from which crystals are grown tends to influence packing [7].

Several previous investigations utilized a single ligand and varied the counteranion or solvent to better understand the relationship between these guiding trends. However, in the current study, we have kept these variables constant and, instead, studied three similar ligands in an attempt to better understand how differences in ligand structure affect how a nitrate anion coordinates to silver. Specifically, we have investigated a trio of 2,3-bis(thienyl) quinoxalines with the expectation that we were likely to see some combinations of chelating or bridging (N, S) bonding modes.

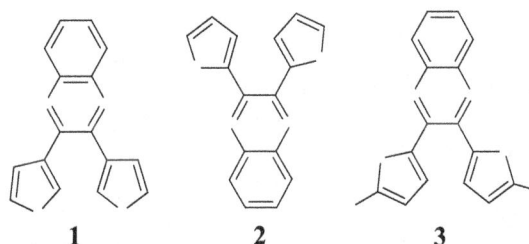

1   **2**   **3**

Previous studies have shown that mono- or bis-silver(I) complexes of pyridylquinoxalines having nitrate counter ions often involve bidentate (N, N) bonding from the nitrogen on the pyridyl substituent in addition to the quinoxaline nitrogen [13]. The nitrates involved in the metal bonding in these complexes show three variations: the nitrate may bind one oxygen to one silver [8], two oxygen to one silver [9,13], or two oxygen on one nitrate group may bridge two silvers on separate molecules [7]. The silver geometry is either linear or bent with the perturbation to non-linear geometries caused by the cation-anion interaction. The silver geometries do not seem to be driven by a simple geometric preference since silver coordination adopts linear, bent, trigonal planar, and pseudo-tetrahedral geometries with these complexes. With this depth of knowledge for pyridylquinoxalines at hand, we aimed to use thienylquinoixalines. Thienylquionoxalines **2** and **3** afford the opportunity to bind in either a metal-bridging and/or a bidentate (N, S) fashion by using the thiophene ring sulfur as well as the quinoxaline nitrogen. This (N, S) binding is not far-fetched; a few ligands similar in structure have shown to utilize both the N-containing and S-containing hetereocycle fragments when bound with a variety of metals such as Cu^{+2} [15], Ru^{+3} [16], Ir^{+3} [17,18], Yb^{+2} [19], and a di-Mo^{+2} complex [20]. To better understand the character of the frontier orbitals in these thienylquinoxaline ligands, we investigated the free ligands with photoelectron spectroscopy.

2. EXPERIMENTAL

2.1. Synthesis of the Ag^+ Complexes

The three ligands, 2,3-Bis(3-thienyl) quinoxaline (**1**), 2,3-bis(2-thienyl) quinoxaline (**2**), and 2,3-bis(5-bromo-2-thienyl) quinoxaline (**3**), see the graphic above, were prepared and purified according to literature methods and their single crystal structures have been reported [21-23]. AgNO$_3$ (≥99.0%, ACS Reagent Grade) was purchased from Aldrich and used without purification as was methanol (≥99.0%, ACS Reagent Grade).

Ag(**1**)$_2$NO$_3$. A warm (~50°C) solution of AgNO$_3$ (0.1003 g, 0.5904 mmol) in 10 mL of methanol was added to a warmed solution of **1** (0.3481 g, 1.182 mmol) in 30 mL of methanol. The resulting mixture was wrapped with aluminum foil, covered with a watch glass, kept in the dark, and allowed to cool and evaporate slowly. Over the course of two weeks, small pale white crystals of bis(2,3-dithien-3'-ylquinoxaline-N-)-(nitrito-O)-silver(I), Ag(**1**)$_2$NO$_3$, precipitated that were suitable for single crystal X-ray diffraction studies.

Ag(**2**)$_2$NO$_3$. Ag(**2**)$_2$NO$_3$ was prepared using the same methodology as for Ag(**1**)$_2$NO$_3$. A warm solution of AgNO$_3$ (0.0998 g, 0.5875 mmol) in 10 mL of methanol was added to a warmed solution of **2** (0.3466 g, 1.177 mmol) in 30 mL of methanol. The resulting mixture was wrapped with foil, covered with a watch glass, kept in the dark, and allowed to cool and evaporate slowly. Over the course of two weeks, small pale yellow crystals of bis(2,3-dithien-2'-ylquinoxaline-N-)-bis(nitrito-O,O')-silver(I), Ag(**2**)$_2$NO$_3$,precipitated that were suitable for single crystal X-ray diffraction studies.

Ag(**3**)$_2$NO$_3$. Ag(**3**)$_2$NO$_3$ was prepared using the same methodology as Ag(**1**)$_2$NO$_3$. A warm solution of AgNO$_3$ (0.0978 g, 0.5757 mmol) in 10 mL of methanol was added to a warmed solution of **3** (0.5211 g, 1.152 mmol) in 40 mL of methanol. The resulting mixture was wrapped with foil, covered with a watch glass, kept in the dark, and allowed to cool and evaporate slowly. Over the course of two weeks, small pale yellow crystals of bis(2,3-di-5'-bromothien-2'-ylquinoxaline-N-)-bis(nitrito-O,O')-silver(I), Ag(**3**)$_2$NO$_3$, precipitated that were suitable for single crystal X-ray diffraction studies.

2.2. Single Crystal X-Ray Diffraction Studies of Ag(**1**)$_2$NO$_3$, Ag(**2**)$_2$NO$_3$, and Ag(**3**)$_2$NO$_3$

Single crystal x-ray diffraction studies were performed on crystal samples with well-defined morphologies displaying uniform birefringence. Diffraction data for all silver complexes were collected on a BrukerSmart Apex instrument with CCD detector at 100 K. The diffractometerwas equipped with a graphite monochromator utilizing Mo Ka radiation (l = 0.71073 Å). Unit cell parameters were obtained via least-square refinements. Integrated data was corrected for absorption effects using multi-scan correction methods. Initial phase models were obtained from direct methods and non-hydrogen peaks were identified through difference maps of the initial and subsequent refined models. Flip-disorders on unsubstitutedthienyl rings in Ag(**1**)$_2$NO$_3$ and Ag(**2**)$_2$NO$_3$ were confirmed by the distortion of ideal thienyl ring geometries and disordered atoms from flipped thienyl rings were located from difference maps and modeled [24]. From evaluation of thermal ellipsoids and difference maps, the nitrate groups in **2** were found to be disordered amongst two sites. Hydrogen atoms on all models were placed at ideal locations. Least-squares refinements on F2 were performed with all non-hydrogen atoms refined anisotropically. Basic crystallographic details for Ag(**1**)$_2$NO$_3$, Ag(**2**)$_2$NO$_3$, and Ag(**3**)$_2$NO$_3$ are given in **Table 1**. Cif files containing detailed descriptions of all refinements are available as supplementary material. CCDC files 275310-275313 and also contain supplementary crystallographic data for this paper. These data can be obtained free of charge from The Cambridge Crystallographic Data Centre via www.ccdc.cam.ac.uk/data_request/cif.

2.3. Photoelectron Spectroscopy of Quinoxalines 1, 2, and 3

The HeI and HeII gas-phase photoelectron spectra of the three ligands were recorded using an instrument and general procedures that have been described previously [25]. The samples showed no signs of impurity or decomposition in the gas phase during controlled sublimation at 90°C - 115°C. A small amount of a non-volatile

Table 1. Data collection and refinement parameters for Ag(**1**)$_2$NO$_3$, Ag(**2**)$_2$NO$_3$, and Ag(**3**)$_2$NO$_3$.

	Ag(**1**)$_2$NO$_3$	Ag(**2**)$_2$NO$_3$	Ag(**3**)$_2$NO$_3$
Color/shape	White block	Yellow block	Yellow block
Crystal size (mm)	0.40 × 0.32 × 0.12 mm	0.44 × 0.25 × 0.12 mm	0.45 × 0.25 × 0.08 mm
Chemical formula	C$_{32}$H$_{20}$AgN$_5$O$_3$S$_4$	C$_{32}$H$_{20}$AgN$_5$O$_3$S$_4$	C$_{32}$H$_{16}$AgBr$_4$N$_5$O$_3$S$_4$
Formula weight	758.64 g/mol	758.64 g/mol	1074.25 g/mol
CSD deposition number	CCDC 275310	CCDC 275311	CCDC 275313
Temperature	100 (2) K	100 (2) K	100 (2) K
Crystal system	Monoclinic	Triclinic	Triclinic
Space group	$P2_1/c$	P-1	P-1
Unit cell dimensions	a = 13.7712 (12) Å b = 16.4862 (15) Å c = 14.4051 (13) Å β = 118.2800 (10)°	a = 10.4991 (7) Å b = 11.9510 (8) Å c = 12.5195 (8) Å α = 73.2320 (10)° β = 83.1450 (10)° γ = 77.9090 (10)°	a = 9.5941 (17) Å b = 9.7749 (17) Å c = 16.183 (3) Å α = 104.520 (3)° β = 98.593 (3)° γ = 107.695 (3)°
Z, Volume	4, 2880.1 (4) Å^3	2, 1467.81(17) Å^3	2, 1710.8(4) Å^3
Density (calculated)	1.750 g/cm^3	1.717 g/cm^3	2.085 g/cm^3
F$_{000}$	1528	764	1036
Max. & min. transmission	0.697 and 0.813	0.683 and 0.885	0.251 and 0.642
θ range for data collection	2.83° to 28.32° (Mo K$_\alpha$)	2.43° to 28.29° (Mo K$_\alpha$)	2.32° to 28.29° (Mo K$_\alpha$)
Reflections measured	29,165	15,097	17,108
Index ranges	$-18 \geq h \geq 18$ $-21 \geq k \geq 21$ $-19 \geq l \geq 19$	$-13 \geq h \geq 13$ $-15 \geq k \geq 15$ $-16 \geq l \geq 16$	$-11 \geq h \geq 11$ $-18 \geq k \geq 18$ $-20 \geq l \geq 20$
Independent reflections	7157 [R$_{int}$ = 0.0286]	7234 [R$_{int}$ = 0.0164]	8395 [R$_{int}$ = 0.0418]
Data/restraints/parameters	7157/30/368	7234/210/407	8395/96/430
Goodness-of-fit on F^2	2.028	0.904	1.169
Δ/σ$_{max}$	0.003	0.002	0.004
R indices [$I > 2\sigma(I)$]	R1 = 0.0778, wR2 = 0.2516	R1 = 0.0351, wR2 = 0.0888	R1 = 0.0438, wR2 = 0.1164
R indices (all data)	R1 = 0.0787, wR2 = 0.2530	R1 = 0.0369, wR2 = 0.0902	R1 = 0.0482, wR2 = 0.1200
Largest diff. peak & hole	4.802 and −3.802 eÅ^{-3}	1.604 and −0.802 eÅ^{-3}	1.941 and −2.675 eÅ^{-3}
R.M.S. deviation from mean	0.199 eÅ^{-3}	0.091 eÅ^{-3}	0.168 eÅ^{-3}

solid remained in the sample cell at the end of each experiment. During data collection the instrument resolution (measured using FWHM of the argon $^2P_{3/2}$ peak) was 0.015 - 0.020 eV for HeI and 0.022 - 0.026 eV for HeII.

3. RESULTS AND DISCUSSION

3.1. Crystal Structure Determinations of Ag(**1**)$_2$NO$_3$, Ag(**2**)$_2$NO$_3$, and Ag(**3**)$_2$NO$_3$

ORTEPs of Ag(**1**)$_2$NO$_3$, Ag(**2**)$_2$NO$_3$, and Ag(**3**)$_2$NO$_3$ are shown in **Figure 1**. All internal bond lengths and angles are within expected values. Molecules of Ag(**1**)$_2$NO$_3$ pack in staggered head-to-head layers. One of the two 2,3-bis(3-thienyl) quinoxalines has a 3-thiophene ring that is nearly coplanar with the quinoxaline moiety [Ring S2 at 11.35(10)°]. The quinoxalines are bent against each other at the silver ion, ∠N-Ag-N = 153.85 (17)°; and the

coordinating nitrate oxygen completes the distorted trigonal planar geometry around the silver ion (**Figure 2**). Molecules of Ag(**2**)$_2$NO$_3$ also pack in staggered head-to-head layers. Again one of the two 2,3-bis(2-thienyl)quinoxalines has a 2-thiophene ring that is nearly coplanar with the quinoxaline moiety [Ring S4 at 10.48 (14)°], and the quinoxalines are bent against each other at the silver, ∠N-Ag-N = 151.75 (8)°; two coordinating nitritooxygens complete a distorted square planar geometry around the silver; however the second oxygen is at a distance of 2.761 (3) Å perhaps making the complex more distorted trigonal planar (**Figure 2**). Molecules of Ag(**3**)$_2$NO$_3$ pack in layers wherein each layer contains nitrate groups oriented in the same direction. Again one of the two 2,3-bis(5-bromo-2-thienyl) quinoxalines has a 2-thiophene ring that is nearly coplanar with the quinoxaline moiety [Ring S4 at 1.77 (19)°]. The quinoxalines

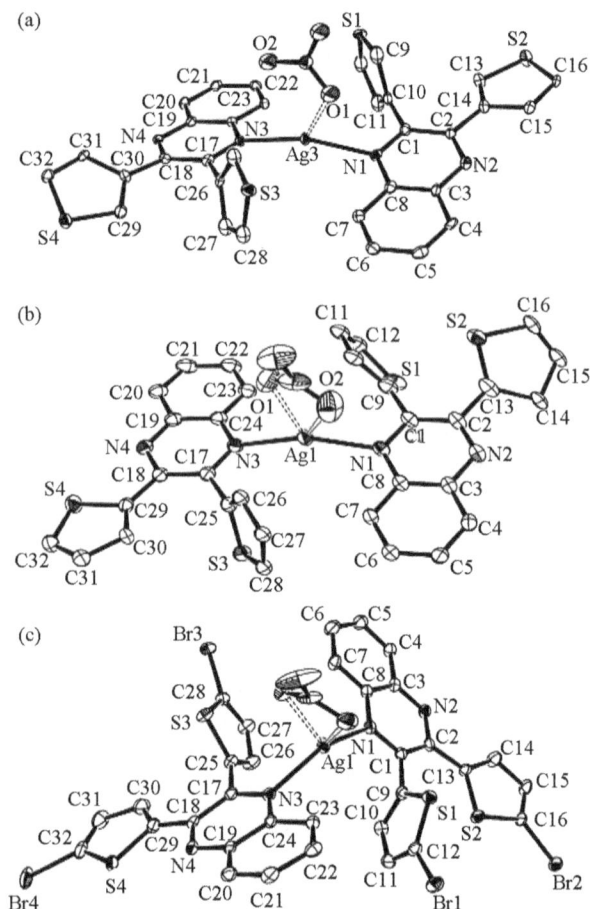

Figure 1. (a) ORTEP of Ag(**1**)$_2$NO$_3$. Displacement ellipsoids are drawn at the 50% probability level. Somethienyl ringsare flip disordered, only the major componentsare shown for clarity. (b) ORTEP of Ag(**2**)$_2$NO$_3$. Displacement ellipsoids are drawn at the 50% probability level. Some thienyl ringsare flip disordered, only the major componentsare shown for clarity. The alternative position of the disordered nitrate is also omitted for clarity. (c) ORTEP of Ag(**3**)$_2$NO$_3$. Displacement ellipsoids are drawn at the 50% probability level.

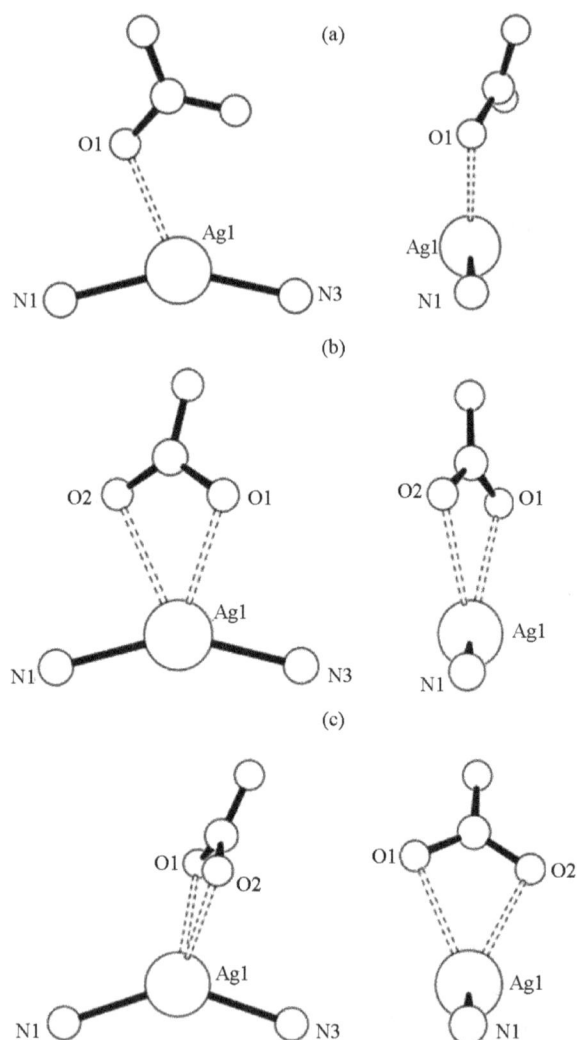

Figure 2. Nitrate group orientation to silver-quinoxaline ligands in Ag(**1**)$_2$NO$_3$ (a), Ag(**2**)$_2$NO$_3$ (b), and Ag(**3**)$_2$NO$_3$ (c).

areslightly bent against each other at the silver, $\angle$N-Ag-N = 141.77 (8)°; and the coordinating nitritooxygens complete a distorted tetrahedral geometry around the silver (**Figure 2**). As the $\angle$N-Ag-N angle become less linear, the quinoxalines twist, the thiophene rings separate, and the nitrate group gets closer to the silver centergoing from a trigonal planar coordination geometry to distorted tetrahedral.

3.2. Thiophene and Nitrate Disorder in Ag(1)$_2$NO$_3$ and Ag(2)$_2$NO$_3$

In the initial structural models, some of the thienyl rings in Ag(**1**)$_2$NO$_3$ and Ag(**2**)$_2$NO$_3$ displayed anomalous bond lengths and angles and had residual electron density peaks suggesting a need to model flip disorder. Specifically for Ag(**1**)$_2$NO$_3$, thienyl-ring flip disorder of 18.4

(4)% for the ring containing S1 and 44.4 (4)% for the ringcontaining S3 was observed. The rings containing S2 and S4 did not show need for disorder modeling. The initial model for Ag(**2**)$_2$NO$_3$ required disorder modeling for three of the four thienyl rings. The percentages of thienyl-ring flip disorder were: 31.2 (3)% for the ring containing S1, 28.9 (3)% for the ring containing S2, and 40.1 (3)% for the ring containing S3. Furthermore, the nitrate group in Ag(**2**)$_2$NO$_3$ is disordered over one additional site at 35.5 (8)%. For Ag(**3**)$_2$NO$_3$, the addition of bromo groups prohibited potential ring flip disorder and none was observed.

3.3. Nitrate Coordination to Silver

The impact of choice of counteranion on extended structure in silver coordination chemistry has been well studied. Preliminary observations have shown that weakly coordinating anions such as PF_6^{-1} and BF_4^{-1} do not tend to be located near the silver [6,10]. However, several structures have seen nitrate-silver interactions that lead to: monodentate O-Ag bonding [8]; bidentate non-bridging O-Ag bonding [9,13]; and bidentate bridging bonding [7]. In Ag(1)$_2$NO$_3$, one nitrito group oxygen is within bonding distance to the silver at a distance of 2.461 (5) Å with the other at a distance of over 3 Å. Of the three complexes, Ag(1)$_2$NO$_3$ is closest to linear (∠N-Ag-N = 153.85 (17)°) and the nitrate group is sandwiched between the nearby thienyl rings. For Ag(2)$_2$NO$_3$ and Ag(3)$_2$NO$_3$, the complexes are bent with ∠N-Ag-N = angles of 151.75 (8)° and 141.78 (10)° respectively and the nitrate groups bind in a non-bridging bidentate fashion. For Ag(2)$_2$NO$_3$, the nitritooxygens bond lengths are 2.558 (2) Å and 2.699 (2) Å and for Ag(3)$_2$NO$_3$, the nitritooxygens bond lengths are 2.482 (3) Å and 2.535 (3) Å. The interesting difference between all three Ag(X)$_2$NO$_3$ structures is the difference in the nitro group coordination. With stronger coordination through the nitrate ions the angle at the silver ion becomes less linear. The nitrate group rotates with respect to the plane created by the quinoxaline nitrogen-silver-quinoxaline nitrogen association (**Figure 2**). The orientation is consistent with the minimization of steric repulsions from the heterocyclic aromatic rings and counteranion.

3.4. C-H ⋯ π and π ⋯ π Interactions in Ag(X)$_2$NO$_3$

The compounds studied also show a variety of intra- and intermolecular "π-stacking" type interactions. Molecules of Ag(1)$_2$NO$_3$ pack in staggered head-to-head layers. This packing may be caused by a number of C-H ⋯ π contacts. Intramolecularly, the hydrogen on C13 in Ag(1)$_2$NO$_3$ interacts with the p-system of the thiophene ring containing S1 at 2.55 Å and the hydrogen on C23 is at 2.80 Å from that same ring. There are intermolecular interactions between molecules as well. C31 is associated with a neighboring thiophene ring containing S2 at 2.97 Å [Symmetry code: −1 + x, 1/2 − y, −1/2 + z]; whereas disordered thiophene ring C12B has a hydrogen that is 2.82 Å away from the centroid of the C3-C8 six-membered ring on a neighboring molecule [Symmetry code: x, 1/2 − y, 1/2 + z]. There are three weak π-π interactions in Ag(1)$_2$NO$_3$. Measurements of ring centroids indicate intermolecular contact between the six-membered ring of C19-C24 with a neighboring six-membered ring containing N3-C24 at a distance of 3.782 (3) Å [Symmetry code: −x, 1 − y, 1 − z]. Additional contacts between ring N1-C8 with two neighboring rings on two different molecules, a C3-C8 ring and a N1-C8, are at 3.952(3) Å and 3.958(3) Å respectively [Symmetry code: 1 − x, 1 − y, 1 − z].

Molecules of Ag(2)$_2$NO$_3$ also pack in staggered head-to-head layers resulting in a number of C-H ⋯ π

contacts. Intramolecularly, hydrogens on carbons C7, C23, and C30 make three hydrogen bonds of length 2.67 Å, 2.86 Å, and 2.67 Å respectively with thiophene rings. However, there are no significant intermolecular C-H ⋯ π interactions. The slipped π ⋯ π interactions between molecules as determined from ring centroid distances are all over 4 Å.

Finally molecules of Ag(3)$_2$NO$_3$ pack in layers wherein each layer contains nitrate groups oriented in the same direction. This packing results in only one C-H ⋯ π contact intramolecularly between the hydrogen on C30 and a thiophene ring at a distance of 2.83 (4) Å. Intermolecular π ⋯ π interactions as determined from ring centroid distances are all over 4 Å and the only observed C-Br ⋯ π-ring centroid distance is 3.7047 (18) Å between Br4 and a neighboring thiophene ring [Symmetry code: 3 − x, 1 − y, −z].

3.5. Photoelectron Spectroscopy of 1, 2, and 3

PE spectra were collected of ligands **1**, **2**, and **3**. The spectra were fitted analytically with asymmetric Gaussian peaks with a confidence limit of peak positions and width deviations generally considered as ±0.02 eV (>3σ level) [26]. For the different complexes, the HeI spectrum was fit first. The number of peaks used in each fit was based on the features of the band profile and the number of peaks necessary for a statistically good fit. For the HeII fits the peak positions and half-widths were fixed with respect to those of the HeI fit. Only the peak amplitudes were allowed to vary to account for changes in photoionization cross-section.

Confidence limits for the relative integrated peak areas are about 5% with the primary source of uncertainty being the determination of the baseline. The baseline arises from electron scattering and is taken to be linear over the small energy range of these spectra. The fitting procedures used are described in more detail elsewhere [26].

The HeI (21.2 eV) PE spectra for the three ligands are shown in **Figure 3**. We observe a noticeable difference in relative energies of the first peak of **1** in comparison with spectra of **2** and **3**. This shift from 7.9 eV (for **2** and **3**) to 8.0 eV (for **1**) may be attributed to either the coplanarity of one thiophene ring with the quinoxaline rings in **2** and **3** which may result in resonance effects for the aromatic ring system's HOMO. A sharp, high peak is seen in **3** (~10.7 eV). This peak is attributed to the lone pair on bromine.

When looking at the highest energy peaks in the spectra of **1** and **2**, a clear growth is observed in the height and area of the peak in the HeII source when compared to HeI (**Figure 4**). For orbitals of predominately nitrogen character, we expect to see a three-fold increase in band-intensity when compared to those of predominately carbon character; for orbitals of predominately sulfur

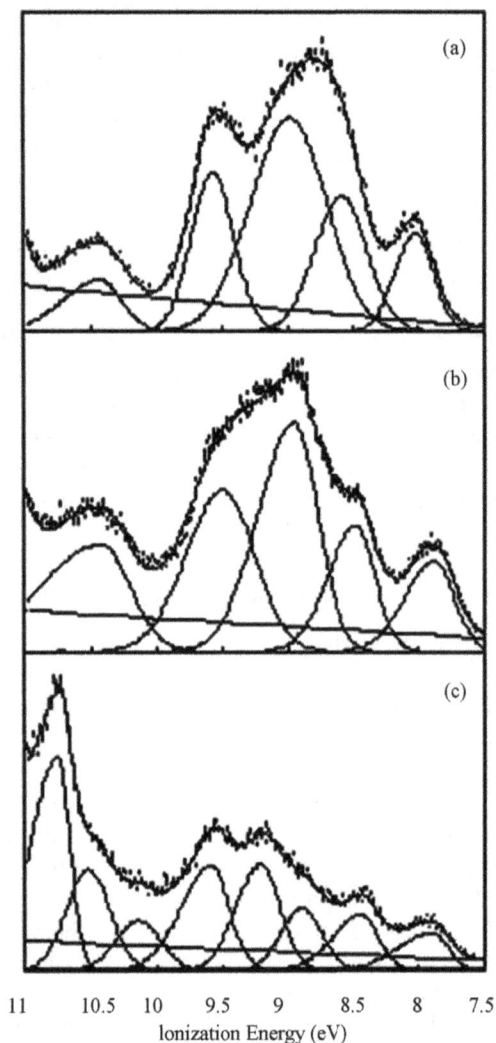

Figure 3. The HeI (21.2 eV) PE spectra for **1** (a), **2** (b), and **3** (c).

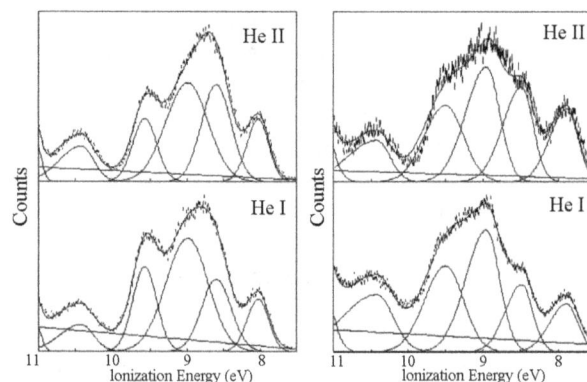

Figure 4. The HeI (bottom, 21.2 eV) and HeII (top, 40.8 eV) PE spectra for **1** (left) and **2** (right). The increasing peaks in the HeII spectra indicate orbitals with increasing N character.

zothiazole ligands [28]. Nitrito coordination differs as the coplanarity of the two quinoxaline moieties changes. The nitrate is a sandwich between thiophene rings on the quinoxalines. Models based upon PE spectra indicate that frontier orbitals are rich in nitrogen character without much sulfur character supporting our observation of quinoxaline nitrogen bonding without any coordination from nearby but misaligned thiophene sulfurs.

ACKNOWLEDGEMENTS

The authors would like to thank Dr. Nadine Gruhn at the Center for Gas Phase Electron Spectroscopy (Managing Director, Center for Enabling New Technologies Through Catalysis, Department of Chemistry, University of Washington, Seattle, WA, 98195) for her assistance in collecting the PES data. The authors also thank Drs. Steven Herron and Katherine Kantardjieffat the W.M. Keck Foundation Center for Molecular Structure (The Department of Chemistry and Biochemistry, California State University, Fullerton, CA, 92831) for initial single crystal XRD datasets on a triclinic polymorph of **1**. Funding for the work was made possible through CSU-AAUP research grants. The diffractometer purchase and operation was funded by NSF grant 0087210, by Ohio Board of Regents grant CAP-491, by ARL grant W911NF-07-1-0642, and by YSU.

character, we expect a slight *decrease* in band intensity when changing source from HeI to HeII [27]. The spectra for both **1** and **2** shows a more than 2-fold increase for bands in the highest energy valence region, suggesting the frontier orbitals contain substantial nitrogen character. There is no evidence of sulfur character in the frontier orbitals of these ligands, consistent with the nitrogen binding observed in the complexes.

4. CONCLUSION

Three 2:1 bis(thienyl) quinoxaline-silver complexes were synthesized and their crystal structures were determined. As seen in other structures containing unsubstituted terminal thienyl rings, crystals containing ligands **1** and **2** contain flip disordered thiophene rings. All three complexes have silver cations coordinated to the nitrate counteranion. This coordination was also recently observed in a silver(I) complex with two 2-(2-thienyl)ben-

REFERENCES

[1] Bi, W.-Y., Chai, W.-L., Lu, X.-Q., Song, J.-R. and Bao, F. (2009) Syntheses and supramolecular structures of silver (I) complexes based upon 2-(2'-pyridyl)-quinoxaline. *Journal of Coordination Chemistry*, **62**, 1928-1938. http://dx.doi.org/10.1080/00958970902736756

[2] Bacchi, A., Bosetti, E., Carcelli, M., Pelagatti, P. and Rogolino, D. (2004) Deconvolution of supramolecular-tectons and analysis of the structural role of the anions for two "wheel-and-axle" silver complexes. *Crystal Engineering Communications*, **6**, 177-183. http://dx.doi.org/10.1039/b405940a

[3] Bhogala, B.R., Thallapally, P.K. and Nangia, A. (2004) 1:2 and 1:1 Ag(I)-Isonicotinamide coordination com-

pounds: Five-fold interpenetrated CdSO₄ network and the first example of (pyridine)N-Ag-O(amide) bonds. *Crystal Growth & Design*, **4**, 215-218. http://dx.doi.org/10.1021/cg034157p

[4] Yaghi, O.M. and Li, H. (1996) T-shaped molecular building units in the porous structure of Ag(4,4'-bpy)·NO₃. *Journal of the American Chemical Society*, **118**, 295-296. http://dx.doi.org/10.1021/ja953438l

[5] Carlucci, L., Ciani, G., Proserpio, D.M. and Sironi, A. (1995) 1-, 2-, and 3-dimensional polymeric frames in the coordination chemistry of AgBF₄ with pyrazino. The first example of three interpenetrating 3-dimensional triconnected nets. *Journal of the American Chemical Society*, **117**, 4562-4569. http://dx.doi.org/10.1021/ja00121a014

[6] Crundwell, G. (2013) Bis[2-(thiophen-2-yl)quinoxaline-kN⁴]silver(I) tetrafluoridoborate. *Acta Crystallographica*, E**69**, m164. http://dx.doi.org/10.1107/S1600536813004510

[7] Yeh, C.-W., Chen, T.-R., Chen, J.-D. and Wang, J.-C. (2009) Roles of anion and solvent in the self-assembly of silver(I) complexes containing 2,3-diphenylquinoxaline. *Crystal Growth & Design*, **9**, 2595-2603. http://dx.doi.org/10.1021/cg800592y

[8] Fitchett, C.M. and Steel, P.J. (2008) Synthesis and X-ray crystal structures of metal complexes of three isomeric bibenzodiazines: Discrete and polymeric assemblies. *Polyhedron*, **27**, 1527-1537. http://dx.doi.org/10.1016/j.poly.2008.01.026

[9] Tong, M.-L., Chen, X.-M., Ye, B.-H. and Ng, S.W. (1998) Helical silver(I)-2,4'-bipyridine chains organized into 2-D networks by metal-counterion or metal-metal bonding. Structures of [Ag(2,4'-bipyridine)]X (X⁻=NO₃⁻ or ClO₄⁻). *Inorganic Chemistry*, **37**, 5278-5281. http://dx.doi.org/10.1021/ic971579d

[10] Kole, G.K., Tan, G.K. and Vittal, J.J. (2012) Photoreactivity of Ag(I) complexes and coordination polymers of pyridyl acrylic acids. *Crystal Growth & Design*, **12**, 326-332. http://dx.doi.org/10.1021/cg201119c

[11] Chen, C.Y., Zeng, J.Y. and Lee, H.M. (2007) Argentophilic interaction and anionic control of supramolecular structures in simple silver pyridine complexes. *Inorganica Chimica Acta*, **360**, 21-30. http://dx.doi.org/10.1016/j.ica.2006.06.013

[12] Patra, G.K., Goldberg, I., De, S. and Datta, D. (2007) Effect of size of discrete anions on the nuclearity of a complex cation. *Crystal Engineering Communications*, **9**, 828-832. http://dx.doi.org/10.1039/b706781j

[13] Raj, S.S.S., Fun, H.-K., Chen, X.-F., Zhu, X.-H. and You, X.-Z. (1999) [2-(2-Pyridyl-N)quinoxaline-Nʲ]silver(I) nitrate. *Acta Crystallographica*, C**55**, 2035-2037. http://dx.doi.org/10.1107/S0108270199010938

[14] Aakerröy, C.B. and Beatty, A.M. (1998) Supramolecular assembly of low-dimensional silver(I) architectures via amide-amide hydrogen bonds. *Chemical Communications*, **10**, 1067-1068. http://dx.doi.org/10.1039/a707919b

[15] Amari, C., Ianelli, S., Pelizzi, C., Pelizzi, G. and Predieri, G. (1993) 2-(2'-Thienyl)pyridine versus 2,2'-bipyridyl binding mode in copper(II) complexes. *Inorganica Chimica Acta*, **211**, 89-94.

http://dx.doi.org/10.1016/S0020-1693(00)82848-0

[16] Patra, S.K. and Bera, J.K. (2006) Axial interaction of the [Ru₂(CO)₄]²⁺ core with the aryl C-H bond: Route to cyclometalated compounds involving a metal-metal-bonded diruthenium unit. *Organometallics*, **25**, 6054-6060. http://dx.doi.org/10.1021/om060774+

[17] McGee, K.A. and Mann, K.R. (2007) Selective low-temperature syntheses of facial and meridionaltris-cyclometalatediridium(III) complexes. *Inorganic Chemistry*, **46**, 7800-7809. http://dx.doi.org/10.1021/ic700440c

[18] Constable, E.C., Henney, R.P.G. and Tocher, D.A. (1991) Different bonding modes in octahedral complexes of 6-(2-thienyl)-2,2'-bipyridine (HL) with transitional metal ions: Crystal structures of [Ru(HL)(py)Cl₃] (py=pyridine) and [Ru(HL)₂Cl][BF₄]·CH₂Cl₂. *Journal of the Chemical Society, Dalton Transactions*, **9**, 2335-2347. http://dx.doi.org/10.1039/dt9910002335

[19] Trifonov, A.A, Shestakov, B.G., Gudilenkov, I.D., Fukin, G.K., Giambastiani, G., Bianchini, C., Rossin, A., Luconi, L., Filippi, J. and Sorace, L. (2011) Steric control on the redox chemistry of (η⁵-C₉H₇)₂Ybᴵᴵ(THF)₂ by 6-aryl substituted iminopyridines. *Dalton Transactions*, **40**, 10568-10575. http://dx.doi.org/10.1039/c1dt10135h

[20] Majumdar, M., Patra, S.K., Kannan, M., Dunbar, K.R. and Bera, J.K. (2008) Role of axial donors in the ligand isomerization processes of quadruply bonded dimolybdenum(II) compounds. *Inorganic Chemistry*, **47**, 2212-2222. http://dx.doi.org/10.1021/ic702298v

[21] deFreitas, J. and Crundwell, G. (2013) 2,3-Bis(thiophene-3-yl)quinoxaline. *Acta Crystallographica*, E**69**, o394. http://dx.doi.org/10.1107/S1600536813004248

[22] Crundwell, G., Sayers, D., Herron, S.R. and Kantardjieff, K. (2003) 2,3-Dithien-2-ylquinoxaline. *Acta Crystallographica*, E**59**, o314-o315. http://dx.doi.org/10.1107/S1600536803002976

[23] Crundwell, G., Linehan, J., Updegraff III, J.B., Zeller, M. and Hunter, A. (2004) 2,3-Bis-(5-bromothien-2-yl) quinoxaline. *Acta Crystallographica*, E**60**, o656-o657. http://dx.doi.org/10.1107/S1600536804006816

[24] Crundwell, G., Sullivan, J., Pelto, R., Kantardjieff, K. (2003) Crystal structures of two thienyl analogs of benzil-1,2-dithien-2-ylethanedione (2,2'-thenil) and 1,2-dithien-3-ylethanedione (3,3'-thenil). *Journal of Chemical Crystallography*, **33**, 239-244. http://dx.doi.org/10.1023/A:1023820908599

[25] Westcott, B.L., Gruhn, N.E. and Enemark, J.H. (1998) Evaluation of molybdenum-sulfur interactions in molybdoenzyme model complexes by gas-phase photoelectron spectroscopy. The "Electronic Buffer" effect. *Journal of the American Chemical Society*, **120**, 3382-3386. http://dx.doi.org/10.1021/ja972674o

[26] Lichtenberger, D. L. and Copenhaver, A.S. (1990) Ionization band profile analysis in valence photoelectron spectroscopy. *Journal of Electron Spectroscopy and Related Phenomena*, **50**, 335-352. http://dx.doi.org/10.1016/0368-2048(90)87076-Z

[27] Yeh, J.J. and Lindau, I. (1985) Atomic subshell photoionization cross sections and asymmetry parameters: 1 ⩽ Z ⩽ 103. *Atomic Data and Nuclear Data Tables*, **32**, 1-155.

http://dx.doi.org/10.1016/0092-640X(85)90016-6

[28] Pereira, G.A., Massabni, A.C., Castellano, E.E., Costa, L.A.S., Leite, C.Q.F., Pavan, F.R. andCuin, A. (2012) A broad study of two new promising antimycobacterial drugs: Ag(I) and Au(I) complexes with 2-(2-thienyl)ben-zothiazole. *Polyhedron*, **38**, 291-296. http://dx.doi.org/10.1016/j.poly.2012.03.016

24

Synthesis and controlled release of cloprop herbicides from cloprop-layered double hydroxide and cloprop-zinc-layered hydroxide nanocomposites

Norhayati Hashim[1,2*], Mohd Zobir Hussein[3], Illyas Md Isa[1,2], Azlan Kamari[1,2], Azmi Mohamed[1,2], Adila Mohamad Jaafar[3], Hafsah Taha[1]

[1]Department of Chemistry, Faculty of Science and Mathematics, Universiti Pendidikan Sultan Idris, Tanjong Malim, Malaysia
[2]Nanotechnology Research Centre, Faculty of Science and Mathematics, Universiti Pendidikan Sultan Idris, Tanjong Malim, Malaysia
[3]Department of Chemistry, Faculty of Science, Universiti Putra Malaysia, Serdang, Malaysia
Email: [*]norhayati.hashim@fsmt.upsi.edu.my

ABSTRACT

Two phenoxyherbicide nanocomposites, namely cloprop-layered double hydroxide and cloprop-zinc-layered hydroxide nanocomposites, have been synthesized by using co-precipitation and direct reaction method. PXRD pattern showed an expansion of interlayer spacing with the value of 21.0 Å and 22.7 Å for cloprop-layered double hydroxide and cloprop-zinc-layered hydroxide nanocomposite, respectively. It is evident from FTIR and elemental analyses that both nanocomposites were successfully intercalated between the interlayers of layered metal hydroxide. Controlled release of cloprop anion from interlayer of nanocomposites for both cloprop-layered double hydroxide and cloprop-zinc-layered hydroxide nanocomposite into phosphate solution was rapid initially and slow thereafter. The percentage of accumulated release of cloprop anion from cloprop-zinc-layered hydroxide nanocomposite was slightly higher than that from cloprop-layered double hydroxide nanocomposite. Kinetic behavior of cloprop release was governed by pseudo-second-order for cloprop-layered double hydroxide nanocomposite while parabolic diffusion for cloprop-zinc-layered hydroxide nanocomposite. Results from this study highlight the potential of both nanocomposites as capsulated material for controlled release of cloprop phenoxyherbicides anion.

KEYWORDS

Cloprop; Cloprop-Layered Double Hydroxide; Cloprop-Zinc-Layered Hydroxide; Nanocomposite; Controlled Release; Kinetic Study

1. INTRODUCTION

Environmental problems associated with the use of highly mobile herbicides are of current concern because of the increasing presence of these agrochemicals in ground and surface waters. Anionic herbicides are of particular concern because they are weakly retained by most of the components of soil sediment, so they remain dissolved in the soil solution and can rapidly move around [1]. One approach to minimizing such transport losses is to use controlled release formulation in which the herbicides are incorporated in a matrix or carrier before application, thereby limiting the amount available for unwanted processes [2]. In this study, LDHs with general formula of

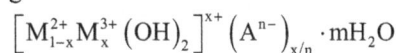

$$\left[M_{1-x}^{2+} M_x^{3+} \left(OH \right)_2 \right]^{x+} \left(A^{n-} \right)_{x/n} \cdot m H_2 O$$

and LMHs with general formula represented by

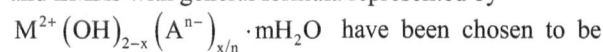

$M^{2+} \left(OH \right)_{2-x} \left(A^{n-} \right)_{x/n} \cdot m H_2 O$ have been chosen to be studied as herbicides controlled release formulation. M^{2+} is the metallic cation such as Mg^{2+}, Zn^{2+}, Co^{2+}, Ni^{2+} and M^{3+} is a trivalent metal ion such as Al^{3+}, Fe^{3+}, Cr^{3+}, Ga^{3+}, Ni^{3+}, Mn^{3+} [3,4]. A is a counterion with n^- charge [5]. This is a safe herbicide system to minimize and prevent the herbicides from direct exposure to environment. Due to these properties, LDH and LMH have a huge opportunity in industrial and environmental research as absorbent [6], chemical sensor [7-9], catalyst [10], drug delivery

[11] and controlled release formulation [12]. Co-precipitation method has been used to prepare cloprop-Zn/Al-layered double hydroxide (cloprop-LDH), meanwhile, cloprop-zinc-layered hydroxide (cloprop-ZLH) is prepared by using a direct reaction method. Cloprop (2(3-chlorophenoxy)propionic acid) is one of phenoxyherbicides widely used in maize, wheat and corn plantations. Controlled release study of cloprop from its nanocomposite was done into various solutions of sodium phosphate concentrations. Four kinetic orders were applied in this study in order to understand the release behavior of cloprop into different aqueous solutions.

2. EXPERIMENTAL

2.1. Preparation of Nanocomposites

All chemicals used in this synthesis were obtained from various chemical suppliers and used without any further purification. All solutions were prepared using deionized water. The preparation of the nanocomposite was carried out using direct co-precipitation method for cloprop-LDH and direct reaction method for cloprop-ZLH. For cloprop-LDH, a mother liquor containing Zn^{2+} and Al^{3+} cations with Zn to Al initial molar ratio R = 3.0 and 0.05 M to 0.3 M cloprop were prepared. The resulting solution was adjusted to pH 7.50 ± 0.02 by drop wise addition of aqueous NaOH (2.0 M). The reaction was carried out under nitrogen atmosphere. The resulting precipitate was aged at 70°C in an oil bath shaker for 18 hours. The synthesized material was then centrifuged, thoroughly washed with deionized water and dried in an oven at 70°C. The resulting nanocomposite was then powdered and stored in a sample bottle for further use and characterizations.

Meanwhile cloprop-ZLH was synthesized by mixing 0.05 g of ZnO (pure commercial zinc oxide, reagent ACS, Acros Organics) into 100 ml deionized water. Solutions of cloprop acid with concentrations of 0.05 M to 0.3 M were added into the ZnO solution. The precipitate was aged at 70°C in an oil bath shaker for 18 hours. The synthesized material was then centrifuged, thoroughly washed with deionized water and dried in an oven at 70°C. The resulting material was then powdered and stored in a sample bottle for further use and characterizations.

2.2. Instrumentation

Powder X-ray diffraction patterns were recorded at 2° - 60° on an ITAL 2000 diffractometer using Cu K_α radiation at 40 kV and 30 mA. FTIR spectra of the materials were recorded over the range 400 - 4000 cm^{-1} on a Perkin-Elmer 1752X Spectrophotometer using KBr disc method. The elemental composition of the samples molar ratio of the resulting nanocomposite were determined by using inductively couple plasma-atomic emission spectrometry (ICP-AES), using a Perkin Elmer Spectrophotometer model Optima 2000DV under standard condition and CHNS analyzer model CHNS-932 (LECO). Thermogravimetric and differential thermogravimetric analyses (TGA/DTG) were carried out using a Mettler Toledo TGA/SDTA851 thermogravimetric analyzer with heating rate of 10°C·min^{-1} between 35°C - 1000°C, under nitrogen flow rate of about 50 ml·min^{-1}.

The release of cloprop from its interlamellae host into the release media, the aqueous solutions at various initial concentrations of sodium phosphate was done by adding about 0.20 g of the nanocomposite into 500 mL of the aqueous solution. The accumulated amount of cloprop released into the solution was measured at preset time at λ_{max} = 217 nm respectively, using a Perkin Elmer UV-visible Spectrophotometer Lambda 35.

3. RESULT AND DISCUSSIONS

3.1. Powder X-Ray Diffraction

Figure 1(a) shows powder XRD patterns of cloprop-LDH nanocomposite prepared by co-precipitation method at R = 3 (ratio zinc to aluminium) with concentration of cloprop 0.05 M to 0.3 M. As shown in the figure, the intercalation started at cloprop concentration of 0.05 M with a small peak at low angle of 2θ below 10° but the peak of unintercalated Zn/Al-LDH remained (peak with basal spacing of 8.9 Å). As the concentration of cloprop is increased from 0.1 M to 0.3 M, the intensity of intercalation peak increased which corresponded with high crystallinity of resulting material when cloprop was introduced. As reported, the mechanism of co-precipitation method relies upon the condensation of hexa-aqua complexes in solution in order to form the brucite-like layers with a distribution of both metallic cations and with solvated interlayer anions [3,13]. As shown in the figure, cloprop intercalated into the layer of LDH shows a pure phase of nanocomposites with sharp and intense peak at low angle of 2θ below 10° at 0.2 M cloprop due to high crystallinity of the materials with the basal spacing (003) of 21.0 Å. The expanded basal spacing was attributed to the bigger size and higher density of cloprop intercalated between the interlayer of brucite-type layer compared to nitrate anions.

Figure 1(b) shows of cloprop-ZLH nanocomposite synthesized by direct reaction method at 0.05 M to 0.3 M cloprop. Generally, all PXRD diffraction of cloprop-ZLH showed intercalation peak at low 2θ angle reflection through the dissociation-deposition mechanism [11,14]. All cloprop-ZLH nanocomposite produced sharp and symmetrical peaks, suggesting a well ordered stacking layer. ZnO phase appeared in all of PXRD diffraction which is due to incomplete reaction. As shown in the

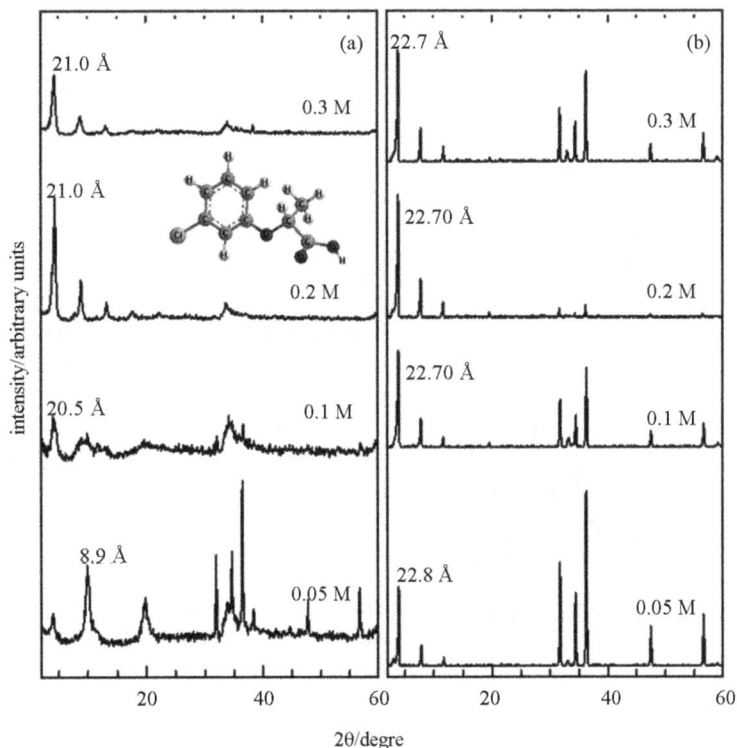

Figure 1. PXRD patterns of cloprop-LDH (a) and cloprop-ZLH (b) nanocomposite at different concentration of cloprop anion. Inset is the molecular structure of cloprop.

figure, cloprop-ZLH synthesized with 0.2 M cloprop showed high crystallinity and the least of ZnO phase compared to others with basal spacing of 22.7 Å. For this reason, cloprop-ZLH at 0.2 M has been chosen for further characterization and elemental analysis.

3.2. FTIR

The presence of the intercalated species in the layered hydroxide materials were also supported by FTIR results (**Figure 2**). FTIR spectra for cloprop showed a broad band at 2959 cm^{-1} which is attributed to the O-H stretching vibration. A sharp band at 1715 cm^{-1} is due to the stretching of C = O and bands at 1465 and 1404 cm^{-1} are attributed to stretching vibrations of aromatic ring C = C. A strong band at 1277 and 1215 cm^{-1} were due to the symmetric and asymmetric stretching modes of C-O-C of the cloprop anions while a sharp band at 746 cm^{-1} is due to C-Cl stretching band of both anions [15].

The FTIR spectrum of cloprop-LDH and cloprop-ZLH synthesized using 0.2 M cloprop are also shown in **Figure 2**. As expected, the spectrum resembles a mixture of both FTIR spectra of cloprop and Zn/Al-LDH, indicating that both functional groups of cloprop and Zn/Al-LDH are present together, except for the peak at 1389 cm^{-1} and 1713 cm^{-1} corresponding to the nitrate anions and carboxylic group. The presence of new peaks at around

1577 cm^{-1} and 1333 cm^{-1} for cloprop-LDH nanocomposite were due to the C = O carboxylate anion, which confirm the presence of cloprop in the anionic form in the interlayer of the Zn/Al-LDH [16]. This shows the cloprop has higher affinity than nitrate towards the inorganic interlamellae, thus occupies the interlamellae space between the inorganic layers and prevents further co-intercalation of nitrate anion.

This is further supported by the CHNS result, which shows the absence of nitrogen in both nanocomposite. FTIR spectrum of cloprop-ZLH nanocomposite (**Figure 2**) shows the characteristic of the bands of pure cloprop. This indicates the cloprop anions have been intercalated into the interlayer galleries of the ZLH. Some of the bands are slightly shifted in position, presumably due to the interaction between cloprop anions and the host interlayer as a result of the intercalation process.

Intense bands at 1583 and 1399 cm^{-1} for cloprop-ZLH nanocomposite are attributed to antisymmetric and symmetric carboxylate stretching of the anion. The bands at 1470 and 1409 cm^{-1} were due to the stretching vibrations of the aromatic ring C = C. A strong band at 869 cm^{-1} is attributed to C-Cl stretching and the bands at 736 and 772 cm^{-1} due to C-H vibration of benzene ring in cloprop-ZLH. This indicates the successful intercalation of the guest anion, cloprop.

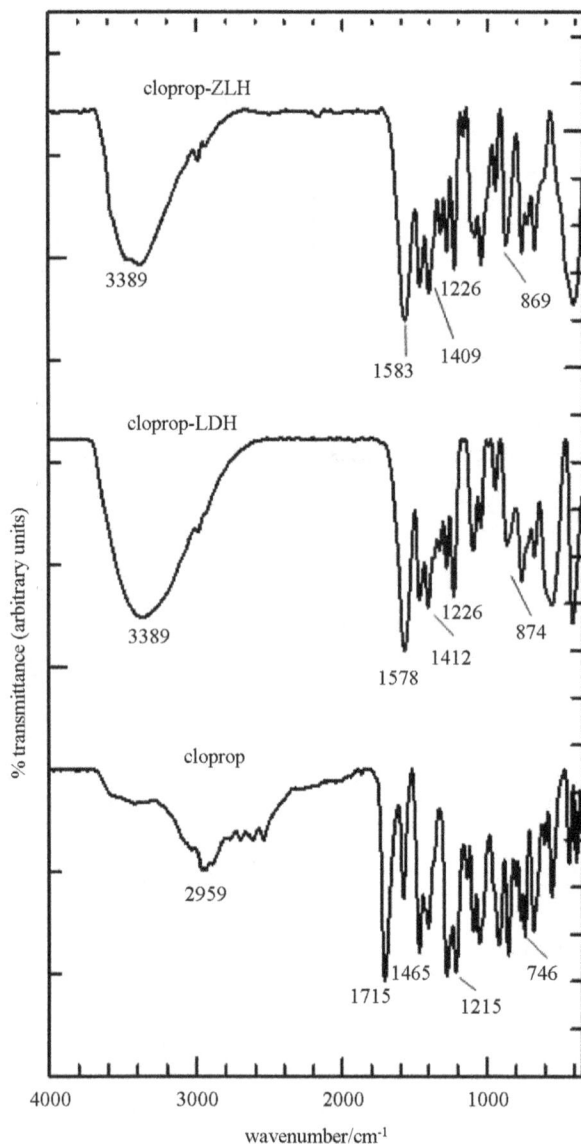

Figure 2. FTIR spectra of cloprop, cloprop-LDH and cloprop-ZLH nanocomposites.

Table 1. Elemental composition of cloprop-LDH and cloprop-ZLH nanocomposites.

Sample	d (Å)	Zn/Al ratio	X_{Al}	N	C	Zn	Cloprop[a]
Cloprop-LDH	21.0	2.9	0.25	-	20.6	-	38.3
Cloprop-ZLH	22.7	-		-	20.8	34.8	38.8

[a]estimated using CHNS analysis.

layer of brucite-type layer for cloprop-LDH and cloprop-ZLH nanocomposite. The elemental analysis shows that cloprop-LDH and cloprop-ZLH contained 20.6% and 20.8% carbon (w/w) and loading percentages of cloprop anion in the nanocomposite is 38.3% and 38.8% (w/w), respectively. The results showed that cloprop anion was successfully intercalated between the interlayer of layered materials.

3.4. Thermal Analysis

TGA/DTG thermogravimetric analysis obtained for the choprop-LDH, cloprop-ZLH and cloprop are reported in **Figures 3(a)-(c)** and **Table 2**. For cloprop, thermal studies show that the maximum temperature was observed at 204°C compared to 350°C and 381°C to the cloprop-LDH and cloprop-ZLH nanocomposites. This indicates that cloprop encapsulated into the inorganic interlamellae is thermally more stable than their counterparts in the sodium salt form.

For cloprop-LDH nanocomposite, three stages of weight loss were observed. Meanwhile, cloprop-ZLH showed two stages of weight loss. The first stage is the weight loss due to the physically adsorbed water and interlayer water which could be seen at around 35°C - 200°C followed by a weight loss due to the removal of interlayer anion and dehydroxylation of the hydroxyl layer which could be observed at 201°C - 600°C. The weight losses for encapsulated organic moiety of the nanocomposite are at 350°C (cloprop-LDH) and 381°C (cloprop-ZLH). The third stage was characterized by combustion of the organic species, leaving only a relatively less volatile, metal oxide.

3.5. Controlled Release Study

Controlled release of cloprop anion from cloprop-LDH, cloprop-ZLH nanocomposite into sodium phosphate had been done at different concentrations of sodium phosphate which is shown in **Figures 4(a)** and **(b)**.

Based on the components of rain water, phosphate solution was chosen as the medium for release of cloprop phenoxyherbicides. Since the research was carried out in the laboratory, the results collected will give us the information and patterns of releasing cloprop phenoxyherbicides from the interlayer of nanocomposite, some adjustments or modifications of the involved method need

3.3. Elemental Analysis

Table 1 shows elemental and organic content of cloprop-LDH and cloprop-ZLH. Elemental analysis shows the molar ratio of Zn to Al in clorrop-LDH 2.9 compared to 3.0 for the value initially prepared for the mother liquor. This indicates that the Zn to Al molar ratio in the resulting synthesized materials was adjusted accordingly for the formation of the positively charged inorganic layers to compensate the negatively charged intercalated anion so that the resulting materials cloprop-LDH is neutrally charged [17].

As shown in the table, the estimated percentage of cloprop intercalated into the interlayer are not much different for both nanocomposites which indicated same amount of cloprop anion intercalated between the inter-

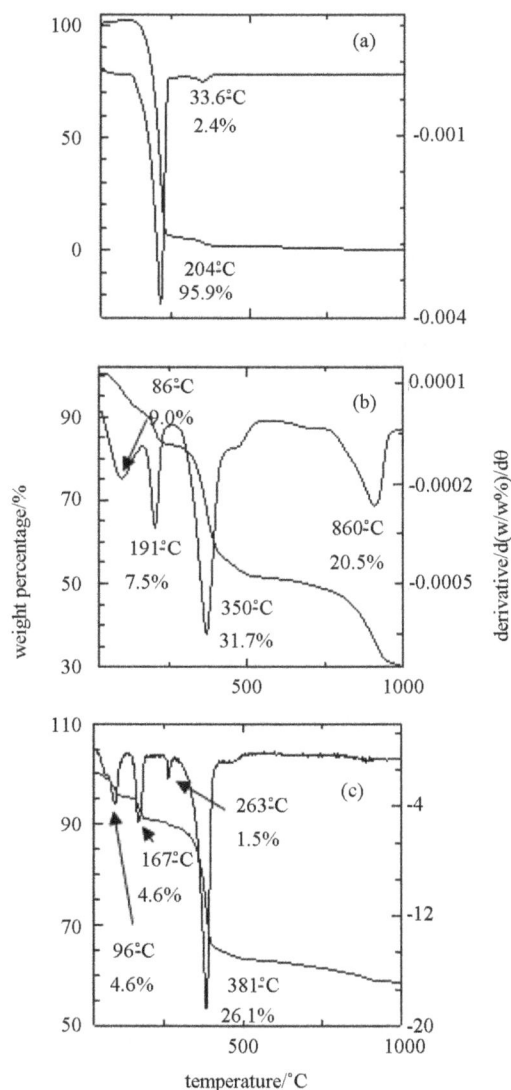

Figure 3. TGA/DTG thermograms of cloprop (a), cloprop-LDH (b) and cloprop-ZLH (c) nanocomposites.

to be done to improve the material before it can be used in plantation. A series of accumulated released profile of cloprop anion from the interlamellae of the nanocomposite into the sodium phosphate aqueous solutions are done at various initial concentrations of 0.002 M, 0.005 M and 0.008 M.

As shown in the figure, a rapid release was found in all solutions at the beginning of time, followed by a more sustained release. The release profile of cloprop from cloprop-LDH and cloprop-ZLH nanocomposite shows an increasing value of accumulated release percentage as the concentration of the release solution was increased. This observation can be due to the many sacrificial anions which act as driving force for the release of cloprop anion from nanocomposite interlayer to aqueous solution. These sacrificial anions have high affinity to-

wards the interlayer nanocomposite compared to the existing counter anion, resulting in ion exchange taking place [12]. At higher concentration of the anions in the aqueous media, more ion exchange process is obtained and a faster rate can be observed. The maximum accumulated release of cloprop (0.008 M) into the medium can be achieved at around 600 and 200 min, respectively for both cloprop from cloprop-LDH, cloprop-ZLH nanocomposite with the percentage of 86% and 91%. A fast release of cloprop anion from its nanocomposite could be due to high density of phosphate anion leading to a high formation of electrostatic interaction between phosphate anion with the positive charged layer of layered hydroxide material in the ion exchange process.

The release profile showed the accumulated release of cloprop from cloprop-LDH is slightly lower than cloprop-ZLH nanocomposite. The differences of the release profile were due to the crystallinity of the synthesized nanocomposite which refers to PXRD pattern for cloprop-ZLH nanocomposite. It shows high crystallinity than cloprop-LDH with the basal spacing of 22.7 Å and 21.0 Å, respectively. High crystallinity material resulted of well-ordered arrangement of anion in the host during the intercalation process. Release of the intercalated guest from nanocomposite with high crystallinity and larger particle size would be slower than the ones with lower crystallinity and smaller particle size. This is due to the stability of the anion arrangement as well the less surface accessibility of the nanocomposite material to be ion exchanged with the incoming anion from medium. This explains the slightly different release rates between the cloprop-ZLH and cloprop-LDH. The method of preparation plays an important role in determining the physico-chemical properties of the resulting nanocomposites, in particular their surface properties and release behaviour of the guest anions [12].

3.6. Kinetic Study

Release kinetic studies of cloprop from its nanocomposites into sodium phosphate solutions were analyzed by applying four different kinetic models in order to study the behaviour of the release process.

It was suggested that the release of guest anion from the nanocomposite could be best described either by dissolution of LDH [18,19] or by diffusion through the LDH [20,21]. The methods used are zeroth order (Equation (1)) [22], first order (Equation (2)) [23], pseudo-second order (Equation (3)) [24] and parabolic diffusion (Equation (4)) [21], for which the equations are given below. The x is the percentage release of herbicides anion at time t, M_i and M_f are the initial and final concentrations of herbicides anions, respectively and C is a constant. The parameter correlation coefficients, r^2 rate

Table 2. TGA-DTG results for cloprop-LDH, cloprop-ZLH and cloprop anion.

	Weight loss (%)			
Temperature	35 - 200 (°C)	201 - 600 (°C)	601 - 1000 (°C)	Total weight loss (%)
Cloprop-LDH	16.5 T_{max} = 191°C	31.7 T_{max} = 350°C	20.5 T_{max} = 860°C	68.7
Cloprop-ZLH	10.7 T_{max} = 167°C C	26.1 T_{max} = 381°C	-	36.8
Cloprop	-	98.3 T_{max} = 204°C	-	98.3

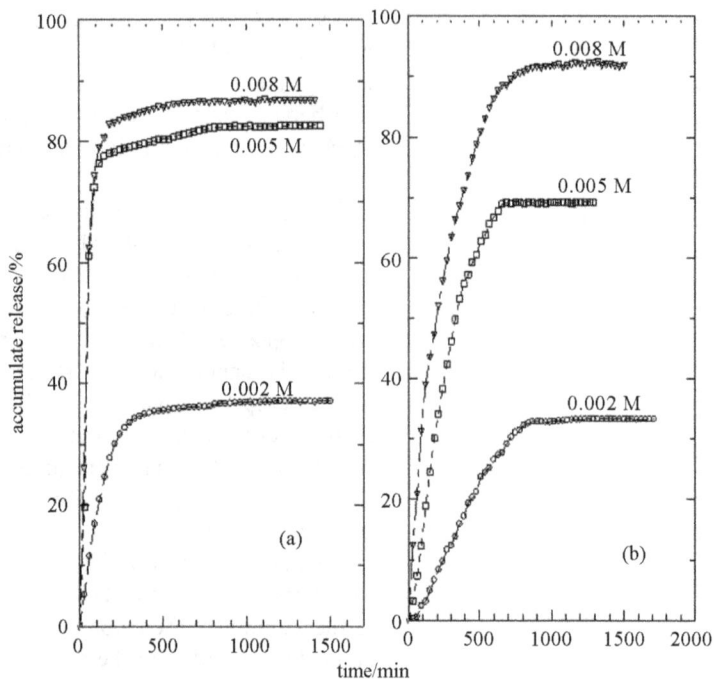

Figure 4. Release profile of clorop from the interlamellae of the cloprop-LDH (a) and cloprop-ZLH (b) nanocomposite into aqueous solutions containing various concentrations of Na$_3$PO$_4$.

constants, k and $t_{1/2}$ values (the time required for 50% of cloprop to be released from nanocomposites) are calculated from the corresponding equations.

$$x = t + c \qquad (1)$$

$$-\log\left(1 - M_i/M_f\right) = t + C \qquad (2)$$

$$t/M_i = 1/M_f^2 + t/M_f \qquad (3)$$

$$M_i/M_f = kt^{0.5} + C \qquad (4)$$

The extent of time determined to gauge the values of cloprop release fitted into the equation was fixed at 0 to 1000 minutes.

Based on the four kinetic models above, the fitting results of cloprop release profiles are given in **Figures 5** and **6** and **Table 3**. It can be seen in **Figure 5** that the best fit is the plot of t/M$_i$ against time which shows an agreement with pseudo-second order kinetic model for cloprop-LDH in all sodium phosphate solution as the evidence of linearization of other models such as zeroth

order, first order and parabolic diffusion do not fit nicely to the experimental data. On the other hand, the release of cloprop from cloprop-ZLH into aqueous solution containing phosphate anion showed a good fitting with parabolic diffusion model as evident by high r^2 values compared to first, zeroth and pseudo-second order (**Figure 6**).

Generally, the parabolic diffusion model elucidates that the release process is controlled by intra particle diffusion or surface diffusion. Therefore, it could be suggested that the results indicate the external surface diffusion or the intra particle diffusion via ion exchange which is the rate-determining step in the release process [25]. These results indicate that LDH and ZLH can be used to host a model guest, cloprop for their with controlled release properties.

For all the nanocomposites, the $t_{1/2}$ values that is the time taken for the herbicides anion concentration to increase to one-half of its initial values decrease as the concentration of the sodium phosphate increases. This is

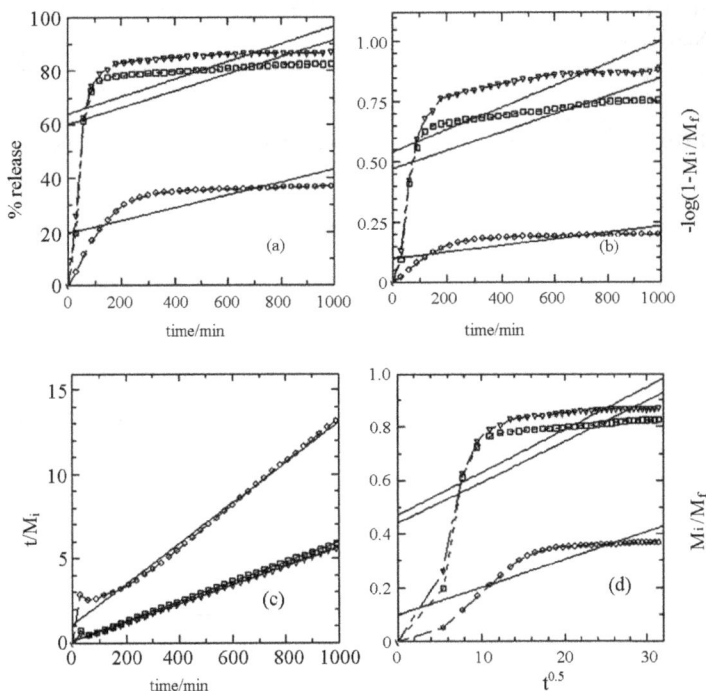

Figure 5. Fitting of the data of cloprop released into aqueous solutions containing various concentrations of Na_3PO_4; 0.002 M (circles), 0.005 M (squares) and 0.008 M (triangles) to the zeroth (a), first (b) and pseudo-second order (c) kinetics, and parabolic diffusion (d) for cloprop-LDH (t = 0 - 1000 min).

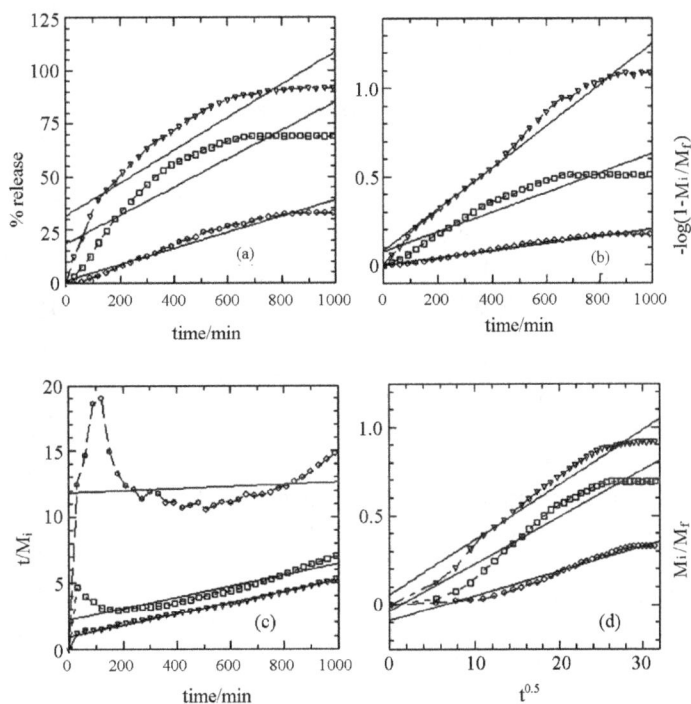

Figure 6. Fitting of the data of cloprop released into aqueous solutions containing various concentrations of Na_3PO_4; 0.002 M (circles), 0.005 M (squares) and 0.008 M (triangles) to the zeroth (a), first (b) and pseudo-second order (c) kinetics, and parabolic diffusion (d) for cloprop-ZLH (t = 0 - 1000 min).

Table 3. Rate constants, half life ($t_{1/2}$) and correlation coefficients obtained from the fitting of the data of cloprop release from cloprop-LDH and cloprop-ZLH into Na_3PO_4 solutions.

Cloprop-LDH nanocomposite						
Na_3PO_4 (mol·L^{-1})	Zeroth Order	First Order	Parabolic Diffusion	Pseudo second Order		
	r^2			r^2	k	$t_{1/2}$
0.002	0.552	0.590	0.765	0.989	0.010	95
0.005	0.290	0.439	0.492	0.998	0.015	25
0.008	0.291	0.454	0.501	0.999	0.015	19
Cloprop-ZLH nanocomposite						
Na_3PO_4 (mol·L^{-1})	Zeroth Order	First Order	Pseudo second Order	Parabolic Diffusion		
	r^2			r^2	k	$t_{1/2}$
0.002	0.961	0.973	0.0067	0.956	0.014	1801
0.005	0.812	0.886	0.758	0.930	0.027	408
0.008	0.825	0.974	0.979	0.956	0.031	207

Unit: k(mg^{-1}Ls^{-1}), $t_{1/2}$(min).

obviously because as the concentration of the phosphate increases, more phosphate anions are available to be ion exchanged with cloprop anion, resulting in lower values of $t_{1/2}$. This pattern is similar to the $t_{1/2}$ values for the release of acetochlor from clay/CMC gel formulations [26]. The results demonstrate the zinc layered hydroxide and layered double hydroxide material are effective inorganic matrix for the herbicides anions storage and possesses controlled delivery at the required time, which is recommendable for the agricultural purposes.

4. CONCLUSION

The intercalation of cloprop phenoxyherbicides anion into Zn/Al-layered double hydroxide and zinc-layered hydroxide using co-precipitation and direct reaction method, respectively, was synthesized. This was proven through the expansion of d spacing in PXRD pattern with value of 21.0 Å (cloprop-LDH) and 22.7 Å (cloprop-ZLH). Percentage release of chlorprop from interlayer of cloprop-ZLH nanocomposite was slightly higher than that from cloprop-LDH nanocomposite in controlled release study due to the high crystallinity of cloprop-ZLH. The release behavior of cloprop from its nanocomposites into phosphate solution followed the pseudo-second-order for cloprop-LDH and parabolic diffusion for cloprop-ZLH. The present study shows that both layered hydroxide materials are compatible inorganic matrix proposed for controlled release formulation of herbicides in agriculture.

ACKNOWLEDGEMENTS

The support of the research by KPM under RACE Grant no. 2012-0151-101-62. NH thanks UPSI for all affords and support in this research.

REFERENCES

[1] Cardoso, L.P. and Valim, J.B. (2006) Study of acids removal by calcined Mg-Al-CO$_3$-LDH. *Journal of Physics and Chemistry of Solid*, **67**, 987-993.
http://dx.doi.org/10.1016/j.jpcs.2006.01.015

[2] Hermosin, M.C., Calderon, M.J., Aguer, J.P. and Cornejo, J. (2001) Organoclays for controlled release of the herbicide fenuron. *Pest Management Science*, **57**, 803-809.
http://dx.doi.org/10.1002/ps.359

[3] Cavani, F., Triffiro, F. and Vaccani, A. (1991) Hydrotalcite-type anionic clays: Preparation, properties and application. *Catalysis Today*, **11**, 173-301.
http://dx.doi.org/10.1016/0920-5861(91)80068-K

[4] Reichle, W.T. (1986) Synthesis of anionic clay minerals (mixed metal hydroxides, hydrotalcite). *Solid State Ionics*, **22**, 135-142.
http://dx.doi.org/10.1016/0167-2738(86)90067-6

[5] Constantino, V.R.L. and Pinnavaia, T.J. (1995) Basic properties of Mg^{2+}$_{1-x}$Al^{3+}$_x$ layered double hydroxide intercalated by carbonate, hydroxide, chloride and sulphate ions. *Inorganic Chemistry*, **34**, 883-892.
http://dx.doi.org/10.1021/ic00108a020

[6] Khan, S.B., Liu, C., Jang, E.S., Akhtar, K. and Han, H. (2011) Encapsulation of organic UV ray absorbents into layered double hydroxide for photochemical properties. *Materials Letters*, **65**, 2923-2926.
http://dx.doi.org/10.1016/j.matlet.2011.03.107

[7] Isa, I.M., Dahlan, S.N.A., Hashim, N., Ahmad, M. and Ghani, S.A. (2012) Electrochemical sensor for cobalt(II) by modified carbon paste electrode with Zn/Al-2(3-Chlorophenoxy) propionate Nanocomposite. *Internation-*

al Journal of Electrochemical Science, **7**, 7797-7808.

[8] Isa, I.M., Sohaimi, N.M., Hashim, N. Kamari, A., Mohamed, A., Ahmad, M., Ghani, S.A. and Suyanta (2013) Determination of salicylate ion by potentiometric membrane electrode based on zinc aluminium layered double hydroxides-4(2,4-dichlorophenoxy)butyrate nanocomposite. *International Journal of Electrochemical Science*, **8**, 2112-2121.

[9] Li, J., Li, Y. and Dong, H. (2008) Controlled release of herbicide acetochlor from clay/carboxylmethylcellulose gel formulations. *Journal of Agricultural and Food Chemistry*, **56**, 1336-1342.

[10] Cordeiro, C.S., Arizaga, G.G.C., Ramos, L.P. and Wypych, F. (2008) A new zinc hydroxide nitrate heterogeneous catalyst for the esterification of free fatty acids and the transesterification of vegetable oils. *Catalysis Communications*, **9**, 2140-2143. http://dx.doi.org/10.1016/j.catcom.2008.04.015

[11] Mohsin, S.M.N., Hussein, M.Z., Sarijo, S.H., Fakurazi, S., Arulselvan, P. and Hin, T.Y.Y. (2013) Synthesis of (cinnamate-zinc layered hydroxide) intercalation compound for sunscreen application. *Chemistry Central Journal*, **7**, 26. http://dx.doi.org/10.1186/1752-153X-7-26

[12] Hussein, M.Z., Hashim, N., Yahaya, A.H. and Zainal, Z. (2009) Controlled release formulation of pesticide agrochemical based on 4-(2,4-dichlorophenoxy)butyrate nanohybrid. *Journal of Nanoscience and Nanotechnology*, **9**, 2140-2147. http://dx.doi.org/10.1166/jnn.2009.445

[13] He, J., Wei, M., Li, B., Kang, Y., Evans, D.G. and Duan, X. (2006) *Layered Double Hydroxide*. Structure and bonding: Preparation of layered double hydroxides. Springer-Verlag, Berlin, Heidelberg. http://dx.doi.org/10.1007/430_006

[14] Xu, S. and Wang, Z.L. (2011) One-dimensional ZnO nanostructures: Solution growth and functional properties. *Nano Research*, **4**, 1013-1098. http://dx.doi.org/10.1007/s12274-011-0160-7

[15] Ragavan, A., Khan, A.I. and O'Hare, D. (2006) Intercalation and controlled release of 2,4-dichlorophenoxyacetic acid using rhombohedral [LiAl$_2$(OH)$_6$]Cl.xH$_2$O. *Journal of Physics Chemistry and Solids*, **67**, 983-968. http://dx.doi.org/10.1039/b610766d

[16] Hussein, M.Z., Sarijo, S.H., Yahaya, A.H. and Zainal, Z. (2007) Synthesis of 4-chlorophenoxyacetate-zinc-aluminium layered double hydroxide nanocomposite: Physicochemical and controlled release properties. *Journal of Nanoscience and Nanotechnology*, **7**, 2852-2862. http://dx.doi.org/10.1166/jnn.2007.613

[17] Hussein, M.Z., Jubri, Z.B., Zainal, Z. and Yahaya, A.H. (2004) Pamoate intercalated Zn-Al layered double hydroxide for the formation of layered organic-inorganic intercalate. *Materials Science-Poland*, **22**, 57-67.

[18] Ambrogi, V., Fardella, G. and Grandolini, G. (2001) Intercalation compounds of hydrotalicite-like anionic clays with anti flammatory agents: I. Intercalation and *in vitro* release of ibufuren. *International Journal of Pharmacy*, **220**, 23-32. http://dx.doi.org/10.1016/S0378-5173(01)00629-9

[19] Hussein, M.Z., Zainal, Z., Yahaya, A.H. and Foo, D.W.V. (2002) Controlled release of plant growth regulator α-naphthalane-acetate from the lamella of Zn-Al-layered double hydroxide nanocomposite. *Journal of Controlled Release*, **82**, 417-427. http://dx.doi.org/10.1016/S0168-3659(02)00172-4

[20] Gunawan, P. and Xu, R. (2008) Direct control of drug release behavior from layered double hydroxides through particle interactions. *Journal of Pharmaceutical Sciences*, **97**, 4367-4378. http://dx.doi.org/10.1002/jps.21321

[21] Kodama, T., Harada, Y., Ueda, M., Shimizu, K., Shuto, K. and Komarneni, S. (2001) Selective exchange and fixation of strontium ions with ultrafine Na-4-mica. *Langmuir*, **17**, 4881-4886. http://dx.doi.org/10.1021/la001774w

[22] Costa, P. and Lobo, J.M.S. (2001) Modeling and comparison of dissolution profiles. *European Journal of Pharmaceutical Sciences*, **13**, 123-133. http://dx.doi.org/10.1016/S0928-0987(01)00095-1

[23] Qiu, H., Lv, L., Pan, B.C., Zhang, Q.J., Zhang, W.M. and Zhang, Q.X. (2009) Critical review in adsorption kinetic models. *Journal of Zhejiang University Science A*, **10**, 716-724. http://dx.doi.org/10.1631/jzus.A0820524

[24] Ho, Y.S. (2006) Review of second-order models for adsorption systems. *Journal of Hazardous Materials*, **B136**, 681-689. http://dx.doi.org/10.1016/j.jhazmat.2005.12.043

[25] Kong, X., Shi, S., Han, J., Zhu, F., Wei, M. and Duan, X. (2010) Preparation of Glycy-l-Tyrosine intercalated layered double hydroxide film and its *in vitro* release behavior. *Chemical Engineering Journal*, **157**, 598-604. http://dx.doi.org/10.1016/j.cej.2010.01.016

[26] Li, W., Yan, D., Gao, R., Lu, J., Wei, M. and Duan, X. (2013) Recent advances in stimuli-responsive photofunctional materials based on accommodation of chromophore into layered double hydroxide nanogallery. *Journal of Nanomaterials*, **2013**, 14 p. http://dx.doi.org/10.1155/2013/586462

25

Synthesis, characterization and biocidal properties of platinum metal complexes derived from 2,6-diacetylpyridine (bis thiosemicarbazone)

Monika Tyagi, Sulekh Chandra

Department of Chemistry, Zakir Husain Delhi College, University of Delhi, New Delhi, India
Email: schandra_00@yahoo.com, mnk02tyg@yahoo.co.in

ABSTRACT

The coordination compounds of Pd^{II}, Pt^{II}, Rh^{III} and Ir^{III} metal ions with a Schiff base ligand (L) *i.e.* 2,6-diacetylpyridine bis(thiosemicarbazone) have been synthesized and characterized by elemental analyses, molar conductance, magnetic susceptibility measurements, IR, NMR and electronic spectral studies. On the basis of molar conductance and elemental analyses the complexes were found to have composition $[M(L)]Cl_2$ and $[M'(L)Cl]Cl_2$, where M = Pd(II), Pt(II) and M' = Rh(III), Ir(III). The spectral studies reveal that the complexes possess monomeric composition. Complexes of Pd^{II} and Pt^{II} were found to have four coordinated square planar geometry whereas the complexes of Rh^{III} and Ir^{III} posses six coordinated octahedral geometry. The ligand field parameters were calculated using various energy level diagrams. *In vitro* synthesized compounds and metal salts have been tested against some species of plant pathogenic fungi and bacteria in order to assess their antimicrobial properties.

Keywords: Schiff Base; Platinum Metal Complexes; Spectral Studies; Biological Screening

1. INTRODUCTION

Schiff's bases are widely studied because of increasing recognition in biological systems [1]. Schiff's bases and their complexes are used in some chemical processes as catalysts. They are also used in biological models to understand the structure of bio-molecules [2]. Schiff bases derived from thiosemicarbazide and their metal complexes are of great significance for their pharmacological properties such as antibacterial, antifungal, antitumoral, antiviral and anticancer [3-9]. After the discovery of the chemotherapeutically active cisplatin large number of metal complexes with thiosemicarbazide derivatives were

synthesized [10]. It is well known that several metal ions enhance the biological activities of thiosemicarbazone particularly the metals of platinum groups. The chemistry of rhodium and iridium has been receiving considerable current attention largely because of the interesting chemical properties exhibited by the complexes of these two metals [11]. Though neither of these two metals are bio-essential elements but their compounds have useful applications in the biological field [12]. Due to the growing interest of palladium, platinum, rhodium and iridium metal complexes of thiosemicarbazones, here we report the synthesis, characterization and biological activities of Pd(II), Pt(II), Rh(III) and Ir(III) complexes of Schiff base ligand derived from 2,6-diacetylpyridine and thiosemicarbazide. The results obtained from antimicrobial activities were compared with standard antifungal drug: Amphotericin-B and antibiotic: Streptomycin.

2. CHEMISTRY

All the chemicals used were of Anala R grade and procured from Sigma-Aldrich and Fluka. Metal salts were purchased from E. Merck and were used as received. They include $PdCl_2$, $PtCl_2$, $RhCl_2$, $RuCl_2$, 2,6-diacetylpyridine and thiosemicarbazide. The solvents used were ethanol, deutrated dimethyl sulfoxide (d6-DMSO), acetone and benzene.

2.1. Synthesis of the Ligand (L)

The ligand L was prepared by the mixing of hot ethanolic solution (20 mL) of thiosemicarbazide (1.82 g, 0.02 mol) and an ethanolic solution (20 mL) of 2,6-diacetylpyridine (1.63 g, 0.01 mol) with constant stirring in the presence of few drops of conc. HCl. This mixture was refluxed for 2 h at 80°C and then allowed to cool overnight at 0°C. The isolated light yellow colored precipitate was filtered. It was washed with cold EtOH and dried under *vacuum* over P_4O_{10}. Yield (67%), mp 265°C. Element chemical analysis data is shown in **Table 1**. The scheme of synthesis of ligand is shown in **Figure 1**.

2.2. Synthesis of Pd(II) and Pt(II) Complexes

Hot ethanolic solution (20 mL) of ligand (1 mmol) and hot ethanolic solution (20 mL) of the corresponding metal salts (1 mmol) was mixed with continuous stirring. The resulting solution was refluxed for 6 - 8 h at 85°C. On cooling the colored product was precipitated out. It was filtered, washed with cold ethanol and dried under *vacuum*. Purity of the complexes were checked by TLC.

2.3. Synthesis of Rh(III) and Ir(III) Complexes

A hot ethanolic solution (20 mL) the ligand (1 mmol) was added slowly to the hot ethanolic solution (20 mL) of the corresponding metal salts (1 mmol) with continuous stirring. The resultant colored solution was refluxed for 10 - 12 h at 80°C. On cooling the colored precipitate of complex was obtained, which was filtered, washed thoroughly with cold ethanol and dried under *vacuum*. Purity of the complexes were checked by TLC.

2.4. Physical Measurements

C, H and N contents were analyzed on a Carlo-Erba 1106 elemental analyzer. NMR spectra were recorded with a model Bruker Avance DPX-300 spectrometer operating at 300 MHz using DMSO-d_6 using as a solvent and TMS as an internal standard. IR spectra were recorded as CsI discs on FT-IR spectrum BX-II spectrophotometer. The electronic spectra were recorded on Shimadzu UV mini-1240 spectrophotometer using DMSO as a solvent. The molar conductance of complexes was measured on the ELICO (CM82T) conductivity bridge. Magnetic susceptibility was measured at room temperature on a Gouy balance using $CuSO_4 \cdot 5H_2O$ as a calibrant.

3. PHARMACOLOGY

3.1. Test Microorganisms

Some fungal species *i.e. Aspergillus niger, Aspergillus fumigates, Fusarium odum* and bacterial species *i.e. Staphyloccocus aureus, Escherichia coli* were screened for antifungal and antibacterial activities of the synthesized compounds.

3.2. Medium

Two solid media namely Potato Dextrose Agar (PDA) (for fungus) and Nutrient agar (NA) (for bacteria) were used for antimicrobial assay.

Table 1. Elemental analyses and molar conductance for Pd(II), Pt(II), Rh(III) and Ir(III) complexes with investigated ligand.

Compounds	Tentative formula	m.p. °C	Color	% Found (Calculated)				Molar conduct. $\Omega^1 \cdot cm^2 \cdot mol$
				C	H	N	M	
Ligand (L)	$C_{11}H_{15}N_7S_2$	265	Light yellow	42.68 (42.72)	4.82 (4.85)	31.74 (31.71)	-	-
[Pd(L)]Cl$_2$	PdC$_{11}$H$_{15}$N$_7$S$_2$Cl$_2$	280	Reddish green	27.22 (27.16)	3.05 (3.08)	20.13 (20.16)	21.85 (21.81)	65
[Pt(L)]Cl$_2$	PtC$_{11}$H$_{15}$N$_7$S$_2$Cl$_2$	290	Light brown	22.92 (22.96)	2.52 (2.60)	17.09 (17.04)	33.86 (33.91)	60
[Rh(L)Cl]Cl$_2$	RhC$_{11}$H$_{15}$N$_7$S$_2$Cl$_3$	286	Rusty brown	25.50 (25.48)	2.86 (2.89)	19.00 (18.92)	19.93 (19.88)	57
[Ir(L)Cl]Cl$_2$	IrC$_{11}$H$_{15}$N$_7$S$_2$Cl$_3$	275	Black	21.79 (21.75)	2.53 (2.47)	16.09 (16.14)	31.60 (31.63)	62

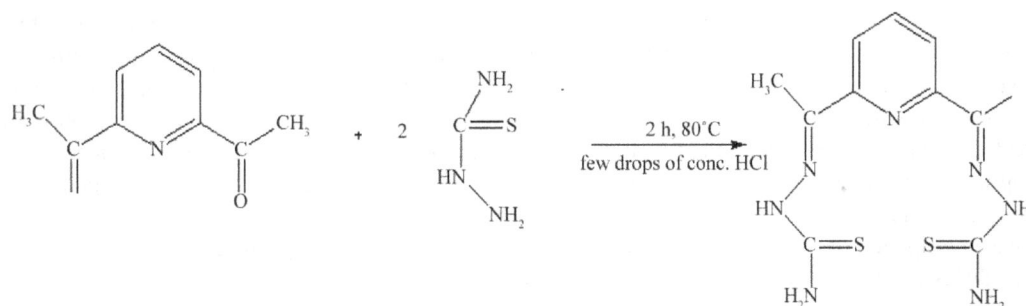

Figure 1. Synthesis of ligand (L).

3.3. *In Vitro* Antimicrobial Activity

3.3.1. Antifungal Screening

The preliminary fungitoxicity screening of the compounds at different concentrations were performed using Food Poison Method [13,14]. Stock solutions of compounds were prepared by dissolving the compounds in DMSO. Amphotericin-B used as antifungal drug and DMSO served as control. Potato dextrose agar medium was prepared by using potato, dextrose, agar-agar and distilled water. Appropriate quantities of the compounds in DMSO was added to potato dextrose agar medium in order to get a concentrations of 100, 200 and 300 ppm of compound in the medium. The medium was poured into a set of two petriplates under aseptic conditions in a laminar flow hood. When the medium in the plates was solidified, a mycelial discs of 0.5 cm in diameter cut from the periphery of the 7 days old culture and it was aseptically inoculated upside down in the centre of the petriplates. These treated petriplates were incubated at 26°C ± 1°C until fungal growth in the control petriplates was almost complete.

The mycelial growth of fungi (mm) in each petriplates was measured diametrically and growth inhibition (I) were calculated by using the formula:

$$I(\%) = \frac{C - T}{C} \times 100$$

where 90 is the diameter (mm) of the petriplates, and C is the growth of the fungus (mm) in control and T is the growth of test compounds.

3.3.2. Antibacterial Screening

The antibacterial activities were evaluated by the Disc Diffusion Method using nutrient agar medium [15,16]. Nutrient agar medium was prepared by using peptone, beef extract, NaCl, agar-agar and distilled water. The test compounds in measured quantities were dissolved in DMSO to get a concentrations of 100, 200 and 300 ppm of compounds. 25 mL nutrient agar media (NA) was poured in each petriplates. After solidification 0.1 mL of test bacteria spreads over the medium using a spreader. The disc of Whatmann no 1 filter paper having the diameter 5.00 mm each containing (1.5 mg·cm^{-1}) of compounds were placed at 4 equidistant places at a distance of 2 cm from the center in the inoculated petriplates. Filter paper disc treated with DMSO served as control and Streptomycin used as a standard drug. All determination were made in duplicate for each of the compounds. Average of two independent readings for each compounds was recorded. These petriplates were kept in refrigerator for 24 hrs for Pre-diffusion. Finally petriplates were incubated for 26 - 30 h 28°C ± 2°C. The zone of inhibition was calculated in mm carefully.

4. RESULTS AND DISCUSSION

The stoichiometric data of complexes reveal that the complexes have 1:1 metal to ligand ratio. The formation of metal complexes is expressed by the following chemical equation:

$$L + MCl_2 \xrightarrow{\text{Reflux}} LMCl_2$$
$$L + M'Cl_3 \xrightarrow{\text{Reflux}} LM'Cl_3$$

where L = 2,6-diacetylpyridine bis(thiosemicarbazone) and M = Pd(II), Pt(II); M' = Rh(III) and Ir(III). On the basis of elemental analysis, the complexes were found to have the composition as given in **Table 1**. The analytical data are in a good agreement with the proposed stoichiometry of the complexes. These complexes are air-stable, insoluble in H_2O and most of organic solvents, but soluble in DMSO and DMF. Molar conductance values of Pd(II) and Pt(II) complexes in DMSO solution correspond to 1:2 electrolytic nature whereas Rh(III) and Ir(III) complexes correspond 1:1 electrolyte nature [17]. Thus these complexes may be formulated as $[M(L)]Cl_2$ and $[M'(L)Cl]Cl_2$.

4.1. Mass Spectrum

The mass spectrum of ligand displays the parental ion peak (M$^+$) at m/z = 309 (78%) and a weak peak at m/z = 310 due to ^{13}C and ^{15}N isotopes. The peak at m/z = 79 (100%) is due to pyridine ring. The other positive ions give the peaks at 295, 248, 193, 170, 121, 85, 70 and 16 mass numbers. The intensities of these peaks give the idea of the stabilities of the fragments.

4.2. NMR Spectra

The ^1H NMR spectra of Schiff base ligand and its metal complexes were recorded in DMSO-d_6 and are given in **Table 2**. The resonance for two methyl groups appeared as a singlet at δ 2.27 ppm in the ligand and no significant change was observed in complexes s [18]. In the ligand a sharp singlet at δ 2.48 ppm is due to –NH$_2$ group. The position of this singlet remains unchanged in the complexes. It shows that the –NH$_2$ group is not taking part in complexation. Significant azomethine proton signal due to HC=N was observed at δ 8.02 ppm in Schiff base ligand. On complexation the position of this signal is shifted to δ 8.18 - 8.30 ppm. It indicates that azomethine nitrogen involved in coordination. The proton peak of N–H group at δ 10.8 - 11.2 ppm remains at same position in ligand and in the complexes which suggests that deprotonation do not occurred [19].

The ^{13}C NMR spectra revealed the presence of expected number of signals corresponding to different types of carbon atoms present in the compounds. The Schiff base ligand shows the signal at δ 21.42 ppm due to carbon atoms of methyl groups. On complexation no change has been observed. The spectra of the Schiff base ligand exhibits a strong band at δ 179.2 ppm due to C=S group. On complex formation the position of this band undergoes upfield shift to δ 171.9 - 172.5 ppm. This indicates that sulphur is involved in coordination.

4.3. IR Spectra

The IR absorption bands, which provide information about the formation of Schiff base ligand and the mode of coordination in its complexes, are given in **Table 3**. The ligand displays an intense band at 1590 cm^{-1} corresponding to v(C=N) stretching vibration. On complex formation, the position of v(C=N) band shows the negative shift which indicates that the nitrogen atom of azomethine group is coordinated to metal ion [20,21]. The spectum of the ligand shows a strong band at 820 cm^{-1} due to v(C=S) group. In the spectra of the complexes the position of this band suffers a downward shift suggesting coordination of the metal ion through sulphur atom of thiosemicarbazone group [22].

Moreover, the Schiff base ligand shows IR spectral bands at 1505, 595 and 475 cm^{-1} due to pyridine ring-stretching, in-plane-ring-bending and out-of-plane-ring-bending vibrations, respectively. In the IR spectra of Rh(III) and Ir(III) complexes, these bands show substantial positive shift with fairly low intensity indicating the involvement of nitrogen atom of pyridine entity present in the ligand to central metal atom. On the other hand, in the IR spectra of Pd(II) and Pt(II) complexes the position of this band remain unchanged which indicates the presence of uncoordinated pyridine ring. This discussion

reveals that the Schiff base ligand possesses five potential coordination sites (SNNNS) for Rh(III) and Ir(III) complexes and four potential coordination sites (SNNS) for Pd(II) and Pt(II) complexes. The complexes show the new bands in the range 345 - 405 cm^{-1} and 309 - 320 cm^{-1} which may be due to v(M-N) and v(M-S) stretching vibrations, respectively [23]. The chloro complexes show the bands in the region 310 - 326 cm^{-1} corresponding v(M-Cl) [24].

4.4. Magnetic Moments and Electronic Spectra

4.4.1. Pd(II) and Pt(II) Complexes

The complexes are diamagnetic as expected for square planar d^8 systems. The electronic spectra of the complexes under study display two bands at 357 - 381 nm and 523 - 537 nm. These bands may be assigned to $^1A_{1g} \rightarrow {}^1A_{2g}(v_1)$ and $^1A_{1g} \rightarrow {}^1B_{1g}(v_2)$ transitions, respectively. The electronic spectra of these complexes indicate the square planar geometry around the Pd(II) and Pt(II) ion [25,26]. By assuming a value of $F_2 = 10F_4 = 600$ for Slater-Condon interelectronic repulsion parameters for both Pd and Pt, it is possible to calculate the value of Δ_1 (**Table 4**) from the first spin allowed d-d transition. The splitting parameter increases in the expected order Pt > Pd.

4.4.2. Rh(III) and Ir(III) Complexes

The complexes are also diamagnetic. The electronic spectrum of Rh(III) complex displays bands at 357 nm, 478 nm and 572 nm. These bands may be assigned $^1A_{1g} \rightarrow {}^3T_{1g}$, $^1A_{1g} \rightarrow {}^1T_{1g}$ and $^1A_{1g} \rightarrow {}^1T_{2g}$, transitions respectively corresponding to an octahedral geometry [27]. The

Table 2. NMR spectral data (δppm) of Pd(II), Pt(II), Rh(III) and Ir(III) complexes with investigated ligand (L).

Compounds	^{1}H NMR				^{13}C NMR		
	δ(NH$_2$)	δ(NH)	δ(HC=N)	δ(CH$_3$)	δ(C=N)	δ(C=S)	δ(CH$_3$)
Ligand (L)	2.48	10.9	8.02	2.27	165.1	179.2	21.42
[Pd(L)]Cl$_2$	2.49	11.2	8.25	2.29	160.5	171.9	21.40
[Pt(L)]Cl$_2$	2.48	11.0	8.18	2.26	161.2	172.5	21.42
[Rh(L)Cl]Cl$_2$	2.47	10.8	8.22	2.29	160.8	172.1	21.41
[Ir(L)Cl]Cl$_2$	2.49	10.9	8.30	2.30	161.6	171.3	21.40

Table 3. IR band assignment (cm^{-1}) of Pd(II), Pt(II), Rh(III) and Ir(III) complexes with investigated ligand (L).

Compounds	v(C=N)	v(C=S)	Pyridine ring	v(M-N)	v(M-S)
Ligand (L)	1590 ms	820 s	1505 vs, 595 mw, 475 mw	-	-
[Pd(L)]Cl$_2$	1565 s	802 ms	1505 vs, 595 mw, 475 mw	390 sh	315 w
[Pt(L)]Cl$_2$	1558 s	795 m	1505 vs, 595mw, 475 mw	345 ms	309 m
[Rh(L)Cl]Cl$_2$	1571 mw	812 sh	1552 s, 613 w, 530 mw	405 w	317 sh
[Ir(L)Cl]Cl$_2$	1660 m	790 m	1529 w, 630 w, 513 m	370 m	320 mw

m = medium, ms = medium strong, vs = very strong, sh = sharp, w = weak, s = strong , mw = medium weak.

B and Dq values are calculated from the positions of their electronic bands using the following equations:

$$\nu_1 = 10Dq - 4B + \frac{86(B)^2}{10Dq}, \quad \nu_2 = 10Dq + 12B + \frac{2(B)^2}{10Dq}.$$

The decrease in B values from the free ion value suggests that there is a considerable orbital overlap with strong covalency in the metal ligand σ bond [28].

The electronic spectrum of the iridium(III) complex displays bands at 292 nm and 324 nm, which may be assigned to $^1A_{1g} \rightarrow {}^1T_{1g}(\nu_1)$ and $^1A_{1g} \rightarrow {}^1T_{2g}(\nu_2)$ transitions in order of increasing energy. These transitions are used to evaluate the ligand field parameters (**Table 4**). The values of these ligand field parameters are compared with those reported for other iridium(III) complexes involving similar donor atoms [29].

4.5. Suggested Structure of the Complexes

On the basis of above mentioned discussions using elemental analysis, mass, NMR and IR spectral studies the following (**Figure 2**) structure may be suggested for the complexes.

4.6. Biological Results and Discussion

The antimicrobial screening data show that the metal chelates exhibit a higher inhibitory effect than the free ligand and metal salts (**Tables 5** and **6**). The increased activity of the metal chelates can be explained based on the chelation theory [30]. The chelation reduces the polarity of the metal atom mainly because of the partial sharing of its positive charge with the donor groups and possible p electron delocalization with in the whole chelation ring. The chelation ring increases the lipophillic nature of the central atom which subsequently favours its permeation through the lipid layer of the cell memberane [31]. The enhanced activity of the complexes can also be explained on the basis of their high solubility fineness of the particles, size of the metal ion and the presence of bulkier organic moieties. The mode of action may involve the formation of a hydrogen bond through the azomethane nitrogen atom with the active centers of the cell constituents, resulting in interference with the normal cell process. The variation in the effectiveness of different compounds against different organisms depend either on the impermeability of the cells of the microbes or difference in ribosomes of microbial cells. It has also been proposed that concentration plays a vital role in increasing the degree of inhibition; as the concentration increases, the activity increases [32].

The results of fungicidal screening (**Figure 3**) show that Pd(II) complex is highly active as compared to free

Table 4. Electronic spectral bands (cm^{-1}) and ligand field parameters of the complexes.

Complexes	λ_{max} (nm)	Dq (cm^{-1})	B (cm^{-1})	C (cm^{-1})	B	ν_2/ν_1	Δ_1 (cm^{-1})
[Pd(L)]Cl$_2$	537,357	-	-	-	-	-	20421
[Pt(L)]Cl$_2$	523,381	-	-	-	-	-	20920
[Rh(L)Cl]Cl$_2$	572,478,357	2265	444	1775	0.62	1.34	-
[Ir(L)Cl]Cl$_2$	324,292	3085	212	850	0.32	1.11	-

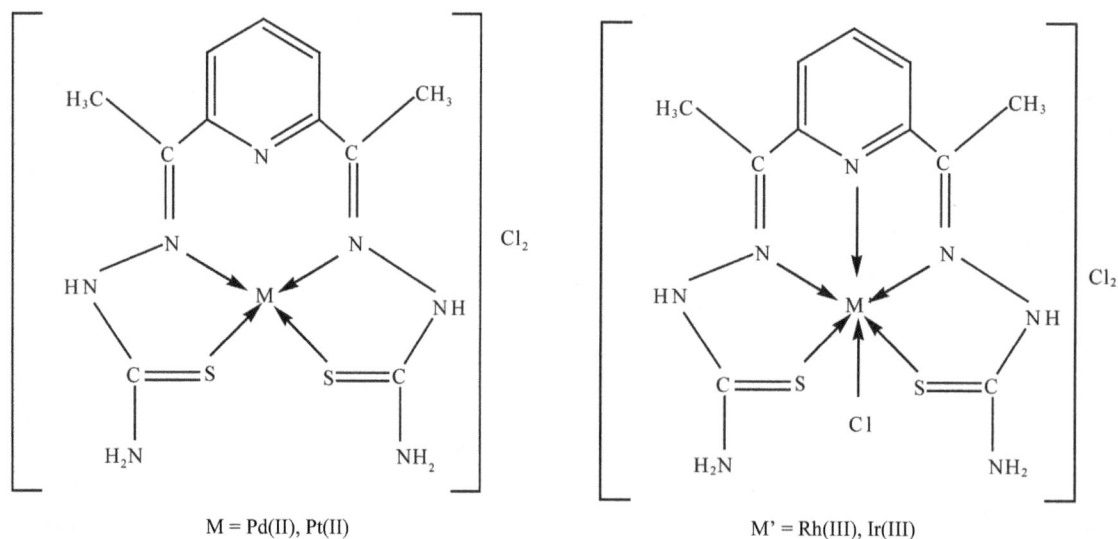

M = Pd(II), Pt(II)   M' = Rh(III), Ir(III)

Figure 2. Suggested structure of the complexes.

Table 5. Antifungal screening data of investigated ligand (L) and its Pd(II), Pt(II), Rh(III), Ir(III) complexes.

Compound no.	Compounds	Fungal Inhibition (%) (conc. in $\mu g \cdot ml^{-1}$)								
		Aspergillus niger			*Aspergillus fumigates*			*Fusarium odum*		
		100	200	300	100	200	300	100	200	300
1.	Ligand (L)	32	46	58	38	57	68	29	35	49
2.	[Pd(L)]Cl₂	51	67	85	56	78	89	49	64	75
3.	[Pt(L)]Cl₂	49	62	80	54	75	87	45	65	72
4.	[Rh(L)Cl]Cl₂	46	59	77	50	67	82	40	59	70
5.	[Ir(L)Cl]Cl₂	48	57	71	52	70	85	39	49	66
6.	PdCl₂	45	50	65	46	62	78	35	43	59
7.	PtCl₂	42	54	62	49	60	76	35	40	57
8.	RhCl₃	39	47	64	39	55	69	30	41	52
9.	IrCl₃	34	45	59	41	59	73	27	39	50
10.	Amphotericin-B	55	72	89	59	81	92	50	69	80

Table 6. Antibacterial screening data of investigated ligand (L) and its Pd(II), Pt(II), Rh(III), Ir(III) complexes.

Compund no.	Compounds	Diameter of Inhibition Zone (mm) (conc. in $\mu g \cdot ml^{-1}$)					
		Staphyloccocus aureus			*Escherichia coli*		
		100	200	300	100	200	300
1.	Ligand (L)	19	25	31	20	29	33
2.	[Pd(L)]Cl₂	25	29	35	24	31	38
3.	[Pt(L)]Cl₂	28	31	39	27	34	41
4.	[Rh(L)Cl]Cl₂	20	26	33	22	29	35
5.	[Ir(L)Cl]Cl₂	22	27	32	19	26	30
6.	PdCl₂	17	22	26	15	19	25
7.	PtCl₂	18	21	25	17	22	29
8.	RhCl₃	18	24	29	19	24	31
9.	IrCl₃	19	22	28	18	21	26
10.	Streptomycin	26	30	36	29	35	40

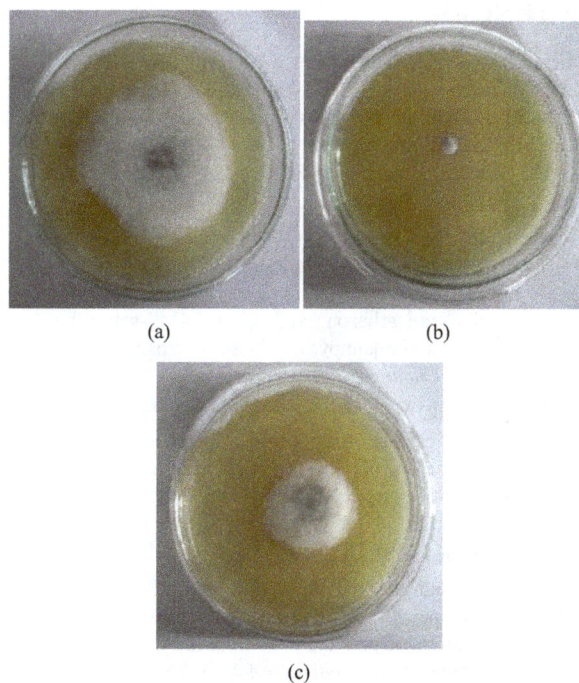

(a) (b)

(c)

Figure 3. Antifungal activities of compounds against *Aspergillus fumigatus* (a) Ligand; (b) [Pd(L)]Cl₂; (c) PdCl₂.

ligand and metal salt against all the fungal species. The order of the fungal and bacterial growth inhibition of the compounds is shown in **Figures 4** and **5**.

5. CONCLUSION

On the basis of NMR, IR and UV palladium(II) and platinum(II) complexes were found with square planar geometry while the rhodium(III) and iridium(III) complexes were found with octahedral geometry. The antifungal and antibacterial screening of the compounds indicates that the Pd(II) complex is highly active while the other metal complexes show moderate activity in comparison of the free ligand and metal salts.

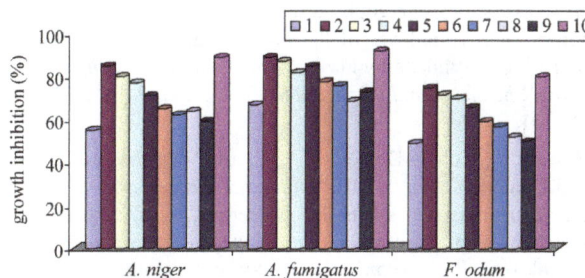

Figure 4. Graphical representation of antifungal activities of the compounds.

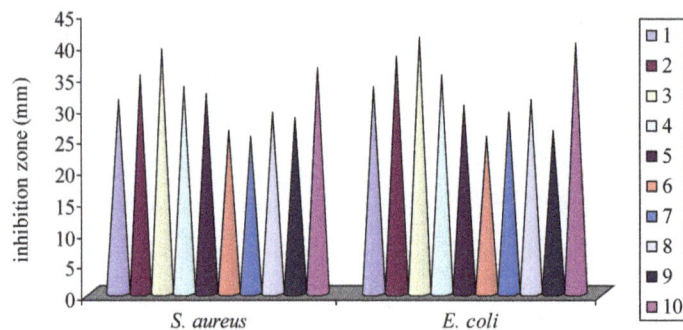

Figure 5. Graphical representation of antibacterial activities of the compounds.

6. ACKNOWLEDGEMENTS

Authors are thankful to the DRDO New Delhi for financial support and Delhi University for recording IR spectra.

REFERENCES

[1] Mishra, D., Naskar, S., Drew, M.G.B. and Chattopadhyay, S.K. (2006) Synthesis, spectroscopic and redox properties of some ruthenium(II) thiosemicarbazone complexes: Structural description of four of these complexes. *Inorganica Chimica Acta*, **359**, 585-592. doi:10.1016/j.ica.2005.11.001

[2] Agarwal, R.K. and Prasad, S. (2005) Synthesis and spectral investigations of some platinum metals ions coordination compounds of 4[N-(furan-2'-carboxalidene)amino] antipyrine thiosemicarbazone and 4[N-(3',4',5'-trime-thoxybenzalidene) amino]antipyrine thiosemicarbazone. *Turkish Journal of Chemistry*, **29**, 289-297.

[3] Patil, S.A., Naik, V.H., Kulkarni A.D. and Badami P.S. (2010) DNA cleavage, antimicrobial, spectroscopic and fluorescence studies of Co(II), Ni(II) and Cu(II) complexes with SNO donor coumarin Schiff bases. *Spectrochimica Acta A*, **75**, 347-354. doi:10.1016/j.saa.2009.10.039

[4] Sharma, K., Singh, R., Fahmi, N. and Singh, R.V. (2010) Microwave assisted synthesis, characterization and biological evaluation of palladium and platinum complexes with azomethines. *Spectrochimica Acta A*, **75**, 422-427. doi:10.1016/j.saa.2009.10.052

[5] Al-Amiery, A.A., Al-Majedy, Y.K., Abdulreazak, H. and Abood, H. (2011) Synthesis, characterization, theoretical crystal structure, and antibacterial activities of some transition metal complexes of the thiosemicarbazone (Z)-2-(pyrrolidin-2-ylidene)hydrazinecarbothioamide. *Bioinorganic Chemistry and Applications*, 2011, 1-6, Article ID 483101.

[6] Wiecek, J., Kovala-Demertzi, D., Ciunik, Z., Zervou, M. and Demertzis, M.A. (2010) Diorganotin complexes of a thiosemicarbazone, synthesis: Properties, x-ray crystal structure, and antiproliferative activity of diorganotin complexes. *Bioinorganic Chemistry and Applications*, 2010, 1-9, Article ID 867195.

[7] Ferraz, K.O., Wardell, S.M.S.V., Wardell, J.L., Louro,

S.R.W. and Beraldo, H. (2009) Copper(II) complexes with 2-pyridineformamide-derived thiosemicarbazones: Spectral studies and toxicity against *Artemia salina*. *Spectrochimica Acta A*, **73**, 140-145. doi:10.1016/j.saa.2009.02.020

[8] Graminha, A.E., Batista, A.A., Mendes, I.C., Teixeira, L.R. and Beraldo, H. (2008) Ruthenium(II) complexes containing 2-pyridineformamide- and 2-benzoylpyridine-derived thiosemicarbazones and PPh₃: NMR and electrochemical studies of *cis-trans*-isomerization. *Spectrochimica Acta A*, **69**, 1277-1282. doi:10.1016/j.saa.2007.07.005

[9] Raja, N. and Ramesh, R. (2010) Mononuclear ruthenium(III) complexes containing chelating thiosemicarbazones: Synthesis, characterization and catalytic property. *Spectrochimica Acta A*, **75**, 713-718. doi:10.1016/j.saa.2009.11.044

[10] Bakir, J.A. and Dissouky, E.I. (2005) Synthesis, spectroscopic and the biological activity studies of thiosemicarbazones containing ferrocene and their copper(II) complexes. *Journal of Coordination Chemistry*, **58**, 1029-1038. doi:10.1080/00958970500096975

[11] Chandra, S., Tyagi, M. and Agrawal, S. (2010) Spectral and antimicrobial studies on tetraaza macrocyclic complexes of Pd^II, Pt^II, Rh^III and Ir^III metal ions. *Journal of Saudi Chemical Society*, **15**, 49-54. doi:10.1016/j.jscs.2010.09.005

[12] Chandra, S. and Kumar, A. (2007) Spectral studies on Co(II), Ni(II) and Cu(II) complexes with thiosemicarbazone(L¹) and semicarbazone(L²) derived from 2-acetyl furan. *Spectrochimica Acta A*, **66**, 1347-1351. doi:10.1016/j.saa.2006.04.047

[13] Kumar, U. and Chandra, S. (2010) Biological active cobalt(II) and nickel(II) complexes of 12-membered hexaaza [N₆] macrocyclic ligand: Synthetic and spectroscopic aspects. *E-Journal of Chemistry*, **7**, 1238-1245. doi:10.1155/2010/518723

[14] Chandra, S., Verma, S., Dev, U. and Joshi, N. (2009) Tetraaza macrocyclic complexes: Synthesis, spectral and antimicrobial studies. *Journal of Coordination Chemistry*, **62**, 1327-1335. doi:10.1080/00958970802521076

[15] Refat, M.S., Chandra, S. and Tyagi, M. (2010) Spectroscopic, thermal and biocidal studies on Mn(II), Co(II), Ni(II) and Cu(II) complexes of tridentate ligand having

semicarbazone moieties. *Journal of Thermal Analysis and Calorimetry*, **61**, 261-267.
doi:10.1007/s10973-009-0397-5

[16] Agarwal, R.K., Singh, L. and Sharma, P. (2006) Synthesis, spectral and biological properties of copper(II) complexes of thiosemicarbazones of schiff bases derived from 4-aminoantipyrine and aromatic aldehyde. *Bioinorganic Chemistry and Applications*, 2006, 1-10, Article ID 59509.

[17] Chandra, S., Raizada, S., Tyagi, M. and Sharma, P. (2008) Spectroscopic and biological approach of Ni(II) and Cu(II) complexes of 2-pyridinecarboxald-ehyde thiosemicarbazone. *Spectrochimica Acta A*, **69**, 816-821.
doi:10.1016/j.saa.2007.05.033

[18] Sharma, K., Singh, R. and Fahmi, N. (2010) Microwave assisted synthesis, characterization and biological evaluation of palladium and s with azomethines. *Spectrochimica Acta A*, **75**, 422-427. doi:10.1016/j.saa.2009.10.052

[19] Wehrli, F.W., Marchand, A.P. and Wehrli, S. (1988) Interpretation of Carbon-^{13}NMR Spectra. 2nd Edition, Wiley, New York.

[20] Deepa, K.P. and Aravindakshan, K.K. (2000) Synthesis, characterizationand thermal studies of thiosemicarbazones of N-methylandN-ethylacetoacetanilide. *Synthesis and Reactivity in Inorganic, Metal-Organic, and Nano-Metal Chemistry*, **30**, 1601-1616.

[21] Joseph, M., Sreekanth, A., Suni, V. and Kurup, M.R.P. (2006) Spectral characterization of iron(III) complexes of 2-benzoylpyridineN(4)-substituted thiosemicarbazones. *Spectrochimica Acta A*, **64**, 637-641.
doi:10.1016/j.saa.2005.07.067

[22] Singh, R.V., Biyala, M.K. and Fahmi, N. (2005) Important properties of sulfur-bonded organoboron(III) complexes with biologically potent ligands. *Phosphorus, Sulfur and Silicon and the Related Elements*, **180**, 425-434.
doi:10.1080/104265090509225

[23] Chandra, S. and Gupta, L.K. (2005) EPR, masss, IR, electronic and magnetic studies on copper(II) complexes of semicarbazones and thiosemicarbazones. *Spectrochimica Acta A*, **61**, 269-275.

[24] Bailey, R.A., Kozak, S.L., Michelson, T.W. and Mills, W.N. (1971) Infrared spectra of complexes of the thio ions. *Coordination Chemistry Reviews*, **6**, 407-445.

[25] Singh, S., Bharti, N., Naqvi, F. and Azam, A. (2004) Synthesis, characterization and *in vitro* antiamoebic activity of 5 nitrothiophene-2-carboxaldehyde thiosemicarbazones and their Palladium(II) and Ruthenium(II) complexes. *European Journal of Medicinal Chemistry*, **39**, 459-465.

[26] Chandra, S., Raizada, S. and Rani, S. (2008) Structural and spectral studies of palladium(II) and platinum(II) complexes derived from *N,N,N,N*-tetradentate macrocyclic ligands. *Spectrochimica Acta A*, **71**, 720-724.

[27] Jorgensen, C.K. (1964) Absorption Spectra and Chemical Bonding in Complexes. Pergamon Press, London.

[28] Chandra, S. and Singh, R. (1988) Pd(II), Pt(II), Rh(III), Ir(III) and Ru(III) complexes of some nitrogen-oxygen donor ligands. *Indian Journal of Chemistry*, **27**, 417-420.

[29] Chandra, S., Kumar, S. and Rani, S. (2011) Synthesis, structural, spectral, thermal and antimicrobial studies of palladium(II), platinum(II), ruthenium(III) and iridium(III) complexes derived from *N,N,N,N*-tetradentate macrocyclic ligand. *Spectrochimica Acta A*, **78**, 1507-1514.

[30] El-Behery, M. and El-Twigry, H. (2007) Synthesis, magnetic, spectral, and antimicrobial studies of Cu(II), Ni(II) Co(II), Fe(III), and UO$_2$(II) complexes of a new Schiff base hydrazone derived from 7-chloro-4-hydrazinoquinoline. *Spectrochimica Acta A*, **66**, 28-36.

[31] Tyagi, M. and Chandra, S. (2012) Chromium(III) and manganese(II) complexes of macrocyclic ligand containing thiosemicarbazide moiety: Spectroscopic and biological studies. *Journal of the Indian Chemical Society*, **89**, 147-154.

[32] Chandra, S. and Tyagi, M. (2008) Ni(II), Pd(II) and Pt(II) complexes with ligand containing thiosemicarbazone and semicarbazone moiety: Synthesis, characterization and biological investigation. *Journal of the Serbian Chemical Society*, **73**, 727-734.

26

Synthesis, spectral characterization and biological activities of Organotin(IV) complexes with *ortho*-vanillin-2-hydrazinopyridine (VHP)

Norrihan Sam, Md Abu Affan[*]**, Md Abdus Salam**[*]**, Fasihuddin B. Ahmad, Mohd Razip Asaruddin**

Department of Chemistry, Faculty of Resource Science and Technology, Universiti Malaysia Sarawak, Kota Samarahan, Malaysia.
Email: [*]maaffan@yahoo.com, [*]salambpx@yahoo.com

ABSTRACT

Five new organotin(IV) complexes of *ortho*-vanillin-2-hydrazinopyridine hydrazone with formula $[R_nSn-Cl_{4-n}(VHP)]$ [R = Me_2, n = 2 (2); R = Ph_2, n = 2 (3); R= nBu_2, n = 2 (4); R = nBu, n = 2 (5) and R = 1, n = 0 (6)] have been synthesized by direct reaction of *ortho*-vanillin-2-hydrazinopyridine hydrazone [(VHP), (1)], base and organotin(IV) chloride(s) in absolute methanol. The hydrazone ligand [(VHP), (1)] and its organotin(IV) complexes (2-6) have been characterized by UV-Visible, FT-IR and ^1H NMR spectral studies. Spectroscopic data suggested that in the complexes (2-4), the ligand (1) acted as a neutral bidentate ligand and is coordinated to the tin(IV) atom *via* the azomethine nitrogen and pyridyl nitrogen atoms, whereas the ligand (1) acted as a uninegative tridentate ligand and coordinated to the tin(IV) atom through phenolic-O, azomethine-N and pyridyl-N atoms in complexes (5-6). The toxicity of the ligand (1) and its organotin (IV) complexes (2-6) were determined against *Artemia salina*. Organotin(IV) complexes showed moderate activity against *Artemia salina*. The ligand (1) and its organotin(IV) complexes (2-6) were also tested against four types of bacteria namely *Bacillus cereus*, *Staphylococcus aureus*, *Escherichia coli* and *Enterobacter aerogenes*. All organotin(IV) complexes and the free ligand (1) showed better antibacterial activities against bacteria. Among the organotin(IV) complexes (2-6), diphenyltin(IV) complex (3) showed higher activity against the four types of bacteria.

Keywords: Hydrazone; Organotin(IV) Complexes; Spectral Analyses; Toxicity; Antibacterial Activity

1. INTRODUCTION

Ortho-vanillin is an organic compound which can be found in the extracts and essential oils of many plants (**Figure 1**) [1]. This type of vanillin is differing from ordinary vanillin (4-hydroxy-3-methoxybenzaldehyde) where the hydroxyl group is in the *para*-position.

Many researches have been used 4-hydroxy-3-methoxybenzaldehyde (vanillin) to synthesize of transition metal complexes with hydrazone ligands but less research using *ortho*-vanillin. The *ortho*-vanillin Schiff base derivative and its Cu(II) complexes were conducted by Nives Galić *et al.* [2]. They studied the tautomeric and protonation equilibria of *ortho*-vanillin Schiff base derivative and its Cu(II) complexes. Vanillin-thiosemicarbazone and its organotin(IV) complexes have been synthesized by Singh *et al.* [3]. The author found that all organotin(IV) complexes showed higher activity toward tested bacteria (*Bacillus cereus*, *Nocardia sp.* and *Enterobacter aerogenes*) than the free ligand. Thiyagarajan *et al.* [4] also were synthesized novel hydrazones from piperidine-4-carboxylic acid methyl ester coupled with 2-chloro pyrimidine along with other vanillin derivatives. They stated that the hydrazone derivatives of vanillin possess antibacterial activities.

To the best of our knowledge, no work has been done on the synthesis of organotin(IV) complexes with *ortho*-vanillin-2-hydrazinopyridine ligand. Therefore, the authors are interested to synthesize, characterize and also to study the biological activities of organotin(IV) complexes of *ortho*-vanillin-2-hydrazinopyridine derivatives against *Artemia salina* and different types of bacteria.

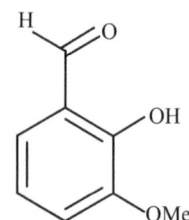

Figure 1. Structure of *ortho*-vanillin (2-hydroxy-3-methoxy- benzaldehyde).

---

[*]Corresponding author.

2. EXPERIMENTAL

2.1. Materials and Methods

2-hydrazinopyridine, 2-hydroxy-3-methoxybenzaldehyde (*ortho*-vanillin) and organotin(IV) salts were purchased from Fluka, Aldrich, Merck and used without further purification. All solvents were purified according to standard procedures [5]. The melting point was measured using open capillary in Stuart MP3. UV-Vis spectra studies were measured using Perkin Elmer Lambda 25 ranging 200 - 800 in DMF. The molar conductance values of all compounds were measured using Jenway 4510 conductivity meter. The FT-IR spectra were obtained on KBr discs using a Perkin Elmer Spectrum GX Fourier-Transform spectrometer (4000 - 370 cm^{-1}). ^{1}H NMR spectra were recorded in DMSO-d_6 solution on a JEOL 500 MHz NMR spectrophotometer.

2.2. Synthesis of *Ortho*-vanillin-2-hydrazinopyri-dine [$C_{13}H_{13}N_3O_2$] (1)

2-hydrazinopyridine (0.546 g, 5 mmol) was dissolved in 20 mL of absolute ethanol before mixing it with 20 mL of ethanolic solution of *ortho*-vanillin (0.761 g, 5 mmol). Then, 2 - 3 drops of glacial acetic acid was added in the reaction mixture. The mixture was heated under reflux for 5 h (**Scheme 1**). The solution was allowed to cool to room temperature for 30 minutes. White light precipitate formed was filtered off and washed several times using absolute ethanol. The white precipitates obtained were purified by recrytallization from hot ethanol and dried *in vacuo* over silica gel. Yield: 1.95 g, 75%, mp: 165°C - 166°C; UV-Visible (DMF) λ_{max}: 335 nm; IR (KBr, cm^{-1}) ν_{max}: 3446 (br, OH), 3191 (s, NH), 1603 (s, C=N), 991 (w, N-N), 728 (s, pyridine in plane); ^{1}H NMR (DMSO-d_6) δ: 10.89 (s, 1H, OH), 9.98 (s, 1H, NH), 8.29 (s, 1H, HC=N), 8.12 (d, 1H, Py-H6), 7.65 (t, 1H, Py-H5), 6.77 - 7.20 (m, 5H, pyridine-H/aromatic-H), 3.80 (s, 3H, CH3) ppm.

2.3. Synthesis of [$Me_2SnCl_2(C_{13}H_{13}N_3O_2)$] (2)

The ligand [(VHP), (1)] (0.486 g, 2 mmol) was dissolved

in 20 mL of absolute methanol in a Schlenk round bottom flask. Then, 10 mL of methanolic solution of potassium hydroxide (0.11 g, 2 mmol) was added dropwise and the colour of the solution changed to light yellow. The resulting solution was refluxed for 1 h under a nitrogen atmosphere. Then, a methanolic solution of dimethyltin(IV) dichloride (0.440 g, 2 mmol) was added dropwise. The solution colour change from light yellow to darker yellow and immediately formed yellow precipitate. The resulting solution was refluxed for 3 hours (**Scheme 2**) and allowed to cool to room temperature. The yellow precipitates obtained were filtered off, washed with pentane, and dried *in vacuo* over silica gel. Yield: 0.62 g, 60% mp: 128°C - 130°C; Molar conductance (DMF) 2.56 Ω-1 cm^2 mol^{-1}; UV-Visible (DMF) λ_{max}: 335, 470, 497 nm; IR (KBr, cm^{-1}) ν_{max}: 3370 (br, OH), 3193 (m, NH), 1604 (s, C=N), 1011 (w, N-N), 730 (s, pyridine in plane), 497 (m, Sn-N); ^{1}H NMR (DMSO-d_6) δ: 10.90 (s, 1H, OH), 10.00 (s, 1H, NH), 8.29 (s, 1H, HC=N), 8.14 (d, 1H, Py-H6), 7.63 (t, 1H, Py-H5), 6.75 - 7.20 (m, 5H, pyridine-H/aromatic-H), 3.80 (s, 3H, CH3), 1.034 (s, 3H, Sn-CH3) ppm.

The other complexes (3-4) were synthesized using a similar procedure to organotin(IV) complexes (2) using appropriate organotin(IV) chloride (s).

2.4. Synthesis of [$Ph_2SnCl_2(C_{13}H_{13}N_3O_2)$] (3)

Yield: 0.83 g, 65% mp: 223°C - 225°C; Molar conductance (DMF) 13.36 Ω^{-1}·cm^2·mol^{-1}; UV-Visble (DMF) λ_{max}: 339, 459, 484 nm; IR (KBr, cm^{-1}) ν_{max}: 3445 (br, OH), 3217 (m, NH), 1623 (s, C=N), 1009 (m, N-N), 733 (s, pyridine in plane), 457 (s, Sn-N); ^{1}H NMR (DMSO-d_6) δ: 10.90 (s, 1H, OH), 9.97 (s, 1H, NH), 8.30 (s, 1H, HC=N), 8.15 (d, 1H, Py-H6), 7.64 (t, 1H, Py-H5), 6.56 - 7.54 (m, 15H, aromatic-H/pyridine-H/Sn-C6H5 protons), 3.80 (s, 3H, CH3) ppm.

2.5 Synthesis of [$nBu_2SnCl_2 (C_{13}H_{13}N_3O_2)$] (4)

Yield: 0.59 g, 58% mp: 151°C - 153°C; Molar conductance (DMF) 16.67 Ω^{-1}·cm^2·mol^{-1}; UV-Visible (DMF) λ_{max}: 338, 485 nm; IR (KBr, cm^{-1}) ν_{max}: 3445 (m, OH), 3188 (m, NH), 1627 (s, C=N), 1013 (w, N-N), 732

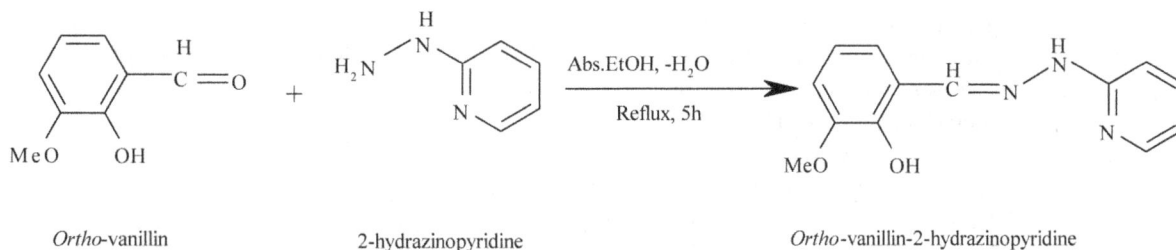

Scheme 1. Synthesis pathway of hydrazone ligand (1).

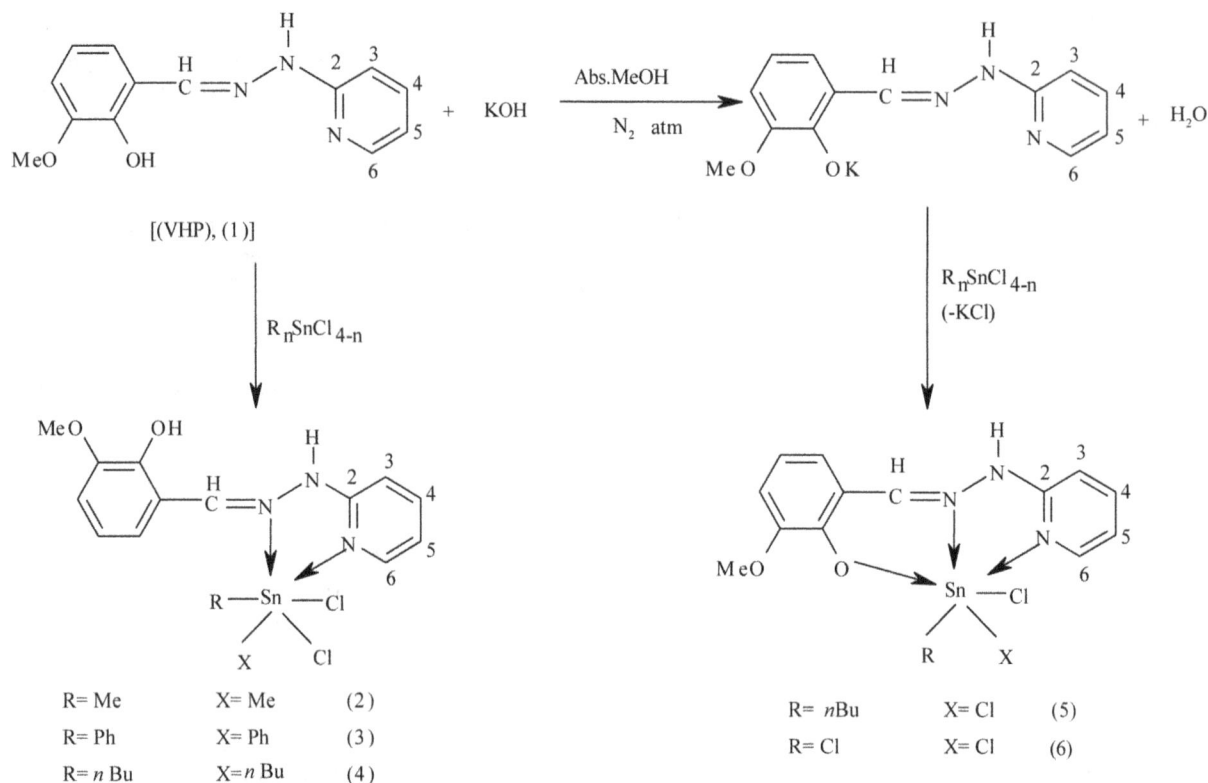

Scheme 2. Synthesis pathway of organotin(IV) complexes (2-6).

(s, pyridine in plane), 499 (w, Sn-N); ^{1}H NMR (DMSO-d_6) δ: 10.86 (s, 1H, OH), 9.95 (s, 1H, NH), 8.39 (s, 1H, HC=N), 8.19 (d, 1H, Py-H6), 7.61 (t, 1H, Py-H5), 6.73 - 7.17 (m, 5H, pyridine-H/aromatic-H), 3.78 (s, 3H, CH3), 0.82 - 1.61 (m, 18H, nBu$_2$-Sn) ppm.

2.6 Synthesis of [nBuSnCl$_2$(C$_{13}$H$_{12}$N$_3$O$_2$)] (5)

The ligand [(VHP), (1)] (0.486 g, 2 mmol) was dissolved in 20 mL of distilled methanol in a Schlenk round bottom flask. Then, 10 mL of methanolic solution of potassium hydroxide (0.11 g, 2 mmol) was added dropwise and the colour of the solution changed to light yellow. The resulting solution was refluxed for 1 h under a nitrogen atmosphere. A solution o f butyltin(IV) trichloride (0.564 g, 2 mmol) in distilled methanol (10 mL) was added dropwise. The solution colour change from light yellow to darker yellow and immediately formed yellow precipitate. The resulting solution was refluxed for 3 hours and allowed to cool to room temperature. The precipitated potassium chloride was removed via filtration and the filtrate was evaporated to dryness. The yellow microcrystals formed were filtered off, washed with pentane, and dried *in vacuo* over silica gel. Yield: 0.75 g, 65% mp: 241°C - 243°C; Molar conductance (DMF) 10.65 $\Omega^{-1}\cdot$cm$^2\cdot$mol^{-1}; UV-Visible (DMF) λ_{max}: 341, 452, 480 nm; IR (KBr, cm^{-1}) ν_{max}: 3449 (br, lattice H$_2$O/OH),

3217 (m, NH), 1626 (s, C=N), 1012 (w, N-N), 733 (s, pyridine in plane), 562 (m, Sn-O), 439 (m, Sn-N); ^{1}H NMR (DMSO-d_6) δ: 9.86 (s, 1H, NH), 8.38 (s, 1H, HC=N), 8.19 (d, 1H, Py-H6), 8.00 (t, 1H, Py-H5), 6.74 - 7.10 (m, 5H, pyridine-H/aromatic-H), 3.81 (s, 3H, CH3), 0.82 - 1.94 (m, 9H, nBu-Sn) ppm.

The complexes (6) were synthesized using a similar procedure to organotin(IV) complexes (5) using stannic(IV) chloride.

2.7 Synthesis of [SnCl$_3$(C$_{13}$H$_{12}$N$_3$O$_2$)] (6)

Yield: 0.78 g, 69% mp: 318°C - 320°C; Molar conductance (DMF) 5.04 $\Omega^{-1}\cdot$cm$^2\cdot$mol^{-1}; UV-Visble (DMF) λ_{max}: 352, 450, 479 nm; IR (KBr, cm^{-1}) ν_{max}: 3436 (br, lattice H$_2$O/OH), 3221 (m, NH), 1630 (s, C=N), 1020 (m, N-N), 730 (s, pyridine in plane), 563 (w, Sn-O), 451 (m, Sn-N); ^{1}H NMR (DMSO-d_6) δ: 9.73 (s, 1H, NH), 8.47 (s, 1H, HC=N), 8.42 (d, 1H, Py-H6), 8.10 (t, 1H, Py-H5), 6.40 - 7.89 (m, 5H, pyridine-H/aromatic-H), 3.82 (s, 1H, CH3) ppm.

2.8. Brine Shrimp Bioassay

The procedures for the brine shrimp bioassay were followed the established method [6] with some modifications. A pinch of brine shrimp eggs (*Artemia salina*) were hatched in treated seawater collected from Pantai

Puteri (Kuching, Sarawak) using a beaker by incubation under a lamp, providing direct light and warmth (24°C - 26°C). The seawater was filtered, placed in autoclave and the salinity measured is 20 psu for hatching.

2.8.1 Sample Preparation

Samples were prepared by dissolving 6 mg of the compounds (1-6) in 6 mL of methanol (stock solution). Mycotoxin solutions were prepared with different concentrations: 1, 5, 10, 50, 100, 150 300 and 500 (μg/mL) by transferring 2, 10, 20, 100, 200, 300, 600 and 1000 μL of stock solution into multiwell plates and air-dried overnight for 24 hours. 50 μL of DMSO and 1 mL of treated seawater were added in each well. About 20 nauplii was pipetted into each well. Then, t he multiwell plates were incubated for 24 hours under direct light at 24°C -26°C. There were one negative control and three replicates per concentration for each of HDMDP ligand (1) and its organotin(IV) complexes (2-6).

2.8.2 LC$_{50}$ Determination

After 24 hours of incubation, a number of dead nauplii in each well were counted. The percentage of death for each concentrations and controls (DMSO, methanol and treated seawater) were determined. If the control death occurred, the mortality was corrected using Abbott's formula [4]. The LC$_{50}$ was determined for each samples from a plot of log samples concentrations versus percentage of death.

2.9 Antibacterial Test

The antibacterial activity was determined using the agar well diffusion method [7]. Grampositive bacteria (*Bacillus cereus* and *Staphylococcus aureus*) and Gramnegative bacteria (*Escherichia coli* and *Enterobacter aerogenes*) were cultivated in nutrient agar on petri dishes. The well was dug in the media with a sterile borer and 18 - 24 h bacterial inoculums containing 0.168 OD was spread on the surface of the nutrient agar using a sterile cotton swab. The samples in the concentration of 200 mg/mL in DMSO, was introduced into respective wells. Other wells containing DMSO and the reference antibacterial drug (Doxycycline) served as negative and positive controls respectively. The plates were incubated immediately at 37°C for 18 - 24 h. The activity was determined by measuring the diameter of inhibition zone (in mm). The results were compared with the control (Doxycycline).

3. RESULTS AND DISCUSSIONS

The VHP ligand (1) was synthesized by the condensation reaction of *ortho*-vanillin and 2-hydrazinopyridine in absolute ethanol (**Scheme 1**). Organotin(IV) complexes

(2-6) have also been synthesized by the direct reaction of the ligand, KOH and organotin(IV) chloride(s) in 1:1:1 (ligand:KOH:metal) mole ratio in absolute methanol (**Scheme 2**). The addition of KOH in the reaction mixture was used for the deprotonation of the ligand (1). The physical and analytical data of the ligand (1) and its organotin(IV) complexes (2-6) are given in the experimental section. The molar conductance values of the organotin(IV) complexes are in the range 2.56 - 8.81 ohm^{-1}·cm^2·mol^{-1} indicating non-electrolytic nature [8]. The compounds are stable in N$_2$ atmosphere, soluble in common organic solvents. In organotin(IV) complexes (2-6), the central tin(IV) atom is six coordinated.

3.1 Electronic Absorption Spectra

The electronic spectra analyses of hydrazone ligand (1) and its organotin(IV) complexes were carried out in DMF (1×10^{-4} M) at room temperature. VHP ligand (1) showed one band at 335 nm which is assigned to the π-π* transition of imino (>C=N) group. After complexation, the λ_{max} value of the imino group were shifted to 338 - 352 nm, due to the coordination of ligand (1) to the tin(IV) ion. New peak in the range 450 - 497 nm in the complexes (2-6) is attributed to the n-π* transition of band which is referred to ligand metal charge transfer (LMCT) [9].

3.2 Infrared Spectroscopy

Several characteristics bands were observed in the free VHP ligand (1) at 3446, 3191, 1603, 991 and 728 cm^{-1} which assigned to v(OH), v(NH), v(C=N), v(N-N) and v(pyridine in plane), respectively. The OH group of the ligand (1) was absent in the complexes (5-6) due to the deprotonation of the ligand (1). This shows that the phenolic oxygen is coordinated to the Sn(IV) ion after deprotonation. However, the OH group was still present in the complexes (2-4) indicating that the phenolic oxygen is not coordinated to the tin(IV) ion. Furthermore, the stretching vibration of azomethine (C=N) value is shifted to higher frequency in all the complexes spectra (2-6) indicating that azomethine nitrogen is involved in the coordination with the Sn(IV) ion [10]. The v(N-N) stretching vibration also shifted to higher frequency which is 1009 - 1020 cm^{-1} compared to the free VHP ligand (1) further supporting that azomethine nitrogen is coordinated to Sn(IV) ion. The infrared spectrum of the free ligand (1) showed band at 728 cm^{-1} which is assigned to the v(pyridine in plane). This band is shifted to the higher frequencies at 729-733 cm^{-1} in all the organotin(IV) complexes (2-6) [11], indicating that the pyridyl ring nitrogen is coordinated to the Sn(IV) ion. A new band observed at 447 - 499 cm^{-1} in the IR spectra is attributed to the v(Sn-N) [12]. This observation indicated that the free ligand (1) is coordinated to the Sn(IV) ion via azomethine nitrogen in all the complexes (2-6). Another a new band at 562 - 563 cm^{-1} in the complexes (5-6) is assigned to the ν(Sn-O) indicating that the phenolic oxygen of ligand (1) is coordinated to the Sn(IV) ion.

3.3. ^{1}H NMR Spectra

The ^{1}H NMR data of the VHP ligand (1) and its all organotin(IV) complexes (2-6) were recorded in DMSO-d_6 solution and interpreted based on the atom-labeling in **Scheme 2**. The ligand (1) showed the resonance signals at 10.89, 9.98, 8.29, 8.11 - 8.12, 6.77 - 7.65, 3.80 ppm are attributed to OH, NH, HC=N, Py-H6, Py-H5, Py-H/aromatic-H and CH$_3$ protons, respectively. The absence of OH proton signal in the ^{1}H NMR spectra of the organotin(IV) complexes (5-6) indicated that the phenolic oxygen is coordinated to the Sn(IV) atom after deprotonation [13]. However, the OH proton signal is still present in the diorganotin(IV) complexes (2-4) due to the less Lewis acidic character of diorganotin(IV) chlorides compare to the monoorganotin(IV) chlorides. For this reason, the diorganotin(IV) chlorides are less reactive compare to the monoorganotin(IV) chlorides. Therefore, the phenolic oxygen is not coordinated to the central Sn(IV) ion in the complexes. The NH resonance signal for the complexes (2-6) shifted to the upfield region (9.86 - 9.95 ppm) compared to the free ligand (1), indicating that the complexation of C=N-NH nitrogen atom to the Sn(IV) ion.

After complexation, the HC=N resonance signal is shifted slightly downfield to 8.29 - 8.47 ppm in all the organotin(IV) complexes (2-6) indicating that the azomethine nitrogen is coordinated to the Sn(IV) ion [14]. The pyridine-H6 proton signal is shifted to downfield at 8.14 - 8.42 ppm in all the complexes (2-6) compared to the free ligand (1) indicating that the pyridyl ring nitrogen atom is also coordinated to the Sn(IV) ion. The multiplet signals of the complexes (2-6) were at 6.73 - 7.60 ppm are due to pyridine-H protons, aromatic-H protons and SnC6H5 protons (3), respectively. The O-CH$_3$ group proton is assigned at 3.78 - 3.82 ppm in all the organotin(IV) complexes (2-6). The ^{1}H NMR spectra analysis also supported the IR spectra analysis of the organotin(IV) complexes (2-6) and attempts were made to grow the single crystals but were unsuccessful.

3.4. Brine Shrimp Bioassay

Brine shrimp bioassay is a preliminary test to screen the potential antitumor activity of hydrazone ligand (1) and its organotin(IV) complexes (2-6). The toxicity of VHP (1) and its organotin(IV) complexes (2-6) are listed in **Table 1**.

All organotin(IV) complexes (2-6) are more toxic than the free ligand (1). Among the organotin(IV) complexes

(2-6), the diphenyltin(IV) complex (3) shows higher toxicity compare to the other complexes. This might be due to the presence of bulky phenyl groups can dissociate the complex to form ionic compounds and increase the permeability of the compound into the cell [15].

3.5. Antibacterial Activity

The antibacterial test was carried out using the disc diffusion method [7]. Two types of Gram-positive bacteria, *Bacillus cereus* and *Staphylococcus aureus*, as well as two types of Gram-negative bacteria, *Escherichia coli* and *Enterobacter aerogenes*, were used to test the ligand (1) and its organotin(IV) complexes (2-6). The antibacterial activity of VHP (1) and its organotin(IV) complexes (2-6) are listed in **Table 2**.

All organotin(IV) complexes (2-5) show moderate and higher activity against four types of bacteria *Bacillus cereus*, *Staphylococcus aureus*, *Escherichia coli* and *Enterobacter aerogenes*. Diameter of inhibition zone which is less than 10 mm are considered as weak; larger than 10 mm but less than 16 mm are considered as moderate and finally larger than 16 mm and above are active [16]. Based on the results, the ligand (1) shows moderate activity towards the bacteria. Among the four complexes, diphenyltin(IV) complex (3) and butyltin(IV) complex (5) were more active than the others. The diphenyltin(IV) complex was most active towards *Staphylococcus aureus*. This might be due to the phenyl ring and the phenolic-OH group because the hydrogen of the phenolic group can enhance the toxicants to combine with constituents of living tissues [17] and the presence of the phenyl ring in the complex (3) bonded with the tin

Table 1. The LC$_{50}$ of the VHP ligand (1) and its complexes (2-6).

Complexes	LC$_{50}$ (ppm)
VHP (1)	138.03
[Me$_2$SnCl$_2$(HDMDP)](2)	100.00
[Ph$_2$SnCl$_2$(HDMDP)](3)	36.31
[nBu$_2$SnCl$_2$(HDMDP)(4)	95.50
[nBuSnCl$_2$(HDMDP)](5)	70.79
[SnCl$_3$(HDMDP)](6)	97.72

Table 2. Results of antibacterial test.

Bacteria	Diameter of Inhibition Zone (mm)					
	1	2	3	4	5	Doxycycline
Bacillus cereus	12.7	9.7	11.7	-	12.3	17.0
Staphylococcus aureus	15.0	-	18.0	11.0	12.3	30.0
Escherichia coli	12.3	9.3	13.7	10.3	11.7	14.7
Enterobacter aerogenes	9.0	-	9.7	-	-	14.3

atom can raise the antibacterial activity [18]. The Cl ion in the complex can enhances the antibacterial activity due to the killing mi- crobes or inhibiting their multiplication by blocking their active site [19].

4. CONCLUSION

Ortho-vanillin-2-hydrazinopyridine ligand [(**VHP**), (1)] and its organotin(IV) complexes (2-6) have been synthesized and fully characterized. The ligand (1) acted as a neutral bidentate nature in complexes (2-4) whereas acted as a mononegative tridentate nature in complexes (5-6). All organotin(IV) complexes (2-6) showed better toxicity compare to the free ligand (1) against *Artemia salina*. However, the free ligand and all the organotin(IV) complexes (2-5) showed moderate and high activity against four types of bacteria (*Bacillus cereus*, *Staphylococcus aureus*, *Escherichia coli* and *Enterobacter aerogenes*). Diphenyltin(IV) complexes (3) shows higher biological activity towards *Artemia salina* and the four types of bacteria compare to the free ligand (1) and other complexes.

REFERENCES

[1] Zeid, A.H.A. and Sleem, A.A. (2002) Natural and stress constituents from *Spinacia oleracea* L. leaves and their biological activities. *Bulletin of the Faculty of Pharmacy Cairo University*, **40**, 153-167.

[2] Galić, N., Cimerman, Z. and Tomišić, V. (2008) Spectrometric study of tautomeric and rotonation equilibria of *o*-vanillin Schiff base derivatives and their complexes with Cu(II). *Spectrochimica Acta Part A*, **71**, 1274-1280. doi:10.1016/j.saa.2008.03.029

[3] Singh, H.L., Singh, J.B. and Sharma, K.P. (2011) Synthesis, structural and antimicrobial studies of organotin(IV) complexes of semicarbazone, thiosemicarbazone derived from 4-hydroxy-3-metoxybenzaldehyde. *Springer Science*, **38**, 53-65.

[4] Thiyagarajan, G., Anjana, P., Nithya, P. and Ashutosh, P. (2011) Synthesis, characterization and antibacterial activity of biologically important vanillin related hydrazone derivatives. *International Journal of Organic Chemical*, **38**, 71-77. doi:10.4236/ijoc.2011.13012

[5] Armarego, W.L.F. and Perrin, D.D. (1996) Purification of laboratory chemicals. 4th Edition, Butterworth-Heinemann Linacre House, Oxford.

[6] Meyer, B.N., Ferrigni N.R., Putnam, J.E., Jacobsen, L.B., Nichols, D.E. and McLaughlin, J.L. (1982) Brine shrimp: A convenient general bioassay for active plant constituents. *Planta Medica*, **45**, 31-34. doi:10.1055/s-2007-971236

[7] Win, Y.F., Teoh, S.G., Vikneswaran, M.R., Ha, S.T. and Pazilah, I. (2010) Synthesis and characterization of organotin(IV) complexes derived of 4-(diethylamino) benzoic acid: In vitro antibacterial screening activity. *International Journal of the Physical Sciences*, **5**, 1263-1269.

[8] Geary, W.G. (1971) The use of conductivity measure-ments in organic solvents for the characterization of coordination compounds. *Coordination Chemical Review*, **7**, 81-122. doi:10.1016/S0010-8545(00)80009-0

[9] Leovac, V.M., Bogdanović, G.A., Češljević, V.I., Jova-nović, L.S., Novaković, S.B. and Vojinović-Ješić, L.S. (2007) Synthesis and characterization of salicylaldehyde Girard-T hydrazone complexes. *Structural Chemistry*, **18**, 113-119. doi:10.1007/s11224-006-9136-8

[10] Dey, D.K., Lycka, A., Mitra, S. and Rosair, G.M. (2004) Simplified synthesis, ^{1}H, ^{13}C, ^{15}N, ^{119}Sn NMR spectra and X-ray structures of diorganotin(IV) complexes containing the 4-phenyl-2,4-butanedionebenzoylhydrazone(2-) ligand. *Journal Organometallic Chemistry*, **689**, 88-95. doi:10.1016/j.jorganchem.2003.09.035

[11] Labib, L., Khalil, T.E., Iskander, M.F. and Refaat, L.S. (1996) Organotin(IV) complexes with tridentate ligand-IV organotin(IV) complexes with N-salicylidene- and N-pyridoxylideneacylhydrazines. *Polyhedron*, **15**, 3697-3707. doi:10.1016/0277-5387(96)00125-8

[12] Affan, M.A., Liew, Y.Z., Ahmad, F.B. Shamsuddin, M.B. and Yamin, B.M. (2007) Synthesis, spectroscopic characterization and antibacterial activity of organotin(IV) complexes containing hydrazone ligand. *Indian Journal of Chemistry*, **46**, 1063-1068.

[13] Hingorani, S. and Agarwala, B.V. (1993) Structural elucidation of *o*-vanillin isonicotinoyl hydrazone and its metal complexes. *Transition Metal Chemistry*, **18**, 576-578. doi:10.1007/BF00191126

[14] Yin, H., Xu, H., Li, K. and Li, G. (2005) Tin(IV) polymers: Part 1. Synthesis of diorganotin esters of thiosalicylicacid: X-ray crystal structure of polymeric thiosalicylatodiorganotin. *Journal of Inorganic and Organometallic Polymers and Materials*, **15**, 319-325.

[15] Affan, M.A., Foo, S.W., Jusoh, I., Hanapi, S. and Tiekink, E.R.T. (2009) Synthesis, characterization and biological studies of organotin(IV) complexes with hydrazone ligand. *Inorganica Chimica Acta*, **362**, 5031-5037. doi:10.1016/j.ica.2009.08.010

[16] Chohan, Z.H., Arif, M. Akhtar, M.M. and Supuran, C.T. (2006) Synthesis, characterization and in vitro biological evaluation of Co(II), Cu(II), Ni(II) and Zn(II) complexes with amino acid-derived compounds. *Bioinorganic Chemistry and Application*, **2006**, 1-13. doi:10.1155/BCA/2006/83131

[17] Rehman, W., Baloch, M.K., Muhammad, B., Badshah, A. and Khan, K.M. (2004) Characteristics spectral studies and *in vitro* antifungal activity of some Schiff bases and their organotin(IV) complexes. *Chinese Science Bulletin*, **49**, 119-122. doi:10.1360/03wb0174

[18] Jamil, K., Bakhtiar, M., Khan, A.R., Rubina, F., Rehana, R., Wajid, R., Qaisar, M., Khan, A.F., Khan, A.K., Danish, M., Awais, M., Bhatti, Z.A., Rizwan, M. Naveed, A., Hussani, M. and Pervez, A. (2011) Synthesis, characterization and antimicrobial activities of noval organotin compounds. *African Journal of Pure and Applied Chemistry*, **3**, 66-71.

[19] Rehman, W., Baloch, M.K. and Badshah, A. (2008) Synthesis, spectral characterization and bio-analysis of some organotin(IV) complexes. *European Journal of Medicinal Chemistry*, **43**, 2380-2385. doi:10.1016/j.ejmech.2008.01.019

27

Structural studies and conductivity of [Fe(O$_3$C$_4$)(COO)]·H$_2$O based H$_4$btec (H$_4$btec = 1,2,4,5-benzenetetracarboxylic acid)

Manel Halouani[1*], Mohamed Abdelhedi[1], Mohamed Dammak[1], Nathalie Audebrand[2], Lilia Ktari[1]

[1]Laboratoire de Chimie Inorganique, Faculté des Sciences de Sfax, Université de Sfax, Sfax, Tunisia
[2]Laboratoire de Matériaux Inorganiques: Chimie Douce et Réactivité, UMR 6226 Sciences Chimiques de Rennes, Université Rennes 1, Rennes, France
Email: [*]manel_halouani@yahoo.fr, m_abdelhedi2002@yahoo.fr, meddammak@yahoo.fr, nathalie.audebrand@univ-rennes1.fr, ktarililia@yahoo.fr

ABSTRACT

A new metal-organic hybrid compound [Fe(O$_3$C$_4$)(COO)]·H$_2$O I has been hydrothermally synthesized and characterized by single-crystal X-ray diffraction. Rust crystals crystallize in the monoclinic system, space group I2/a, a = 6.9651(2) Å, b = 8.12630(10) Å, c = 19.4245(2) Å, β = 92.6600(10)°; V = 1098.25(4) Å^3; Z = 2 and D$_x$ = 3.63 g/cm^3. The refinement converged into R = 0.042; R$_w$ = 0.058. The structure, determined by single crystal X-ray diffraction, consists of a network of FeO$_6$ centers, octahedral coordinated by btec (btec = 1,2,4,5-benzenetetracarboxylic acid) anions giving rise to a two-dimensional sheet structure. In the compound I, [Fe(O$_3$C$_4$)(COO)]·H$_2$O, the FeO$_6$ group bridged by the 1,2,4,5-benzenetetracarboxyl anion exist in a unit cell, with each anion lying about an inversion centre. One of the FeO$_2$ a distance [1.965(2)] significantly corresponds to the shortest distance as the other and the distances found in the axial direction of compound I.

Keywords: Hydrothermal Synthesis; X-Ray Diffraction; Crystal Structure

1. INTRODUCTION

During the past decade, the design and synthesis of crystalline material constructed from molecular clusters linked by extended groups have attracted great attention. Most notable fields are metal-organic frameworks (MOFs), [1-7] in which polyatomic inorganic metal-containing clusters are connected by polytopic linkers. The direct driving forces originate not only from their fascinating topological structures but also from their versatile applications in gas adsorption and separation, catalytic activities, optoelectronic material, luminescence, magnetism, and so on [8-19]. Thus, structural design or modification of the coordination polymers has become a very active field in crystal engineering [20-24]. In this process, judicious selection of ligands as basic building blocks is of great importance because slight structural changes in the organic building blocks such as length, flexibility, and symmetry can dramatically change the structural motifs of coordination polymers. It is well known that organic ligands play a rather important role in the construction of MOFs, and multicarboxylate ligands are frequently chosen owing to their rich coordination modes, coordinating to metal ions through complete or partial deprotonation of their carboxy groups, and their metal binding ability [25-30]. Especially interesting are the complexes formed by H$_2$bdc and H3btc having one-dimensional polymeric chain or brick-wall structures and their selective guest binding abilities [31-46].

In spite of the rich coordination chemistry exhibited by H$_2$bdc and H$_3$btc (Chart 1) in the presence of auxiliary ligands and coordinated solvents, barring a few sporadic reports, which mainly concentrate on the direct interaction between the metal ion and the ligand, there have been no serious attempts to prepare metal-organic polymeric or supramolecular structures based on 1,2,4,5-benzene tetracarboxylic acid (H$_4$btec) [47-65]. It may be argued that, due to steric reasons, all the four carboxyl groups of H$_4$btec are unlikely to take part in coordination to the metal. However, even the presence of free -COOH groups (especially in the vicinity of coordinated

---

*Corresponding author.

water molecules, donor solvents and added amines) would lead to formation of new extended structures aided by hydrogen bonding inter-actions. Moreover, the presence of a large number of uncoordinated water molecules within the lattice, may in turn lead to interesting possibilities for the preparation of porous solids. 1,2,4, 5-Benzenetetracarboxylic acid (H$_4$btec) has been well known to be an ideal ligand to encapsulated metal nodes, forming coordination compounds with unique structures and interesting properties [66-70], which may have the following advantages: 1) Higher symmetry of the ligand may cause the generation of regular structures; 2) The rigidity of the ligand may reduce the possibility of lattice interpenetration in the products [71,72]; 3) The multidentate carboxylate is known to be essential in chelating metal ions to form chain-like units with M-O-M connectivity; 4) The four carboxylic groups of this ligand may chelate to metal ions by using various coordination modes to form fascinating multidimensional compounds. In consideration of the varieties of topologies and properties, incorporating functional moieties into MOFs is often a popular method used in crystal engineering.

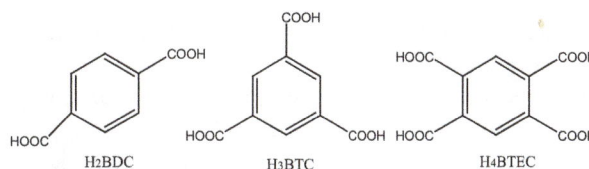

2. EXPERIMENTAL

2.1. Synthesis and Initial Characterization

The title compound was synthesized under hydrothermal conditions in the presence of tetramethylammonium nitrate. In a typical synthesis, 0.1158 g of pyromellitic acid (Acros Organics) was dispersed in 9 ml of H$_2$O. To this, 0.1975 g of iron nitrate monohydrate (Prolabo) was added under constant stirring. Finally, we add 0.062 g of tetramethylammonium nitrate (Alfa Aesar) and the mixture was homogenized for 15 min at room temperature, was sealed in a 23-ml PTFE-lined stainless steel autoclave and heated at 120°C for 60 h. Then the product obtained is filtered and washed with a small amount of distilled water. The chemical purity of the product was tested by EDAX measurements. [**Figure 1(a)**] presents

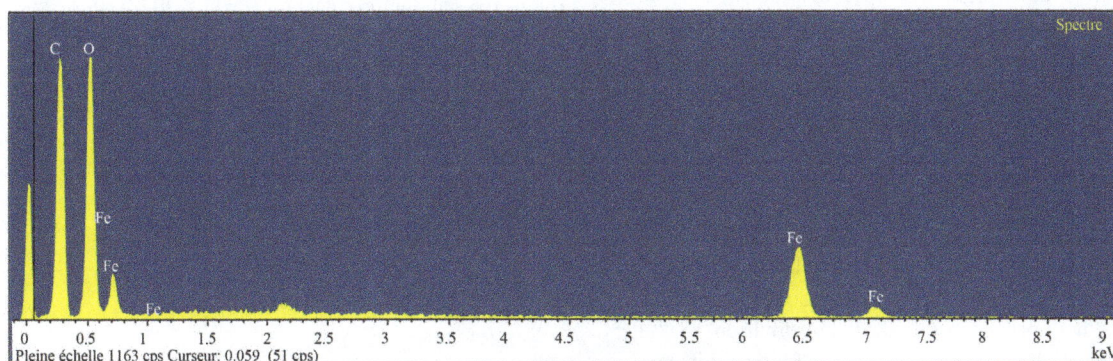

(a)

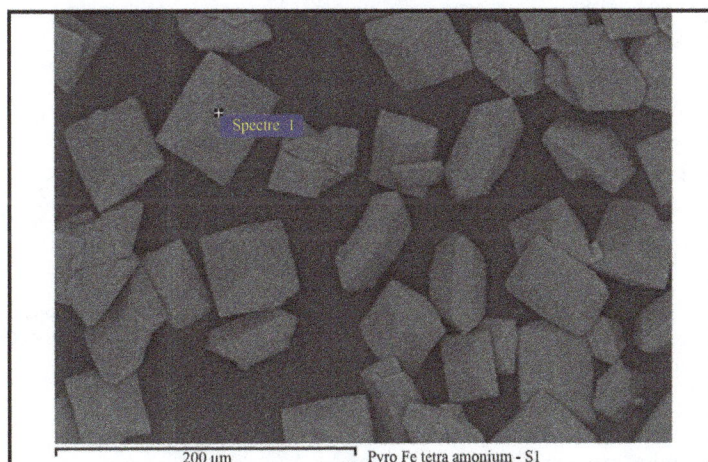

(b)

Figure 1. (a) Typical EDAX spectrum of [Fe(O$_3$C$_4$)(COO)]·H$_2$O showing the presence of Fe, O and C; (b) Scanning electron microscopic image of the [Fe(O$_3$C$_4$)(COO)]·H$_2$O.

the EDAX spectrum of $[Fe(O_3C_4)(COO)]\cdot H_2O$ which reveals the presence of all non-hydrogen atoms: Fe, C and O. Elemental analysis give these results: for observed we have C 34.10%, O 43.52%, Fe 2.38%; whereas for calculated we find C 43.93%, O 47.08%, Fe 8.99%. The amine used in this synthesis does not appear in the reaction product and its role remains unexplained. The [**Figure 1(b)**] shows the photograph of scanning electron microscopy (SEM) of the samples $[Fe(O_3C_4)(COO)]\cdot H_2O$ at room temperature.

2.2. Single Crystal Structure Determination

The unit-cell dimensions were refined using X-ray diffraction data collected with a Kappa CCD Enraf Noninus diffractometer using Mo Kα radiation .The structure, $[Fe(O_3C_4)(COO)]\cdot H_2O$, was analyzed with the crystallographic CRYSTALS program [73]. The structure was solved by conventional Patterson and difference-Fourier techniques. The chemical crystal data, the parameters used for X-ray diffraction data collection and strategy used for the crystal structure determination and their results, are listed in **Table 1**. **Table 2** shows the atomic coordinates and equivalent isotropic displacement. The anisotropic displacement parameters are listed in **Table 3**. Selected bond distances and angles are given in **Table 4**. Structural graphics were created by the DIAMOND program [74]. The asymmetric unit is shown in (**Figure 2**).

3. RESULTS AND DISCUSSION

The structure of compound I is representatively described in detail here. In an asymmetrical unit of 2, The structure of $[Fe(O_3C_4)(COO)]\cdot H_2O$, Compound I, is formed of a network of octahedral coordination by Fe carboxylate units btec. It also contains 11 non-hydrogen atoms. The Fe atom is octahedrally coordinated by six oxygen atoms from four different carboxylic groups. The Fe-O distances are in the range 1.965(11) - 2.038(4) (av.2.013 Å) and the O-Fe-O angles are in the range 86.34(16) - 179.994° (av.102.856°). Bond valence sum calculations 23 indicated that the valence states of the Fe, C and O in II were +2, +4 and −2 respectively. Their selected bond lengths and bond angles are listed in **Table 4**.

The bonded oxygen atoms of the carboxylate groups have C-O distances in the range 1.276(5) - 1.290(5)Å. The O-C-O bond angles have an average value of 125.16°. The present compound adds another example to this family of compounds. The various structural parameters observed in the present compound are in agreement with those observed before. The present compoun d adds another example to this family of compounds. The various structural parameters observed in the present compound are in agreement with those observed before. The structure of $[Fe(O_3C_4)(COO)]\cdot H_2O$, compound I, consists of a network of octahedral Fe cen-

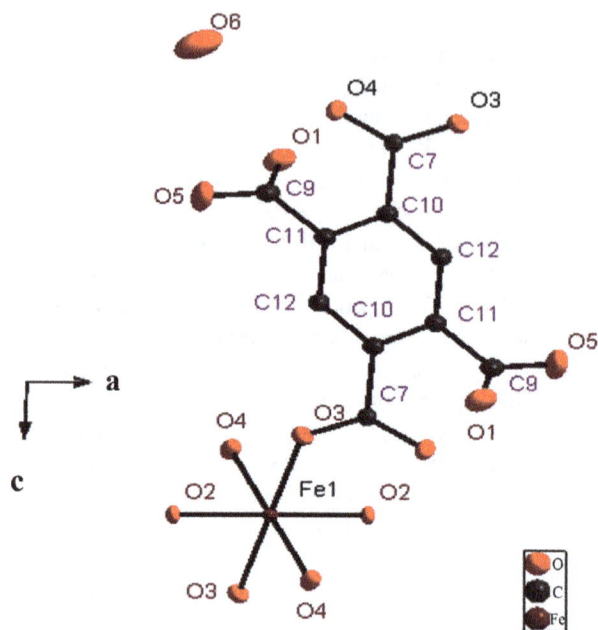

Figure 2. Asymmetric unit of $[Fe(O_3C_4)(COO)]\cdot H_2O$. Thermal ellipsoids are given at 50% probability.

Table 1. Crystallographic data for $[Fe(O_3C_4)(COO)]\cdot H_2O$.

Chemical formula $[Fe(O_3C_4)(COO)]\cdot H_2O$
Formula weight = 209.845 $g\cdot mol^{-1}$
Crystal system: Monoclinic
Space group: $I2/a$
a = 6.9651 (2) Å
b = 8.1263 (1) Å
c = 19.4245 (2) Å
β = 92.660 (1)°
V = 1098.25 (4) $Å^3$
Z = 2
$2\theta°\ max$ = 27.5° with Mo Kα
T = 293 K
D_x = 3.631 $Mg\cdot m^{-3}$
$-9 \le h \le 9$
$-10 \le k \le 10$
$-25 \le l \le 25$
μ = 1.43 mm^{-1}
Data collection instrument: Nonius Kappa CCD diffractometerRadiation, monochromator graphite λ = 0.71073 Å
Measured reflections: 17.777
Unique reflections : 8559
R = 0.050 and R_W = 0.058

ters coordinated by the btec carboxylate units (**Figure 2**). The connectivity between these units gives rise to a two dimensional hybrid layered structure in the ac and bc planes as shown in (**Figures 3** and **4**). The structure re-

Table 2. Fractional atomic coordinates and equivalent isotropic displacement for [Fe(O$_3$C$_4$)(COO)]·H$_2$O.

	x	y	z	$U_{iso}{}^*/U_{eq}$	Occupancy
Fe1	0.5000	0.0000	1.0000	0.0092	1.000
O2	0.7500	−0.1120 (7)	1.0000	0.0122	1.000
O3	0.5712 (5)	0.1353 (5)	0.9167 (2)	0.0155	1.000
O4	0.3897 (5)	−0.1677 (5)	0.9309 (2)	0.0155	1.000
C7	0.7342 (7)	0.1644 (7)	0.8953 (3)	0.0124	1.000
C9	1.0515 (8)	0.3736 (9)	0.8420 (3)	0.0201	1.000
C10	0.7395 (9)	0.2115 (7)	0.8197 (3)	0.0132	1.000
C11	0.8953 (8)	0.3027 (7)	0.7952 (3)	0.0161	1.000
C12	0.5959 (8)	0.1636 (7)	0.7741 (3)	0.0142	1.000
O1	1.0256 (5)	0.4966 (13)	0.87741 (17)	0.0245	1.000
O6	0.2500	0.4444 (8)	0.0000	0.0475	1.000
O11	1.2191 (6)	0.3011 (8)	0.8392 (3)	0.0361	1.000

Table 3. Anisotropic displacement parameters (Å^2).

	U^{11}	U^{22}	U^{33}	U^{12}	U^{13}
Fe1	0.0104 (5)	0.0123 (6)	0.0048 (5)	0.0015 (6)	−0.0002 (3)
O2	0.0080 (8)	0.0157 (14)	0.0128 (16)	0.0000	0.0005 (6)
O3	0.0123 (8)	0.0187 (12)	0.0158 (8)	−0.0009 (9)	0.0038 (8)
O4	0.0133 (9)	0.0142 (11)	0.0185 (10)	−0.0012 (10)	−0.0047 (9)
C7	0.0086 (8)	0.0150 (17)	0.0136 (9)	0.0024 (12)	0.0007 (6)
C9	0.0138 (11)	0.032 (2)	0.0145 (13)	−0.0064 (13)	−0.0006 (12)
C10	0.0091 (14)	0.0180 (18)	0.0127 (10)	0.0015 (15)	0.0021 (7)
C11	0.0099 (13)	0.025 (2)	0.0137 (10)	−0.0012 (15)	0.0004 (9)
C12	0.0079 (15)	0.020 (2)	0.0142 (10)	0.0020 (15)	0.0000 (9)
O1	0.0276 (18)	0.0208 (18)	0.0243 (19)	−0.009 (2)	−0.0070 (14)
O6	0.088 (4)	0.007 (3)	0.046 (4)	0.0000	−0.020 (4)
O11	0.0110 (13)	0.061 (3)	0.035 (3)	−0.0020 (18)	−0.0051 (15)

Table 4. Main interatomic distances (Å) and bonds angles (deg).

Atoms	distance	atoms	distance
Fe1—O4[i]	2.038 (4)	C7—C10	1.520 (8)
Fe1—O3[i]	2.037 (4)	C9—C11	1.501 (8)
Fe1—O2[ii]	1.965 (2)	C9—O1	1.231 (11)
Fe1—O2	1.965 (2)	C9—O11	1.311 (8)
Fe1—O3	2.037 (4)	C10—C11	1.416 (8)
Fe1—O4	2.038 (4)	C10—C12	1.361 (9)
O3—C7	1.249 (6)	C11—C12[iii]	1.377 (8)
O4—C7[ii]	1.257 (7)		
O4i—Fe1—O3i	86.34 (16)	Fe1—O3—C7	128.7 (4)
O4i—Fe1—O2ii	89.85 (14)	Fe1—O4—C7ii	129.5 (4)
O3i—Fe1—O2ii	90.09 (13)	O4v—C7—O3	126.5 (5)
O4i—Fe1—O2	90.15 (14)	O4v—C7—C10	117.9 (4)
O3i—Fe1—O2	89.91 (13)	O3—C7—C10	115.6 (5)
O2ii—Fe1—O2	179.994	C11—C9—O1	122.0 (5)
O4i—Fe1—O3	93.66 (16)	C11—C9—O11	115.1 (6)
O3i—Fe1—O3	179.994	O1—C9—O11	122.8 (5)
O2ii—Fe1—O3	89.91 (13)	C7—C10—C11	120.7 (5)
O2—Fe1—O3	90.09 (13)	C7—C10—C12	120.5 (5)
O4i—Fe1—O4	179.994	C11—C10—C12	118.8 (6)
O3i—Fe1—O4	93.66 (16)	C9—C11—C10	122.8 (5)
O2ii—Fe1—O4	90.15 (14)	C9—C11—C12iii	116.9 (5)
O2—Fe1—O4	89.85 (14)	C10—C11—C12iii	120.2 (6)
O3—Fe1—O4	86.34 (16)	C11iii—C12—C10	121.0 (5)
Fe1—O2—Fe1iv	124.8 (3)		

Symmetry codes: (i) $-x + 1, -y, -z + 2$; (ii) $x - 1/2, -y, z$; (iii) $-x + 3/2, -y + 1/2, -z + 3/2$; (iv) $-x + 3/2, y, -z + 2$; (v) $x + 1/2, -y, z$.

sembles a 4-connected network in which each Fe atom is connected to four btec anions and each btec anion is linked to two Fe^{2+} ions. The water molecules also occupy the interlamellar region.

4. CONCLUSION

In this work, we report a novel metal–organic complex $[Fe(O_3C_4)(COO)]\cdot H_2O$ (I), which is prepared by the hydrothermal synthesis route. It crystallizes in the mono-

clinic symmetry, space group $I2/a$. Compound I exhibits a novel bi-dimensional network constructed from bridging btec ligand. The successful isolation of compound I not only confirms that such metal–organic compounds may be designed and synthesized according to the inherent stereo and interactive information stored in the organic ligands and metal ions [75], but also further proves the strong capability of hydrothermal reactions in preparing novel metal–organic materials with mixed or-

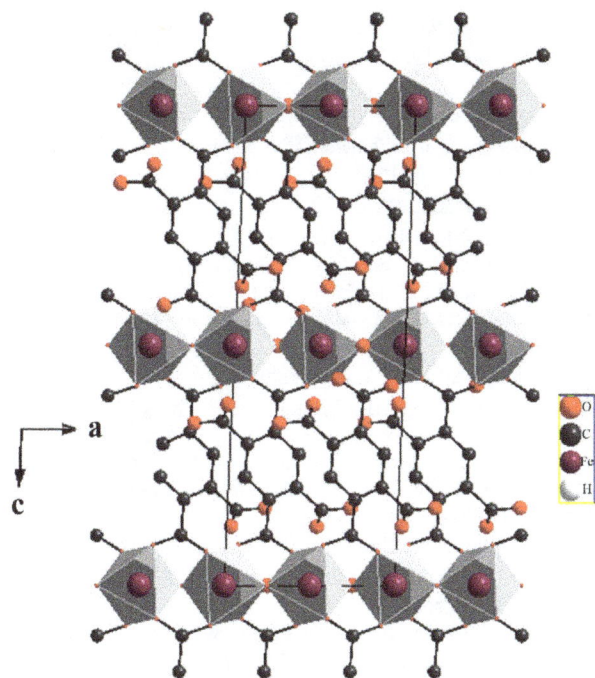

Figure 3. The projection structure of [Fe(O$_3$C$_4$)(COO)]·H$_2$O, in the *ac* plane showing a single layer.

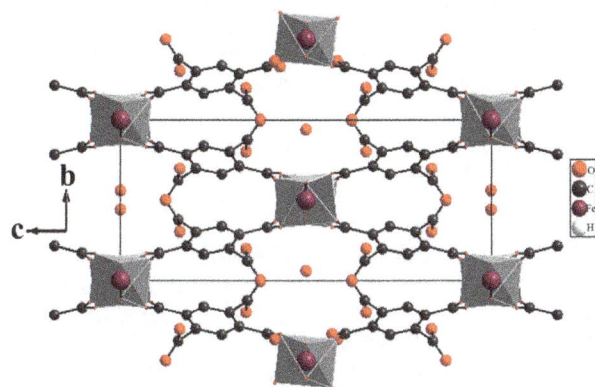

Figure 4. The projection structure of [Fe(O$_3$C$_4$)(COO)]·H$_2$O, in the *bc* plane showing a single layer.

ganic ligands.

5. ACKNOWLEDGEMENT

This work is supported by the minister of superior education and research.

REFERENCES

[1]   O'Keeffe, M., Peskov, M. A., Ramsden S. J. and Yaghi O. M. (2008) The reticular chemistry structure resource (RCSR) database of, and symbols for, crystal nets. *Accounts of Chemical Research*, **41**, 1782-1789. http://dx.doi.org/10.1021/ar800124u

[2]   Wang, Z. and Cohen, S.M. (2009) Postsynthetic modification of metal-organic frameworks. *Chemical Society Reviews*, **38**, 1315-1329. http://dx.doi.org/10.1039/b802258p

[3]   Yaghi, O.M., O'Keeffe, M., Ockwing, N.W., Chae, H.K., Eddaoudi, M. and Kim, J. (2003) Reticular synthesis and the design of new materials. *Nature*, **42**, 3705-3714. http://dx.doi.org/10.1038/nature01650

[4]   Yang, W.B., Lin, X., Jia, J.H., Blake, A.J., Wison, C., Hubberstey, P., Champness, N.R. and Schröder, M. (2008) A biporous coordination framework with high H$_2$ storage density. *Chemical Communications*, 359-361. http://dx.doi.org/10.1039/b712201b

[5]   Lee, Y.G., Moon, H.R., Cheon, Y.E. and Suh, M.P. (2008) A comparison of the H$_2$ sorption capacities of isostructural metal-organic frameworks with and without accessible metal sites: [{Zn$_2$(abtc)(dmf)$_2$}$_3$] and [{Cu$_2$(abtc)(dmf)$_2$}$_3$] versus [{Cu2(abtc)}3]. *Angewandte Chemie International Edition*, **47**, 7741-7745. http://dx.doi.org/10.1002/anie.200801488

[6]   Liu, Y.L., Eubank, J.F., Cairns, A.J., Eckert, J., Kravtsow, V.C., Luebke, R. and Eddaoudi, M. (2007) Assembly of metal-organic frameworks (MOFs) based on indium-trimer building blocks: A porous MOF with soc topology and high hydrogen storage. *Angewandte Chemie International Edition*, **46**, 3278-3283. http://dx.doi.org/10.1002/anie.200604306

[7]   McManus, G.J., Wang, Z.Q., Beauchamp, D.A. and Zaworotko, M.J. (2007) A novel metal-organic ternary topology constructed from triangular, square and tetrahedral molecular building blocks. *Chemical Communications*, 5212-5213. http://dx.doi.org/10.1039/b712905j

[8]   Perry IV, J.J., Perman, J.A. and Zaworotko, M.J. (2009) Design and synthesis of metal-organic frameworks using metal-organic polyhedra as supermolecular building blocks. *Chemical Society Reviews*, **38**, 1400-1417. http://dx.doi.org/10.1039/b807086p

[9]   Qiu, Y.C., Deng, H., Yang, S., Mou, J., Daiguebonne, C., Kerbellec, N., Guillou, O. and Batten, S.R. (2009) Syntheses, crystal structures, and gas storage studies in new three-dimensional 5-aminoisophthalate praseodymium polymeric complexes. *Inorganic Chemistry*, **48**, 3976-3981. http://dx.doi.org/10.1021/ic8020518

[10] Li,B. K., Olson, D.H., Lee, J.Y., Bi, W., Yuen, T., Xu, Q. and Li, J. (2008) Multifunctional microporous MOFs exhibiting gas/hydrocarbon adsorption selectivity, separation capability and three-dimensional magnetic ordering. *Advanced Functional Materials*, **18**, 2205-2214. http://dx.doi.org/10.1002/adfm.200800058

[11] Cheng, P., Yan, S.P., Xie, C.Z., Zhao, B., Chen, X.Y., Liu, X.W., Li, C.H., Liao, D.Z., Jiang, Z.H. and Wang, G.L. (*2004*) Ferromagnetic and antiferromagnetic polymeric complexes with the macrocyclic ligand 1,4,7-triazacyclononane. *European Journal of Inorganic Chemistry*, **2004**, 2369-2378. http://dx.doi.org/10.1002/ejic.200300791

[12] Liu, H.Y., Zhao, B., Shi, W., Zhang, Z.J., Cheng, P., Liao, D.Z. and Yan, S.P. (2009) A chiral metal-organic framework based on heptanuclear zinc cores. *European Journal of Inorganic Chemistry*, **2009**, 2599-2602. http://dx.doi.org/10.1002/ejic.200900185

[13] Janiank, C. (2003) Engineering coordination polymers towards applications. *Dalton Transactions*, 2781-2804. http://dx.doi.org/10.1039/b305705b

[14] Kitagawa, S. and Uemura, K. (2005) Dynamic porous properties of coordination polymers inspired by hydrogen bonds. *Chemical Society Reviews*, **34**, 109-119. http://dx.doi.org/10.1039/b313997m

[15] Shen, W.Z., Chen, X.Y., Cheng, P., Yan, S.P., Zhai, B., Liao, D.Z. And Jiang, Z.H. (2005) A structural and magnetic investigation of ferromagnetically coupled copper(II) isophthalates. *European Journal of Inorganic Chemistry*, **2005**, 2297-2305. http://dx.doi.org/10.1002/ejic.200401022

[16] Zhao, X.Q., Zuo, Y., Gao, D.L., Zhao, B., Shi, W. and Cheng, P. (2007) Pillared 3d-4f frameworks with rare 3D architecture showing the coexistence of ferromagnetic and antiferromagnetic interactions between gadolinium ions. *Crystal Growth & Design*, **7**, 851-853. http://dx.doi.org/10.1021/cg0608231

[17] Zhao, B., Cheng, P., Dai, Y., Cheng, C., Liao, D.Z., Yan, S.P., Jiang, Z.H. and Wang, G.L. (2003) A nanotubular 3d coordination polymer based on a 3d-4f heterometallic assembly. *Angewandte Chemie International Edition*, **42**, 934-936. http://dx.doi.org/10.1002/anie.200390248

[18] Wang, R.H., Hong, M.C., Luo, J.H., Cao, R. and Weng, J.B. (2003) A new type of three-dimensional framework constructed from dodecanuclear cadmium(II) macrocycles. *Chemical Communications*, **8**, 1018-1019. http://dx.doi.org/10.1039/b212425d

[19] Vaidhyanathan, R., Natarajan, S. and Rao, C.N.R. (2002) Three-dimensional open-framework neodymium oxalates with organic functional groups protruding in 12-member channels. *Inorganic Chemistry*, **41**, 4496-4501. http://dx.doi.org/10.1021/ic020197r

[20] Cerdeira, A.C., Sim, D., Santos, I.C., Machado, A. and Pereira, L.C.J. (2008) (*n*-Bu₄N)[Fe(cbdt)₂]: Synthesis, crystal structure and magnetic characterisation of a new FeIII bisdithiolene complex. *Inorganica Chimica Acta*, **361**, 3836-3840. http://dx.doi.org/10.1016/j.ica.2008.02.068

[21] Yamada, T. and Kitagawa, H. (2009) Protection and deprotection approach for the introduction of functional groups into metal-organic frameworks. *Journal of the American Chemical Society*, **131**, 6312-6313. http://dx.doi.org/10.1021/ja809352y

[22] Fang, M., Zhao, B., Zuo, Y., Chen, J., Shi, W., Liang, J. and Cheng, P. (2009) Unique two-fold interpenetration of 3D microporous 3d-4f heterometal-organic frameworks (HMOF) based on a rigid ligand. *Dalton Transactions*, 7765-7770. http://dx.doi.org/10.1039/b903737c

[23] Mavrandonakis, A., Klontzas, E., Tylianakis, E. and Froudakis, G.E. (2009) Enhancement of hydrogen adsorption in metal-organic frameworks by the incorporation of the sulfonate group and Li cations. A multiscale computational study. *Journal of the American Chemical Society*, **131**, 13410-13414. http://dx.doi.org/10.1021/ja9043888

[24] Manna, S.C., Zangrando, E., Ribas, J. and Chaudhuri, N.R. (2007) Cobalt(II)-(dpyo)-dicarboxylate networks: unique H-bonded assembly and rare bridging mode of dpyo in one of them [dpyo = 4,4'-dipyridyl N,N'-dioxide].

Dalton Transactions, 1383-1391. http://dx.doi.org/10.1039/b617278d

[25] Wang, X. S., Ma, S., Sun, D. F., Parkin, S. and Zhou, H. C. (2006) A Mesoporous metal-organic framework with permanent porosity. *Journal of the American Chemical Society*, **128**, 16474-16475. http://dx.doi.org/10.1021/ja066616r

[26] Chung, H. T., Tsai, H. L., Yang, E. C., Chien, P. H., Peng, C. C., Huang, Y. C. and Liu, Y. H. (2009) A new manganese coordination polymer containing 1,2,4-benzenetricarboxylic acid. *European Journal of Inorganic Chemistry*, 3661-3666. http://dx.doi.org/10.1002/ejic.200900375

[27] Shekhah, O., Wang, H., Zacher, D., Fischer, R.A. and Woll, C. (2009) Growth mechanism of metal-organic frameworks: Insights into the nucleation by employing a step-by-step route. *Angewandte Chemie International Edition*, **48**, 5038-5041. http://dx.doi.org/10.1002/anie.200900378

[28] Ma, S., Simmons, J.M., Sun, D.F., Yuan, D.Q. and Zhou, H.C. (2009) Porous metal-organic frameworks based on an anthracene derivative: Syntheses, structure analysis, and hydrogen sorption studies. *Inorganic Chemistry*, **48**, 5263-5268. http://dx.doi.org/10.1021/ic900217t

[29] Hong, S., Oh, M., Park, M., Yoon, J.W., Chang, J.S. and Lah, M.S. (2009) Metal-organic frameworks (MOFs) constructed from ZnII/CdII-2,2'-bipyridines and polycarboxylic acids: Synthesis, characterization and microstructural studies. *Chemical Communications*, 5397-5399. http://dx.doi.org/10.1039/b909250a

[30] Prajapati, R., Mishra, L., Kimura, K. and Raghavaiah, P. (2009) Metal-organic frameworks (MOFs) Constructed from ZnII/CdII-2,2'-bipyridines and polycarboxylic acids: Synthesis, characterization and microstructural studies. *Polyhedron*, **28**, 600-608. http://dx.doi.org/10.1016/j.poly.2008.11.059

[31] Li, H., Eddaoudi, M., Groy, T.L. and Yaghi, O.M. (1998) Establishing microporosity in open metal-organic frameworks: Gas sorption isotherms for Zn(BDC) (BDC = 1,4-benzenedicarboxylate). *Journal of the American Chemical Society*, **120**, 8571-8572. http://dx.doi.org/10.1021/ja981669x

[32] Reineke, T.M., Eddaoudi, M., Fehr, M., Kelley, D. and Yaghi, O.M. (1999) From condensed lanthanide coordination solids to microporous frameworks having accessible metal sites. *Journal of the American Chemical Society*, 121, 1651-1657. http://dx.doi.org/10.1021/ja983577d

[33] Groeneman, R.H., MacGillivray, L.R. and Atwood, J.L. (1998) Aromatic inclusion within a neutral cavitycontaining rectangular grid. *Chemical Communications*, 2735-2736. http://dx.doi.org/10.1039/a807738j

[34] Groeneman, R.H., MacGillivray, L.R. and Atwood, J.L. (1999) One-dimensional coordination polymers based upon bridging terephthalate ions. *Inorganic Chemistry*, **38**, 208-209. http://dx.doi.org/10.1021/ic9807137

[35] Yaghi, O.M., Li, H. and Groy, T.L. (1999) Construction of porous solids from hydrogen-bonded metal complexes of 1,3,5-benzenetricarboxylic acid. *Journal of the American Chemical Society*, **118**, 9096-9101 http://dx.doi.org/10.1021/ja960746q

[36] Yaghi, O.M., Li, G. and Li, H. (1995) Selective binding and removal of guests in a microporous metal-organic framework. *Nature*, **378**, 703-706. http://dx.doi.org/10.1038/378703a0

[37] Yaghi, O.M., Davis, C.E., Li, G. and Li, H. (1997) Selective guest binding by tailored channels in a 3-d porous zinc(II)-benzenetricarboxylate network. *Journal of the American Chemical Society*, **119**, 2861-2868. http://dx.doi.org/10.1021/ja9639473

[38] Kepert, C.J. and Rosseinsky, M.J. (1998) A porous chiral framework of coordinated 1,3,5-benzenetricarboxylate: Quadruple interpenetration of the (10,3)—A network. *Chemical Communications*, 31-32. http://dx.doi.org/10.1039/a705336c

[39] Platers, M.J., Howie, R.A. and Roberts, A.J. (1997) Hydrothermal synthesis and X-ray structural characterisation ofcalcium benzene-1,3,5-tricarboxylate. *Chemical Communications*, 893-894. http://dx.doi.org/10.1039/a700675f

[40] Choi, H.J. and Suh, M.P. (1998) Self-assembly of molecular brick wall and molecular honeycomb from nickel (II) macrocycle and 1,3,5-benzenetricarboxylate: Guest-dependent host structures. *Journal of the American Chemical Society*, **120**, 10622-10628. http://dx.doi.org/10.1021/ja980504l

[41] Choi, H.J., Lee, T.S. and Suh, M.P. (1999) Self-assembly of a molecular floral lace with one-dimensional channels and inclusion of glucose. *Angewandte Chemie International Edition*, **38**, 1405-1408. http://dx.doi.org/10.1002/(SICI)1521-3773(19990517)38:10<1405::AID-ANIE1405>3.0.CO;2-H

[42] Yaghi, O.M. and Li, H. (1996) T-shaped molecular building units in the porous structure of Ag(4,4'-bpy)·NO3. *Journal of the American Chemical Society*, **118**, 295-296. http://dx.doi.org/10.1021/ja953438l

[43] Deakin, L., Arif, A.M. and Miller, J.S. (1999) Observation of ferromagnetic and antiferromagnetic coupling in 1-D and 2-D extended structures of copper(II) terephthalates. *Inorganic Chemistry*, **38**, 5072-5077. http://dx.doi.org/10.1021/ic990400r

[44] St. Foreman, J.M.R., Gelbrich, T., Hursthouse, M.B. and Plater, M.J. (2000) Hydrothermal synthesis and characterization of lead(II) benzene-1,3,5-tricarboxylate [Pb3BTC2]·H2O: A lead(II) carboxylate polymer. *Inorganic Chemistry Communications*, **3**, 234-238. http://dx.doi.org/10.1016/S1387-7003(00)00064-2

[45] Li, H., Davis, C.E., Groy, T.L., Kelley, D.G. and Yaghi, O.M. (1998) Coordinatively unsaturated metal centers in the extended porous framework of Zn3(BDC)3·6CH3OH (BDC=1,4-benzenedi-carboxylate). *Journal of the American Chemical Society*, **120**, 2186-2187. http://dx.doi.org/10.1021/ja974172g

[46] Chen, B., Eddaoudi, M., Reineke, T.M., Kampf, J.W., Keeffe, M.O. and Yaghi, O.M. (2000) Cu2(ATC)·6H2O: Design of open metal sites in porous metal—Organic crystals (ATC: 1,3,5,7-adamantane tetracarboxylate). *Journal of the American Chemical Society*, **122**, 11559-11560. http://dx.doi.org/10.1021/ja003159k

[47] Robl, C. and Hentschel, S. (1991) On dicobaltpyromelli-

tate-octadecahydrate (Co2[C6H2(COO)4·18H2O)—A compound with the novel chain-like polyanion Co(H2O)4 [C6H2(COO)4]2n-n. *Materials Research Bulletin*, **26**, 1355-1362. http://dx.doi.org/10.1016/0025-5408(91)90152-C

[48] Poleti, D. and Karanovic, L. (1989) Structure of hexaaquacobalt(II) catena-tetraaqua [1,2,4,5-benzenetetracarboxylato(4-)]-cobaltate(II) 7.36-hydrate. *Acta Crystallographica Section C*, **45**, 1716-1718. http://dx.doi.org/10.1107/S0108270189002489

[49] Ward, D.L. and Leuhrs, D.C. (1983) Hexaaquacobalt(II) dihydrogen 1,2,4,5-benzenetetracarboxylate, [Co(H2O)6] [C10H4O8]. *Acta Crystallographica Section C*, **39**, 1370-1372. http://dx.doi.org/10.1107/S0108270183008550

[50] Robl, C. (1992) Water clustering in the zeolite-like channel structure of Na2Zn[C6H2(COO)4]·9H2O. *Materials Research Bulletin*, **22**, 99-107. http://dx.doi.org/10.1016/0025-5408(92)90047-4

[51] Poleti, D., Stojakovic, D.R., Prelesnik, B.V. and Herak, R.M. (1988) Structure of binuclear hexaaqua[1,2,4,5-benzenetetracarboxylato(4-)]-bis(ethylenediamine)dinickel(II) tetrahydrate. *Acta Crystallographica Section C*, **44**, 242-245. http://dx.doi.org/10.1107/S0108270187009764

[52] Jessen, S.M. and Kuppers, H. (1990) The structure of dilithium dihydrogen 1,2,4,5-benzenetetracarboxylate tetrahydrate (dilithium dihydrogen pyromellitate tetrahydrate). *Acta Crystallographica Section C*, **46**, 2351-2354. http://dx.doi.org/10.1107/S0108270190004863

[53] Chu, D., Xu, J., Duan, L., Wang, T., Tang, A. and Ye, L. (2001) Hydrothermal synthesis of a two-dimensional coordination polymer [Fe(phen)(μ6-bta)1/2]n (bta = benzene-1,2,4,5-tetracarboxylate, phen=1,10-phenanthroline). *European Journal of Inorganic Chemistry*, **2001**, 1135-1137. http://dx.doi.org/10.1002/1099-0682(200105)2001:5<1135::AID-EJIC1135>3.0.CO;2-G

[54] Rochon, F.D. and Massarweh, G. (2001) Study of the aqueous reactions of metallic ions with benzenetetracarboxylate ions: Part 2. Crystal structures of compounds of the type M(H2O)5(μ-C6H2(COO)4)M(H2O)5 (M=Mn and Co) and a novel mixed-metallic Mn-Co dimeric compound. *Inorganica Chimica Acta*, **314**, 163-171. http://dx.doi.org/10.1016/S0020-1693(01)00307-3

[55] Plater, M.J., St. Foreman, J.M.R., Howie, R.A., Skakle, J.M.S. and Slawin, A.M.Z. (2001) Hydrothermal synthesis of polymeric metal carboxylates from benzene1,2,4,5-tetracarboxylic acid and benzene-1,2,4-tricarboxylic acid. *Inorganica Chimica Acta*, **315**, 126-132. http://dx.doi.org/10.1016/S0020-1693(01)00318-8

[56] Usubaliev, B.T., Shnulin, A.N. and Mamadow, H.S. (1982) Molecular and crystal structure of tetra-aquobis(pyridine)-tetra(p-nitrobenzoato)dicopper (II). *Journal of Structural Chemistry September*, **23**, 760-764.

[57] Robl, C. (1987) Komplexe mit aromatischen carbonsäuren. I Darstellung und struktur von zinkpyromellitat-heptahydrat. *Zeitschrift für anorganische und allgemeine Chemie*, **554**, 79-86. http://dx.doi.org/10.1002/zaac.19875541109

[58] Zou, J.Z., Liu, Q., Xu, Z., You, X.Z. and Huang, X.Y. (1998) TCB bridged binuclear and polynuclear copper(II)

complexes: A novel three dimension-network structure complex [(Cudien)$_2$(Cudien·H$_2$O)TCB (ClO4)$_2$·H$_2$O]n (TCB= tetracarboxylatobenzene, dien=3-azapentane-1,5-diamine). *Polyhedron*, **17**, 1863-1869. http://dx.doi.org/10.1016/S0277-5387(97)00524-X

[59] Jessen, S.M. Kuppers, H. and Luehrs, D.C. (1992) Hydrogen-bonding in acid litetrabutylammonium, nitetrabutylammonium, tetrabutylammonium, and ammonium-salts of benzene-1,2,4,5-tetracarboxylic acid (pyromellitic acid). *Zeitschrift Fur Naturforschung Section B—A Journal of Chemical Sciences*, **47**, 1141-1153.

[60] Karanovic, L., Poleti, D., Bogdanovic, G.A. and Spasojevic-de Bire, A. (1999) Disodium hexaaquacobalt (II) bis [1,2,4,5-dihydrogénophosphate benzenetetracarboxylate (2-)] tétrahydraté. *ACTA Crystallographica Section de Communications C—Structure Cristalline*, **55**, 911-913.

[61] Zhang, B., Jin, Z.S., Wang, Y.L., Wei, G.C. and Chen, W.Q. (1992) Syntheses and crystal structures of benzyl- and cyclopentyl-cyclopentadienyl sodium tetra-hydrofuranate complexes. *Jiegou Huaxue*, **8**, 15-20.

[62] Chaudhuri, P., Oder, K., Wieghardt, K., Gehring, S., Haase, W., Nuber, B. and Weiss, J. (1988) Moderately strong intramolecular magnetic exchange interaction between the copper(II) ions separated by 11.25 .ANG. in [L$_2$Cu$_2$(OH2)$_2$(.eta.-terephthalato)](ClO$_4$)$_2$ (L = 1,4,7-trimethyl-1,4,7-triazacyclononane). *Journal of the American Chemical Society*, **110**, 3657-3658. http://dx.doi.org/10.1021/ja00219a051

[63] Jaber, F., Charbonnier, F. and Faure, R. (1997) Crystal structure of a silver(I) complex with the 1,2,4,5-benzenetetracarboxylic acid (pyromellitic acid). *Journal of Chemical Crystallography*, **27**, 397-400. http://dx.doi.org/10.1007/BF02576473

[64] Chen, W., Heng Tioh, N., Zou, J.-Z., Xu, Z. and You, X.-Z. (1996) A Tetracarboxylatobenzene-Bridged Binuclear Copper(II) Complex. *Acta Crystallographica Section C*, **52**, 43-45. http://dx.doi.org/10.1107/S0108270195010444

[65] Luehrs, D.C. and Day, C.S. (1988) The crystal structure of thallium(I) 1,2,4,5-benzenetetracarboxylate. *Inorganica Chimica Acta*, **142**, 201-202. http://dx.doi.org/10.1016/S0020-1693(00)81560-1

[66] Vitillo, J.G., Regli, L., Chavan, S., Ricchiardi, G., Spoto, G., Dietzel, P.D.C., Bordiga, S. and Zecchina, A. (2008) Role of exposed metal sites in hydrogen storage in MOFs.

Journal of the American Chemical Society, **130**, 8386-8396. http://dx.doi.org/10.1021/ja8007159

[67] Wang, Y.B., Zhuang, W.J., Jin, L.P. and Lu, S.Z. (2005) New lanthanide coordination polymers of 1,2,4,5-benzenetetracarboxylic acid and 4,4'-bipyridine with 1D channels. *Journal of Molecular Structure*, **737**, 165-172. http://dx.doi.org/10.1016/j.molstruc.2004.10.062

[68] Cheng, D., Khan, M.A. and Houser, R.P. (2002) Novel sandwich coordination polymers composed of cobalt(II), 1,2,4,5-benzenetetracarboxylato ligands, and homopiperazonium cations. *Crystal Growth & Design*, **2**, 415-420. http://dx.doi.org/10.1021/cg025547z

[69] Ding, M.T., Wu, J.Y., Liu, Y.H. and Lu, K.L. (2009) Dissolution/reorganization toward the destruction/construction of porous cobalt(II)- and nickel(II)-carboxylate coordination polymers. *Inorganic Chemistry*, **48**, 7457-7465. http://dx.doi.org/10.1021/ic900037g

[70] Caskey, S.R. and Matzger, A.J. (2008) Selective metal substitution for the preparation of heterobimetallic microporous coordination polymers. *Inorganic Chemistry*, **47**, 7942-7944. http://dx.doi.org/10.1021/ic8007427

[71] Cheng, D., Khan, M.A. and Houser, R.P. (2003) Nickel(II) and manganese(II) 1D chain coordination polymers with 1,2,4,5-benzenetetracarboxylato anions. *Inorganica Chimica Acta*, **351**, 242-250. http://dx.doi.org/10.1016/S0020-1693(03)00122-1

[72] Xiao, D.R., Wang, E.B., An, H.Y., Li, Y.G., Su, Z.M. and Sun, C.Y. (2006) A bridge between pillared-layer and helical structures: A series of three-dimensional pillared coordination polymers with multiform helical chains. *Chemistry*, **12**, 2680-2691. http://dx.doi.org/10.1002/chem.200501308

[73] Betteridge, P.W., Carruthers, J.R., Cooper, R.I., Prout, K. and Watkin, D.J. (2003) Software for guided crystal structure analysis. *Journal of Applied Crystallography*, **36**, 1487. http://dx.doi.org/10.1107/S0021889803021800

[74] Brandenburg, K. and Berndt, M. (1999) Diamond. Crystal impact Gb R, Bonn, Germany, Version 2.1.b.

[75] Cao, R., Sun, D.F., Liang, Y.C., Hong, M.C., Tatsumi, K., Shi, Q. (2002) Syntheses and characterizations of three-dimensional channel-like polymeric lanthanide complexes constructed by 1,2,4,5-benzenetetracarboxylic acid. *Inorganic Chemistry*, **41**, 2087-2094. http://dx.doi.org/10.1021/ic0110124

28

Preparation of N,N,N',N'-tetrakis-(2-benzimidazolyl-methyl)-1,2-ethanediamine and crystal assemblies of the relative complexes

Shanling Tong[1], Yahong Wu[1], Zhaoyong Tian[2], Sheng Hu[1], Yan Yan[1*]

[1]College of Light Industry & Chemical Engineering, Guangdong University of Technology, Guangzhou, China
[2]Logistics Group, Huainan Normal College, Huainan, China
Email: *yanyan600716@hotmail.com

ABSTRACT

N,N,N',N'-tetrakis-(2-benzimidazolylmethyl)-1,2-etha-nediamine (TBIMEDA), was prepared by reaction of ethylenediamine tetra-acetic acid disodium salt (EDTA) with 1,2-diaminobenzene in a refluxed glycol solution, and furthermore, three allomeric complexes [(M^{II} TBIMEDA) $SO_4 \cdot 5H_2O$, M = Cd, Co, Ni] were self-assembled by solvothermal method based on reaction of this ligand with the relative sulfates respectively. These allomeric complexes were characterized by elemental analysis and IR spectroscopy and their crystal structures were determined by single crystal X-ray structural analysis. In the crystal architecture of these complexes, every metal(II) ion is chelated by one neutral TBIMEDA ligand to form an octahedral core with configuration of five heterocyclic rings (five-member ring). These cores then were linked together by multi hydrogen bond interactions with sulfate ions and water molecules to construct their 3D crystal architectures.

Keywords: Benzoimidazolmethyl Ethane Diamine; Solvothermal; Complex; Crystal Architecture

1. INTRODUCTION

Azaheterocyclic compounds, with strong coordination ability during complex preparations and as better acceptors in hydrogen bonding formation, were more frequently selected in coordination research [1-4]. After self-assemblies, the conjugated π bonds in the heterocyclic ligands provided abounding information on electromagnetism and photoelectrochemistry, and these special complex's aggregation was well known as functionalized material in future applications [5-9]. Meanwhile EDTA (ethylenediamine tetra-acetic acid) was regard as the

most useful multidentate ligand using in analytic chemistry. In this work, based on reaction of EDTA with benzene-1,2-diamine, a multidentate ligand with four benzoimidazole groups, N,N,N',N'-tetrakis-(2-benzim-idazolyl-methyl)-1,2-ethanediamine(TBIMEDA), was prepared according to the reported method (**Scheme 1**) [10-13]. Furthermore, by self-assembly of TBIMEDA reacting with different sulfates, three single-nuclear complexes with same crystal configuration were constructed by sovothermal methods, and their crystal structures were well defined by X-ray analysis.

2. EXPERIMENTAL

2.1. Materials and Physical Measurements

All chemicals and solvents were of analytical reagent grade and used as received. Elemental analysis was performed in a Perkin-Elmer 240 elemental analyzer. IR spectrum was obtained using a Nicolet IR200 infrared spectrometer. The fluorescence spectra were taken on an Edinburgh Instruments FLS920 fluorescence spectrumeter respecttively.

2.2. Synthesis of Ligand

Preparation of the ligand, N,N,N',N'-tetrakis-(2-benzi-midazolylmethyl)-1,2-ethanediamine (TBIMEDA) was described as below [10-13]: A solution of EDTA (2.92 g, 0.01 mol) and 1,2-diaminobenzene (4.32 g, 0.04 mol) in 100 mL of glycol was heated to boiling and kept in reflux for 16 h. After cooling down to room temperature, the mixture was added to water (ac. 400 mL) for precipitation over night. The crude product was separated by filtration, purified by recrystallization with small volume of ethanol for three times and dried in air. A white powder product in yield of 70% was obtained with m. p. = 156°C - 158°C. IR data (KBr, cm^{-1}): 3186 vs (ν_{N-H}), 2966 w, 2823 w, 1617 s, 1535 s, 1486 w(σ_{N-H}), 1454 s, 1433 vs, 1348 s, 1311 m, 1274 vs, 1245 m, 1217 w, 1094

---

*Corresponding author.

s, 1049 m, 1021 m, 996 m, 963 w, 845 m, 747 vs, 616 w, 485 m. Calc. (found) for $C_{34}H_{32}N_{10}$: C 72.88 (72.90), H 3.84 (3.82), N 7.08 (7.12). This IR data were similar to the reported values. [10] The preparation of TBIMEDA was described as **Scheme 1**.

2.3. Synthesis of Complexes

2.3.1. Co(TBIMEDA)·SO₄·5H₂O(1)

Keeping under stirring, TBIMEDA (0.0706 g, 0.1 mmol), $CoSO_4·7H_2O$ (0.0422 g, 0.15 mmol) are mixed with water (8 mL). The mixture was sealed in an autoclave and the autoclave was placed in an oven at 120°C for 60 h. After cooling down to room temperature at rate of 5°C/h, filtrating and washing, several large and transparent orange crystals were collected (in yield of 34%). Found (Calc.) for $CoC_{34}H_{42}N_{10}O_9S$: C, 49.44(49.45); H, 5.11 (5.13); N, 16.98(16.96)%. IR data(KBr, cm^{-1}): 3415 s, 3101 w, 3056 w, 2917 w, 2774 w, 2643 w, 1621 m, 1540 m, 1470 m, 1450 vs, 1392 m, 1274 vs, 1119 vs, 1029 m, 939 m, 910 w, 910 w, 857 w, 743 vs, 620 vs, 555 w, 514 w.

2.3.2. Ni(TBIMEDA)·SO₄·5H₂O(2)

Keeping under stirring, TBIMEDA (0.0706 g, 0.1 mmol), $NiSO_4·6H_2O$ (0.0262 g, 0.1 mmol) were mixed with water (8 mL). The mixture was sealed in an autoclave and the autoclave was placed in an oven at 140°C for 60 h. After cooling down to room temperature at rate of 5°C/h, filtrating and washing, several large and transparent blue crystals were collected (in yield of 45%). Found (Calc.) for $NiC_{34}H_{42}N_{10}O_9S$: C, 46.48(46.47); H, 5.13(5.11); N, 16.97(16.99)%. IR data (KBr, cm^{-1}): 3415 s, 3105 m, 3064 w, 2913 w, 2765 w, 2639 w, 1544 m, 1405 vs, 1392 s, 1331 vs, 1278 vs, 1221 w, 1115 vs, 1029 s, 988 w, 939 s, 906 m, 849 m, 743 vs, 616 vs, 545 w, 518 w.

2.3.3. Cd(TBIMEDA)·SO₄·5H₂O(3)

Similar to the assembly of **1**, when $NiSO_4·6H_2O$ was replaced by $CdSO_4·8H_2O$ (0.0385 g, 0.05 mmol), large and transparent blue crystals were obtained in yield of 45% by the same hydrothermal method. Found(Calc.) for $CdC_{34}H_{42}N_{10}O_9S$:C, 49.44(49.45); H, 5.07(5.09); N, 15.95(15.93)%. IR data (KBr, cm^{-1}): 3395 s, 3096 m, 2761 w, 1621 m, 1540 m, 1446 vs, 1384 s, 1335 m, 1278 s, 1094 vs, 1025 s, 755 s, 620 m.

2.4. Single Crystal X-Ray Diffraction Analysis

The X-ray data collections and structure determinations were performed on a Bruker SMART CCD. The data were collected using graphite-monochromatic Mo-Kα radiation (λ = 0.71073 Å). The crystal structure was solved by direct methods and refined by full-matrix least-square calculation on F^2 with SHELX-97 program package.

[14]. All non-hydrogen atoms were treated anisotropically. Hydrogen atoms were placed in calculated positions. The crystallographic data for **1-3** are summarized in **Tables 1** and **2**.

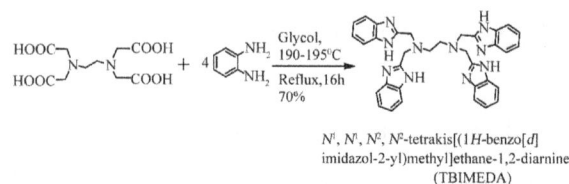

Scheme 1. Preparation of N,N,N,N-tetrakis-(2-benzimidazolylmethyl)-1,2-ethanediamine(TBIMEDA).

Table 1. Crystallographic data for **1-3**.

Crystalline data empirical formula	1	2	3
	$C_{34}H_{42}CdN_{10}O_9S$	$C_{34}H_{40}CoN_{10}O_8S$	$C_{34}H_{40}N_{10}NiO_8S$
Mr/g mol⁻¹	879.24	807.75	807.53
T/K	293(2)	293(2)	293(2)
Crystal system	Orthorhombic	Orthorhombic	Orthorhombic
Space group	$P2_12_12_1$	$P2_12_12_1$	$P2_12_12_1$
a/Å	11.2807(16)	11.366(2)	11.3360(13)
b/Å	14.980(2)	14.872(3)	14.2121(17)
c/Å	23.081(3)	22.655(5)	23.605(3)
V/Å³	3900.2(10)	3829.7(14)	3802.9(8)
Z	4	4	4
ρ_c/g·cm⁻³	1.497	1.401	1.410
μ/mm⁻¹	0.679	0.565	0.628
S	1.042	0.969	1.026
Reflections collected	24451	21215	21583
Unique collected	7645	7492	7453
R_{int}	0.0303	0.0925	0.0375
R_1[a] (I > 2 (I))	0.0484,	0.0814	0.0590
wR_2[b] (all data)	0.1357	0.2417	0.1663

[a]$R_1 = \sum \|F_o\| - |F_e\| / \sum |F_o|$,

[b]$wR_2 = \left[\sum w \left(F_o^2 - F_e^2 \right)^2 / \sum w \left(F_o^2 \right)^2 \right]^{1/2}$.

3. RESULTS AND DISCUSSION

3.1. Crystal Analyses for 1 - 3

Complexes **1** - **3** are alloisomers with each other, **1** is therefore selected as a typical example for discussing their crystal configurations. The view of the single core of **1** is depicted in **Figure 1**. Some selected bond distances and angles are listed in **Table 1**, and the hydrogen bonding data are listed in **Table 2**. **1** adopted orthorhombic system with $P2_12_12_1$ space group (unit cell parameters: a = 11.2807(16) Å, b = 14.980(2) Å, c = 23.081(3) Å, β = 90.00°, V = 3900.2(10) Å^3, Z = 4, Dc = 1.497 mg/cm^3). Cadmium(II) ion in the core of complex **1** is coordinated by one neutral TBIMEDA ligand, forming five of five-member rings (**Figure 1**). The octahedral coordination core around cadmium(II) ion includes six N atoms, two of them originate from the chain of ethanediamine and the others originate from the unsaturated N atoms in imidazole rings. All distances of Cd-N described above are near to or shorter than the sum of Van der Waals radii for Cd and N, and the effective chelations result in distorted octahedron around Cd(II) ions. However, the saturated N atoms in imidazoles, include N3, N5, N11 and N12, are protonated and without any coordination contribution.

3.2. Hydrogen Bond Interactions in 1

In the crystal structure of complex **1**, the Cd(II) coordination cores are linked through intermolecular N-H⋯O and O-H⋯O hydrogen bonds with sulfate ions and water molecules (**Figure 2**), forming a complicate 3D architecture. If all water molecules are omitted, the Cd(II)

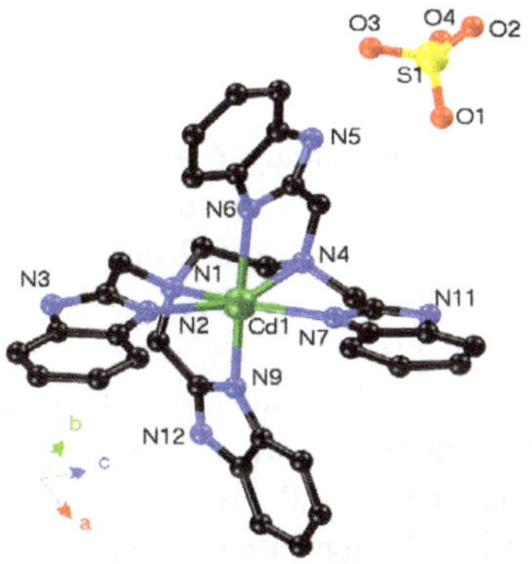

Figure 1. Diagram of single core in crystal architecture of **1**. All H atoms and 5 water molecules (O5-O9) are omitted for clarity. Just like the chelation of EDTA with metal ion, Cd^{II} in **1** adopting an octahedral configuration is chelated by five of 5-membered rings, where the two positive charges are counteracted by the sulfate anion.

Table 2. Select bond lengths (Å) and angles (°) for **1** - **3**.

1			
Cd(1)-N(7)	2.234(5)	Cd(1)-N(2)	2.251(5)
Cd(1)-N(9)	2.312(5)	Cd(1)-N(1)	2.480(5)
Cd(1)-N(6)	2.311(5)	Cd(1)-N(4)	2.526(4)
N(7)-Cd(1)-N(2)	141.56(19)	N(7)-Cd(1)-N(6)	91.0(2)
N(7)-Cd(1)-N(9)	92.50(19)	N(2)-Cd(1)-N(9)	93.55(18)
N(7)-Cd(1)-N(1)	143.9(2)	N(2)-Cd(1)-N(1)	73.72(18)
N(9)-Cd(1)-N(1)	73.5(2)	N(7)-Cd(1)-N(4)	72.87(18)
N(6)-Cd(1)-N(4)	71.84(17)	N(9)-Cd(1)-N(4)	90.07(18)
N(2)-Cd(1)-N(6)	96.09(17)	N(2)-Cd(1)-N(4)	144.97(18)
N(6)-Cd(1)-N(9)	159.6(2)	N(1)-Cd(1)-N(4)	74.01(17)
N(6)-Cd(1)-N(1)	92.1(2)		

2			
Co(1)-N(3)	2.142(7)	Co(1)-N(8)	2.148(7)
Co(1)-N(1)	2.169(7)	Co(1)-N(10)	2.292(8)
Co(1)-N(6)	2.165(6)	Co(1)-N(9)	2.337(6)
N(3)-Co(1)-N(8)	131.9(3)	N(3)-Co(1)-N(6)	91.1(3)
N(3)-Co(1)-N(1)	92.7(3)	N(8)-Co(1)-N(1)	93.4(3)
N(3)-Co(1)-N(10)	151.1(3)	N(8)-Co(1)-N(10)	76.0(3)
N(1)-Co(1)-N(10)	93.1(3)	N(3)-Co(1)-N(9)	75.2(3)
N(6)-Co(1)-N(9)	93.4(2)	N(1)-Co(1)-N(9)	76.5(2)
N(8)-Co(1)-N(6)	92.7(2)	N(10)-Co(1)-N(9)	78.7(3)
N(6)-Co(1)-N(1)	167.8(3)	N(10)-Co(1)-N(9)	78.7(3)
N(10)-Co(1)-N(9)	78.7(3)		

3			
Ni(1)-N(7)	2.078(4)	Ni(1)-N(1)	2.085(4)
Ni(1)-N(3)	2.098(4)	Ni(1)-N(9)	2.158(4)
Ni(1)-N(5)	2.088(4)	Ni(1)-N(10)	2.173(4)
N(9)-Ni(1)-N(10)	82.91(15)	N(9)-Ni(1)-N(10)	82.91(15)
N(9)-Ni(1)-N(10)	82.91(15)	N(9)-Ni(1)-N(10)	82.91(15)
N(9)-Ni(1)-N(10)	82.91(15)	N(9)-Ni(1)-N(10)	82.91(15)
N(9)-Ni(1)-N(10)	82.91(15)	N(9)-Ni(1)-N(10)	82.91(15)
N(9)-Ni(1)-N(10)	82.91(15)	N(9)-Ni(1)-N(10)	82.91(15)
N(9)-Ni(1)-N(10)	82.91(15)	N(9)-Ni(1)-N(10)	82.91(15)
N(9)-Ni(1)-N(10)	82.91(15)	N(9)-Ni(1)-N(10)	82.91(15)
N(9)-Ni(1)-N(10)	82.91(15)	N(9)-Ni(1)-N(10)	82.91(15)
N(9)-Ni(1)-N(10)	82.91(15)		

coordination water molecules are omitted, the Cd(II) coordination cores and sulfate ions will be linked by hydrogen bonding interactions to form a metal organic framework (MOF, as describing in **Figure 3**), and this MOF can be simplified as a topologic diagram with characters of 3×7^2 around Cd cores and $3^2 \times 7^5 \times 8^3$ around sulfate cores (**Figure 4**). Meanwhile, if all Cd cores were ignored, the sulfate ions and water molecules will be linked together to construct a 2D chain running along the a axis (**Figure 5**).

3.3. Photoluminescence

Excitation under $\lambda_{Ex} = 341$ nm, ligand TBIMEDA gave a emission at $\lambda_{Em} = 375$ nm, which was contributed from its conjugated configuration with π-π^* electron transition (**Figure 6**). After crystal assembly using the ligand with the metal ions with d^{10} configuration, such as Cd(II) in **1**, its crystal state also presented a strong emission peaks at $\lambda_{Em} = 417$ nm ($\lambda_{Ex} = 339$ nm). Obviously, compared with

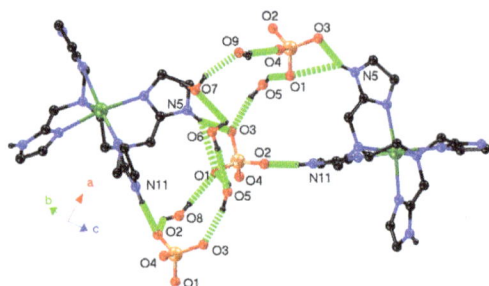

Figure 2. Hydrogen bonding interactions in **1**'s crystal architecture.

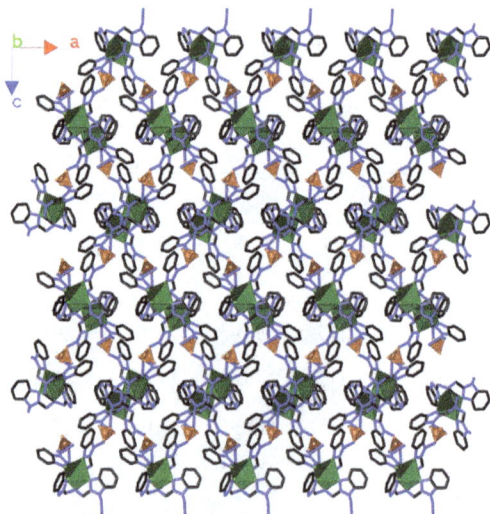

Figure 3. Crystal architecture of **1** constructed by hydrogen bonding interactions. Cd^{II} cores and SO$_4^{2-}$ ions are linked together to form a network stretched in the ac plane. All H$_2$O molecules and H atoms without hydrogen bonding contributions are omitted for clarity.

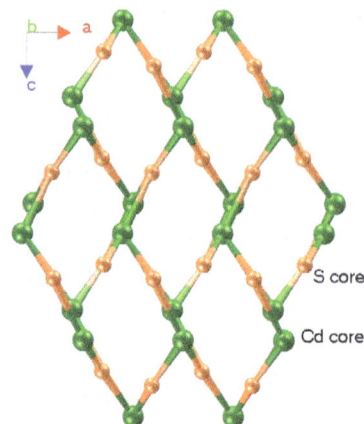

Figure 4. Simplified graph of for complex **1** with topologic characters of $3 \times 7 \times 8$ around Cd(II) cores and $3^2 \times 7^4 \times 8^4$ around sulfate anions.

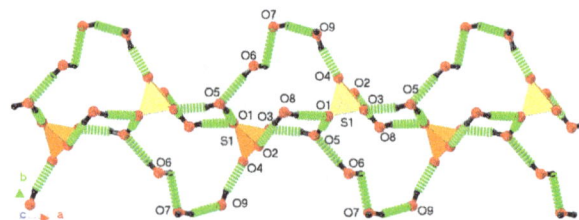

Figure 5. Simplified graph of for **1** only with water molecules and sulfate anions.

the ligand's emission, the fluorescent peaks of **1** emerged with red shift. This photoluminescence mechanism originated from ligand-metal charge transition (LMCT) [15], and its emission peak position was decided by the coordination situation of ligand with metal ion [16].

4. CONCLUSION

Ligands of diamine with imidazole group exhibited potentially application in biodegradation. In this work, after preparation of N,N,N',N'-tetrakis-(2-benzimidazolylmethyl)-1,2-ethanediamine(TBIMEDA), and using it as ligand reacting with different salts, three crystal architectures were self-assembled under solvothermal conditions. In crystal assembly, beside coordination between metal and ligands, the crystal architectures were also sustained by multi hydrogen bonding interactions from the complex cores with sulfate anions and water molecules. Three crystal architectures of M(TBIMEDA) SO$_4 \cdot$5H$_2$O(M = Cd^{II}, Co^{II}, Ni^{II}) all adopted orthorhombic crystalline with $P2_12_12_1$ space group. The metal center was chelated by three TBIMEDA to construct an octahedral configuration with five 5-numbered chelating rings, where the coordinating atoms were unsaturated N in imidazole ring and saturated N from ethylenediamine chain. These complexes can be used as a model to study

Table 3. Typical hydrogen bonding interactions in **1 - 3**.

1				
Donor-H···Acceptor	D-H/Å	H···A/Å	D-A/Å	D-H-A/°
N11-H11···O2	0.86	1.83	2.68	171
N5-H5···O3	0.86	1.98	2.77	152
O5-H5a···O1	0.85	2.00	2.81	159
O8-H8e···O1	0.85	1.98	2.81	166
O5-H5b···O3	0.85	1.91	2.74	166
O7-H7a···O9	0.85	1.97	2.78	158

2				
Donor-H···Acceptor	D-H/Å	H-A/Å	D-A/Å	D-H-A/°
N4-H4···O4	0.86	1.97	2.86	153
O2w-H2wa···O1	0.85	1.96	2.81	176
N7-H7a···O2w	0.86	1.91	2.77	177
O3w-H3wa···O3	0.85	1.90	2.74	168
O5w-H5wa···O4w	0.85	1.60	2.45	179

3				
Donor-H···Acceptor	D-H/Å	H-A/Å	D-A/Å	D-H-A/°
N4-H4···O2	0.86	1.86	2.70	162
N2-H2···O3	0.86	1.85	2.70	172
O1w-H1wa···O3	0.85	1.97	2.80	165
O1w-H1wb···O4w	0.85	1.99	2.82	164
O3w-H3wa···O1	0.85	1.87	2.72	178
O3w-H3wb···O4	0.85	2.33	3.18	178
N8-H8···O1w	0.86	1.87	2.72	166

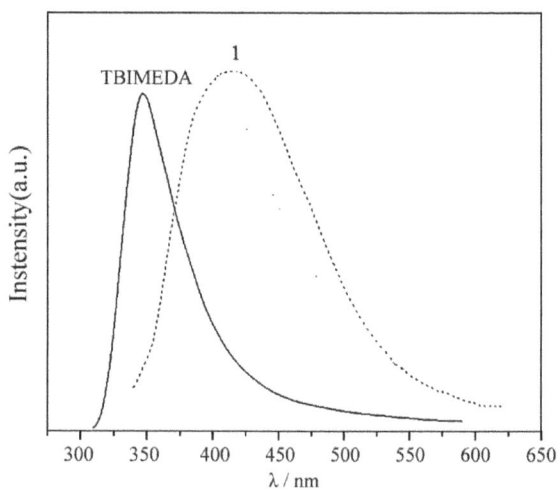

Figure 6. Emission spectra of ligand TBIMEDA (λ_{exa} = 341 nm) and **1** (λ_{exa} = 339 nm).

the effect of stereochemistry on the coordination polyhedron of M(II) ions. Although these M(II) ions were coordinated by six atoms, their coordination bonds were different in length, and therefore they may use different hybrid obits to form inner orbital or outer orbital coordination compounds. Obviously, these conclusions should be supported by magnetic determination and thermal gravimetric analysis. Further study should be intensively investigated.

5. ACKNOWLEDGEMENTS

This work was financially supported by the National Natural Science Foundation of China (No. 20771073); by the "211" Project of Guangdong Province (the third time): The Key and Common Technologies in Fine Chemical Engineering; and by The Key Project of Education Office from Guangdong Province, China.

REFERENCES

[1]   Huang, X.-C., Zhang, J.-P. and Chen, X.-M. (2004) A new route to supramolecular isomers via molecular templating: Nanosized molecular polygons of copper(I) 2-methylimidazolates. *Journal of the American Chemical Society*, **126**, 13218-13219. doi:10.1021/ja0452491

[2]   Moon, D., Kang, S., Park, J., Lee, K., John, R.P., Won, H., Seong, G.H., Kim, Y.S., Kim, G.H., Rhee, H. and Lah, M.S. (2006) Face-driven corner-linked octahedral nanocages: M_6L_8 cages formed by C_3-symmetric triangular facial ligands linked via C_4-symmetric square tetratopic Pd^{II} ions at truncated octahedron corners. *Journal of the American Chemical Society*, **128**, 3530-3531. doi:10.1021/ja060051h

[3]   Luo, F., Zheng, J.-M. and Batten, S.R. (2007) Unprecedented (3,4)-connected metal-organic frameworks (MOFs) with 3-fold interpenetration and considerable solvent-accessible void space. *Chemical Communications*, **36**, 3744-3746. doi:10.1039/B706177C

[4]   Britt, D., Tranchemontagne, D. and Yaghi, O.M. (2008) Metal-organic frameworks with high capacity and selectivity for harmful gases. *PNAS*, **105**, 11623-11627. doi: 10.1073/pnas.0804900105

[5]   Wu, A.J., Hahn, J.E.P. and Pecoraro, V.L. (2005) Structural, spectroscopic, and reactivity models for the manganese catalase. *Chemical Reviews*, **104**, 903-938. doi: 10.1021/cr020627v

[6]   Maji, T.K., Ohba, M. and Kitagawa, S. (2005) Transformation from a 2D stacked layer to 3D interpenetrated framework by changing the spacer functionality: Synthesis, structure, adsorption, and magnetic properties. *Inorganic Chemistry*, **44**, 9225-9231. doi: 10.1021/ic050835g

[7]   Lopez, X., Benard, M. and Rohmer, M.M. (2007) Influence of electron-attractor substituents on the magnetic properties of Ni(II) string complexes. *Inorganic Chemistry*, **46**, 5-7. doi: 10.1021/ic061705q

[8]   Belloni, M., Kariuki, B.M., Manickam, M., Wilkie, J. and Preece, J.A. (2005) Design of potentially photorefractive

liquid crystalline materials: Derivatives of 3,6-disubstituted carbazole. *Crystal Growth & Design*, **5**, 1443-1450. doi: 10.1021/cg049580s

[9] Kjlo, J. and Okamoto, Y. (2002) Organo lanthanide complexes for electroluminescent materials. *Chemical Reviews*, **102**, 2357-2368. doi: 10.1021/cr010448y

[10] Hendriks, H.M.J., Bokkel, H.W.O. and Reedijk, J. (1979) Synthesis, characterisation and complex formation of N,N,N',N'-tetrakis-(2-benzimidazolylmethyl)-1,2-ethane diamine, a new hexadentate ligand. *Recueil des Travaux Chimiques des Pays-Bas*, **98**, 499-500.

[11] Vishweshwar, P., Nangia, A. and Lynch, V.M. (2002) Recurrence of carboxylic acid-pyridine supramolecular synthon in the crystal structures of some pyrazinecarboxylic acids. *The Journal of Organic Chemistry*, **67**, 556-565. doi: 10.1021/jo0162484

[12] Addison, A.W., Hendriks, H.M.J., Reedijk, J. and Thompson, L.K. (1981) Copper complexes of the "tripod" ligand tris(2-benzimidazolylmethyl)amine: Five- and six-coordinate copper(II) derivatives and some copper(I) derivatives. *Inorganic Chemistry*, **20**, 103-110.

doi: 10.1021/ic50215a024

[13] Wahlgren, C.G. and Addison, A.W. (1989) Synthesis of some benzimidazole-, pyridine- and imidazole-derived chelating agents. *Journal of Heterocyclic Chemistry*, **26**, 541-543. doi:10.1002/jhet.5570260303

[14] Sheldrick G.M. (1997) SHELXL-97, program for the refinement of the crystal structures. University of Göttingen, Göttingen.

[15] Guo, C.Y., Wang, Y.Y., Xu, K.Z., Zhu, H.L., Liu, P., Shi, Q.Z. and Peng, S.M. (2007) Crystal structures, bioactivities and fluorescent properties of four diverse complexes with a new symmetric benzimidazolic ligand. *Polyhedron*, **27**, 3529-3536. doi:10.1016/j.poly.2008.08.018

[16] Wang, X.L., Bi, Y.F., Liu, G.C., Lin, H.Y., Hu, T.L. and Bu, X.H. (2008) Zn(II) coordination architectures with mixed ligands of dipyrido [3,2-d:2',3'-f] quinoxaline/2,3-di-2-pyridylquinoxaline and benzenedicarboxylate: Syntheses, crystal structures, and photoluminescence properties. *CrystEngComm*, **10**, 349-356. doi:10.1039/B706071H

29

Spectroscopic and molecular structure characterization of the bis(2-aminophenol)(5,10,15,20-tetraphenylporphyrinato) cobalt(II) complex

Yassin Belghith[1], Anissa Mansour[1], Jean-Claude Daran[2], Habib Nasri[1]

[1]Laboratoire de Physico-chimie des Matériaux, Faculté des Sciences de Monastir, Université de Monastir, Monastir, Tunisia
[2]Laboratoire de Chimie de Coordination, CNRS UPR 8241, Toulouse, France
Email: hnasri1@gmail.com, Habib.Nasri@fsm.rnu.tn

ABSTRACT

The reaction of the cobalt(II) *meso*-tetraphenylporphyrin (TPP) starting material with an excess of 2-aminophenol (Hon) in organic solvents, yields the cobalt(II) porphyrin species $[Co^{II}(TPP)(Hon)_2]$ (1). This compound has been characterized by UV-vis, IR, MSI MS and ^1H NMR spectroscopy. The UV-vis data and especially the proton NMR results, for the isolated product, indicated that complex 1 is a Co(II) *meso*-porphyrin derivative. The X-ray molecular structure of the title compound bis(2-aminophenol) (tetraphenyl-porphyrinato) cobalt(II) has been determined. This structure is the first one reported of a metalloporphyrin with a 2-aminophenol axial ligand species. The central metal is hexacoordinated by the four nitrogen atoms of the pyrrole rings and the nitrogen atoms of the two Hon trans axial ligands.

Keywords: *Meso*-Tetraphenylporphyrin Cobalt(II); 2-Aminophenol Cobalt(II) Complex; X-Ray Molecular Structure

1. INTRODUCTION

Complexes with noble metal such as rhenium and technetium have been used for decades in medicine specially as potential radiotherapeutic agents for cancer [1] and many cobalt(II) phthalocyaninetetrasodiumsulfonate has been found to enhance the rate of oxidation of 2-aminophenol with dioxygen in water to 2-amino-phenoxazin-3-one [2]. Thus, the main goal of the present work is to prepare a metalloporphyrin complex with much cheaper metal in particular transition metals such as cobalt, which is coordinated to a 2-aminophenol (Hon) axial ligand [3]. Such species can eventually replace the rhenium and technetium as potential radiotherapeutic agents and also can be used as catalysts in many organic reactions. It is noteworthy that no structures of 2-aminophenol Co(II)

porphyrin derivative were reported up to day. The only reported cobalt(II) with a substituted 2-aminophenol is the polymer$\{[Co^{II}(\mu_2\text{-}AT)]\}_n$(where μ_2-AT = μ_2-3-amino-(S)-tyrosinato) [4]. In order to gain more insight into the physico-chemistry properties of this new Co(II)-porphyrin derivative. We report in this paper the results of the spectroscopy and the X-ray molecular structure investigation on the 2-aminophenol cobalt(II) *meso*-tetraphenyl-porphyrin complex $[Co^{II}(TPP)(Hon)_2]$.

2. EXPERIMENTAL

2.1. Materials and Methods

Ultraviolet-visible (UV-vis) spectra were measured on a SHIMADZU UV-2401 spectrometer and IR spectra were recorded from pure products using a PerkinElmer Spectrum 100 FT-IR equipped with a single-bounce diamond attenuated total reflectance (ATR) sampling accessory. Proton magnetic resonance spectra were measured at room temperature on a Bruker 300 Ultrashield spectrometer. Mass spectra were recorded on spectrometers MS/MS API-365 (Perkin Elmer Sciex) equipped with electrospray source (ESI). All chemicals were purchased from SIGMA-ALDRICH Co. LLC. They were used as received without further purification except the dichloromethane which was distilled under CaH_2 and freshly used.

Synthetic Methods

The free base 5,10,15,20-tetraphenylporphyrin and the corresponding cobalt(II) derivative $[Co^{II}(TPP)]$ were synthesized by literature methods [5,6].

2.2. Synthesis of $[Co^{II}(TPP)(Hon)_2]$

$[Co^{II}(TPP)]$ [6] (100 mg, 0.149 mmol) and (190 mg, 2.231 mmol) of 2-aminophenol in 25 mL of dichloromethane were stirred overnight at room temperature. The color of the solution turns from red-orange to dark-red and single crystals of the complex were prepared by slow diffusion of the hexanes into the CH_2Cl_2 solution. Typical yields were 55% - 65%. UV-vis [λ_{max} (nm) in CH_2Cl_2, (logε)]: 434 (4.46), 521 (3.54), 555 (3.41). IR (KBr):ν, cm^{-1} 3411, 3385, 3354, 3329, 3298, 3027, 2324, 1888, 1811, 1622, 1595, 1500, 1440, 1378 , 1362, 1350, 1271, 1229, 1203, 1175, 1156, 1147, 1071, 1000, 993, 954, 928, 834, 796, 747, 701. ^{1}H NMR ($CDCl_3$): δ, ppm, 16.07 (s, 8 H), 13.20 (s, 8 H), 9.97 (s, 8 H), 9.76 (s, 4 H), 5.30 (s, H), 3.70 (s, 2 H). ESI(+) MS m/z calc for (C_6H_7NO) [M]$^+$ 109.05, found 109.20; m/z calc for ($C_{44}H_{28}N_4Co$) [M]$^+$ 671.16, found 671.40; m/z calc for ($C_{50}H_{35}CoN_5O$) [M-H]$^+$ 779.21 found 779.50.

2.3. X-Ray Diffraction

A dark purple crystal of the cobalt(II) 2-aminophenol porphyrin derivative with approximate dimensions of 0.44 × 0.40 × 0.27 mm^3 was mounted under inert perfluoropolyether at the tip of glass fiber and cooled in the cryostream of an Oxford Diffraction XCALIBUR.

The data were collected at 180 K using the monochromatic Mo $K\alpha$ radiation (λ = 0.71073). Lattice parameters were obtained by least-squares fit to the optimized setting angles of the 4447 collected unique reflections in the full theta range data collectioon 3.02° < θ < 26.02°. Intensity data were recorded using ω scans. Data reduction was done using CrysAlisPro, Oxford Diffraction Ltd. [7]. Empirical absorption correction using sphereeeeeeecal harmonics, implemented in SCALE3 ABSPACK scaling algorithm. The minimum and the maximum transmission factors values are 0.8592 and 1.000, respectively. The structure was solved by direct methods using SIR-2004-1.0 [8] and refined by full-matrix least squares on |F|^2 using the SHELXL-97 program [9]. The asymmetric unit of the structure contains half [Co^{II}(TPP)(Hon)$_2$] molecule where the Co(II) is located on an inversion center. The two hydrogens of the NH_2 group and the H of the OH group of the axial ligand were found in the difference Fourier map and were included in the refinement using restraints (N-H = 0.92(1) Å; O-H = 0.85(1) Å) with U_{iso}(H) = 1.2U_{eq}(N_3, O1). In the last refinement cycles, the H of the NH_2 group were fixed and treated as riding on their parent N atom. The H attached to O1 was kept refining. The positions of H atoms at- tached to C atoms were calculated and they were treated as riding on their parent C atoms with C-H = 0.95Å and U_{iso}(H) = 1.2 U_{eq}(C). Refinement was then carried on to convergence with anisotropic thermal parameters for all non hydrogen atoms. Crystal data and experimental parameters used for the intensity data collection are summarized in **Table 1**. Complete crystallographic details are available from the CCDC (see Supporting information section).

Table 1. Crystal data and refinement parameters for [Co^{II}(TPP)(Hon)$_2$].

Empirical formula	$C_{56}H_{42}N_6O_2Co$
Formula weight, (g·mol^{-1})	889.89
Crystal system	triclinic
Space group	*P*-1
Lattice constants	
a (Å)	10.7139(3)
b (Å)	11.0463(4)
c (Å)	11.6128(4)
α (°)	114.290(3)
β (°)	107.973(3)
γ (°)	103.296(3)
Volume *V* (Å^3)	1085.92(9)
Z	1
D$_{cal}$, g/cm^3	1.361
Absorption coefficient, μ (mm^{-1})	0.448
F(000)	463
Crystal size (mm^3)	0.44 × 0.40 × 0.27
T (K)	180(2)
θ-range for data collection	3.02 - 26.02
Limiting indices	$-13 \leq h \leq 13, -13 \leq k \leq 13$
	$-14 \leq l \leq 14$
Completeness (%)	99.9
T_{min}/T_{max}	0.859/1.000
Reflections collected unique	21397/4447
R(int)/*R*(sigma)	0.0245/0.0142
Data/restraints/parameters	3975/1/298
Goodness-of-fit on F^2	1.361
Final R indices [I > 2σ(I)]	R$_1$ = 0.0492, wR$_2$ = 0.1442
R indices (all data)	R$_1$ = 0.0546, wR$_2$ = 0.1491
Largest peak and hole (eÅ^{-3})	1.30 and –0.62

3. RESULS AND DISCUSSION

3.1. UV-Visible, IR and Mass Spectrometry

The electronic spectra of complex **1** and the [Co^{II}(TPP)] starting compound are represented in **Figure 1**. The Soret bands of the related species [Co^{II}(TpivPP)(OAc)]$^-$ [10], and other Co(II) *meso*-porphyrin derivatives (**Table 2**), are in the range [438 - 441 nm]. This data confirm complex creation of a new Co(II)-*meso*-porphyrin coordination compound by displaying red shifted Soret band of complex **1**. The IR spectrum of [Co^{II}(TPP)(Hon)$_2$] (**Figure 2**) shows the existence of the coordinated 2-aminophenol ligand. Thus, the frequencies of the O-H(Hon) and the N-H(NH$_2$) stretching are 3386 and 3332 cm^{-1} respectively.

The ESI MS experiments recorded in the positive ion mode show the presence of the three fragments: [Hon]$^+$ with a m/z value of 190.20, the [Co^{II}(TPP)]$^+$ and the [Co^{II}(TPP)(on)]$^+$ with a m/z values of 671.40 and 779.50 respectively. For the latter fragment, it corresponds to [M-H]$^+$ where the deprotonation of the OH group occurs on one 2-aminophenol axial ligand.

3.2. Proton NMR Spectroscopy

The paramagnetic starting material [Co^{II}(TPP)] species (with the ground state configuration 3d^7) presents downfield chemical shifts of the β-pyrrole protons (Hβ-pyrr) = 15.75 ppm) as shown in **Figure 3**. For the diamagnetic cobalt(III) porphyrin derivatives (with the ground state configuration 3d^6), the β-pyrrole protons resonate in the normal regions of the free base TPP porphyrin (8.1 ppm < δ(Hβ-pyrr) < 9.1 ppm) [11,12] (**Table 3**). Complex **1**

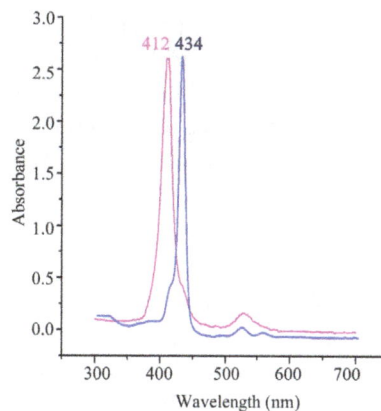

Figure 1. UV-vis spectra of [Co^{II}(TPP)] starting material (pink line) and [Co^{II}(TPP)(Hon)$_2$] (blue line). The spectra were taken in CH$_2$Cl$_2$ at room temperature.

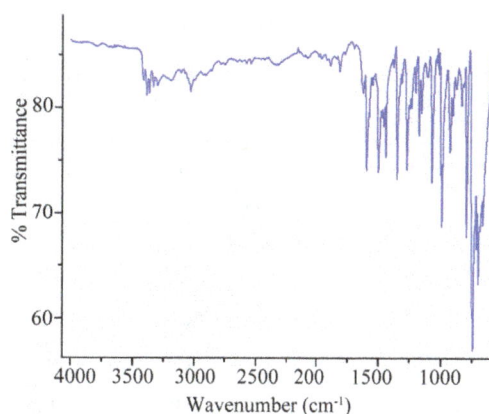

Figure 2. IR spectrum complex [Co^{II}(TPP)(Hon)$_2$] (solid neat).

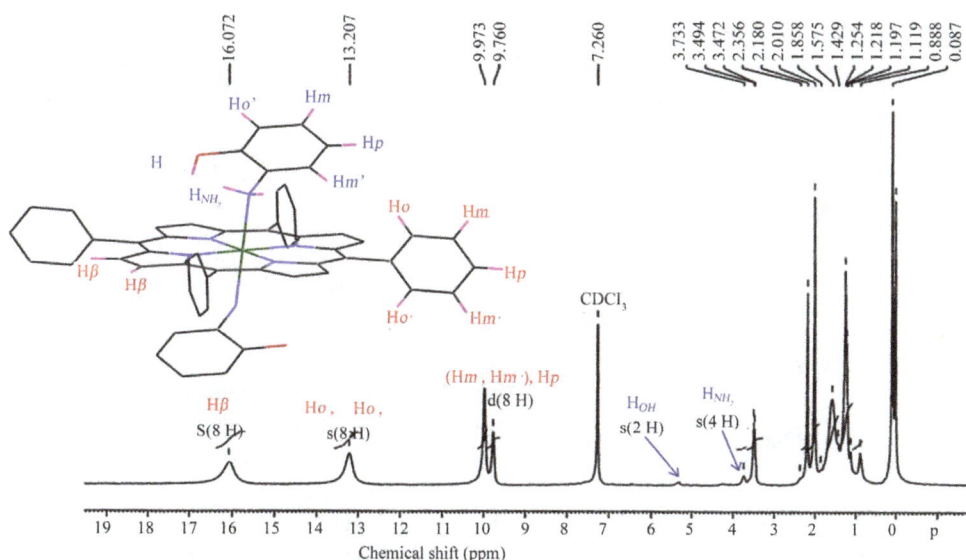

Figure 3. Proton NMR spectra of the [Co^{II}(TPP)(Hon)$_2$] complex taken in CDCl$_3$ solution at room temperature.

presents a peak at 16.07 attributed to the β-pyrrole protons. This is an indication that our derivative is a paramagnetic cobalt(II) *meso*-porphyrin species [13]. The ^{1}H NMR peaks for the NH$_2$ and the OH protons of the two Hon ligands appear at $\delta \approx 3.7$ ppm and $\delta \approx 5.3$ ppm respectively.

3.3. Crystal Structure

The structure of [Co^{II}(TPP)(Hon)$_2$] compound was determined at 180 K and **Figure 4** is an ORTEP diagram [15] of this complex. The Co(II) is coordinated to the four nitrogens of the porphyrin ring and the nitrogen at oms of the NH$_2$ groups of the two 2-aminophenol trans axial ligands. Selected bond distance (Å) angles (deg) for the [Co^{II}(TPP)(Hon)$_2$] complex are represented in **Table 4**. The preference for the amino nitrogen is understandable since Co(II) would have a lower affinity for the more electronegative oxygen donor atom, which prefers

Table 2. Electronic spectra data for selected cobalt(II) *meso*-porphyrins.

Complex	λ_{max} (nm)		Ref.
	Soret band	$\alpha\,\beta$ bands	
[Co^{II}(TpivPP)]^a	412	524	-	[10]
[Co^{II}(TPP)]^b	412	528	-	t.w.^{c}
[Co^{II}(TPP)(Hon)$_2$]^b	434	521	555	t.w.
{[Co^{II}(TPP)(μ-4,4'-bipy)].2bipy}$_n^b$	438	545	-	[10]
[Co^{II}(TPP)(OAC)]$^{-a}$	441	553	-	[10]

^{a}In chlorobenzene. ^{b}In dichloromethane. ^{c}t.w: This work.

Table 3. ^{1}H NMR^a data for the free-base TPP porphyrin and a selected cobalt(II) tetraphenylporphyrin complexes.

Complex	Hβ-pyrrol.	H-Phenyl.^{b}	Ref.
TPP	8.84	8.23; 7.91; 7.67; 7.26	t.w.^{c}
[Co^{II}(TPP)]	15.75	13.10; 9.80; 8.95	t.w.
{[Co^{II}(TPP)(μ-4,4'-bipy)]-2bipy}$_n$	13.45	9.98; 8.82; 8.42; 7.64	[10]
[Co^{II}(TPP)(py)]	12.5	8.50; 8.33; 7.82	[14]
[Co^{II}(TPP)(Hon)$_2$]	16.07	13.21; 9.97; 9.76	t.w.

^{a}Data are recorded in CDCl$_3$ solvent at room temperature. ^{b}Designate respectively the protons Ho, Ho', Hm, Hm' and Hp of the phenyl rings of the porphyrin TPP. ^{c}t.w: This work.

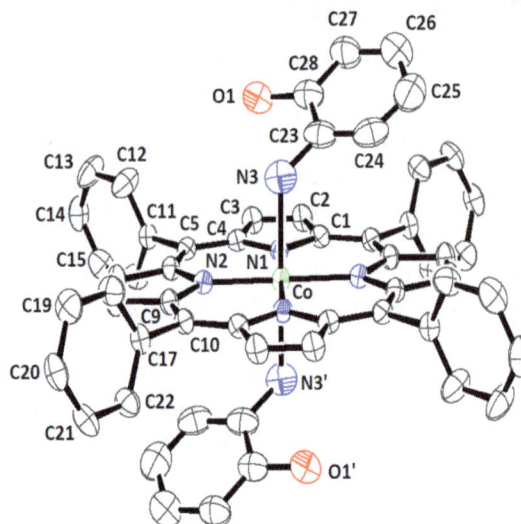

Figure. 4. ORTEP drawing [15] of [Co^{II}(TPP)(Hon)$_2$] complex showing thermal ellipsoids at 50% probability level. The H atoms have been omitted for clarity.

Table 4. Selected bond distances (Å) and angles (deg) for the [Co^{II}(TPP)(Hon)$_2$] complex.

Cobalt coordination polyhedron			
Co-N1	1.973(2)	N1-Co-N3	92.88(9)
Co-N2	1.975(2)	N2-Co-N3	91.30(9)
Co-N3	2.391(3)	N1-Co-N2	89.97(8)
Co-N$_p^a$	1.974(2)		
2-aminophenol ligand			
N3-C23	1.413(4)	O1-C28-C23	115.9(3)
C28-O1	1.405(4)	N3-C23-C28	118.5(3)
C23-C28	1.386(5)	C23-N3-Co	120.9(2)

^{a}Co-N$_p$ represents the average cobalt-porphyrinato nitrogen atoms bond usually known as the equatorial bond.

ionic interaction with the metal. In the Cambridge Structural Database (CSD, version 5.33 of November 2011 [16]) few structures of transition-metal-Hon structures were reported among them only one structure of a Co(II) substituted 2-aminophenol ligand was mentioned [4]. For all these published structures the metal ion is either bonded only to the nitrogen of the amino group or to the NH$_2$ group and the deprotonated oxygen (**Table 5**). **Figure 5** illustrates the geometry of the transition metal (M) and the amino group of one coordinated Hon ligand. The Co^{II}-NH$_2$(Hon) bond length for **1** [2.391(3) Å] is slightly longer than the one of the related Co^{II}-bis[2-μ_2-3amino-(S)-tyrosinato] polymer derivative [4] with values of 2.311(7) and 2.258(7) Å. The only reported structures of complexes with metal coordinating Hon ligand are those with the rhenium transition metal [17,18] (**Table 6**) where the M-NH$_2$ distance is in the range [2.048(4) - 2.270(4) Å]. The C$_I$-NH$_2$ and C$_{II}$-OH of the Hon ligand for complex 1 and the reported complexes are within the normal values: [1.40 - 1.45 Å] and [1.35 - 1.39 Å] respectively. The C$_I$-N-M angle is 120.9(2)° for our species and the related Co(II) derivatives are in the range [115.4° -

121.0°]. The complex [Co^{II}(TPP)(Hon)$_2$] presents an intramolecular hydrogen-bond (N$_3$-H.....O$_I$) with a distance of 2.677(5) Å which falls in the range [2.639(5) - 2.788(3) Å] for related species (**Table 5**). It has been noticed [19] for cobalt-porphyrin complexes that ruffling of the porphyrin core always results in a shortening of the porphyrin equatorial cobalt pyrrole nitrogen M-N$_p$ bonds. Thus, for the very ruffled structure [Co^{II}(TPP)] [6], the Co-N$_p$ bond length value is 1.923(4) Å. while the practically planar porphyrin core of the dimer

Table 5. Selected bond lengths (Å) and angles (°) for several 2-aminophenol and substituted 2-aminophenol complexes.

Complexe	M-NH$_2$	C$_I$-NH	C$_{II}$-OH	<C$_I$-N-M>	NH...O	<N-H...O>	Ref
[Co^{II}(TPP)(Hon)$_2$]	2.391(3)	1.413(1)	1.405(4)	120.9(2)	2.677(5)	110	t.w.^{e}
{[Co^{II}(μ$_2$-AT)]}n^a	2.311(7)	1.40(1)	1.35(1)	121.0(6)	2.72(1)	107	[4]
	2.258(7)	1.45(1)	1.39(1)	118.9(6)	2.72(1)		
[Pd^{II}(adtbp)(noo')]^{b,c}	2.066(2)	1.441(4)	1.382(4)	115.4(2)	2.788(3)		
Fac-[Re(CO)$_3$(opa)(Hon)]^d	2.248(4)	1.426(6)	1.346(5)	119.4(3)	2.639(5)	105	[17]
[Re(CO)$_3$(opa)(Hon)]^d	2.270(4)	1.441(5)	1.365(6)	117.9(4)	2.703(4	147	[18]

^{a}μ$_2$-AT = μ$_2$-3-amino-(S)-tyrosinato, ^{b}adtbp = (2-amino-4,6-di-t-butylphenol-*N*), ^{c}noo' = (2,4-di-t-butyl 6) (salicylideneamino) phenolato-N-O-O', ^{d}opa = 2-aminophenolato-N,O. ^{e}t.w: This work.

Table 6. Selected intermolecular hydrogen interactions.

	D-H^a (Å)	H...A^b (Å)	D...A (Å)	D-H...A (°)
C19-H19...Cg1^{iv}	0.95	2.82	3.655(3)	148
C13-H13...Cg2^i	0.95	2.71	3.500(4)	141
C14-H14...Cg8^{ii}	0.95	2.92	3.851(4)	166
C15-H15...Cg9^{iii}	0.95	2.85	3.636(4)	141

^{a}D: designate the donor atom , ^{b}A: designate the acceptor atom; Cg1 is the centroid of the N1-C1-C2-C3- C4 five membered ring; Cg2 is the centroid of the N2-C6-C7-C8-C9 five membered ring; Cg8 is the centroid of the C17-C18-C19-C20-C21-C22 six membered ring; Cg9 is the centroid of the C23-C24-C25-C26-C27-C28 six membered ring; Symmetry codes : (i) -x,1-y,1-z°; (ii) -x,-y,1-z°; (iii) x, y, 1+z°; (iv) 1-x, -y, 1-z.

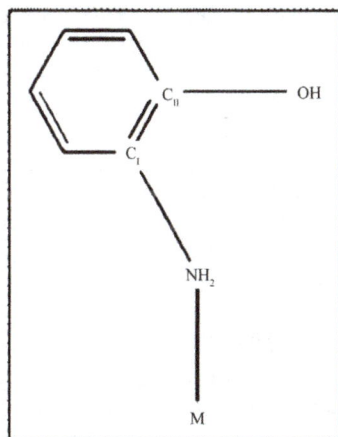

Figure 5. Drawing showing the ion metal M coordinated to the amino group of one 2-aminophenol (Hon) axial ligand.

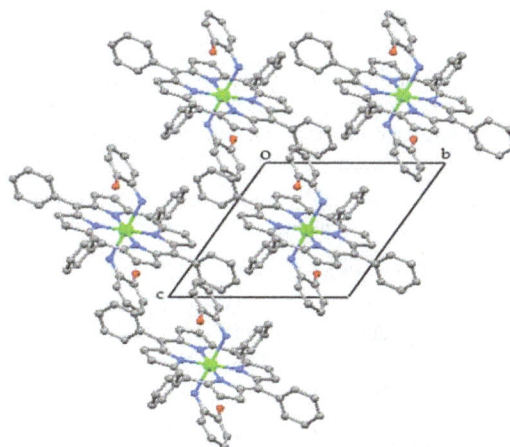

Figure 6. Drawing showing the packing in lattice of [Co^{II}(TPP) (Hon)$_2$], viewed down the *a* axis.

$\{[Co^{II}(TPP)(\mu\text{-}4,4'\text{-bipy})].2bipy\}_n$ [10] presents a Co-N_p distances of 1.985(5) and 1.993(1) Å. The Co-N_p value for our complex $[Co^{II}(TPP)(Hon)_2]$ is 1.974(2) Å which is an indication that **1** presents a slightly ruffled porphyrin core. On the other hand, the average displacements of the *meso* carbons above and below the porphyrin mean plan are small [−0.01 and 0.034 Å] which indicate a moderate ruffling.

The crystal structure of **1** is stabilized by intermolecular interactions type O-H.....Cg and C-H.....Cg where Cg is the centroid of pyrrole five member ring or phenyl six member ring. A selection of these intermolecular distances is summarized in **Table 6**. **Figure 6** is a drawing showing the packing in lattice of $[Co^{II}(TPP)(Hon)_2]$, viewed down the *a* axis.

4. SUPPORTING INFORMATION

Crystallographic data (excluding structure factors) for the $[Co^{II}(TPP)(Hon)_2]$ compound have been deposited with the Cambridge Crystallographic Data Centre as supplementary publication number CCDC-878829. Copies of the data can be obtained free of charge on application to CCDC, 12 Union Road, Cambridge CB2 1EZ, UK or email: deposit@ccdc.cam.ac.uk

5. ACKNOWLEDGEMENTS

The authors gratefully acknowledge financial support from the Ministry of Higher Education and Scientific Research of Tunisia.

REFERENCES

[1] Gerber, T.I.A., Luzipo, D. and Mayer, P. (2004) The reaction of the *cis*-dioxorhenium(V) core with N,O-donor ligands: Mono- and bidentate coordination of 3-methyl-2-aminophenol. *Inorganica Chimica Acta* , **357**, 429-435. doi:10.1016/j.ica.2009.02.037

[2] Hassenein, M., Abdo M., Gerges, S. and El-Khalafy, S. (2008) Study of the oxidation of 2-aminophenol by molecular oxygen catalyzed by cobalt(II) phthalocyaninetetrasodiumsulfonate in water. *Journal of Molecular Catalysis A*, **287**, 53-56. doi:10.1016/j.molcata.2006.11.045

[3] Abbreviations used: TPP: dianion of *meso*-tetraphenylporphyrin; Hon: 2-aminophenol (or ortho-aminophenol); OAc: acetato $\left(CH_3CO_2^-\right)^-$; μ_2-AT: μ^2-3-amino-(S)-tyrosinato; adtpb: (2-amino-4,6-di-t-butylphenol-N); noo'(2,4-di-t-butyl-6) (salicylideneamino) phenolato-N-O-O'; d: opa2-aminophenolato-N,O; PhCN: benzonitrile; BAP: tetrabutylammonium perchlorate.

[4] Gao, X.-Y., Xing, L.-X., Chen, Z.-F., Bao, Y.-C., Yang, J.-H., Wu, M.-F., Xiong, R.-G., You, X.-Z. and Fun, H.-K. (2003). Synthesis and crystal structure of a biologically relevant homochiral two-dimensional metal-organic herring-bone molecular grid: Bis(3-amino-(S)-ty-

rosinato) cobalt(II). *Chinese Journal of Inorganic Chemistry*, **19**, 802.

[5] Alder, A.D., Longo, F.R, Finarelli, J.D., Gildmacher, J., Assour, J. and Korsakoff, L. (1967) A simplified synthesis for meso- tetraphenyl porphine. *Journal of Organic Chemistry*, **32**, 476-476. doi:10.1021/jo01288a053

[6] Mansour, A., Belkhiria , M.S., Daran, J.C. and Nasri, H. (2010) (5,10,15,20-tetraphenylporphyrinato) cobalt(II)-18-crown-6 (1/1). *Acta Crystallographica   Section E*, **66**, m509-m510. doi:10.1107/S1600536810012080

[7] Oxford Diffraction (2009) Chrysalis promotions. Oxford Diffraction Ltd., Abingdon, Oxfordshire, England. doi:10.1021/ic50166a056

[8] Burla, M.C., Caliandro, R., Camalli, M., Carrozzini, B., Cascarano, G.L., De Caro, L., Giacovazzo, C., Polidori, G. and Spagna, R. (2005) *SIR*2004: An improved tool for crystal structure determination and refinement. *Journal of Applied Crystallography*, **38**, 381-388. doi:10.1107/S002188980403225X

[9] Sheldrick, G.M. (2008) A short history of *SHELX*. *Acta Crystallographica Section E*, **64**, 112-122. doi.10.1107/S0108767307043930

[10] Mansour, A. and Nasri, H. (2012) Synthesis, crystal structure and spectroscopic characterization of 1D Co(II) coordination polymers $\{[Co^{II}(\alpha,\beta,\alpha,\beta\text{-TpivPP})(\mu\text{-}4,4'\text{-bp} y)_2]\}_n$ and $\{[Co^{II}(TPP)(\mu\text{-}4,4'\text{-bpy})2]\}_n$ with the ($\alpha,\beta,\alpha,\beta$-tetrakis(o-pivalamidophenyl)porphyrin) $\alpha,\beta,\alpha,\beta$-TpivPP) and the tetraphenylporphyrin (TPP). In press.

[11] Hoshino, M., Sonoki, H., Miyazaki, Y., Iimura, Y. and Yamamoto, K. (2000) Structure and photo chemistry of dicyanocobalt(III) tetraphenyl porphyrin. Photochromic reaction caused by photodissociation of axial ligand. *Inorganic Chemistry*, **39**, 4850-4857. doi: 10.1021/ic000408x

[12] Shirazi, A. and Goff, H.M. (1982) Carbon-13 and proton NMR spectroscopy of four- and five-coordinate cobalt(II) porphyrins: Analysis of NMR isotropic shifts. *Inorganic Chemistry*, **21**, 3420-3425. doi: 10.1021/ic00139a030

[13] La Mar, G.N. and Walker, F.A. (1973) Proton nuclear magnetic resonance and electron spin resonance investigation of axial solvation in planar, low-spin cobalt(II) porphyrin complexes. *Journal of American Chemical Society*, **95**, 1790-1796. doi:10.1021/ja00787a017

[14] Choi, I.-K., Lin, Y., Wei, Z. and Rayan, M.D. (1997) Reactions of hydroxylamine with metal porphyrins. *Inorganic Chemistry*, **36**, 3113-3118. doi:10.1021/ic9605783

[15] Farrugia, L.J. (1997) *ORTEP*-3 for windows a ver sion of *ORTEP*-III with a graphical user interface (GUI). *Journal of Applied Crystallography*, **30**, 565. doi:10.1107/S0021889897003117

[16] Allen, F.H. (2002) The Cambridge Structural Data base: A quarter of a million crystal structures and rising. *Acta Crystallographica Section B*, **58**, 380-388. doi:10.1107/S0108768102003890

[17] Mayer, P., Potgieter, K.C. and Gerber, T.I.A. (2010) Rhenium (I), (III) and (V) complexes of tridentate ONX (X=O, N, S)-donor Schiff bases. *Polyhedron*, **29**, 1423-

1430. doi: 10.1016/j.poly.2010.01.013

[18] Gerber, T.I.A., Betz, R., Booysen, I.N., Potgieter, K.C. and Mayer, P. (2011) Coordination of bidentate aniline derivatives to the *fac*-[Re(CO)$_3$]$^+$ core. *Polyhedron*, **30**, 1739-1745. doi: 10.1016/j.poly.2011.04.019

[19] Iimura, Y., Sakurai, T. and Yamamoto, K. (1988) The crystal and molecular structure of aquachloro (*meso*-tetra-phenylporphyrinato) cobalt(III). *Bulletin of the Chemical Society of Japan*, **61**, 821-826. doi: 10.1246/bcsj.61.821.

Permissions

All chapters in this book were first published in OJIC, by Scientific Research Publishing; hereby published with permission under the Creative Commons Attribution License or equivalent. Every chapter published in this book has been scrutinized by our experts. Their significance has been extensively debated. The topics covered herein carry significant findings which will fuel the growth of the discipline. They may even be implemented as practical applications or may be referred to as a beginning point for another development.

The contributors of this book come from diverse backgrounds, making this book a truly international effort. This book will bring forth new frontiers with its revolutionizing research information and detailed analysis of the nascent developments around the world.

We would like to thank all the contributing authors for lending their expertise to make the book truly unique. They have played a crucial role in the development of this book. Without their invaluable contributions this book wouldn't have been possible. They have made vital efforts to compile up to date information on the varied aspects of this subject to make this book a valuable addition to the collection of many professionals and students.

This book was conceptualized with the vision of imparting up-to-date information and advanced data in this field. To ensure the same, a matchless editorial board was set up. Every individual on the board went through rigorous rounds of assessment to prove their worth. After which they invested a large part of their time researching and compiling the most relevant data for our readers.

The editorial board has been involved in producing this book since its inception. They have spent rigorous hours researching and exploring the diverse topics which have resulted in the successful publishing of this book. They have passed on their knowledge of decades through this book. To expedite this challenging task, the publisher supported the team at every step. A small team of assistant editors was also appointed to further simplify the editing procedure and attain best results for the readers.

Apart from the editorial board, the designing team has also invested a significant amount of their time in understanding the subject and creating the most relevant covers. They scrutinized every image to scout for the most suitable representation of the subject and create an appropriate cover for the book.

The publishing team has been an ardent support to the editorial, designing and production team. Their endless efforts to recruit the best for this project, has resulted in the accomplishment of this book. They are a veteran in the field of academics and their pool of knowledge is as vast as their experience in printing. Their expertise and guidance has proved useful at every step. Their uncompromising quality standards have made this book an exceptional effort. Their encouragement from time to time has been an inspiration for everyone.

The publisher and the editorial board hope that this book will prove to be a valuable piece of knowledge for researchers, students, practitioners and scholars across the globe.

List of Contributors

Olga Kovalchukova, Ali Sheikh Bostanabad and Svetlana Strashnova
Peoples' Friendship University of Russia, Moscow, Russia

Vladimir Sergienko and Irina Polyakova
Institute of General and Inorganic Chemistry, Russian Academy of Sciences, Moscow, Russia

Igor Zyuzin
Institute of Problems of Chemical Physics, Russian Academy of Sciences, Chernogolovka, Russia

Oluwayemi O. Esther Onawumi, Idowu O. Adeoye and Florence A. Oluwafunmilayo Adekunle
Department of Pure and Applied Chemistry, Ladoke Akintola University of Technology, Ogbomoso, Nigeria

Yichen Ye, Congcong Chen, Hui Feng, Jin Zhou, Juanjuan Ma, Jianrong Chen, Junhua Yuan, Lichun Kong and Zhaosheng Qian
College of Chemistry and Life Science, Zhejiang Normal University, Jinhua, China

Lin Lin and Bin Zhao
National Engineering Research Center for Nanotechnology, Shanghai, China

Yuchao Chai
School of Material Science and Engineering, Shanghai Jiao Tong University, Shanghai, China

Wei Wei
State Key Laboratory of Bioreactor Engineering, New World Institute of Biotechnology, East China University of Science and Technology, Shanghai, China

Dannong He
National Engineering Research Center for Nanotechnology, Shanghai, China
School of Material Science and Engineering, Shanghai Jiao Tong University, Shanghai, China

Belin He and Qunwei Tang
Institute of Materials Science and Engineering, Ocean University of China, Qingdao, China

S. N. Azizi, A. R. Samadi-Maybodi and M. Yarmohammadi
Analytical Division, Faculty of Chemistry, University of Mazandaran, Babolsar, Iran

Tengiz Machaladze
Rafiel Agladze Institute of Inorganic Chemistry and Electrochemistry, I. Javakhishvili Tbilisi State University, Tbilisi, Georgia

Madona Samkharadze, Nino Kakhidze and Maia Makhviladze
Akaki Tsereteli State University, Kutaisi, Georgia

Zhaoyan Deng, Yongheng Xing, Na Xing and Liting Xu
College of Chemistry and Chemical Engineering, Liaoning Normal University, Dalian, China

Fengying Bai
College of Life Science, Liaoning Normal University, Dalian, China

Xiaoli Tang and Longguan Zhu
Department of Chemistry, Zhejiang University, Hangzhou, China

Mostafa M. H. Khalil, Eman H. Ismail and Ahmed Badr
Chemistry Department, Faculty of Science, Ain Shams University, Cairo, Egypt

Gehad G. Mohamed and Ehab M. Zayed
Chemistry Department, Faculty of Science, Cairo University, Cairo, Egypt

Yunsheng Hu, Weidong Zhuang Xiaowei Huang and Huaqiang He
National Engineering Research Center for Rare Earth Materials, General Research Institute for Nonferrous Metals, and Grirem Ad-vcanced Materials Co., Ltd., Beijing, China

Jianhua Hao
Department of Applied Physics, The Hong Kong Polytechnic University, Hong Kong, China

Jing Chen
College of Chemistry and Chemical Engineering, Anyang Normal University, Anyang, China
School of Chemical Engineering and Energy, Zhengzhou University, Zhengzhou, China

You-Juan Zhang
College of Chemistry and Chemical Engineering, Anyang Normal University, Anyang, China

Kun-Tao Huang and Qiang Huang
School of Chemical Engineering and Energy, Zhengzhou University, Zhengzhou, China

Rudik Hamazaspi Grigoryan
Institute of Structural Macrokinetics and Materials Science, Russian Academy of Sciences, Moscow, Russia

Levoh Hamazaspi Grigoryan
Institute of Problems of Chemical Physics, Russian Academy of Sciences, Moscow, Russia Academician Semenova av. 1, Chernogolovka, Moscow, Russia

Anjali Goel and Neetu Rani
Department of Chemistry, Kanya Gurukul Mahavidyalaya, Gurukul Kangri University, Hardwar, India

Jia Li, Gao Zhang, Yan-Tao Li, Xin Wang, Jian-Qi Zhu and Yun-Qi Tian
Institute of Chemistry for Functionalized Materials, College of Chemistry and Chemical Engineering, Liaoning Normal University, Dalian, China

Bikash Kumar Panda
Department of Chemistry, Jangipur College, Murshidabad, India

Shawnt Tosonian, Charles J. Ruiz, Andrew Rios, Elma Frias and Jack F. Eichler
Department of Chemistry, University of California, Riverside, Riverside, USA

Jing Chen
College of Chemistry and Chemical Engineering, Anyang Normal University, Anyang, China
School of Chemical Engineering and Energy, Zhengzhou University, Zhengzhou, China

Jun-Jie Wang and You-Juan Zhang
College of Chemistry and Chemical Engineering, Anyang Normal University, Anyang, China

Kun-Tao Huang and Qiang Huang
School of Chemical Engineering and Energy, Zhengzhou University, Zhengzhou, China

Papiya Biswas, Kotikalapudi Rajeswari, Somasani Chaitanya and Roy Johnson
Centre for Ceramic Processing, International Advanced Research Centre for Powder Metallurgy and New Materials, Hyderabad, India

Swapnil A. Prabhudesai, Veerendra K. Sharma, Subhankur Mitra and Ramaprosad Mukhopadhyay
Solid State Physics Division, Bhabha Atomic Research Centre, Trombay, India

Sulekh Chandra
Department of Chemistry, Zakir Husain Delhi College, University of Delhi, New Delhi, India

Poonam Pipil
Department of Chemistry, Rajdhani College, University of Delhi, New Delhi, India

S. A. Stepanchikova and R. P. Biteykina
Institute of Geology and Mineralogy, Siberian Division of Russian Academy of Sciences, Novosibirsk, Russia

A. A. Sava
Novapharm Ltd., Sydney, Australia

Imen Ketata, Lassad Mechi and RachedBen Hassen
Unité de Recherche de Chimie des Matériaux et de l'Environnement, Université de Tunis El Manar, Tunis, Tunisia

Taïcir Ben Ayed
Unité de Recherche de Chimie des Matériaux et de l'Environnement, Université de Tunis El Manar, Tunis, Tunisia
INSAT, Université de Carthage, Tunis, Tunisia

Michal Dusek and Vaclav Petricek
Institute of Physics, Academy of Sciences of the Czech Republic, Praha, Czech Republic

Olalere G. Adeyemi and Umaru Salami
Department of Chemical Sciences, Redeemer's University, Redemption City, Nigeria

Guy Crundwell, Stefanie Cantalupo, Paul D. C. Foss, Brian McBurney, Kristen Kopp and Barry L. Westcott
Department of Chemistry & Biochemistry, Central Connecticut State University, New Britain, USA

James Updegraff III, Matthias Zeller, Allen D. Hunter
YSU Structure Center, Department of Chemistry, Youngstown State University, Youngstown, USA

Norhayati Hashim, Illyas Md Isa, Azlan Kamari and Azmi Mohamed
Department of Chemistry, Faculty of Science and Mathematics, Universiti Pendidikan Sultan Idris, Tanjong Malim, Malaysia
Nanotechnology Research Centre, Faculty of Science and Mathematics, Universiti Pendidikan Sultan Idris, Tanjong Malim, Malaysia

Adila Mohamad Jaafar and Mohd Zobir Hussein
Department of Chemistry, Faculty of Science, Universiti Putra Malaysia, Serdang, Malaysia

Hafsah Taha
Department of Chemistry, Faculty of Science and Mathematics, Universiti Pendidikan Sultan Idris, Tanjong Malim, Malaysia

Monika Tyagi and Sulekh Chandra
Department of Chemistry, Zakir Husain Delhi College, University of Delhi, New Delhi, India

Norrihan Sam, Md Abu Affan, Md Abdus Salam, Fasihuddin B. Ahmad and Mohd Razip Asaruddin
Department of Chemistry, Faculty of Resource Science and Technology, Universiti Malaysia Sarawak, Kota Samarahan, Malaysia

Manel Halouani, Mohamed Abdelhedi, Mohamed Dammak and Lilia Ktari
Laboratoire de Chimie Inorganique, Faculté des Sciences de Sfax, Université de Sfax, Sfax, Tunisia

Nathalie Audebrand
Laboratoire de Matériaux Inorganiques: Chimie Douce et Réactivité, UMR 6226 Sciences Chimiques de Rennes, Université Rennes 1, Rennes, France

Shanling Tong, Yahong Wu, Sheng Hu and Yan Yan
College of Light Industry & Chemical Engineering, Guangdong University of Technology, Guangzhou, China

Zhaoyong Tian
Logistics Group, Huainan Normal College, Huainan, China

Yassin Belghith, Anissa Mansour and Habib Nasri
Laboratoire de Physico-chimie des Matériaux, Faculté des Sciences de Monastir, Université de Monastir, Monastir, Tunisia

Jean-Claude Daran
Laboratoire de Chimie de Coordination, CNRS UPR 8241, Toulouse, Francew

www.ingramcontent.com/pod-product-compliance
Lightning Source LLC
Chambersburg PA
CBHW080534200326
41458CB00012B/4431